PUZZLES, PARADOXES, and PROBLEM SOLVING

An Introduction to Mathematical Thinking

PUZZLES, PARADOXES, and PROBLEM SOLVING

An Introduction to Mathematical Thinking

Marilyn A. Reba

Clemson University
South Carolina, USA

Douglas R. Shier

Clemson University
South Carolina, USA

CRC Press
Taylor & Francis Group
Boca Raton   London   New York

CRC Press is an imprint of the
Taylor & Francis Group, an **informa** business

A CHAPMAN & HALL BOOK

CRC Press
Taylor & Francis Group
6000 Broken Sound Parkway NW, Suite 300
Boca Raton, FL 33487-2742

© 2015 by Taylor & Francis Group, LLC
CRC Press is an imprint of Taylor & Francis Group, an Informa business

No claim to original U.S. Government works

Printed on acid-free paper
Version Date: 20141103

International Standard Book Number-13: 978-1-4822-2753-6 (Hardback)

Visit the Taylor & Francis Web site at
http://www.taylorandfrancis.com

and the CRC Press Web site at
http://www.crcpress.com

MR: For Michael, Charles, Katherine, Elijah, and Omar (MIAZPXEECJKJYBX)

DS: For Joan, with appreciation for her patient support during the writing of this book

Contents

Preface

> JW: I am afraid that to me, mathematics for enjoyment sounds rather like a contradiction in terms.
>
> CD: Oh, I think there is a certain fun to be had in questions of logic and probability, and in the surprises and paradoxes that can arise.[1]

The above fictional exchange between Dr. John Watson (companion of Sherlock Holmes) and the Reverend Charles Dodgson (better known as Lewis Carroll) captures the guiding theme of this textbook. Namely, that the use of puzzles and paradoxes can provide a natural and satisfying avenue for introducing certain basic principles of mathematical thought. We believe it is the best approach to offer to non-science, non-technical majors both in terms of motivation and development of critical-thinking skills. Students are introduced to puzzles and decision-making situations that are of sufficient difficulty to require creative problem-solving approaches. The same structural problem can appear in various disguises throughout the text and a set of common solution approaches emerge as students progress from recreational problems to important contemporary applications. A list of such applications appears in Appendix A.

Rather than presenting a diverse, but loosely connected, set of topics, this text strives to present a unified approach that logically connects the subject matter and employs a recurring set of solution approaches. Specifically, a common set of mathematical representations (e.g., graphs, diagrams, tables, sets, and symbols) is encountered throughout the various chapters. This allows diverse problems to be viewed as familiar friends. As well, common solution strategies (e.g., applying brute force, enumeration, problem decomposition, and well-defined rules) can then be attempted, resulting in the application of specific solution algorithms (e.g., systematic search techniques, heuristic methods, and sequential/recursive procedures). The solutions produced can give exact or only approximate answers to the problem being studied. This classification into common representations, strategies, and algorithms appears throughout.

The material covered in this text is clustered into five units: graphs, logic, probability, voting, and cryptography. Within these units, other subject areas are discussed, including operations research, game theory, number theory, combinatorics, statistics, and circuit design. Every area is injected with numerous games and puzzles (e.g., mazes, jug-pouring problems, word puzzles, logic puzzles, chess, cards, sports) as well as numerous contemporary applications (e.g., genetics, legal forensics, business, politics, medicine, computer security).

Features

- Any mathematical tools beyond arithmetic and basic algebra are gently presented as needed in each chapter. We use non-threatening, non-technical prose, but do not sacrifice a mathematically precise development of concepts.

- New problem-solving skills are explained through a progression of numbered *examples*. Examples and their solutions are clearly delineated, with the symbol ◁ being used to indicate when a solution is complete. The chapter exercises related to specific examples

---

[1] Bruce, C. *Conned again, Watson! Cautionary tales of logic, math, and probability.* Cambridge, MA: Perseus, 2001.

are listed as well; students will then find it easy to locate related problems they may need to review.

- Boxed definitions, summaries of algorithms, and hundreds of detailed graphs, diagrams, and tables are interspersed throughout the text. These elements reinforce and summarize the important concepts introduced within each of the chapters. *Historical comments* are also provided to place the material in context.

- *Chapter summaries* are used to organize the fundamental ideas and approaches presented in the chapter. At the end of each chapter, we also provide a *classification table* that lists the various applications encountered as well as the common representations, strategies, and algorithms employed. (Appendix B explains in further detail these *representations*, *strategies*, and *algorithms*.)

- Extensive *exercises* at the end of each chapter are a strong feature of the text. They are both varied and graduated in terms of difficulty. In addition, they are carefully organized into groups and cross-referenced to the corresponding chapter material.

- *Projects* are suggested for further exploration at the end of each chapter. Also listed are references, both printed and online, that are accessible to the students.

- An Instructor's Solution Manual is available that provides detailed solutions to each problem, as well as a Student's Solution Key that lists answers to all odd-numbered problems. A website is forthcoming that will house additional resources for instructors.

Audience

This is an undergraduate text designed for students in liberal arts mathematics courses and for students in general education courses fulfilling a quantitative or a critical-thinking requirement. It can also be used in an honors problem-solving course, suitable for both non-science and science undergraduates. There are no mathematical prerequisites, so anyone interested in becoming familiar with puzzles, paradoxes, and their logical underpinnings might use this text for self-study. Not all chapters of the text are intended to be covered in a single course. For example, sections containing more complex examples can be excluded to suit the pace of the course or the intended audience.

There are numerous liberal arts mathematics textbooks available, in both the quantitative literacy and quantitative appreciation genres. This text is an improved resource for students, in terms of challenges, thoroughness, and critical-thinking learning outcomes. We have developed this text through its classroom use over a number of years. Whether the audience consists of liberal arts students enrolled in the single mathematics course of their college career, or science and engineering students enrolled in an honors problem-solving seminar, students achieve a better understanding of what creative mathematical thinking really is. Computer science students can also benefit from the material presented, both in terms of its wide range of applications and its unified treatment of a recurring set of solution approaches.

Acknowledgments

We thank the Honors College at Clemson University for contributing to the development of this text by allowing this course to be taught for seven years and to the honors students for their enthusiastic participation and feedback. We also thank Clemson's Department of Mathematical Sciences for encouraging the use of a preliminary version of this text in sections of its liberal arts mathematics course. We are grateful to several graduate students, Michael Dowling and Kara Stasikelis, who contributed ideas to exercises and helped with the preparation of the material. We also wish to acknowledge the assistance of Patrick Buckingham and Drew Lipman for their contributions to the solutions manual for this text. We are especially grateful to Roland Minton (Roanoke College), Kevin Hutson (Furman University), and Ralph Grimaldi (Rose-Hulman Institute of Technology) for their insightful comments as reviewers of this text.

List of Figures

List of Tables

Unit I

Graphs: Puzzles and Optimization

Chapter 1

Graphical Representation and Search

TOSCA: I want a safe conduct, so that he and I can flee the state together.
SCARPIA: Your wish shall be granted. And which road do you prefer?
TOSCA: The shortest!

—Giacomo Puccini, Italian composer, 1858–1924

Chapter Preview

This chapter focuses on the power of graphical representation, illustrated in a number of different contexts. Even though the problems explored may seem unrelated, their commonality is revealed by exhibiting an underlying graph structure. Then we can apply an appropriate solution method to produce an answer to the original problem. Specifically, a number of puzzles can be represented as graphs in which we wish to start at an initial vertex (or state) s and arrive at a desired final vertex (state) t. Finding a path in the graph from vertex s to vertex t then reveals a solution to the puzzle. A breadth-first or depth-first search can aid us in discovering such a path when the underlying graph is large or complex. At times we may wish to find not just any path from s to t, but rather a most efficient (shortest) such path.

1.1 Mazes and Graphs

Mazes (and labyrinths) have a long history as pure visual art, as architectural decoration, and as cultural and religious artifacts. At one time the word "amaze" connoted being "lost in thought," which can happen to us when trying to navigate a maze. Today mazes are found in puzzle books or as paid attractions constructed for people to walk through. Visual examples of such mazes can be seen in [OR1].

We typically use trial-and-error to navigate a maze. For example, in the maze of Figure 1.1 (found at Hampton Court Palace, in England), we are asked to start at a location (A) in the center of the diagram, and then find our way to the outside (Z). Along the way, we need to make decisions at each intersection and possibly backtrack if we encounter a dead-end.

It is not hard to find a solution to this simple maze, such as the one shown in Figure 1.2. However, given a more complicated maze (such as the one shown in Figure 1.3), you would

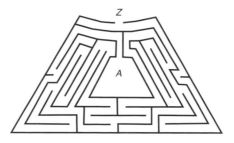

FIGURE 1.1: The hedge maze at Hampton Court

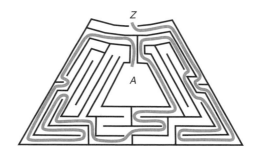

FIGURE 1.2: A solution to the Hampton Court maze

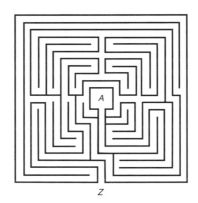

FIGURE 1.3: A more challenging maze

not want to use trial-and-error. We will develop a more systematic and general approach. Our key strategy is to represent a maze as a *graph*, composed of *vertices* and *edges*. The next example illustrates these concepts in the context of social networks.

Example 1.1 Represent a Social Network as a Graph

A group of five students is selected from a large lecture class to work on a class project. A few of them are close friends with one another, while others are not. Specifically, Anand is friends with Brittany and with Claire; Dexter is also friends with Brittany and Claire; in addition, Claire is a friend of Ethan. Represent this situation as a graph, with vertices representing the students and an edge indicating friendship between two students. Then determine two ways in which a message can be passed from Anand to Ethan.

Solution

In Figure 1.4, we place a circle for each vertex (labeled by the initial letter of the student's name), and we join two vertices by an edge if they are mutual friends. We see that C is *adjacent* to A, D, and E. The sequence $A \rightarrow C \rightarrow E$ represents a *path* of *length* 2 from Anand to Ethan. A second (longer) path is $A \rightarrow B \rightarrow D \rightarrow C \rightarrow E$, of length 4. ◁

(Related Exercises: 1–5)

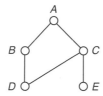

FIGURE 1.4: A small social network

DEFINITION **Graph**
A diagram consisting of points (vertices) that are connected by lines (edges).

DEFINITION **Adjacent Vertices**
Two vertices that are connected by an edge.

DEFINITION **Path**
A succession of adjacent vertices.

DEFINITION **Length of a Path**
The number of edges in the path.

In the context of social networks, a huge graph of mutual friendships can be constructed and we can measure how far away any two people are, based on the shortest path length between them (in Example 1.1, Anand and Ethan are just two steps away). The phrase "six degrees of separation" refers to the observation that any two people are typically (at most) six steps away, measured by these "friend of a friend" steps [OR2]. A recent Facebook report concluded the world may be even smaller, with only an average of 4.7 degrees of separation between any two of us [OR3].

Graphs also provide the underlying framework for a number of commonly encountered systems in our modern world: for example, airline networks, road networks, and computer networks. In addition, graphs have proved useful in crime prevention and interdiction, as a way to represent the structure of drug-trafficking organizations and terrorist networks. The formal study of graphs is a subspecialty within mathematics and computer science.

DEFINITION **Graph Theory**
The study of graphs as mathematical structures used to model relations between pairs of objects within a defined set of objects.

We now return to see how graphs can be used to distill the essential ingredients of a maze, and to aid in solving such mazes. Namely, we place vertices at all of the decision points in the maze, including dead-ends. If you can travel from one decision point to another in the maze without having to make a decision in between, then an edge joins these two vertices. The next example shows how the underlying structure of a maze can be concisely represented by a graph.

Example 1.2 Use a Graph to Solve a Maze

Construct the graph associated with the maze displayed in Figure 1.5, where the objective is to find a way from the outside (A) to a designated location in the middle (M). Identify a path from A to M in your graph and then interpret it as a solution to the maze.

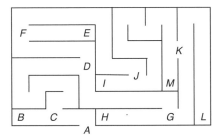

FIGURE 1.5: A simple maze

Solution

Place a vertex for each decision point in the maze—the starting location, the ending location, each intersection, and each dead-end—and label it with a letter. Then join two vertices by an edge if they can be reached directly in the maze. This produces the graph in Figure 1.6. Using this graph depiction it is easy to identify two paths from location A to location M: one uses the top path $A \to C \to D \to F \to J \to K \to M$ of length 6 and one uses the bottom path $A \to C \to D \to G \to K \to M$ of length 5. Since the bottom path is shorter than the top path, a highly methodical rat trapped in this maze would be well advised to take the bottom path. This path of length 5 translates into the solution of the original maze shown in Figure 1.7. ◁

(Related Exercises: 7 8)

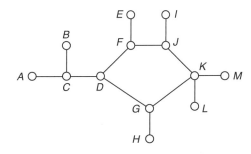

FIGURE 1.6: Graph for the simple maze

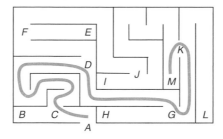

FIGURE 1.7: A solution for the simple maze

1.2 Systematic Search Methods

A common project given to computer science or engineering students is to program a robot that can navigate its way through a maze, and this is aided by carrying out a systematic search of the associated graph. Computer scientists, when programming computers to retrieve information, distinguish a *breadth-first search* (BFS) from a *depth-first search* (DFS). Both are systematic methods for exploring a graph. We can apply these methods to the task of finding our way out of a maze.

Example 1.3 Use a Breadth-First Search in the Graph of Example 1.2

Carry out a breadth-first search of the graph in Figure 1.6, beginning at vertex A and exploring *all* edges leading from it to any adjacent vertices. In turn from each of these vertices, explore all adjacent vertices not yet encountered. Continue this branching out process until the goal vertex M is reached.

Solution

There is only one vertex C adjacent to A, so we explore (and mark) the edge (A, C). Then from C, we encounter the new adjacent vertices B and D via edges (C, B) and (C, D). As there are no new vertices adjacent to B, we continue with vertex D and encounter the new adjacent vertices F and G via edges (D, F) and (D, G). We can keep track of this search by using Table 1.1, in which Current indicates the vertex currently being processed and Adjacent lists all of its adjacent vertices not already encountered. For example, when our current vertex is J, we only list the adjacent vertex I but not the adjacent vertex K, since it was encountered earlier when processing G as the current vertex. Note that the order of processing the current vertices is the order in which they are encountered, given by their order of appearance in Adjacent. We can stop the search when the goal vertex M is first encountered.

TABLE 1.1: Carrying out a BFS

Current	A	C	B	D	F	G	E	J	H	K
Adjacent	C	B, D		F, G	E, J	H, K		I		L, M

The set of marked edges are just those of the form (x, y) where x is a **Current** vertex and y is one of the entries in the **Adjacent** list for x. These edges are shown in bold in Figure 1.8. Notice that the edges marked during the BFS form what is called a *tree*. ◁

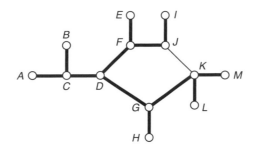

FIGURE 1.8: Tree found during the BFS

DEFINITION **Breadth-First Search (BFS)**
A method of exploring a graph in which each vertex of the graph is processed in the order it is encountered. At each vertex we explore all edges leading from it to adjacent vertices not already seen.

DEFINITION **Tree**
A graph in which there is a unique path between any two vertices.

A BFS mimics the spread of a rumor or a disease. One person may start the rumor by mentioning it to several of his friends, who in turn spread it to several of their own friends, and so on. There is an important feature of a BFS: following the (unique) path in the BFS tree from a starting vertex to an ending vertex produces a solution path in the graph that has the *fewest* number of edges. In the context of the maze of Example 1.2, the unique path in the BFS tree from M to M in Figure 1.8 is $A \to C \to D \to G \to K \to M$, the bottom shorter path discovered in Figure 1.6.

(Related Exercise: 6)

We can also apply a different type of search to find a solution path in a maze.

Example 1.4 Use a Depth-First Search (DFS) in the Graph of Example 1.2

Carry out a DFS of the graph in Figure 1.6 by beginning at the starting vertex A and exploring a *single* edge leading from it to an adjacent vertex. From this vertex, select a single edge leading to some adjacent vertex not yet encountered. If necessary, you may have to retreat or backtrack if you are unable to progress further. Continue this process until the goal vertex M is reached.

Solution

To make the search process well defined, we establish the convention that when there are several vertices adjacent from a vertex just reached, we select the one with the (alphabetically) smallest label. We begin our search from vertex A; there is only one vertex C adjacent to A, so we move to C and add C to our current **Path**. At vertex C, there are two adjacent vertices B and D, so we select B (having the smaller label) and add it to **Path**. As there are no new vertices adjacent to B, we retreat back to vertex C and select the adjacent vertex D. At D, there are two new vertices F and G, so we select F which has the smaller label, and so on. We can keep track of this search by using Table 1.2, which shows each new vertex explored as the depth-first search wends its way through the graph; underscored vertices indicate that we needed to retreat from that vertex and return to a previous vertex to continue the path. We stop the search when the goal vertex M is first encountered.

TABLE 1.2: Carrying out a DFS

	Path											
Vertex	A	C	$\underline{B}$	D	F	$\underline{E}$	J	$\underline{I}$	K	$\underline{G}$	$\underline{H}$	$\underline{L}$	M

This table shows that the DFS makes six retreats before eventually finding the path $A \to C \to D \to F \to J \to K \to M$. This path is defined by the non-underlined vertices appearing in Table 1.2, and in fact corresponds to the top path previously discovered in Figure 1.6. While we are able to produce a solution to the maze, it is not one of smallest length. ◁

DEFINITION **Depth-First Search (DFS)**
A method of exploring a graph in which you move forward to a single new adjacent vertex whenever possible, progressing deeper and deeper until the desired goal is found or you are unable to move further. In the latter case, you retreat to the first vertex with an option not yet explored.

Unlike a BFS, a DFS cannot guarantee using the fewest number of edges. For this reason, we prefer using a BFS in navigating our way through a maze. Here is a summary of our graphical approach to solving mazes:

Graphical Solution of Maze Problems

1. Place a letter at each decision point in the maze.
2. Construct the graph with vertices for letters and edges between adjacent letters.
3. Identify a (shortest) path from the initial vertex to the final vertex: for example, by using a BFS.
4. Trace out the corresponding path in the maze.

We now apply this approach to a previously seen maze.

Example 1.5 Solve the Hampton Court Maze Using the Graphical Approach

The hedge maze at the Hampton Court Palace is shown in Figure 1.1. Our objective is to find our way from the interior of the maze (A) to the outside (Z) using the fewest number of decisions. (a) Create the graph of this maze. (b) Use a BFS of this graph to find a shortest path from A to Z. (c) Interpret this solution path as an escape route in the maze.

Solution

(a) We begin by adding labels to all decision points in the maze, as is done in Figure 1.9. Joining adjacent vertices then produces the graph representation displayed in Figure 1.10.

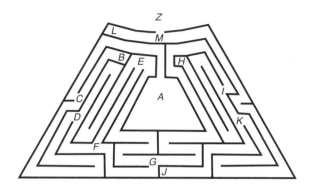

FIGURE 1.9: Labels for the Hampton Court maze

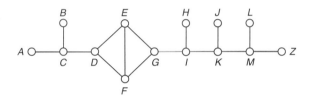

FIGURE 1.10: Graph for the Hampton Court maze

(b) Performing a BFS of this graph yields Table 1.3. The edges marked during the BFS are found by consulting each column of this table and are shown in bold in Figure 1.11. A shortest path from A to Z is obtained by following the bold edges within the BFS tree: $A \to C \to D \to E \to G \to I \to K \to M \to Z$.

TABLE 1.3: Carrying out a BFS for the Hampton Court maze

Current	A	C	B	D	E	F	G	I	H	K	J	M
Adjacent	C	B, D		E, F	G		I	H, K		J, M		L, Z

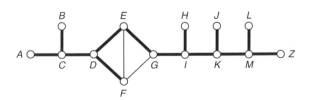

FIGURE 1.11: BFS tree for the Hampton Court maze

(c) The path found in (b) has length 8, and it involves the fewest number of decisions needed to escape the maze. When interpreted back in the original maze, this path produces the routing shown in Figure 1.12. ◁

(Related Exercises: 11–12)

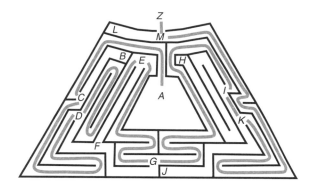

FIGURE 1.12: A solution for the Hampton Court maze

Notice that there is another shortest path from A to Z in Figure 1.10, one that uses edge (F, G) instead of (E, G): $A \to C \to D \to F \to G \to I \to K \to M \to Z$. In other words, there is not a unique shortest routing through the Hampton Court maze. Moreover, there are two other longer paths, of length 9, in Figure 1.10. Can you identify them?

(Related Exercises: 9–10)

This general strategy—creating an appropriate graph, carrying out a search to identify a solution path, and then interpreting this solution path in the original context—will be followed in the next section, where we explore another type of recreational puzzle.

1.3   Jug-Pouring Problems

Jug-pouring problems are puzzles that involve pouring water between jugs with known capacities in order to measure out a specific quantity of water. As there are no markings on the jugs, the only operations allowed are filling a jug to capacity, entirely emptying a jug, or transferring the contents of one jug to another. Variations of jug-pouring problems have been around since the 13th century. In the 1995 movie *Die Hard with a Vengeance*, the protagonists John McClain (Bruce Willis) and Zeus (Samuel L. Jackson) are challenged by the evil Simon Gruber (Jeremy Irons) to measure out exactly 4 gallons from 5- and 3-gallon jugs in just a few minutes in order to disarm a bomb [OR4]. This is an example of a *two-jug problem*.

DEFINITION **Two-Jug Problem**
You are given two empty jugs of known capacities and an unlimited supply of liquid. At each step, you can fill a jug to capacity or completely empty it. Also, you can transfer the contents of one jug to another, stopping once either one jug is filled to capacity or the other is empty. The objective is to find a sequence of operations to measure out a specified quantity, typically using the fewest number of pouring operations.

Example 1.6 Solve the *Die Hard* Puzzle by Trial-and-Error

Using two unmarked 5- and 3-gallon jugs, obtain exactly 4 gallons in one of these jugs by filling, pouring, and emptying operations. Assume there is an endless supply of water.

Solution

One possible solution is depicted in Table 1.4, which records the contents of the two jugs and the associated operation. We achieve success after eight pourings, with 4 gallons measured out in the large jug. ◁

TABLE 1.4: Solution for the two-jug problem

5-Gallon Jug	3-Gallon Jug	Pouring Operation
0	0	both empty
0	3	fill small jug
3	0	pour small jug into large jug
3	3	fill small jug
5	1	pour small jug into large jug
0	1	empty large jug
1	0	pour small jug into large jug
1	3	fill small jug
4	0	pour small jug into large jug

The solution in Example 1.6 was found by trial-and-error. Is there a more systematic way to find this solution? Is it possible to find another solution involving fewer pourings? We investigate these questions by first creating an appropriate graphical representation and then applying a BFS.

Example 1.7 Create a BFS Tree for the *Die Hard* Puzzle

As suggested by Table 1.4, it is helpful to represent the *states* of the system by the respective amounts in the 5- and 3-gallon jugs. *Vertices* are then ordered pairs of amounts, such as $(0,0), (5,0), (0,3), (5,3)$. *Edges* with arrows represent transitions between states via a single action (adding water to a jug, transferring water from one jug to another, or emptying a jug). These vertices and edges define a *directed graph*, which can be searched in a breadth-first manner. Display the associated BFS tree that you obtain.

Solution

We start with the initial state $(0,0)$ and then find all adjacent states: namely, $(0,3)$ and $(5,0)$. The corresponding two edges directed from $(0,0)$ are seen in Figure 1.13. Next, we find any new states adjacent to $(0,3)$—namely, states $(3,0), (5,3)$—and these produce two more edges in the BFS tree. We then process state $(5,0)$ and identify the adjacent states

$(5, 3)$ and $(2, 3)$. Since we have already encountered $(5, 3)$, we do not add a second edge to it from $(5, 0)$. However, $(2, 3)$ has not been seen before, so this adds a new directed edge to the BFS tree. Continuing in this fashion, we obtain the BFS tree shown in Figure 1.13. ◁

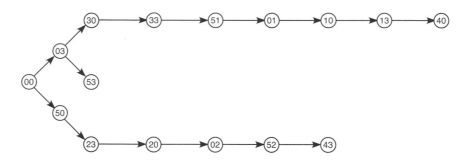

FIGURE 1.13: BFS tree for the *Die Hard* puzzle

DEFINITION **Directed Graph**
A graph with arrows on the edges that indicate the allowable direction of progress between vertices.

Example 1.8 Use the BFS Tree to Identify Solutions to the *Die Hard* Puzzle

Identify all solution paths revealed by the BFS tree in Figure 1.13 and interpret these as pouring sequences that solve the *Die Hard* puzzle.

Solution

Observe that the vertices in Figure 1.13 are arranged in vertical layers, according to their distance from the initial state $(0, 0)$. This shows that we can get to state $(4, 0)$ from the origin in eight moves (pourings) along the top path. However, it is more efficient to use the bottom path to state $(4, 3)$, which involves only six moves. Once we have this path $(0, 0) \rightarrow (5, 0) \rightarrow (2, 3) \rightarrow (2, 0) \rightarrow (0, 2) \rightarrow (5, 2) \rightarrow (4, 3)$, we can interpret it in the context of the original problem as the following operations: fill the 5-jug, pour the 5-jug into the 3-jug, empty the 3-jug, pour the 5-jug into the 3-jug, fill the 5-jug, and pour the 5-jug into the 3-jug. The longer top path $(0, 0) \rightarrow (0, 3) \rightarrow (3, 0) \rightarrow (3, 3) \rightarrow (5, 1) \rightarrow (0, 1) \rightarrow (1, 0) \rightarrow (1, 3) \rightarrow (4, 0)$ corresponds to the sequence of operations described in Table 1.4. ◁

(Related Exercises: 13–18)

Notice that in contrast to maze problems, which can be represented using (undirected) graphs, we choose to use directed graphs to model jug-pouring problems. This is especially important for general jug-pouring problems with more than just two jugs. For example, if jug capacities are 5, 3, and 2, then it is possible to move directly from state $(4, 2, 0)$ to state $(3, 3, 0)$ but not the reverse. The next example provides an instance of a *three-jug problem*.

DEFINITION **Three-Jug Problem**

You are given three jugs of known capacities, some of which are already full. At each step, you can transfer the contents of one jug to another, stopping once either one jug is filled to capacity or the other is empty. The objective is to find a sequence of operations to measure out specified quantities, typically using the fewest number of pouring operations.

Example 1.9 Solve a Three-Jug Problem Using a BFS

Suppose that Jug A has a capacity of 8 pints, Jug B has a capacity of 5 pints, and Jug C has a capacity of 3 pints. Initially, there are 8 pints in the largest jug. By pouring back and forth between the jugs, find a way to get 3 pints in two of the jugs. (Notice that we use only the 8 pints of liquid with which we started; only transfers between pairs of jugs are allowed.) Create an associated BFS tree and use it to identify a solution having the minimum number of pourings.

Solution

Starting with 8 pints in Jug A, our goal is to reach two jugs with 3 pints each. As before, we can label the vertices to represent the amounts in each jug. To begin, the initial vertex is labeled $(8, 0, 0)$ to indicate that all the liquid is in Jug A, while nothing is in Jug B or Jug C. Remember that the three jugs have capacities 8, 5, and 3, respectively. Using a BFS starting at $(8, 0, 0)$, we can find all new arrangements reachable in one pouring: namely, $(5, 0, 3)$ and $(3, 5, 0)$. Then we can look at all new arrangements reachable from those arrangements, and so on. This produces the BFS tree shown in Figure 1.14.

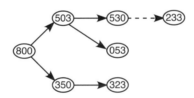

FIGURE 1.14: BFS tree for the three-jug problem

It is now clear from this search tree that we can get from $(8, 0, 0)$ to $(3, 2, 3)$ in just two pourings, using the path $(8, 0, 0) \to (3, 5, 0) \to (3, 2, 3)$. Namely, we pour the contents of Jug A into Jug B, and then pour from Jug B to Jug C. There is no solution with fewer pourings. By contrast, if we had continued the BFS further (indicated by the dashed edge in Figure 1.14), we would have found a solution path from $(8, 0, 0)$ to $(2, 3, 3)$; however, this involves three pourings not two. ◁

(Related Exercises: 19–25)

1.4 Other Graph Search Applications

We now present two additional puzzles that can be usefully represented as graph problems and solved using appropriate search techniques. Recall that in the jug-pouring problems, we found a shortest path in an associated graph: that is, a path involving the fewest number of pourings. We now investigate a problem in which edges can have different lengths and we seek again a shortest path. This topic will be explored again in Chapter 3.

Example 1.10 Solve an Egg-Timer Puzzle Using a BFS

Suppose that we have on hand two egg timers. Each is an hourglass-shaped device (filled with sand) used to count down time. The first can count down 5 minutes and the second can count down 3 minutes. However, our recipe calls for measuring out exactly 4 minutes. Is it possible to use these two egg timers to do this? Specifically, we want to describe a sequence of actions so that we will have exactly 4 minutes left on one of the timers (namely, the 5-minute timer). Define appropriate states and transitions for this problem so that it becomes a graph search problem. Then use a BFS to identify a solution.

Solution

Each egg timer measures the time remaining until it runs out. This suggests construction of a directed graph with vertices representing possible states of the egg timers. For example, state $(3, 1)$ would indicate that there are 3 minutes remaining in the 5-minute timer and only 1 minute remaining in the 3-minute timer.

In terms of state transitions, we can use an edge to indicate the passage of time until one of the timers runs out; these (dashed) edges are labeled with the elapsed time. Also, each timer can be turned over (flipped) to measure the time already passed. For example, if there are 3 minutes remaining in the 5-minute timer, it can be turned over to measure 2 minutes. So, there are other (solid) edges to indicate flipping each of the timers individually or both at the same time; no elapsed time is associated with such edges. These vertices and edges define a graph that can be searched in a breadth-first manner. Figure 1.15 shows the BFS tree obtained. As done earlier, we do not show edges that lead to states previously found in the BFS. For example, there is no dashed edge directed from state $(0, 3)$ to $(0, 0)$ nor a solid edge from $(0, 3)$ to $(5, 3)$, as we have seen these states already. Similarly, $(5, 2)$ produces the previously seen state $(3, 0)$ after 2 minutes have elapsed.

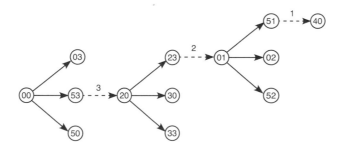

FIGURE 1.15: BFS tree for the egg-timer puzzle

Figure 1.15 reveals a path of length $3 + 2 + 1 = 6$ from the starting state to $(4, 0)$, meaning that after 6 minutes of elapsed time we can get exactly 4 minutes remaining in the 5-minute timer. We can interpret this path $(0, 0) \to (5, 3) \to (2, 0) \to (2, 3) \to (0, 1) \to (5, 1) \to (4, 0)$ in the original context as follows:

> Turn over both timers; after the second timer runs out, flip it over again; wait for the first timer to run out and then flip it over. Finally, wait for the second timer to run out and you will have 4 minutes remaining on the first timer to use for the recipe.

We note that this solution involves the smallest amount of total elapsed time. ◁

(*Related Exercises: 26–28*)

Example 1.11 Solve a Chessboard Puzzle Using a Graph Representation

In the chessboard shown in Figure 1.16, a knight piece is to be moved from square A to square B using the minimum number of (valid) moves. The knight can jump over, but cannot land, on any shaded square. A knight moves in an L-shaped manner: two squares in one direction and one in the other direction. (a) Represent this problem using a graph with vertices defined by grid coordinates and edges defined by valid moves. For example, $A = (5, 1)$ and $B = (5, 3)$, using vertical and horizontal coordinates, relative to the top left corner of the board. (b) Using this graph, identify a way to move the knight from A to B using the fewest number of moves. In how many different ways can this be done?

FIGURE 1.16: A chessboard with forbidden (shaded) squares

Solution

(a) First, we construct the associated graph, with vertices representing the unshaded squares and edges representing valid knight moves. For example, the only valid move from square $A = (5, 1)$ is to $(4, 3)$, located right above square B. Figure 1.17 shows the resulting graph.

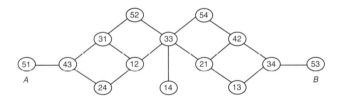

FIGURE 1.17: Graph for the chessboard puzzle

(b) One solution path in the graph is the topmost path $(5, 1) \rightarrow (4, 3) \rightarrow (3, 1) \rightarrow (5, 2) \rightarrow (3, 3) \rightarrow (5, 4) \rightarrow (4, 2) \rightarrow (3, 4) \rightarrow (5, 3)$, involving eight moves of the chess piece. In fact, there are nine such paths involving only eight moves. To see this, notice that all paths from $(5, 1)$ to $(5, 3)$ must pass through $(3, 3)$. Since there are three paths from $(5, 1)$ to $(3, 3)$ and symmetrically three paths from $(3, 3)$ to $(5, 3)$, we have $3 \times 3 = 9$ paths in total. We will encounter another systematic way of counting paths in Section 3.1. ◁

(*Related Exercises: 29–31*)

Chapter Summary

The essential features of several unrelated problems can be represented using a graph having vertices for different locations or states in the problem and edges for the transitions between these locations or states. The graph provides a model of the original problem that helps us visualize the connections between problem elements and devise an appropriate strategy to support a solution method. Here we presented a number of constructive strategies, which build a path from an initial vertex (state) s to a desired final vertex (state) t. All examples in this chapter illustrate a common approach: (1) the problem is modeled using a graph, (2) constructive strategies are used to produce a solution (path) in the graph, and (3) the solution from the graph is translated into a solution to the original problem. While the examples provided draw mainly from the literature of puzzles, subsequent chapters address important applied situations, again using appropriate graph representations and solution techniques.

Applications, Representations, Strategies, and Algorithms

Applications			
Society/Politics (1)		Time Measurement (10)	

Representations			
Graphs			Tables
Weighted	Undirected	Directed	Decision
10	1–5, 11	7–10	6

Strategies	
Brute-Force	Solution-Space Restrictions
1–2, 8	3–7, 9–11

Algorithms				
Exact	Inspection	Sequential	BFS	DFS
1–11	1–2, 8, 11	6	3, 5, 7, 9–10	4

Exercises

Graph Representation of Puzzles

1. You are friends with Aliana, Bob, and Chen. Dashaun is friends with Bob, Elmer, Forrest, and Hayes. Ilya is friends with Elmer, Gillian, and Zeke. Bob is friends with Aliana, Chen, and Forrest. Gillian is friends with Hayes and Zeke, and Forrest is friends with Hayes.

(a) Graphically represent this social network.

(b) Describe a way to send a message to Zeke in the most efficient way. Is there more than one way to do this?

2. You are on a scavenger hunt with the following succession of clues. The initial clue A leads to location B. The clue at location B could lead to location C or location D. There is no further clue at location C. The clue at location D could lead to location E or location F, both of which have clues. The clue at location E points to location G. The clue at location F could lead to location J or location H. There is no further clue at location J, but there is a clue at location H. The clue at location H could lead to location G or location M, both of which have clues. Location G could lead to location L or the prize. Location L has no further clue. The clue at location M also leads to the prize. You are in charge of the scavenger hunt and want to sort out this information.

(a) Draw a directed graph to represent the scavenger hunt.

(b) How many paths are there to the prize (assuming we never visit a location twice)?

(c) Identify a shortest path to the prize.

3. In the grid below, you can move any adjacent square into the blank square (effectively moving the blank square either horizontally or vertically). As in Example 1.11, each square can be given coordinates relative to the top left corner; for example, $(3, 3)$ denotes the current position of the blank square, which we would like to move to $(2, 1)$.

(a) Draw the graph representing this problem.

(b) How many different paths are there from $(3, 3)$ to $(2, 1)$, assuming you cannot go through a vertex twice?

(c) What is the length of a shortest path to the desired final configuration?

6	2	7
4	1	3
8	5	

4. Consider the following word puzzle. You begin with the four-letter word OOIIS and you are allowed to transpose any two adjacent letters: the first and second, or the second and third, or the third and fourth. For example, from OOHS you can obtain the two new words OHOS and OOSH. The objective is to transform OOHS into SHOO by a sequence of such transpositions.

(a) Draw a graph representing all words you can obtain by this process; two words are adjacent if they can be obtained from one another by a single transposition.

(b) Find a shortest path from OOHS to SHOO in this graph and list the corresponding word transformations that achieve this.

(c) Find a sequence of transpositions that start at OOHS and visit all vertices of the graph exactly once, ending at SHOO.

5. A knight piece is placed at position A in the partial chessboard shown next. We wish to move it to position B using the fewest number of moves. Recall that a knight can move two squares in one direction and one square in the other direction; here all moves must take place on the eleven squares of this partial chessboard.

(a) Draw a graph in which vertices represent the squares of the chessboard and edges correspond to valid moves of the knight on the chessboard.

(b) Identify a shortest path from vertex A to vertex B in this graph.

(c) How many moves are needed to move the knight from position A to position B?

6. Alicia, Claudia, and Gerry are all mutual Facebook friends, as are Bev, Ethan, and Gerry. Ethan is also friends with Claudia, Devi, and Francisco; Devi is also friends with Claudia and Francisco.

(a) Draw a graph G showing the friendship relationship between these seven individuals.

(b) Starting from Alicia, carry out a BFS of G and construct an associated BFS tree.

(c) Suppose that Alicia posts a juicy piece of gossip on her page. Assuming that it takes 15 minutes for this message to spread between any pair of Facebook friends, how long will it be before every one of the seven individuals knows this gossip item?

Maze Problems

7. In the maze shown below, we wish to start at location A and end at location Z.

(a) Label the remaining decision points in the maze with letters.

(b) Draw the graph representing the maze.

(c) Identify a shortest path from A to Z, and then display this solution on the maze.

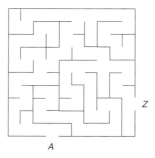

8. In the maze shown next, we wish to start at location A and end at location Z.

(a) Label the remaining decision points in the maze with letters.

(b) Draw the graph representing the maze.

(c) Identify a shortest path from A to Z, and then display this solution on the maze.

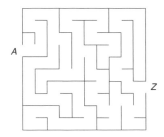

9. A second (lower) path of length 8 from A to Z, using edge (F, G), is seen in Figure 1.10. Interpret this path as a solution to the Hampton Court maze and display this solution on the maze.

10. Figure 1.10 contains two paths of length 9 from A to Z. Identify both paths and display each as a solution to the Hampton Court maze.

11. In the maze shown below, we wish to start at location A and end at location Z.

   (a) Label the remaining decision points in the maze with letters.

   (b) Draw the graph representing the maze.

   (c) Construct a BFS tree from vertex A and use this to determine a shortest path from A to Z.

   (d) Display your solution in part (c) on the maze.

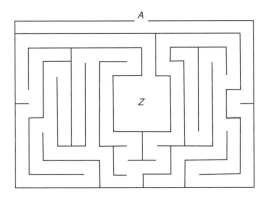

12. In the maze of Figure 1.3, we wish to start at location A and end at location Z.

   (a) Label the remaining decision points in the maze with letters.

   (b) Draw the graph representing the maze. You may need to use multiple edges between some pairs of adjacent vertices.

   (c) Construct a BFS tree from vertex A and use this to determine a shortest path from A to Z.

   (d) Display your solution in part (c) on the maze.

Two-Jug Problems

In the following problems, we are given empty jugs of known capacities. Allowable actions are filling, emptying, and transferring liquid between jugs.

13. We have two empty jugs of capacities 9 pints and 4 pints, respectively. We would like to measure out exactly 3 pints, even though there are no markings on the jugs.

    (a) Extend the BFS tree for this problem, partially shown below, in order to provide a solution to this problem.

    (b) What is the minimum number of pourings needed? Show the corresponding solution path on your graph.

    (c) Interpret this solution in the context of the original problem, as a sequence of pourings.

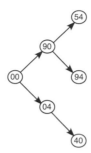

14. We have two empty jugs of capacities 8 pints and 3 pints, respectively. We would like to measure out exactly 1 pint, even though there are no markings on the jugs.

    (a) Create a BFS tree for this problem, extended far enough to solve this problem.

    (b) What is the minimum number of pourings needed? Show the corresponding solution path on your graph.

    (c) Interpret this solution in the context of the original problem, as a sequence of pourings.

15. We have two empty jugs of capacities 9 gallons and 5 gallons, respectively. We would like to measure out exactly 6 gallons, even though there are no markings on the jugs.

    (a) Create a BFS tree for this problem, extended far enough to solve this problem.

    (b) What is the minimum number of pourings needed? Show the corresponding solution path on your graph.

    (c) Interpret this solution in the context of the original problem, as a sequence of pourings.

16. We have two empty jugs of capacities 9 gallons and 5 gallons, respectively. We would like to measure out exactly 2 gallons, even though there are no markings on the jugs.

    (a) Create a BFS tree for this problem, extended far enough to solve this problem.

    (b) What is the minimum number of pourings needed? Show the corresponding solution path on your graph.

    (c) Interpret this solution in the context of the original problem, as a sequence of pourings.

17. We have two empty jugs of capacities 7 gallons and 5 gallons, respectively. We would like to measure out exactly 1 gallon, even though there are no markings on the jugs.

    (a) Create a BFS tree for this problem, extended far enough to solve this problem.

(b) What is the minimum number of pourings needed? Show the corresponding solution path on your graph.

(c) Answer part (b) if you now want to measure out exactly 6 gallons.

18. Suppose that evil Simon Gruber had given John McClain and Zeus a 6-gallon jug and a 9-gallon jug to measure out exactly 5 gallons. Would they have been able to disarm the bomb? Justify your answer.

Three-Jug Problems

In the following problems, we are given jugs of known capacities that contain specified amounts of liquid. The only allowable action is transferring liquid between the jugs.

19. Example 1.9 discovered a solution that involved just two pourings. It also indicated a solution path that uses three pourings. Interpret this path in the context of the original problem, as a sequence of pourings between the jugs.

20. Consider the three-jug problem of Example 1.9, with jugs of capacities of 8, 5, and 3 pints, respectively. Starting with 8 pints in the largest jug, we want to measure out exactly 1 pint.

    (a) Extend the BFS tree in Figure 1.14 enough to provide a solution to this problem.

    (b) Using this graph, find a solution involving the minimum number of pourings.

21. Consider the three-jug problem of Example 1.9, with jugs of capacities of 8, 5, and 3 pints, respectively. We want to divide the original 8 pints in the largest jug equally into two jugs having 4 pints each.

    (a) Extend the BFS tree in Figure 1.14 enough to provide a solution to this problem.

    (b) Using this graph, find a solution involving the minimum number of pourings.

22. Suppose we have three jugs with capacities of 10, 9, and 5 gallons, respectively. The 10-gallon jug is empty and the other two are completely filled with water. Suppose that our goal is to leave 6 gallons in one jug and 3 gallons in another.

    (a) Create a BFS tree for this problem, extended far enough to solve this problem.

    (b) Using your graph, find a solution involving the minimum number of pourings.

    (c) Identify a second solution using the minimum number of pourings.

23. Suppose we have three jugs with capacities of 9, 8, and 4 pints, respectively. We begin with the smaller two jugs completely full, giving a total of 12 pints. We want to measure out exactly 7 pints.

    (a) Create a BFS tree for this problem, extended far enough to solve this problem.

    (b) Is it possible to measure out exactly 7 pints? If so, how can this be done? Are there several solutions using the minimum number of pourings?

    (c) Answer the questions in part (b) for measuring out exactly 6 pints.

24. Suppose we have three jugs with capacities of 9, 8, and 2 pints, respectively. We begin with the smaller two jugs completely full, giving a total of 10 pints. We want to divide these 10 pints so that there are two jugs having 4 pints each. That is, we wish to reach the final configuration of $(4, 4, 2)$.

(a) Create a BFS tree for this problem, extended far enough to solve this problem.

(b) Using this graph, find a solution involving the minimum number of pourings.

25. Suppose we have three jugs with capacities of 9, 8, and 2 pints, respectively. We begin with the smaller two jugs completely full, giving a total of 10 pints. We want to divide these 10 pints equally so that there are two jugs having 5 pints each. That is, we wish to reach the final configuration of $(5, 5, 0)$.

(a) Create a BFS tree for this problem, extended far enough to solve this problem.

(b) Using this graph, find a solution involving the minimum number of pourings.

Other Graph Search Problems

26. We have egg timers that measure out 7 minutes and 4 minutes, respectively. A recipe requires that we be able to measure out exactly 6 minutes.

(a) Using a graph representation, determine if it is possible to do this.

(b) If so, describe the sequence of steps to measure out exactly 6 minutes. How long does it take?

27. We have egg timers that measure out 7 minutes and 4 minutes, respectively. A recipe requires that we be able to measure out exactly 5 minutes.

(a) Using a graph representation, determine if it is possible to do this.

(b) If so, describe the sequence of steps to measure out exactly 5 minutes. How long does it take?

28. A baker in ancient Greece has two hourglasses. One measures 15 minutes and the other measures 7 minutes. A recipe requires him to bake his baklava for 9 minutes.

(a) Using a graph representation, determine if it is possible for him to use these hourglasses to do this.

(b) If so, describe the sequence of steps he should take so that he has exactly 9 minutes left (on the 15-minute timer).

29. In the chessboard of Example 1.11, we now want to move a knight from square $(5, 1)$ to square $(4, 2)$ using valid knight moves and not landing on any shaded square.

(a) What is the fewest number of moves the knight will need to reach square $(4, 2)$?

(b) In how many ways can the knight reach square $(4, 2)$ using the minimum number of moves?

30. In the chessboard shown next, we want to move a knight starting from square A.

(a) What is the fewest number of valid moves needed to reach square B? Show such a solution on the chessboard.

(b) What is the fewest number of valid moves needed to reach square C? Show such a solution on the chessboard.

(c) Are there any squares that cannot be reached from square A? If so, which ones?

Project

31. Write a computer program to determine the fewest number of valid knight moves needed to progress from a given starting square to a designated ending square in a chessboard with forbidden (shaded) squares. Then test out your code on the sample chessboard below for two cases: finding a shortest sequence from square A to square B, and a shortest sequence from square A to square C.

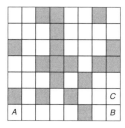

Bibliography

[Barabasi 2002] Barabási, A.-L. 2002. *Linked: The new science of networks.* Cambridge, MA: Perseus.

[Easley 2011] Easley, D., and Kleinberg, J. 2010. *Networks, crowds, and markets.* New York: Cambridge University Press.

[Fisher 2006] Fisher, A. 2006. *The amazing book of mazes.* New York: Harry N. Abrams.

[Higgins 2007] Higgins, P. M. 2007. *Nets, puzzles, and postmen,* Chapter 9. New York: Oxford University Press.

[Watts 2004] Watts, D. 2004. *Six degrees: The science of a connected age.* New York: W. W. Norton.

[OR1] http://gwydir.demon.co.uk/jo/maze (accessed September 25, 2013).

[OR2] http://abcnews.go.com/Primetime/story?id=2717038 (accessed September 25, 2013).

[OR3] http://www.nytimes.com/2011/11/22/technology/between-you-and-me-4-74-degrees.html?_r=0 (accessed December 9, 2013).

[OR4] http://www.youtube.com/watch?v=1Z64IR2bz5o (accessed September 25, 2013).

Chapter 2

Greedy Algorithms and Dynamic Programming

Greed is right, greed works. Greed clarifies, cuts through, and captures the essence of the evolutionary spirit. Greed, in all of its forms; greed for life, for money, for love, knowledge has marked the upward surge of mankind.

—Gordon Gekko, protagonist, *Wall Street*, 1987 film

Chapter Preview

Problems occurring in finance, simple change-making, or TV game shows can all be viewed as sequential decision-making problems. Such situations typically involve a number of stages, each composed of states. A decision at a given state leads us to a state in the next stage, which again requires a decision. We can represent these decision problems using both tables and graphs. We first discuss a greedy approach, which successively takes the best option at each stage to obtain an answer. While quick, it need not produce the best (or optimal) overall solution. We then explore the technique of dynamic programming, which breaks up a formidable decision problem into several smaller subproblems. By using a "best of the best" approach (incorporating the best solution to each subproblem), we can then optimally solve the original problem. Heuristic algorithms such as the greedy approach are useful if we are interested in quickly obtaining a good, though not necessarily optimal, answer.

2.1   Investments and the Knapsack Problem

You have a fixed amount of money to invest and several mutual funds from which to choose. There are costs and benefits associated with each fund. You want to maximize the financial return earned on your investments. Which investments should you select?

This type of problem is an instance of what is called a *knapsack problem*. The name derives from the predicament of a traveller who has to pack a fixed-size knapsack or backpack, yet can't take everything. A choice must be made about which items to pack and which items to leave behind. The traveller must then *optimize* the choices made, packing the most useful items, yet respecting the size constraints imposed by the knapsack.

DEFINITION **Knapsack Problem**

This type of optimization problem requires making decisions about which items to include or exclude in order to maximize the overall utility (value or benefit), taking into account limited resources or financial constraints.

DEFINITION **Optimization Problem**

In mathematics and computer science, an optimization problem involves finding the best option from among all feasible choices—for example, one that is fastest, cheapest, or most desirable.

In attempting to solve a knapsack problem, we might intuitively use a strategy where we begin by choosing an item with the greatest value that our constraints allow. At each subsequent decision point (or stage of the problem), we select an item with the greatest value that our constraints allow. Repeatedly applying this method is referred to as using a *Greedy Algorithm*. For example, suppose Cristina is packing a knapsack (that can hold 12 lb) for a trail bike ride, and she must choose from five items whose weights and importance (value) are given in Table 2.1. Here we assume that no more than one of any item can be chosen: there is no need to take along two video cameras, for example.

TABLE 2.1: An illustrative knapsack problem

Item	Weight (lb)	Value
A	4	14
B	7	29
C	2	9
D	5	17
E	3	11

Using a Greedy Algorithm, we would choose item B having the greatest value 29 first; this takes up 7 of the 12 allowable pounds, which then leaves us with 5 pounds. Among the remaining items, D has the largest value of 17, and it completely uses up the 5 pounds available in the knapsack. The greedy solution consists then of selecting items B and D with total value $29 + 17 = 46$. However, this is not the highest value that can be achieved; using B, C, E uses up the 12 allowable pounds, but gives a higher total value $29 + 9 + 11 = 49$. We see here that using a Greedy Algorithm doesn't deliver the best solution. An online tool for finding optimal solutions to knapsack problems can be found at [OR1].

(Related Exercises: 1–2)

DEFINITION **Algorithm**

A systematic recipe, or set of computational steps, used to produce an answer to problems of a certain type.

DEFINITION **Greedy Algorithm**

A procedure that takes the best choice at each stage of a problem, hoping that concentrating on the best local choice will lead to an optimal solution of the problem. Choices taken in one stage are not revised or reconsidered at a later stage.

Example 2.1 Solve an Investment Problem Using a Greedy Algorithm

Jordan has $50,000 to invest in five possible bond funds with different returns (see Table 2.2), but with the same 3-year maturity. In the interest of diversifying his portfolio, Jordan will select at most one of each bond fund. He wants to fully invest the $50,000 and earn the maximum return. This is a knapsack problem: we are trying to maximize the total return of the investment, subject to a budget constraint of $50,000. (a) Apply the Greedy Algorithm to produce an investment strategy. (b) Investigate whether this investment strategy is optimal for Jordan.

TABLE 2.2: An investment problem with five choices

Fund	Denomination	Return
A	$10,000	$2300
B	$20,000	$3400
C	$30,000	$5400
D	$10,000	$900
E	$20,000	$3200

Solution

(a) Fund C has the largest return and is selected first, leaving us with $50,000 − $30,000 = $20,000. Among the remaining choices, fund B has the largest return and it completely uses up the $20,000 available. Application of the Greedy Algorithm then results in selecting funds B and C for a total return of $3400 + $5400 = $8800.

(b) The Greedy Algorithm is quick, but again it doesn't deliver the best result. To see this, we develop Table 2.3, which lists all possible selections that add up to $50,000, using 1 to mean invest and 0 to mean don't invest. The optimal solution occurs in the second row (select investments A, B, E), producing $2300+$3400+$3200 = $8900. So here again greed doesn't pay! ◁

(*Related Exercises: 3–5, 29*)

TABLE 2.3: List of all choices for the investment problem

A ($10,000)	B ($20,000)	C ($30,000)	D ($10,000)	E ($20,000)	Total Return
1	0	1	1	0	$8600
1	1	0	0	1	$8900
0	1	0	1	1	$7500
0	1	1	0	0	$8800
0	0	1	0	1	$8600

Unlike the Greedy Algorithm, constructing a table of all feasible choices is guaranteed to identify the best overall solution. However, it requires considerable work, especially for more complex problems. It would be nice if we could enumerate all the relevant choices in a systematic way. It turns out that we can—by using a directed graph representation (see Section 1.3), one in which the edges allow travel in a specified direction.

We illustrate this idea by constructing a directed graph for the investment situation in Example 2.1. This construction is not something you need to learn to do by hand, but rather to appreciate. A computer can be easily programmed to carry out this task for problems with many investment options. Here are the steps in constructing the graph.

1. Vertices are labeled with two elements: for example, $(A,5), (B,5), (B,4)$. The letter represents the investment, and the number represents the current budget (in \$10,000 units). For example, $(B,4)$ indicates that we are about to make a decision whether to choose investment B and that we have \$40,000 available at that point.

2. Each directed edge indicates an investment choice: horizontal edges indicate no investment and diagonal edges indicate an investment. For example, the horizontal edge directed from $(A,5)$ to $(B,5)$ means that we don't choose investment A, so \$50,000 remains to be used and we can move on to considering investment B. The diagonal edge directed from $(A,5)$ to $(B,4)$ means that we choose investment A (using up \$10,000), now leaving us with \$40,000 and a decision to make about investment B.

3. Each edge is labeled with the value produced by the corresponding investment choice: for example, \$0 if A is not chosen and \$2300 if A is chosen.

4. Dashed edges mean that at this point there is no way to invest the entire \$50,000 in view of previous choices. These edges lead to a dead-end, as in a maze (Section 1.1).

5. Starting from vertex $(A,5)$, our objective is then to reach the final vertex $(F,0)$, indicating that we have allocated the entire \$50,000.

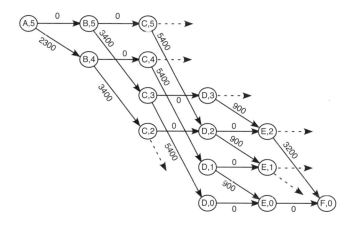

FIGURE 2.1: Directed graph for the investment problem

The directed graph in Figure 2.1 is composed of several (vertical) stages as we move from left to right, with each stage representing an investment decision. The important thing to note is that each path from the starting vertex $(A,5)$ to the final vertex $(F,0)$ represents a possible investment strategy; the overall value of a path is obtained by adding up the values along its edges.

Example 2.2 Identify Paths and Investment Decisions

(a) Interpret the path $(A,5) \to (B,5) \to (C,5) \to (D,2) \to (E,2) \to (F,0)$ in Figure 2.1 as a particular selection of investments. Compute the value of this path. (b) Interpret the selection of investments B, D, E as a particular path from $(A,5)$ to $(F,0)$ in Figure 2.1. Compute the value of this path.

Solution

(a) In the path $(A,5) \to (B,5) \to (C,5) \to (D,2) \to (E,2) \to (F,0)$, the edge from $(A,5)$ to $(B,5)$ indicates that we did not choose A since no budget was consumed; similarly, the edge from $(B,5)$ to $(C,5)$ shows that we did not choose B. However, the edge $(C,5)$ to $(D,2)$ shows that we did choose C, reducing our available budget to $\$50{,}000 - \$30{,}000 = \$20{,}000$. Finally, the edges from $(D,2)$ to $(E,2)$ and from $(E,2)$ to $(F,0)$ indicate that D was not used but E was used. That is, we selected investments C and E only. The value of this path is $\$0 + \$0 + \$5400 + \$0 + \$3200 = \$8600$.

(b) The selection of investments B, D, E is represented by the path $(A,5) \to (B,5) \to (C,3) \to (D,3) \to (E,2) \to (F,0)$. This selection entirely uses the $\$50{,}000$ and provides a return of $\$0 + \$3400 + \$0 + \$900 + \$3200 = \$7500$. ◁

(Related Exercises: 6–7)

There are in fact five possible investment paths from vertex $(A,5)$ to vertex $(F,0)$. These are shown in Table 2.4, along with their associated values. We can see that the best path, the one with highest value, is $(A,5) \to (B,4) \to (C,2) \to (D,2) \to (E,2) \to (F,0)$. This verifies that the best investment strategy is to select A, B, E for a total return of $\$8900$.

TABLE 2.4: Paths and investment decisions

Path	Investments	Total Return
$(A,5) \to (B,4) \to (C,4) \to (D,1) \to (E,0) \to (F,0)$	A, C, D	$\$8600$
$(A,5) \to (B,4) \to (C,2) \to (D,2) \to (E,2) \to (F,0)$	A, B, E	$\$8900$
$(A,5) \to (B,5) \to (C,3) \to (D,3) \to (E,2) \to (F,0)$	B, D, E	$\$7500$
$(A,5) \to (B,5) \to (C,3) \to (D,0) \to (E,0) \to (F,0)$	B, C	$\$8800$
$(A,5) \to (B,5) \to (C,5) \to (D,2) \to (E,2) \to (F,0)$	C, E	$\$8600$

To summarize, the investment problem can be viewed as an "optimal path" problem, in which we seek a path from a starting vertex to an ending vertex that has the maximum value. (In the maze and jug problems of Chapter 1 we also searched for an optimal path, one having the minimum number of edges.)

A natural question is how to find all these paths systematically. Or if there are too many possible paths, is there an efficient way to find the best path? We will return to these issues in Section 3.1, where finding the best path arises in both DNA sequencing and GPS routing.

2.2 Change-Making Problems

Suppose we want to give someone change for 31 cents and we want to do it using as few coins as possible. This can be viewed as a type of knapsack problem: now we have a budget of 31 cents (rather than a budget of 12 lb) and desire to *minimize* the number of coins (rather than maximize the utility value). In this case, there is a known set of allowable denominations for the coins and (unlike our previous investment problem) we are able to choose more than one coin of each denomination in order to make the required change. In general, *change-making problems* ask us to determine which coins should be chosen in order to minimize the number of coins required to make specific change.

DEFINITION **Change-Making Problem**
Given a set of coins of various denominations, the goal is to use the fewest coins that sum to a specified amount.

Let's use trial-and-error to solve the problem of making change for 31 cents using U.S. coinage of 1, 5, 10, and 25 cents. Table 2.5 lists several possibilities for making change along with the number of coins used. Clearly, using the three coins $1, 5, 25$ is the optimal solution.

TABLE 2.5: Making change for 31 cents

Coins Used	Number of Coins
$5 + 5 + 5 + 5 + 5 + 5 + 1$	7
$10 + 10 + 5 + 5 + 1$	5
$10 + 10 + 10 + 1$	4
$25 + 5 + 1$	3

If we had used the Greedy Algorithm in the above situation, we would have first chosen the largest coin denomination of 25 cents, leaving us with the need to make change for $31 - 25 = 6$ cents. The largest coin denomination (not exceeding 6 cents) is 5 cents. Choosing this as our second coin leaves us with just 1 cent, so we select 1 cent as our third coin. We see that the Greedy Algorithm produces the optimal solution for this particular problem. It turns out that the Greedy Algorithm is always optimal for U.S. coinage. The reason for this has to do with the fact that any denomination of a U.S. coin is a multiple of another U.S. coin with a smaller denomination. A general description of the Greedy Algorithm for change-making problems is as follows:

Greedy Algorithm for Making Change

1. Select a coin with the largest denomination not exceeding the change amount; subtract its value from the change amount.

2. Continue to choose a coin of largest denomination that does not exceed the amount left to be paid; then subtract this coin's value from the amount remaining.

3. Repeat Step 2 until the selected coins sum to the change required.

Example 2.3 Solve Change-Making Problems Using the Greedy Algorithm

Suppose we have non-U.S. coin denominations of 1, 5, 9, and 16 cents. (a) Give change to a customer for 22 cents using the Greedy Algorithm. (b) Give change to another customer for 42 cents using the Greedy Algorithm.

Solution

(a) Applying the Greedy Algorithm with 22 cents, we use the largest denomination 16 cents, leaving us with $22 - 16 = 6$ cents. The largest denomination coin that can now be used is 5 cents, leaving us with $6 - 5 = 1$ cent. At this point we use the 1-cent coin, giving the following three coins to the first customer: $16 + 5 + 1 = 22$.

(b) Applying the Greedy Algorithm with 42 cents, we first select 16 cents, leaving $42 - 16 = 26$ cents. We can again use the largest denomination 16 cents, now leaving us with $26 - 16 = 10$ cents. The largest coin that can be used to make change for 10 cents is the 9-cent coin, leaving us with $10 - 9 = 1$ cent. At this point, we use a 1-cent coin, giving the following four coins to the second customer: $16 + 16 + 9 + 1 = 42$. ◁

(Related Exercises: 8–9)

In the situations encountered in Example 2.3, the Greedy Algorithm is in fact optimal—it produces change using the fewest coins. However, this need not always be the case.

Example 2.4 Improve a Solution Found by the Greedy Algorithm

With non-U.S. coin denominations of 1, 5, 10, 12, and 25 cents, give change for 41 cents to a customer using the Greedy Algorithm. Then, see if you can identify a better solution.

Solution

Applying the Greedy Algorithm results in selecting the following six coins:

$$41 = 25 + 12 + 1 + 1 + 1 + 1.$$

Is this the fewest number of coins? Using trial-and-error, we list in Table 2.6 several combinations of coins that add to 41. For example, the first row indicates that we use one 25-cent coin, one 12-cent coin, and four 1-cent coins. The solutions shown reduce the number of coins used from 6 to 5 and then to 4. It turns out that the solution indicated with four coins (25, 10, 5, 1) is in fact optimal. So, this example shows that a greedy approach does not always produce the fewest number of coins. ◁

(Related Exercises: 10–15)

TABLE 2.6: Several ways to make 41 cents

25	12	10	5	1	Total	Number of Coins
1	1	0	0	4	41	6
1	0	0	3	1	41	5
1	0	1	1	1	41	4

2.3 Dynamic Programming

In Example 2.4, applying the Greedy Algorithm resulted in using six coins, which is not an optimal solution. If we knew in advance that our earlier choices would ultimately force us to use four 1-cent coins, we would have proceeded differently. To remedy this short-sighted approach (which assumes that making the best local decision at each step will lead to the best result overall), we use a *dynamic programming* approach—one that is guaranteed to find an optimal solution to any change-making problem. The dynamic programming philosophy is a more sophisticated greedy approach to problems that proceed in stages: you make the best local decision at the current stage, based on the best solutions available at other stages. In this way, the original optimization problem is broken up into smaller, more manageable, optimization problems. This technique is commonly used in applied problem-solving domains, such as in *Operations Research.*

DEFINITION **Dynamic Programming**
This technique involves breaking an optimization problem into smaller subproblems and then combining solutions of those subproblems to obtain an optimal solution to the larger problem. The term "programming" does not refer to computer programming, but rather to the formulation and solution of optimization (planning/programming) problems.

DEFINITION **Operations Research**
An interdisciplinary approach that uses methods of mathematics, statistics, and computer science to arrive at optimal or near-optimal solutions to complex decision problems.

Thai 21

Before revisiting change-making problems, we illustrate the ideas behind dynamic programming using a game called *Thai 21*, featured in the TV reality show *Survivor*. *Thai 21* is a variation of an ancient game called *Nim* [OR1], in which two players take turns removing objects from distinct heaps and the winner is the person who takes the last object.

Example 2.5 Solve the *Thai 21* Immunity Challenge Game

There are 21 flags in total and the two teams alternate in selecting either 1, 2, or 3 flags. The team that removes the last flag wins the game. What is the best strategy for a team to pursue?

Solution

We can consider the number of flags n remaining to be the *state* of the system when a team is about to make its decision. We classify each state as either being a winning (W) state or a losing (L) state. Rather than trying to decide what to do in state $n = 21$, we begin small. That is, we find out the best strategy for small values of n and then use that information to make optimal decisions for larger values of n.

If a team is presented with $n = 1$, $n = 2$, or $n = 3$ flags, then all of these are winning states since we simply remove all the flags and we win. We denote this by $W(1), W(2), W(3)$.

Now suppose $n = 4$. If we choose 1 flag, then we leave the opposing team in state $W(3)$—a winning state for them. Likewise, choosing 2 flags leaves $W(2)$ and choosing 3 flags leaves $W(1)$. So in any case we leave the opponent with a winning state, meaning state 4 is a losing state $L(4)$ for us.

When $n = 5$, selecting 1, 2, 3 flags leaves the respective states $L(4), W(3), W(2)$ for the opposing team. Then our best strategy is to select 1 flag so the other team is left in a losing state. Similarly, when $n = 6$, we should select 2 flags so the other team is left in the losing state $L(4)$. When $n = 7$, we should select 3 flags so the other team is also left in the losing state $L(4)$. We summarize these results by writing $W(5), W(6), W(7)$.

Now consider $n = 8$. Selecting 1, 2, 3 flags leaves the respective states $W(7), W(6), W(5)$ all of which are winning states. So state 8 is a losing state $L(8)$. On the other hand, states 9, 10, 11 are winning states since there is some choice that leaves the opponent in the losing state $L(8)$. [Check this out!]

Using this same logic, we can develop Table 2.7, which shows winning and losing states. In particular, $n = 21$ is a winning state: taking 1 flag will leave the opponent in the losing state $L(20)$. In general, the optimal strategy is to take enough flags to leave the opponent in a losing state: an integer multiple of 4. In the *Thai 21* game, the first team to select flags is guaranteed a win using this strategy. ◁

(Related Exercises: 16–22, 30)

TABLE 2.7: Winning and losing states

1	2	3	4	5	6	7	8	9	10	11	12	13	14	15	16	17	18	19	20	21
W	W	W	L	W	W	W	L	W	W	W	L	W	W	W	L	W	W	W	L	W

Thai 21 is played so that the person making the last move will win the game. But it can also be played in what is called the "misère" format, where the person making the last move loses. For example, in the game called *21*, two players take turns saying a number and the player forced by the rules to say 21 or more loses. The rules state that the first player must say 1. At each subsequent turn, the player must increase the current number by 1, 2, or 3. To illustrate, Table 2.8 shows the possible alternating moves for the two players. Player says 1, and then Player 2 increases this by 3, saying 4. Suppose Player 1 increases the current number by 2 to 6; then Player 2 responds by saying 8, an increase of 2. Player 2 also responds with 8 if Player 1 says 5 or 7. The winning strategy for this game is to say a multiple of 4; after that, it is guaranteed that the other player will have to say 21 (or more). This game is now biased in favor of the second player, who can get to 4 first and then subsequently control the game.

TABLE 2.8: Possible moves in the game *21*

Player	Possible Moves
1	1
2	4
1	5, 6, 7
2	8
1	9, 10, 11
2	12
1	13, 14, 15
2	16
1	17, 18, 19
2	20

Notice that our analysis proceeds by developing optimal strategies for successively larger subproblems. This is the essential feature of a dynamic programming approach. In Example 2.6 we will apply this same dynamic programming philosophy to the change-making problem for 41 cents, introduced in Example 2.4. Similar to the investment problem, we will represent the states by vertices and draw a directed graph that shows allowable transitions between states.

Example 2.6 Solve a Change-Making Problem Using Dynamic Programming

With coin denominations of 1, 5, 10, 12, and 25 cents, use dynamic programming to determine the fewest number of coins to make change for 41 cents.

Solution

We construct a directed graph in which each vertex (state) is labeled with an ordered pair of numbers (A, L), where A is the amount of change needed and L is the largest denomination coin that is allowed. That is, L restricts the size of the largest coin that we are allowed to use to make change for A. For example, if a vertex is labeled $(41, 12)$, then we are allowed to use 12, 10, 5, and 1, but not 25 cents. Also, each vertex will have directed edges leading away from it, labeled with one of the available denominations.

Stage 1 consists of the initial vertex $(41, 25)$, which has five edges leaving it. These correspond to the five options of using 25, 12, 10, 5, or 1 as the largest denomination coin to make change for the amount 41.

Stage 2 contains the five vertices adjacent to $(41, 25)$. In Figure 2.2, the edge labeled 25 from vertex $(41, 25)$ to vertex $(16, 12)$ represents first choosing 25 cents, leaving $41 - 25 = 16$ cents. The second number 12 appearing in vertex $(16, 12)$ now restricts the largest available denomination to 12, since it is not possible to use 25 cents in making change for 16 cents.

Similarly, the edge labeled 12 from $(41, 25)$ to $(29, 12)$ represents choosing 12 cents as the largest denomination, leaving $41 - 12 = 29$ cents. Since this edge is labeled 12, the second number 12 appearing in $(29, 12)$ now restricts the largest available denomination to 12.

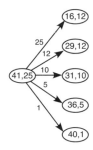

FIGURE 2.2: First two stages for the change-making problem

In order to optimally solve our original first-stage problem $(41, 25)$ we need to optimally solve the smaller subproblems $(16, 12), (29, 12), (31, 10), (36, 5), (40, 1)$ appearing at the second stage. We first identify those subproblems that seem to be easily solved by inspection: where we can convince ourselves (see Exercise 23) that the Greedy Algorithm does indeed use the fewest coins. For example,

- Subproblem $(29, 12)$ is optimally solved using three coins, namely two 12-cent coins and one 5-cent coin.

- Subproblem $(31, 10)$ is optimally solved using four coins, namely three 10-cent coins and one 1-cent coin.

- Subproblem $(36, 5)$ is optimally solved using eight coins, namely seven 5-cent coins and one 1-cent coin. Note that we cannot use three 12-cent coins, since for subproblem $(36, 5)$ we are restricted to using only 1-cent and 5-cent coins.

- Subproblem $(40, 1)$ requires 40 coins, as we are only allowed to use 1-cent coins.

The optimal number of coins 3, 4, 8, 40 needed to solve these four (Stage 2) subproblems are shown in blue in Figure 2.3.

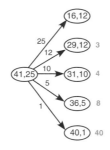

FIGURE 2.3: Optimal solution values for several Stage 2 subproblems

If we apply the Greedy Algorithm to subproblem $(16, 12)$, we would need five coins: 12, 1,1,1,1; however, it is not clear that this greedy solution is the best we can do. So, when in doubt, we develop the next stage (Stage 3) for this subproblem. Specifically, we add edges from $(16, 12)$ to four new subproblems, corresponding to the available denominations 12, 10, 5, 1. Each of these subproblems can be solved by inspection; Figure 2.4 displays as blue the optimal values (fewest coins) for these Stage 3 subproblems.

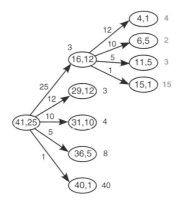

FIGURE 2.4: Expanded Stage 3 subproblems

Given the optimal values for the Stage 3 subproblems of $(16, 12)$, we find the optimal value for $(16, 12)$ by taking an edge leading to the smallest Stage 3 optimal value (namely, 2). The optimal value for $(16, 12)$ is then $2 + 1 = 3$, since we need one additional coin to get from Stage 2 to Stage 3; this optimal value is shown as the label 3 on $(16, 12)$. In other words, three coins is the fewest number needed to make change for subproblem $(16, 12)$. Note that the Greedy Algorithm applied to $(16, 12)$ would have followed the topmost edge, requiring $4 + 1 = 5$ coins to make change for 16 cents. By contrast, the dynamic programming approach says to follow the edge that leads to the *smallest* value at the next stage.

Finally, let's return to the original problem $(41, 25)$, making change for 41 cents, by working backward from Stage 2 to Stage 1. We look at all of the optimal values for each Stage 2 subproblem and pick the smallest one (namely, 3) to obtain the optimal value $3 + 1 = 4$ for the single vertex at Stage 1. This means that four is the fewest coins needed to make change for 41 cents. The optimal value 4 for the original problem is shown in blue in Figure 2.5.

Moreover, this graphical representation enables easy identification of four coins that constitute an optimal solution. Figure 2.5 shows the path $(41, 25) \rightarrow (16, 12) \rightarrow (6, 5)$, decreasing from value 4 to value 3 to value 2. The edges of this path indicate that we chose coin 25 and then coin 10 leading us to vertex $(6, 5)$, where the indicated solution is to use the two coins 5 and 1. Overall, this path corresponds to selecting coins $25 + 10 + 5 + 1 = 41$. ◁

(*Related Exercises: 23–28*)

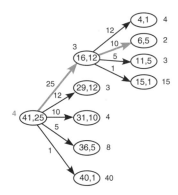

FIGURE 2.5: Optimal solution for Stage 1

By our construction, every path from $(41, 25)$ will have edge weights appearing in nonincreasing order. This results from adding the second component L to each vertex, which ensures that our coins are processed in nonincreasing (denomination) order. Also, notice that there is a second edge in Figure 2.5 from $(41, 25)$ leading to an optimal Stage 2 value of 3: from $(41, 25)$ to $(29, 12)$ using a 12-cent coin. The greedy solution associated with subproblem $(29, 12)$ uses the three coins 12, 12, 5. This produces a second way to make change for 41 cents with four coins: $12 + 12 + 12 + 5 = 41$. Again, this solution is produced in nonincreasing order of coin denominations.

In contrast to the trial-and-error approach used in Example 2.4, the graph representation enables a more systematic approach to find an optimal solution. As a matter of fact, it allows us to identify *all* optimal solutions to the change-making problem. Neither of these optimal solutions would have been found by the Greedy Algorithm. On the other hand, we can apply dynamic programming to identify the best solutions to any change-making problem. The idea is to decompose the original problem into smaller, more manageable subproblems. As needed, this decomposition can be continued until the optimal solutions to all subproblems are obtained. Then we "work backward" to solve our original problem.

Chapter Summary

This chapter introduces several decision problems in which a best solution is desired: for example, an investment strategy yielding the maximum revenue, or a way to make change that requires the fewest coins. We can search for a quick, but not necessarily the best, answer using the Greedy Algorithm. By contrast, optimal solutions can be found by a systematic enumeration of all possibilities (aided by a graphical representation), or by using the technique of dynamic programming. This latter strategy decomposes the original problem into smaller subproblems—whose solution provides stepping stones leading to an answer to the original problem. The dynamic programming approach also enables us to retrieve not just one optimal solution to the given problem, but it facilitates the discovery of all optimal solutions.

Applications, Representations, Strategies, and Algorithms

Applications					
Finance (1–4, 6)					

Representations					
Graphs			Tables		Sets
Weighted	Directed	Tree	Data	Decision	Lists
2, 6	2, 6	6	1	5	1

Strategies		
Brute-Force	Solution-Space Restrictions	Composition/Decomposition
1	1–4	5–6

Algorithms					
Exact	Approximate	Greedy	Enumeration	Sequential	Dynamic Programming
1–2, 4–6	3–4	1, 3–4	1, 4	2	5–6

Exercises

Knapsack Problems and Investment Decisions

1. Luke has a backpack to fill with items and must decide on which ones to select. The backpack has a maximum allowable weight of 27 lb. The weight and value of each item are given in the table below.

   (a) Using the Greedy Algorithm, what items would be selected?

   (b) Can you find a better way to fill the backpack?

Item	Weight (lb)	Value
A	10	30
B	3	8
C	7	16
D	9	20
E	6	12
F	5	9
G	2	3

2. Serena has a prepaid mailing envelope to fill with items and must decide on which ones to place in the envelope. The envelope can have a maximum allowable weight of 25 oz. The weight and value of each item are given in the following table.

   (a) Using the Greedy Algorithm, what items would be selected?

   (b) Can you find a better way to fill the envelope?

Item	Weight (oz)	Value
A	10	20
B	9	16
C	8	15
D	7	12
E	6	10

3. You have $50,000 to invest in five possible bond funds with the same maturity and having the financial details shown in the table below. In the interest of diversifying your portfolio, you want to select at most one of each bond fund. You also want to fully invest the $50,000.

   (a) What investment strategy results if you use the Greedy Algorithm to select the largest return at each step? Determine the associated total return for this selection.

   (b) Create a table listing all selections of funds that fully use the given $50,000 and then compute the total return for each selection.

   (c) Identify from part (b) the optimal investment strategy.

Fund	Denomination	Return
A	$10,000	$2100
B	$10,000	$1300
C	$20,000	$3200
D	$20,000	$3100
E	$30,000	$5100

4. A company has $35,000 to allocate for funding internal research projects. Five projects (A, B, C, D, E) have been proposed, each with the associated cost and benefit listed in the following table. The firm would like to select a group of projects giving the highest total benefit, and that uses the entire $35,000 budget.

   (a) What projects would be funded using the Greedy Algorithm to select the largest benefit at each step? Determine the associated total benefit for this selection.

   (b) Create a table listing all selections of projects that fully use the $35,000 and then compute the total benefit for each selection. What is the optimal funding strategy?

Project	Cost	Benefit
A	$5000	$10,000
B	$10,000	$16,000
C	$10,000	$18,000
D	$15,000	$23,000
E	$20,000	$32,000

5. A homeowner has a line of credit of $25,000 upon which to draw in order to make home improvements (e.g., upgrading the kitchen, extending the screen porch). Five possible improvements (A, B, C, D, E) are shown below. The homeowner wants to select those improvements giving the highest total benefit (addition to the resale value), without exceeding the $25,000 line of credit.

   (a) What improvements would be selected using the Greedy Algorithm? Determine the associated total benefit for this selection.

   (b) Create a table listing all selections of three improvements with total cost at most $25,000 and then compute the total benefit for each such selection. Identify the best set of improvements from this list. Argue why you have found the optimal solution.

Project	Cost	Benefit
A	$12,000	$16,000
B	$9000	$13,000
C	$7600	$11,500
D	$6800	$10,000
E	$3500	$5000

6. Consider the investment problem represented by the directed graph in Figure 2.1.

   (a) Interpret the path $(A, 5) \to (B, 5) \to (C, 3) \to (D, 0) \to (E, 0) \to (F, 0)$ in this graph as a particular selection of investments. Compute the value of this path.

   (b) Interpret the selection of investments A, C, D as a particular path from $(A, 5)$ to $(F, 0)$ in this graph. Compute the value of this path.

7. Explain why there is a dashed horizontal line leaving vertex $(C, 4)$ in Figure 2.1.

Greedy Algorithms and Change-Making Problems

8. In Example 2.3(a), the Greedy Algorithm produced a solution that uses three coins to make change for 22 cents, when the available denominations are $1, 5, 9, 16$ cents. Verify that this solution is optimal—namely, there is no way to make change for 22 cents by using fewer than three coins.

9. In Example 2.3(b), the Greedy Algorithm produced a solution that uses four coins to make change for 42 cents, when the available denominations are $1, 5, 9, 16$ cents. Verify that this solution is optimal—namely, there is no way to make change for 42 cents by using fewer than four coins. [*Hint*: First, argue that a 16-cent coin must be used to make change when using fewer than four coins; that then leaves the task of making change for 26 cents.]

10. Using coin denominations of $1, 5, 10, 12, 25$, we want to make change for 31 cents.

    (a) What solution does the Greedy Algorithm produce?
    (b) Can you find a solution using fewer coins?

11. Using coin denominations of $1, 5, 10, 12, 25$, we want to make change for 32 cents.

    (a) What solution does the Greedy Algorithm produce?
    (b) Can you find a solution using fewer coins?

12. Using coin denominations of $1, 5, 10, 12, 25$, we want to make change for 34 cents.

    (a) What solution does the Greedy Algorithm produce?
    (b) Can you find a solution using fewer coins?

13. Using coin denominations of $1, 5, 10, 12, 25$, we want to make change for 67 cents.

    (a) What solution does the Greedy Algorithm produce?
    (b) Can you find a solution using fewer coins?

14. Using coin denominations of $1, 4, 10, 12$, we want to make change for 31 cents.

    (a) What solution does the Greedy Algorithm produce?
    (b) Can you find a solution using fewer coins?

15. Using coin denominations of $1, 5, 7, 12, 17$, we want to make change for 42 cents.

    (a) What solution does the Greedy Algorithm produce?
    (b) Find all optimal solutions to this problem.

Dynamic Programming and Thai 21

16. Verify the entries given for $n = 8, 9, 10, 11, 12$ in Table 2.7.

17. There are 21 flags and each team must select either 1, 3, or 4 flags. The team that removes the last flag wins the game.

    (a) Fill in the table below with either a W or L to indicate whether the state (number of flags) is a Winning or Losing state.
    (b) Decide whether it is advantageous to go first or second in this game. Explain.

1	2	3	4	5	6	7	8	9	10	11	12	13	14	15	16	17	18	19	20	21
W		W	W																	

18. There are 21 flags and each team must select either 1, 2, or 4 flags. The team that removes the last flag wins the game.

    (a) Fill in the following table with either a W or L to indicate whether the state (number of flags) is a Winning or Losing state.
    (b) Decide whether it is advantageous to go first or second in this game. Explain.

1	2	3	4	5	6	7	8	9	10	11	12	13	14	15	16	17	18	19	20	21
W	W		W																	

19. There are 21 flags and each team must select either 1, 4, or 6 flags. The team that removes the last flag wins the game.

    (a) Fill in the table below with either a *W* or *L* to indicate whether the state (number of flags) is a Winning or Losing state.

    (b) Decide whether it is advantageous to go first or second in this game. Explain.

1	2	3	4	5	6	7	8	9	10	11	12	13	14	15	16	17	18	19	20	21
W			W		W															

20. There are 21 flags and each team must select either 1, 4, or 5 flags. The team that removes the last flag wins the game.

    (a) Fill in the table below with either a *W* or *L* to indicate whether the state (number of flags) is a Winning or Losing state.

    (b) Decide whether it is advantageous to go first or second in this game. Explain.

1	2	3	4	5	6	7	8	9	10	11	12	13	14	15	16	17	18	19	20	21

21. There are 21 flags and each team must select either 1, 2, or 6 flags. The team that removes the last flag wins the game.

    (a) Fill in the table below with either a *W* or *L* to indicate whether the state (number of flags) is a Winning or Losing state.

    (b) Decide whether it is advantageous to go first or second in this game. Explain.

1	2	3	4	5	6	7	8	9	10	11	12	13	14	15	16	17	18	19	20	21

22. There are 21 flags and each team must select either 1, 5, or 6 flags. The team that removes the last flag wins the game.

    (a) Fill in the table below with either a *W* or *L* to indicate whether the state (number of flags) is a Winning or Losing state.

    (b) Decide whether it is advantageous to go first or second in this game. Explain.

1	2	3	4	5	6	7	8	9	10	11	12	13	14	15	16	17	18	19	20	21

Dynamic Programming and Change-Making Problems

23. In Example 2.6, we asserted that certain subproblems are optimally solved by the Greedy Algorithm. Verify this assertion for the following two cases by arguing that the required change cannot be made using fewer coins:

    (a) subproblem $(29, 12)$ is optimally solved using three coins

    (b) subproblem $(31, 10)$ is optimally solved using four coins

24. Suppose the available denominations are $1, 10, 12, 15$ and we want to make change for 34 cents.

    (a) What solution does the Greedy Algorithm produce?

    (b) Use dynamic programming to solve this problem optimally. Begin with the graph shown next.

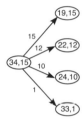

25. Suppose the available denominations are $1, 5, 10, 12$ and we want to make change for 16 cents.

    (a) What solution does the Greedy Algorithm produce?

    (b) Use dynamic programming to solve this problem optimally. [*Hint*: Begin with the initial vertex $(16, 12)$; two stages will be required.]

26. Suppose the available denominations are $1, 5, 10, 12, 25$ and we want to make change for 32 cents. Using dynamic programming, find the fewest number of coins needed to make change for 32 cents. To solve the problem, begin with the following graph and extend to a third stage where necessary.

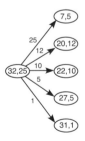

27. Consider the change-making problem for 29 cents, given four denominations of coins: $1, 6, 10, 15$. Using dynamic programming, find all solutions using the fewest number of coins to make change for 29 cents. [*Hint*: Begin with the initial vertex $(29, 15)$; three stages will be required.]

28. Consider the change-making problem for 32 cents, given four denominations of coins: $1, 6, 10, 19$. Using dynamic programming, find all solutions using the fewest number of coins to make change for 32 cents. [*Hint*: Begin with the initial vertex $(32, 19)$; four stages will be required.]

Projects

29. A more sophisticated heuristic for quickly finding a solution to a knapsack problem is the *Greedy Ratio Algorithm*. Rather than selecting at each step an item that fits in the available space and that has the largest benefit, we can select one that has the largest ratio of benefit to weight. The rationale is that selecting items with large benefits will also quickly fill up the available space (weight) in the knapsack. The Greedy Ratio Algorithm ensures that we get the best "benefit per weight" that our constraints allow. Apply this heuristic algorithm to the examples and exercises in this chapter, comparing its behavior to that of the standard Greedy Algorithm.

30. A simple version of *Nim* involves two players who alternate in selecting either 1 or 2 flags from a total of 26 flags. Create a table of winning and losing states for this game. Do similarly when the players can choose 1 or 3 flags; 1 or 4 flags; and finally 1 or 5 flags. Now see if you can conjecture a pattern in the winning and losing states when the players are allowed to select either 1 or k flags, for some integer $k \geq 2$.

Bibliography

[Conway 1996] Conway, J. H., and Guy, R. 1996. *The book of numbers*. New York: Springer-Verlag.

[OR1] `http://coma-coding.com/blog/javascript-knapsack-solver` (accessed April 24, 2014).

[OR2] `http://education.jlab.org/nim/index.html` (accessed October 2, 2013).

Chapter 3

Shortest Paths, DNA Sequences, and GPS Systems

Of all the paths a man could strike into, there is, at any given moment, a best path.

—Thomas Carlyle, Scottish essayist, 1795–1881

Chapter Preview

Finding a best alignment between DNA sequences and finding a desirable routing using a GPS system can be accomplished by determining shortest paths in an associated table or graph. Solutions to both of these problems can be approached by invoking the dynamic programming principle introduced in Chapter 2. Namely, we optimally solve certain smaller subproblems along the way to solving optimally the original problem. In particular, we introduce Dijkstra's Algorithm, a greedy-type approach that is nonetheless guaranteed to produce an optimal path in a weighted graph.

3.1 Finding DNA Sequence Alignments

Geneticists scan the human genome to find patterns and variations that have implications for deciphering and treating complex human diseases, for understanding differences among individuals and populations, for creating a taxonomy of species, and for analyzing forensic evidence in criminal court cases. Improvements in *DNA sequencing* technologies are making this possible and highly reliable.

DEFINITION **DNA Sequencing**
The human genome is composed of approximately three billion base pairs involving four different nucleotides: adenine (A), thymine (T), guanine (G), and cytosine (C). Finding a single, unique sequence of As, Ts, Gs, and Cs within an organism's genome often reveals something biologically important about that organism.

Suppose we are looking for certain genes or proteins in an existing database. We have a specific subsequence (or probe) *Sub* that we want to find within a large main sequence *Main*. If we required an exact match, the search would be relatively straightforward; either it exists somewhere in the main sequence (as it does below) or it does not.

Main: ACGTCAAGT
Sub: AAG

Often scientists have reasons to expect a match that is less than perfect, due to phenomena such as errors or mutations in the genetic code. Below we illustrate a 3-out-of-4 letter match, in which there is a mismatch between the letter A in the main sequence and the letter G in the subsequence:

Main: ACGTC**A**AGT
Sub: C**G**AG

In the next two examples we have a 3-out-of-4 letter match but allow for a *gap*. In the first case below, the gap (−) occurs in the main sequence. In the second case, the gap occurs in the subsequence. Gaps in a sequence could be the result of genetic miscopying of DNA, where a nucleotide has been deleted. A gap mismatch may not be as serious as a direct mismatch because what is missing could have been the desired letter.

Main: ACGTC−AAGT
Sub: C**G**AA

Main: ACGTCA**A**GT
Sub: CA−G

Scientists who are interested in comparing two genetic sequences have devised a scoring scheme to differentiate between true mismatches and possible mismatches. The best alignment is considered to be one with the lowest total penalty score (or cost) in this scheme. An example would be the scoring scheme defined in Table 3.1.

TABLE 3.1: A scoring scheme for alignments

Comparison	Penalty
direct letter match	0
letter gap in main sequence	1
letter gap in subsequence	1
direct letter mismatch	3

Example 3.1 Use a Path in a Table to Represent an Alignment

Construct a table to represent the main sequence GACGATG and the subsequence AGTA. Then identify a path in the table that corresponds to the following sequence alignment:

Main: GA−CGATG
Sub: AGT−A

Solution

We create a table in which each column corresponds to a letter in the main sequence, and each row corresponds to a letter in the subsequence. Consider the path shown in Figure 3.1, which starts at position (A, A) in the table. This path consists of vertical, horizontal, and

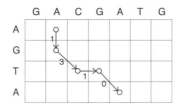

FIGURE 3.1: A path in the alignment table

diagonal edges; each edge has an associated cost, which represents the penalty assigned using Table 3.1. We now explain how this path corresponds to the given alignment of the two sequences.

Since the path in Figure 3.1 begins at vertex (A, A), which is a direct match between letters A and A, there is no penalty at the start of the path. Then the path makes a *vertical* move in column A down to row G, arriving at vertex (G, A). This incurs a penalty of 1 because of a gap in the main sequence; in the given sequence alignment, we have not moved from A in the main sequence, but we have made a move to G in the subsequence. The corresponding edge from (A, A) to (G, A) is assigned the cost 1.

The path now moves *diagonally* from (G, A) to (T, C). Since the two letters T and C are different, we incur the direct mismatch penalty of 3; this cost is associated with the diagonal edge from (G, A) to (T, C).

The path then moves *horizontally* from (T, C) to (T, G), representing a gap in the subsequence. We still remain at T in the subsequence, but move along to the next letter G in the main sequence. The edge from (T, C) to (T, G) is assigned the cost 1, representing the subsequence gap penalty.

The path ends with the diagonal edge from (T, G) to (A, A), representing a direct match of letter A with letter A and incurring penalty 0. Accordingly, the edge from (T, G) to (A, A) is assigned the cost 0. We have now completely exhausted all of the letters of the subsequence. The total cost associated with this path is then $0 + 1 + 3 + 1 + 0 = 5$. This is the total penalty that would be calculated for the given alignment using Table 3.1. ◁

(Related Exercises: 1–4)

As seen in Example 3.1, a vertical edge indicates that there is a gap in the main sequence, a horizontal edge indicates that there is a gap in the subsequence, and a diagonal edge indicates that we have two genuine letters to compare for either a direct match or direct mismatch. It is also important to add the cost of the first vertex in the alignment path to get the correct total penalty for the alignment.

Example 3.2 Show the Alignment Corresponding to a Path

Determine the alignment that is associated with the path displayed in Figure 3.2. Compute the cost of this path and the total penalty score of the corresponding alignment.

Solution

Note that the path starts in the first row of the table and ends in the last row of the table, indicating that all letters of the subsequence are matched with letters or gaps in the main sequence. Initially, the letter A is matched with the letter A. The first diagonal edge corresponds to a mismatch of G with C. The next horizontal edge corresponds to a gap

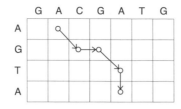

FIGURE 3.2: Another path in the alignment table

in the subsequence, the diagonal edge results from mismatching T with A, and the final vertical edge corresponds to a gap in the main sequence. So the corresponding alignment, along with each penalty, is

Main: GACGA−TG
Sub: AG−TA
Penalty: 0 3 1 3 1

First, A is matched with A at cost 0. Each diagonal edge means a mismatch (cost 3) and each horizontal or vertical edge means a gap (cost 1). So the cost of the path is $0+3+1+3+1 = 8$, agreeing with the total penalty score of 8 calculated using Table 3.1. ◁

(Related Exercises: 5–7)

Since the path illustrated in Example 3.2 has a higher cost than the path in Example 3.1, it is not as desirable as the first path; it produces a less satisfactory alignment of the two given sequences. Since we seek a best alignment, our objective can then be rephrased as finding an optimal (minimum cost) path in the table. First, what are *admissible* paths? Since we need to match up all letters in the subsequence with letters or gaps in the main sequence, we seek a path starting in the first row and ending in the last row of our table. Also, the path is only allowed to make vertical, horizontal, and diagonal moves as it progresses in a south/east/southeast direction.

There is good news and there is bad news. First, let's deal with the bad news—there can be a very large number of paths in our table. To illustrate this, we will see how many paths are admissible in an even smaller table, composed of just four rows and five columns.

Example 3.3 Count the Number of Admissible Paths Starting at the Upper Left Corner

In a 4×5 table, calculate the number of admissible paths that start from the upper left corner. Recall that to be admissible a path must make only vertical, horizontal, or diagonal moves and it must end up in the last row of the table.

Solution

We begin by labeling the upper left cell (A, A) in Figure 3.3 with a 1 to indicate that there is just a single way to get from (A, A) to (A, A): namely, stay where we are! Likewise, we can label all the cells in the first column with a 1 to indicate that there is just one admissible path (vertical down moves) from (A, A) to each of them. All cells in the first row can be labeled with 1 by the same reasoning (only horizontal right moves).

Each remaining cell in the table can be reached in one of three ways. For example, (G, C) can be reached either from (A, C) by a vertical move, from (G, A) by a horizontal move, or from (A, A) by a diagonal move. Since there is one path to each of the preceding cells, there are $1 + 1 + 1 = 3$ ways to get from the starting position to (G, C). We fill in the

	A	C	G	A	T
A	1	1	1	1	1
G	1				
T	1				
A	1				

FIGURE 3.3: 4×5 alignment table

entries of Figure 3.4 from left to right and from top to bottom in the same way. That is, each entry is the sum of the entries in its bordering north, west, and northwest cells. For example, the number of admissible paths from the upper left corner (A, A) to the bottom right corner (A, T) is $41 + 63 + 25 = 129$. By adding all entries in the last row, we get $1 + 7 + 25 + 63 + 129 = 225$ paths from the upper left corner to a location in the last row. ◁

(*Related Exercises: 8–9, 31*)

	A	C	G	A	T
A	1	1	1	1	1
G	1	3	5	7	9
T	1	5	13	25	41
A	1	7	25	63	129

FIGURE 3.4: Counting paths in the alignment table

Even in this small example there are many admissible paths starting at the upper left corner and ending at the bottom row. Moreover, there are other paths (possible alignments) that start elsewhere in the first row. In fact, there are 363 admissible paths from a cell in the top row to a cell in the bottom row (see Exercise 8). So it seems that we are doomed in our quest for an optimal alignment since there are just too many possibilities to consider.

Now for the good news! We don't need to consider all admissible paths, but only require promising paths relative to our penalty scheme. We can again use dynamic programming (Section 2.3) to determine optimal paths in an alignment table; namely, the solutions of smaller subproblems are used in order to solve successively larger problems. Our strategy mimics what was done in the solution of Example 3.3, counting *admissible* paths in an alignment table. Now we extend this technique in order to determine *optimal* paths in an alignment table. As mentioned in the previous paragraph, we need to account for starting our paths anywhere in the first row. The next two examples pursue this line of reasoning.

Example 3.4 Compute Optimal Path Costs in an Alignment Table

Consider the optimal alignment problem for the subsequence AGTA and the main sequence GACGATG. First, construct a 4×7 table, since there are four letters in the subsequence and seven letters in the main sequence. Fill in this table with optimal path costs, beginning with the first column and the first row, and then progressively assign the remaining entries of the table. Remember that vertical and horizontal moves incur a penalty of 1, a mismatch incurs a penalty of 3, and a match incurs no penalty.

Solution

We start by assigning a cost of 3 to the upper left corner since the A and G do not match (we are assuming there is no initial gap). Moving down the first column corresponds to inserting gaps in the main sequence, each with a penalty of 1. This gives costs of $3 + 1 = 4$ for cell (G, G), $4 + 1 = 5$ for cell (T, G), and $5 + 1 = 6$ for cell (A, G).

We now consider the rest of the first row. Cell (A, A) in the first row could be assigned cost $3 + 1 = 4$ if we made a horizontal move from (A, G). But instead, since any entry in the first row can be a starting point for our path, cell (A, A) is assigned cost 0 since we have a direct match. Next, cell (A, C) could either receive cost 3 (as a mismatched starting point) or cost $0 + 1 = 1$ if we make a horizontal move from (A, A). We always choose the smaller cost (the dynamic programming idea!), so cell (A, C) is assigned cost 1. We sequentially fill in the rest of the first row using similar reasoning. Figure 3.5 shows the optimal costs obtained so far.

	G	A	C	G	A	T	G
A	3	0	1	2	0	1	2
G	4						
T	5						
A	6						

FIGURE 3.5: Optimal costs for the first row and column

Just as in the path-counting problem (Example 3.3), we can progressively fill in the remaining cells of the table with the optimal cost to reach that cell from the first row. Again, we simply look at the costs in the bordering north, west, northwest cells and add to these the corresponding costs for a vertical, horizontal, diagonal move. For example, the cost for (G, A) in row 2, column 2 is the smaller of $0 + 1 = 1$ (vertical), $4 + 1 = 5$ (horizontal), $3 + 3 = 6$ (diagonal). So the cost 1 is assigned to cell (G, A). Similarly, the cost for (G, C) in row 2, column 3 is the smaller of $1 + 1 = 2$ (vertical), $1 + 1 = 2$ (horizontal), $0 + 3 = 3$ (diagonal). Notice that there is a tie for the smallest cost 2, which is assigned to cell (G, C). Figure 3.6 displays the optimal costs for the entire table. ◁

(Related Exercises: 10–12, 32)

	G	A	C	G	A	T	G
A	3	0	1	2	0	1	2
G	4	1	2	1	1	2	1
T	5	2	3	2	2	1	2
A	6	3	4	3	2	2	3

FIGURE 3.6: Optimal costs for the entire table

Example 3.5 Compute Optimal Paths and Best Alignments

Using the optimal costs shown in Figure 3.6, determine the smallest penalty score for an alignment. Then identify the alignments that achieve this smallest score.

Solution

Since we are interested in paths ending at the last row, we consult the costs in the last row and find the smallest one, which is 2. This value, the smallest possible penalty of an alignment, appears twice in the last row: at (A, A) in column 5 and at (A, T) in column 6. Let's begin with (A, A) and *trace back* a corresponding path (and alignment). There are three possible ways the cost of cell (A, A) could have been assigned: either vertically, horizontally, or diagonally. The three values used to determine the cost for (A, A) are $2 + 1 = 3$ (vertical), $3 + 1 = 4$ (horizontal), or $2 + 0 = 2$ (diagonal). This means that cell (A, A) was assigned cost 2 by using a diagonal edge from cell (T, G). We now work backward from cell (T, G), with cost 2: it was determined as the smaller of $1 + 1 = 2$ (vertical), $3 + 1 = 4$ (horizontal), or $2 + 3 = 5$ (diagonal). So the cost 2 of cell (T, G) was assigned using the vertical edge from (G, G). Continuing in this way we generate the path $(A, A) \rightarrow (A, C) \rightarrow (G, G) \rightarrow (T, G) \rightarrow (A, A)$ shown in Figure 3.7. Its corresponding alignment is

Main: GACG–ATG
Sub: A–GTA

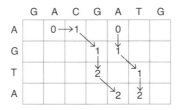

FIGURE 3.7: Optimal paths for the alignment problem

This technique of working backward can also be applied to cell (A, T), which also has cost 2. It produces the second optimal path in Figure 3.7, which starts in row 1, column 5: $(A, A) \rightarrow (G, A) \rightarrow (T, T) \rightarrow (A, T)$. The corresponding alignment is

Main: GACGA–T–G
Sub: AGTA

As a reality check, notice that both optimal alignments create no mismatches and insert just two gaps for a penalty score of $1 + 1 = 2$, agreeing with our optimal cost of 2. ◁

(Related Exercise: 13)

Example 3.6 Identify Alternative Optimal Paths

The values shown in Figure 3.8 represent optimal alignment costs. By tracing backward, find all optimal paths that correspond to the cost 3 in the lower right corner of the table.

	C	A	T	G	G	C
A	3	0	1	2	3	3
G	4	1	2	1	2	3
C	5	2	3	2	3	2
A	6	3	4	3	4	3

FIGURE 3.8: Optimal costs for an alignment problem

Solution

The optimal cost 3 shown in column 6 of the last row is determined by a vertical move from (C, C), and the cost 2 in (C, C) is determined by a diagonal move from (G, G). However, the cost 2 in (G, G) can be obtained by a diagonal move from (A, G) as $2+0 = 2$ or as a horizontal move from (G, G) as $1 + 1 = 2$. Continuing to trace backward leads to the edges shown in Figure 3.9. From this we obtain the two optimal paths $(A, A) \rightarrow (A, T) \rightarrow (A, G) \rightarrow (G, G) \rightarrow (C, C) \rightarrow (A, C)$ and $(A, A) \rightarrow (A, T) \rightarrow (G, G) \rightarrow (G, G) \rightarrow (C, C) \rightarrow (A, C)$. There are additional optimal paths associated with the other two cost values 3 in the last row of the alignment table. $\triangleleft$

(Related Exercises: 14–18)

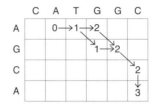

FIGURE 3.9: Alternative optimal paths in an alignment table

We now summarize our approach for obtaining optimal sequence alignments.

Dynamic Programming Algorithm for Optimal Sequence Alignment

1. Create a table, with rows corresponding to the subsequence letters and columns corresponding to the main sequence letters.

2. Fill in the costs of the first column, based on incurring a vertical gap penalty (1).

3. Fill in the costs of the first row, based on whether it is less costly to directly match (0), mismatch (3), or incur a horizontal gap penalty (1).

4. Working from top to bottom, and left to right, determine the remaining costs by choosing the smallest sum of adjacent cell cost plus move penalty from the north, west, and northwest cells.

5. The optimal cost is the smallest value in the last row of the table. For each optimal cost entry, trace back the associated optimal path(s).

6. For each optimal path, find the corresponding optimal sequence alignment.

3.2   GPS Routings and Dijkstra's Algorithm

The ability to quickly compute a shortest path between any two points on a map is an important feature of a GPS device. After choosing the desired destination (address or landmark), GPS software computes a shortest path from the current location to the final destination. The shortest route is then displayed on the map and the application guides the user along the road. If the user takes a different path or requires a detour, then a new shortest path can be recomputed from the current location to the final destination.

DEFINITION **Global Positioning System (GPS)**
This space-based radio-navigation system consists of a group of orbiting satellites that transmit highly accurate time and position signals. By computing the time it takes signals to travel from several satellites, GPS receivers are able to compute a user's position as well as the time.

A shortest path in routing systems can represent the minimum distance path (computed using road lengths), or the minimum time path (assuming that each road is traversed at the posted speed limit). These can produce different routes, and the GPS user is typically given a choice of which cost criterion to use. In fact, some GPS systems give the user the opportunity to weight these criteria, or to take into account road repairs, traffic congestion, tolls, or possibly aesthetics (scenic routes).

Suppose that we wish to find a shortest route between Atlanta and Boston. Given a road map indicating the various streets/highways, how can we efficiently determine such a shortest route? We can model this problem using a *directed network*. Intersections correspond to vertices, and roads joining intersections correspond to edges. A one-way road corresponds to a directed edge, whereas a two-way road can be represented by a pair of oppositely directed edges. Each edge of the network has an associated numerical *cost* (e.g., representing length or travel time). Given an origin vertex s and a destination vertex t, the shortest path problem requires finding a path from s to t having minimum cost.

(Related Exercise: 19)

DEFINITION **Directed Network**
A directed graph in which each edge (i, j) has an associated cost $c_{i,j}$.

DEFINITION **Cost of a Path**
The sum of all edge costs along the path.

The most popular shortest path algorithm used in GPS systems is known as *Dijkstra's Algorithm*. In order to find a shortest path from a specified origin vertex s to a specified destination vertex t, this algorithm systematically expands the set S of vertices with known shortest path distances from vertex s. Again we implement a dynamic programming philosophy, by solving small problems in an optimal way in order to solve larger and larger problems optimally. Dijkstra's Algorithm is also a greedy-type approach in the sense that it expands the current set S by selecting at each step a nearest vertex. Remarkably, this greedy approach is guaranteed to produce an optimal solution.

HISTORICAL CONTEXT **Dijkstra's Algorithm**

This shortest path algorithm was developed in 1956 by E. W. Dijkstra (1930–2002), a famous Dutch computer scientist. He did not consider publishing his method until 1959, since before that time research into efficient algorithms had barely begun and was not considered legitimate scholarship. Given the massive sizes of the transportation and communication networks in today's highly connected world, efficient algorithms are now the holy grail of both mathematicians and computer scientists alike.

Before launching into an example of using Dijkstra's Algorithm, we first point out how this method parallels and extends the approach used to determine optimal sequence alignments (Section 3.1). There we had an *initialization* step (filling out cells in the first row and first column), an *order* for selecting the next cell to process (top to bottom, left to right), a *calculation* step that used cost values on adjacent cells to determine the cost for the selected cell, and a *retracing* step to retrieve an optimal path from the optimal cost values. Here are the analogous steps for Dijkstra's Algorithm, applied to the problem of determining a shortest path from vertex s to vertex t in a network with edge costs $c_{i,j}$. At each step of this algorithm, certain vertices will have labels (or *distances*) that represent the cost of a path (not necessarily optimal) from s to that vertex. When that label is in fact optimal, we call it a *permanent* label; otherwise, it is called *temporary*.

- *Initialization.* Since vertex s is at distance 0 from itself, vertex s receives the permanent label 0 and is added to the set S of permanently labeled vertices. If vertex j is adjacent to s, it receives the temporary label $c_{s,j}$, the cost of edge (s, j).

- *Order.* Rather than using an order dictated by its position in the alignment table, the next vertex processed is one having the smallest temporary label. This amounts to a greedy choice—selecting a vertex that is closest to s and then adding it to S.

- *Calculation.* Any vertex j that is outside S but is adjacent to some vertex i in S receives a temporary label that represents the smallest sum of the permanent label on i plus the edge cost $c_{i,j}$.

- *Retracing.* Rather than waiting until the very end to determine the edges constituting an optimal path (as in the alignment problem), we keep track of the *routing edge* (i, j) that produced the minimum sum in the calculation step.

Dijkstra's Algorithm continues this process (identifying a new permanently labeled vertex and updating labels on vertices not in S) until the destination vertex t has been added to S. At that point we can use the routing edges to reconstruct a shortest path from s to t. In the next example, we illustrate these steps for a very small network. While brute force could be used to find the shortest path, that is not our objective. Rather, we need to appreciate how Dijkstra's Algorithm carries out its work efficiently for very large networks.

Example 3.7 Use Dijkstra's Algorithm to Find Shortest Paths

Figure 3.10 shows a directed network with a cost appearing on each edge. Apply Dijkstra's Algorithm to find a shortest path from the origin vertex 1 to the destination vertex 6.

Solution

To initialize the algorithm, permanently label the origin vertex $s = 1$ with the distance 0 and place vertex 1 in set S. Then label each vertex adjacent to 1 with the cost of the edge from 1 to that vertex. Namely, vertex 2 receives the temporary label 6 while vertex 3

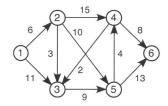

FIGURE 3.10: Example network for Dijkstra's Algorithm

receives the temporary label 11. Since vertex 2 has the smaller temporary label, this label is declared permanent and vertex 2 is added to S, giving $S = \{1, 2\}$. The edge $(1, 2)$ used to label vertex 2 is the routing edge. Figure 3.11 shows the current vertex labels, the set S (within the blue curve), and the routing edge (bold); the accompanying table records each vertex selected to enter S, the routing edge, and the permanent label (distance from s).

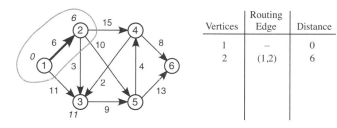

Vertices	Routing Edge	Distance
1	–	0
2	(1,2)	6

FIGURE 3.11: Dijkstra's Algorithm with two permanent labels

Vertices $3, 4, 5$ are adjacent to the set $S = \{1, 2\}$. Each of these vertices is labeled with the smallest sum of permanent label plus edge cost to that vertex. Specifically, vertex 4 receives the label $6 + 15 = 21$ and vertex 5 receives the label $6 + 10 = 16$. For vertex 3, there are two sums to consider: $0 + 11 = 11$ and $6 + 3 = 9$; so vertex 3 is labeled with the smaller value 9, in effect crossing out the previous value 11 appearing on this vertex. Vertex 3 has the smallest temporary label; it is added to the set S and its label 9 becomes permanent. The associated routing edge $(2, 3)$ is added to the table and appears bold in Figure 3.12.

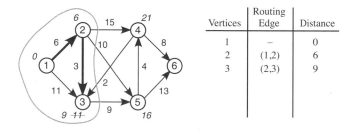

Vertices	Routing Edge	Distance
1	–	0
2	(1,2)	6
3	(2,3)	9

FIGURE 3.12: Dijkstra's Algorithm with three permanent labels

Vertices 4 and 5 are adjacent to the set $S = \{1, 2, 3\}$. Vertex 4 is labeled with $6 + 15 = 21$, the same as its previous label. Vertex 5 receives the label 16, the smaller of $6 + 10 = 16$ and $9 + 9 = 18$. Vertex 5 has the smaller temporary label, so 5 is added to the set S and its label 16 is declared permanent. The associated routing edge $(2, 5)$ is shown in bold in Figure 3.13.

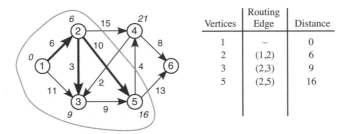

FIGURE 3.13: Dijkstra's Algorithm with four permanent labels

Vertices 4 and 6 are adjacent to the set $S = \{1, 2, 3, 5\}$. Vertex 4 receives the label 20, the smaller of $6 + 15 = 21$ and $16 + 4 = 20$, replacing the previous label 21. Vertex 6 is labeled with $16 + 13 = 29$. Vertex 4 has the smaller temporary label, so 4 is added to the set S and its label 20 is declared permanent. The routing edge $(5, 4)$ is shown in bold in Figure 3.14.

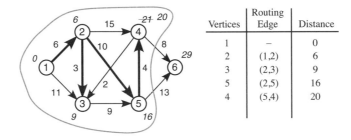

FIGURE 3.14: Dijkstra's Algorithm with five permanent labels

Now vertex 6 is the only vertex adjacent to the set $S = \{1, 2, 3, 4, 5\}$ and it receives the label 28, the smaller of $20 + 8 = 28$ and $16 + 13 = 29$. The destination vertex 6 is now added to the set S with its permanent label 28, and the associated routing edge is $(4, 6)$. This completes the labeling portion of Dijkstra's Algorithm. Figure 3.15 shows all routing edges and the final table. The permanent label (distance) 28 on vertex 6 means that the minimum cost of a path from $s = 1$ to $t = 6$ is 28. Using the final table, we can now trace backward to obtain a corresponding shortest path. Specifically from entry 6, the routing edge is $(4, 6)$; from entry 4, the routing edge is $(5, 4)$; from entry 5, the routing edge is $(2, 5)$; and from entry 2, the routing edge is $(1, 2)$. This identifies the shortest path $1 \rightarrow 2 \rightarrow 5 \rightarrow 4 \rightarrow 6$ with total cost $6 + 10 + 4 + 8 = 28$. ◁

(*Related Exercises: 20–24*)

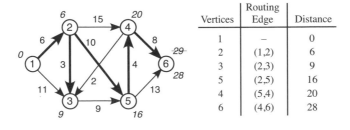

FIGURE 3.15: Final results of Dijkstra's Algorithm

Note that the bold edges in Figure 3.15 form a tree directed from the origin vertex s. Just as in the jug-pouring problems of Section 1.3, an optimal solution is found by following the edges in this solution tree. For example, the tree not only shows us a shortest path from vertex 1 to vertex 6, but it also indicates a shortest path from vertex 1 to vertex 3: $1 \to 2 \to 3$ with total cost $6 + 3 = 9$.

Here are the steps of Dijkstra's Algorithm, which can be applied to any network with nonnegative edge costs $c_{i,j} \geq 0$. An interactive illustration of the algorithm can be found at [OR1].

Dijkstra's Shortest Path Algorithm

1. Initialize the set S to contain the origin vertex s, which receives the permanent label of 0. Assign the temporary label $c_{s,j}$ to each vertex j adjacent to s.

2. Select a vertex with smallest temporary label, making its label permanent and adding it to S. Identify the corresponding routing edge to this vertex.

3. For each vertex j adjacent to S identify all the edges (i, j) leading to it from a vertex i in S. Compute the temporary label for vertex j as the (smallest) sum of the permanent label on a vertex i plus the edge cost $c_{i,j}$.

4. Repeat Steps 2 and 3 until S contains the destination vertex t.

5. Use the routing edges to trace a shortest path from vertex s to vertex t.

Example 3.8 Find Shortest Paths from a Routing Table

Dijkstra's Algorithm has been used to calculate shortest paths from a source vertex $s = 1$ in a network with seven vertices. Table 3.2 shows the final routing information. (a) Trace out a shortest path from vertex 1 to vertex 6 and determine its total cost. (b) Trace out a shortest path from vertex 1 to vertex 7 and determine its total cost. (c) Trace out a shortest path from vertex 1 to vertex 2 and determine its total cost.

TABLE 3.2: Routing table from Dijkstra's Algorithm

Vertex	Routing Edge	Distance
1	—	0
4	$(1, 4)$	8
2	$(1, 2)$	12
5	$(4, 5)$	14
3	$(5, 3)$	21
7	$(5, 7)$	22
6	$(3, 6)$	25

Solution

(a) The distance listed in the table for vertex 6 is 25, so that is the cost of a shortest path from 1 to 6. To trace backward from 6, we consult the routing edge entry $(3, 6)$ for vertex 6. This indicates that the last edge on the shortest path is from 3 to 6. Continuing, the routing edge entry for vertex 3 is $(5, 3)$, for vertex 5 is $(4, 5)$, and for vertex 4 is $(1, 4)$. We have now traced back the shortest path $1 \to 4 \to 5 \to 3 \to 6$.

(b) The distance listed in the table for vertex 7 is 22, so that is the cost of a shortest path from 1 to 7. To trace backward from 7, we start with the routing edge $(5,7)$ for vertex 7. Then the routing edge entry for vertex 5 is $(4,5)$, and for vertex 4 it is $(1,4)$. We have now traced back the shortest path $1 \to 4 \to 5 \to 7$.

(c) From the table the cost of a shortest path from 1 to 2 is 12. To trace backward from 2, we start with the routing edge $(1,2)$ for vertex 2. Since we are now back at the origin vertex 1, we stop. So the shortest path is simply the edge $(1,2)$ from 1 to 2. ◁

(*Related Exercises: 25–26*)

3.3   Other Optimal Path Problems

Even though the discussion in Section 3.2 focused on determining shortest paths from an origin vertex s to a destination vertex t, there are other optimization problems that can be viewed as variants of shortest path problems. We investigate how these problems can be modeled as optimal path problems and solved by slightly modifying Dijkstra's Algorithm.

Example 3.9 Arrange a Business Meeting

Figure 3.16 shows the (two-way) street system of the downtown area of a city. Adrian, Briana, and Carmen need to meet an important client at 9:00 am at the Triumph Tower building. Their respective offices (a, b, c) and the Tower (t) are indicated in Figure 3.16, along with the travel times (in minutes) along each street. What are the latest times that Adrian, Briana, and Carmen can leave their offices to arrive on time at the Triumph Tower?

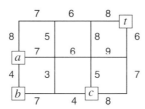

FIGURE 3.16: Grid of city streets

Solution

Because Adrian, Briana, and Carmen want to arrive at the Triumph Tower as soon as possible, they should take shortest paths *to* the destination vertex t, using the travel times as costs. While we previously considered shortest paths *from* an origin vertex s, we can easily modify our solution approach to solve the problem at hand. Another difference here is that the streets are two-way; this does not present any difficulty either, since each undirected edge can be treated as two oppositely directed edges with the same cost.

Dijkstra's Algorithm can be adapted in the following way. First, we label the destination vertex t with cost 0, add it to the set S, and label all vertices i adjacent to t with the cost $c_{i,t}$. We select a vertex not in S with the minimum temporary label to enter S. In this way, the set S "grows" from the northeast corner of the street grid. At each step, we compute temporary labels for all vertices adjacent to the current set S.

For example, the vertex just to the west of t receives the temporary label 8 and the vertex just to the south of t receives the temporary label 6. As this latter vertex has the smaller temporary label, the label 6 now becomes permanent and this vertex enters S. Once all vertices a, b, c have entered S, the processing stops and we have the routing edges shown in bold in Figure 3.17. The permanent labels on a, b, c are also displayed and extraneous edge costs have been removed for clarity. Each of the individuals should follow his or her shortest path (using the bold routing edges). For example, Adrian should leave his office by 8:34 am at the latest, to arrive 26 minutes later at Triumph Tower for the 9:00 am meeting. Similarly, Briana should leave by 8:31 am and Carmen by 8:40 am. ◁

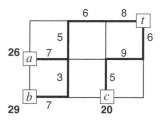

FIGURE 3.17: Shortest path tree for Example 3.9

Example 3.10 Determine an Optimal Rental Policy

Ridha rents out a cottage for visitors to a nearby music festival. He has received a number of bids for differing rental periods over the next nine days. Table 3.3 shows the dates requested and the rental bids offered. For example, the first bid is for $150 to cover check-in on September 1 and check-out on September 3. What is the best rental policy for Ridha?

TABLE 3.3: Rental bids for the cottage

	1	2	3	4	5	6	7	8
Starting Date	9/1	9/1	9/2	9/3	9/5	9/6	9/7	9/8
Ending Date	9/3	9/4	9/5	9/6	9/7	9/9	9/9	9/9
Bid	$150	$240	$300	$180	$170	$280	$150	$130

Solution

First, we construct the network of Figure 3.18, in which the vertices correspond to the rental days and the directed edge (i, j) indicates a rental checking in on day i (of September) and checking out on day j (of September). The value shown on each edge is the rental income for that time period. Notice that a rental for bid 4 can directly follow a rental for bid 1, whereas the rentals for bids 1 and 3 cannot both be accepted (they overlap on day 2). In addition, we have placed edges from day i to day $i + 1$ giving income 0 to allow for the possibility that the cottage is not rented for that time period.

Each path in this network corresponds to a valid set of bids that could be accepted. For instance, the path $1 \to 4 \to 5 \to 7 \to 9$ corresponds to accepting bids 2, 5, 7 and keeping the cottage vacant for day 4, producing a total income of $240 + 0 + 170 + 150 = 560$. Since a set of bids that maximizes the total income is desired, we are in fact looking for a path from vertex 1 to vertex 9 that has *maximum* value. Notice that the edges (i, j) in this network are always moving forward in time: i.e., we have $i < j$. Because of this property, Dijkstra's Algorithm can be modified to find a longest rather than a shortest path in the

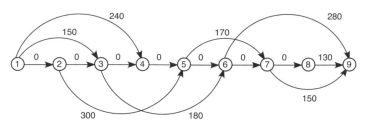

FIGURE 3.18: Network of rental bids

network. Rather than selecting a vertex with minimum temporary label to enter S, we just select the vertices in numerical order 1, 2, ..., 9. Also, when computing the temporary label of a vertex, we find the *maximum* sum of permanent label and edge value. Applying this modified algorithm yields the routing edges in Figure 3.19. Tracing backward from vertex 9 produces the path $1 \rightarrow 2 \rightarrow 5 \rightarrow 7 \rightarrow 9$ of value $0 + 300 + 170 + 150 = 620$. This means that the maximum rental income of \$620 can be obtained by accepting bids 3, 5, and 7. ◁

(*Related Exercises: 27, 30*)

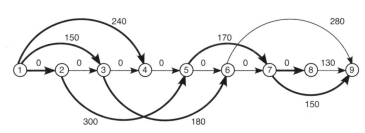

FIGURE 3.19: Optimal selection of rental bids

Example 3.11 Find a Most Reliable Path in a Communication Network

The links of a communication network can occasionally be inoperable because of congestion or outages. Historically one can associate a likelihood that each link will be functioning. These edge *reliabilities* are indicated on the directed edges of the communication network shown in Figure 3.20. Assume that communication links operate independently of one another. What is the best way to send a message from vertex 1 to vertex 7 in this network?

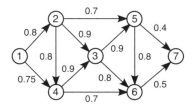

FIGURE 3.20: A communication network with edge reliabilities

Solution

Assuming that edges do not influence one another, each path has an associated reliability given by the product of its constituent edge reliabilities. For example, the path $1 \rightarrow 2 \rightarrow$

$5 \to 7$ has reliability $0.8 \times 0.7 \times 0.4 = 0.224$. Is there a better path along which to send our message? We seek then a path from vertex 1 to vertex 7 having the maximum reliability. While this is not directly a shortest path problem, it is an optimal path problem where it is desired to *maximize* the *product* of edge values, rather than minimize the sum of edge costs. Dijkstra's Algorithm can be modified to handle this problem. We initialize the origin vertex s to have label 1 (perfectly reliable). As in Example 3.10, vertices will be selected to enter S based on the maximum temporary label. To accommodate the product (rather than the sum) measure of a path's reliability, the temporary label on a vertex j not in S is computed using the product of the permanent label on vertex i and the reliability of edge (i, j). With these modifications, Dijkstra's Algorithm produces the routing edges shown in Figure 3.21. Tracing backward from vertex 7 along these routing edges produces the path $1 \to 2 \to 3 \to 6 \to 7$ with maximum reliability $0.8 \times 0.9 \times 0.8 \times 0.5 = 0.228$. ◁

(*Related Exercise: 28*)

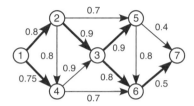

FIGURE 3.21: Most reliable paths in a communication network

Criminal activities (such as smuggling and money laundering) can be tracked from data sources that link various "entities" and their occurrence together. The entities could be people, financial institutions, locations, or vehicles. The associations could arise from meetings or geographical proximity between individuals, from financial transactions, and so forth. An important aspect of uncovering criminal activities is establishing an overall association between two entities (e.g., between two seemingly unrelated individuals). This can be aided by analyzing paths in a criminal network.

Example 3.12 Identify Strong Associations in a Criminal Network

A weight is placed on each edge of a criminal network, indicating the strength of association between the corresponding entities. It can be thought of as the likelihood that the two entities (vertices) i and j are related to one another. While this weight measures the direct association between entities i and j, there may be indirect connections between the two, given by a path of edges between i and j in the criminal network. As in Example 3.11, this indirect association can be measured by the product of the edge weights along the path. We therefore seek the *strongest* association between two particular entities, corresponding to the *largest* weight of a path between these vertices. Determine the strongest association between vertices 1 and 6 in the criminal network shown in Figure 3.22.

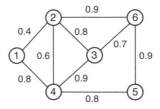

FIGURE 3.22: A criminal network

Solution

To find the largest weight path from 1 to 6 in this undirected network, we apply Dijkstra's Algorithm, modified as in Example 3.11. First, the source vertex 1 is labeled with the weight 1 and all vertices adjacent to 1 (namely, 2 and 4) receive the weight of the edge from vertex 1. This labels vertex 2 with 0.4 and vertex 4 with 0.8; vertex 4 (having maximum label) now enters S, giving the situation in Figure 3.23(a). Next, the vertices $2, 3, 5$ adjacent to S are given the updated labels $0.48, 0.72, 0.64$ shown in Figure 3.23(b). Notice that vertex 2 receives the label 0.48, the larger of $1 \times 0.4 = 0.4$ and $0.8 \times 0.6 = 0.48$. Vertex 3 has the largest temporary label and enters S. This process continues and the algorithm terminates with the routing edges shown in Figure 3.24. The largest association between vertices 1 and 6 is then determined by the path $1 \to 4 \to 5 \to 6$ with weight $0.8 \times 0.8 \times 0.9 = 0.576$. ◁

(*Related Exercise: 29*)

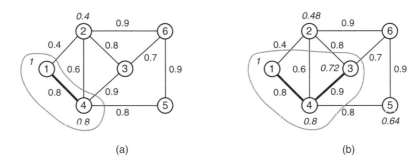

(a)   (b)

FIGURE 3.23: First two iterations of the modified Dijkstra Algorithm

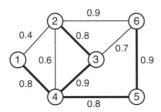

FIGURE 3.24: Strongest association paths in a criminal network

Chapter Summary

This chapter discusses two practical optimization problems, one a biological sequencing problem and another a vehicle routing problem. Both require the determination of least cost paths. In the first problem, a table provides a useful representation for studying the alignment of two sequences. Moreover, a suitable path in this table corresponds to a valid alignment. To determine a minimum penalty alignment, we equivalently seek a minimum cost path in the alignment table. Dynamic programming can again be applied to build up the final solution from the solutions to smaller subproblems. This same approach can be used to solve minimum cost routing problems in networks, such as those representing road systems, airline routes, or communication systems. Specifically, we describe Dijkstra's Algorithm, which builds a shortest path tree by greedily adding a nearest vertex to a successively growing set S of vertices with known minimum distances from the origin vertex s. Even though greedy algorithms do not in general produce optimal results, Dijkstra's Algorithm is indeed guaranteed to produce genuine shortest paths from the origin vertex to all vertices of the network. Other applications of finding optimal paths in a network are also illustrated.

Applications, Representations, Strategies, and Algorithms

Applications					
Biology (1–2, 4–6)		Navigation (7–8)		Scheduling (9)	
Business (10)		Communication (11)		Law Enforcement (12)	

Representations					
Graphs				Tables	
Weighted	Undirected	Directed	Path	Data	Decision
7, 9–12	9, 12	7, 10–11	1–2	5–6, 10	1–4, 8

Strategies	
Rule-Based	Composition/Decomposition
1–2, 5–6	3–4, 7–12

Algorithms		
Exact	Dynamic Programming	Recursive
1–12	3–4, 7, 9–12	5–6, 8

Exercises

DNA Alignments and Paths

1. Construct an alignment table for the main sequence GAGCTGT and the subsequence ACGT. Show in that table the path corresponding to the alignment

Main: GA−GCTGT
Sub: ACG−T

2. Construct an alignment table for the main sequence GCACGTGA and the subsequence ATGG. Show in that table the path corresponding to the alignment

      Main: GCA−CGTGA
      Sub: AT−G−G

3. Construct an alignment table for the main sequence CGATGAC and the subsequence GCGGA. Show in that table the path corresponding to the alignment

      Main: CG−ATGAC
      Sub: GCG−G−A

4. Construct an alignment table for the main sequence AGTCGCA and the subsequence GCTTA. Show in that table the path corresponding to the alignment

      Main: AG−TC−GCA
      Sub: GCT−T−−A

5. Find the alignment corresponding to the following path and calculate its penalty score.

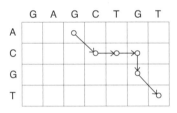

6. Find the alignment corresponding to the following path and calculate its penalty score.

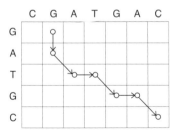

7. Find the alignment corresponding to the following path and calculate its penalty score.

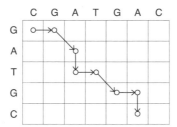

Admissible Paths in an Alignment Table

8. We are interested in counting all admissible paths from the first row to the last row in Figure 3.4. We already know there are 225 admissible paths from the upper left corner of the table to the last row of the table.

   (a) Determine the number of admissible paths from the fifth cell in the first row to a cell in the last row.

   (b) Determine the number of admissible paths from the fourth cell in the first row to a cell in the last row.

   (c) Determine the number of admissible paths from the third cell in the first row to a cell in the last row.

   (d) Determine the number of admissible paths from the second cell in the first row to a cell in the last row.

   (e) Use the above results to verify that there are 363 admissible paths in this table.

9. We are interested in counting all admissible paths from the upper left corner to the last row in a 3×6 alignment table. Fill out the entries of this table, progressing from top to bottom and left to right. Use the entries in the last row of the table to calculate the total number of such admissible paths.

Optimal Paths in an Alignment Table

10. Construct a table of optimal path costs for the problem of aligning the main sequence TGTACTG and the subsequence GATC.

11. Construct a table of optimal path costs for the problem of aligning the main sequence ATGATAC and the subsequence AGTC.

12. Construct a table of optimal path costs for the problem of aligning the main sequence CGAGCAT and the subsequence GCTGA.

13. The table of optimal costs in Figure 3.8 contains the minimum penalty score 3 in the fourth column of the last row. Trace backward to obtain the associated optimal path and then determine the optimal alignment that corresponds to this path.

14. Consider the alignment table with optimal costs displayed next.

    (a) What is the smallest penalty of an alignment?

    (b) Display all optimal paths and determine the corresponding optimal alignments.

	A	G	C	T	A	G
A	0	1	2	3	0	1
G	1	0	1	2	1	0
T	2	1	2	1	2	1
A	3	2	3	2	1	2

15. Consider the alignment table with optimal costs displayed next.

    (a) What is the smallest penalty of an alignment?

    (b) Display all optimal paths and determine the corresponding optimal alignments.

	C	G	A	T	G	A	C
G	3	0	1	2	0	1	2
A	4	1	0	1	1	0	1
T	5	2	1	0	1	1	2
G	6	3	2	1	0	1	2
C	7	4	3	2	1	2	1

16. Using the table produced in Exercise 10, find the smallest penalty of an alignment. Then display all optimal paths and determine the corresponding optimal alignments.

17. Using the table produced in Exercise 11, find the smallest penalty of an alignment. Then display all optimal paths and determine the corresponding optimal alignments.

18. Using the table produced in Exercise 12, find the smallest penalty of an alignment. Then display all optimal paths and determine the corresponding optimal alignments.

Shortest Paths and Dijkstra's Algorithm

19. In the network below, determine a shortest path from vertex 1 to vertex 7 by listing all paths, calculating their cost, and selecting the one with minimum cost.

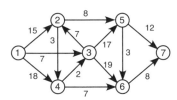

20. Carry out Dijkstra's Algorithm to find a shortest path from vertex 1 to vertex 6 in the following network.

    (a) Display the labels that develop at the vertices (some of which get crossed out and replaced), and show the final table with routing edges and distances.

    (b) Using the routing edges in the table, determine a shortest path from 1 to 6.

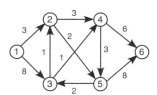

21. Carry out Dijkstra's Algorithm to find a shortest path from vertex 1 to vertex 6 in the following network.

    (a) Display the labels that develop at the vertices (some of which get crossed out and replaced), and show the final table with routing edges and distances.

    (b) Using the routing edges in the table, determine a shortest path from 1 to 6.

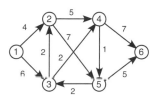

22. Carry out Dijkstra's Algorithm to find a shortest path from vertex 1 to vertex 6 in the network below.

    (a) Display the labels that develop at the vertices (some of which get crossed out and replaced), and show the final table with routing edges and distances.

    (b) Using the routing edges in the table, determine a shortest path from 1 to 6.

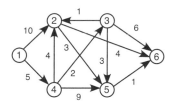

23. Carry out Dijkstra's Algorithm to find a shortest path from vertex 1 to vertex 6 in the network below.

    (a) Display the labels that develop at the vertices (some of which get crossed out and replaced), and show the final table with routing edges and distances.

    (b) Using the routing edges in the table, determine a shortest path from 1 to 6.

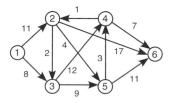

24. Carry out Dijkstra's Algorithm to find a shortest path from vertex 1 to vertex 6 in the network below.

    (a) Display the labels that develop at the vertices (some of which get crossed out and replaced), and show the final table with routing edges and distances.

    (b) Using the routing edges in the table, determine a shortest path from 1 to 6.

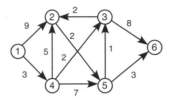

25. Use Dijkstra's Algorithm to find shortest paths from vertex 1 in the network below.

    (a) Display the labels that develop at the vertices (some of which get crossed out and replaced), and show the final table with routing edges and distances.

    (b) Using the routing edges in the table, determine a shortest path from 1 to 5.

    (c) Using the routing edges in the table, determine a shortest path from 1 to 7.

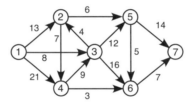

26. Use Dijkstra's Algorithm to find shortest paths from vertex 1 in the network below.

    (a) Display the labels that develop at the vertices (some of which get crossed out and replaced), and show the final table with routing edges and distances.

    (b) Using the routing edges in the table, determine a shortest path from 1 to 4.

    (c) Using the routing edges in the table, determine a shortest path from 1 to 5.

    (d) Using the routing edges in the table, determine a shortest path from 1 to 7.

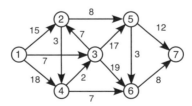

Other Optimal Path Problems

27. Carry out the steps of a modified Dijkstra Algorithm to find a maximum income path from vertex 1 to vertex 9 for the network in Figure 3.18. You should arrive at the solution depicted in Figure 3.19.

28. Carry out the steps of a modified Dijkstra Algorithm to find a most reliable path from vertex 1 to vertex 7 for the network in Figure 3.20. You should arrive at the solution depicted in Figure 3.21.

29. Continue the iterations started in Figure 3.23 to determine a strongest association path from vertex 1 to vertex 6 in Figure 3.22. You should arrive at the solution depicted in Figure 3.24.

30. A software company needs to hire technical support staff to answer questions and troubleshoot issues related to a newly released mobile operating system. Nine applicants are available who possess different amounts of expertise and have different availabilities during the 9 am to 5 pm support hours. The following table lists the hours that can be covered by each applicant, as well as the associated hiring cost. A requirement is that there be at least one support staff on duty throughout the day, so that overlap of shifts is permitted. Formulate this as a shortest path problem on an appropriate network and solve for a minimum cost selection of applicants to cover the entire day. [*Hint*: Let the vertices represent the nine hours of the day and let edges correspond to the shifts. Cost 0 edges $(i, i-1)$ can be used to accommodate the possible overlap of shifts.]

Applicant	1	2	3	4	5	6	7	8	9
Starting Time	9	9	10	12	12	1	2	3	4
Ending Time	1	11	12	3	5	4	5	5	5
Cost	$30	$18	$5	$18	$38	$22	$20	$12	$9

Projects

31. The entries of the table constructed in Figure 3.4 reveal several numerical patterns. The second row contains the odd integers $1, 3, 5, 7, 9$. But what about the next two rows? To get further insight, notice that if we take differences between successive values in the second row we get $2, 2, 2, 2$.

    (a) What do you get if you take differences between successive values in the third row? And what if you take differences between those differences? Can you conjecture a formula for the entries in row three?

    (b) What do you get if you take differences between successive values in the fourth row? And what if you take differences between those differences? And then differences yet again? Can you conjecture a formula for the entries in row four?

32. Develop a spreadsheet (or computer program) that will compute the table of optimal costs for an alignment problem, with a specified main sequence and subsequence.

Bibliography

[Ridley 2006] Ridley, M. 2006. *Genome: The autobiography of a species in 23 chapters*. New York: Harper Perennial.

[Shasha 1998] Shasha, D., and Lazere, C. 1998. *Out of their minds: The lives and discoveries of 15 great computer scientists*, pp. 55–67. New York: Springer-Verlag.

[OR1] `http://www.unf.edu/~wkloster/foundations/DijkstraApplet/DijkstraApplet.htm` (accessed October 13, 2013).

Chapter 4

Routing Problems and Optimal Circuits

It's the Circle of Life, and it moves us all. Through despair and hope, through faith and love. Till we find our place, on the path unwinding, in the Circle. The Circle of Life.

—song from *The Lion King*, lyrics by Tim Rice

Chapter Preview

Eulerian circuits and Eulerian paths arise in puzzles as well as in real-world situations in which service vehicles need to traverse all the streets of a road network. The overall task is to trace out a continuous route, possibly returning to the starting point. In some cases, all of the graph edges have to be traversed exactly once (forming an Eulerian path/circuit). In other cases, repetition is unavoidable in traversing the edges and we wish to minimize these repetitions (Chinese Postman Problem). A more general optimal routing problem is also studied in which the objective is to minimize the total time required to traverse all edges of a weighted graph at least once.

4.1   Eulerian Circuits and Eulerian Paths

A problem frequently found in puzzle books asks you to reproduce a drawing using as few continuous penstrokes as possible. A *unicursal drawing* is one that can be made with a single penstroke—which means it can be drawn without picking up the pen from the drawing surface and without retracing any lines. Several artists have created intricate drawings rendered with a single (unicursal) line [OR1].

It is possible to draw Figure 4.1(a) unicursally, in a single stroke without retracing any line. One way to accomplish this is to follow the sequence of vertices $AFDEBDCEACBA$ in the corresponding graph of Figure 4.1(b). In fact, you can start such an *Eulerian circuit* from any vertex of this graph and you will necessarily end up at the same vertex.

DEFINITION **Circuit**
A path that starts and ends at the same vertex.

DEFINITION **Eulerian Circuit**
A circuit that traverses every edge of the graph exactly once.

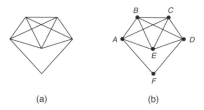

FIGURE 4.1: Unicursal drawing and an Eulerian circuit

It is also possible to draw Figure 4.2 unicursally, but only if you start at either vertex B or vertex E. Moreover, when you start at one of these vertices, you must necessarily end at the other. One possible unicursal drawing follows the sequence $BCDABDEAFE$. This unicursal drawing forms what is termed an *Eulerian path* in the graph.

FIGURE 4.2: Eulerian path in a graph

DEFINITION **Eulerian Path**
A path that traverses all the edges of the graph exactly once, but does not necessarily begin and end at the same vertex.

These special types of circuits and paths are named in honor of Leonhard Euler [OR2], considered by many to be the most influential and prolific mathematician of the 18th century.

HISTORICAL CONTEXT **Leonhard Euler (1707–1783)**
An eminent Swiss mathematician who also made substantial contributions to physics, mechanics, and astronomy. His collected works fill over 70 encyclopedia-sized volumes.

Euler is considered the originator of graph theory because of his novel approach and solution to the famous Königsberg Bridges Problem. The river Pregel divided the city of Königsberg, Prussia into several land masses. The inhabitants of the city wondered if it was possible to take a walk that crossed each of the city's seven bridges exactly once. A map showing the four land areas and the seven bridges is shown in Figure 4.3(a).

Euler converted the map of Königsberg into a graph with four vertices (A, B, C, D), one for each land mass, and seven edges, one for each bridge. Notice that this graph, shown in Figure 4.3(b), contains multiple edges that connect the same two vertices.

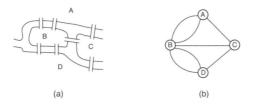

(a)   (b)

FIGURE 4.3: Königsberg map and graph

Now the question becomes whether this graph has an Eulerian path. Euler demonstrated that there can be no such path in the graph, and so it is not possible to construct a continuous walk that crosses each bridge exactly once. The reasoning goes as follows. Suppose there is an Eulerian path starting at vertex v and ending at vertex w. Whenever you enter an intermediate vertex on this path by one edge, you must leave the vertex by a different edge. So apart from vertices v and w, the number of edges touching each vertex must be an even integer; that is, the *degree* of each vertex other than v or w must be even. Since vertices A, B, C, D in Figure 4.3(b) have respective degrees of $3, 5, 3, 3$ (all odd), there can be no Eulerian path and so the Königsberg Bridges Problem has no solution.

DEFINITION **Degree of a Vertex**
The number of edges that are incident to (touching) the vertex.

The following result provides a way to test whether a graph has an Eulerian circuit or, if not, whether it has an Eulerian path.

RESULT **Euler's Theorem**
A connected graph has an Eulerian circuit if and only if it has no vertices of odd degree. If it contains exactly two vertices of odd degree, then there is an Eulerian path in the graph; moreover, the path must begin and must end at these vertices of odd degree.

Example 4.1 Determine Whether a Graph has an Eulerian Circuit

Establish whether each graph in Figure 4.4 has an Eulerian circuit. If it does, describe a circuit in the graph that uses each edge exactly once.

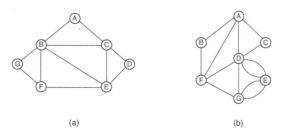

(a)   (b)

FIGURE 4.4: Eulerian circuits in graphs

Solution

(a) The graph in Figure 4.4(a) does not have an Eulerian circuit since it contains two vertices of odd degree: B has degree 5 and F has degree 3. However, it has an Eulerian path; for example, $BACDECBFEBGF$ is an Eulerian path starting at vertex B and ending at vertex F.

(b) All vertices in Figure 4.4(b) have even degree, so the graph has an Eulerian circuit; for example, $ACDEGEDGFDAFBA$ is an Eulerian circuit that starts and ends at vertex A.
◁

(*Related Exercises: 1–3, 29–30*)

Example 4.2 Determine if a Diagram is Unicursal

Is it possible to draw the diagram in Figure 4.5(a) with a single penstroke?

 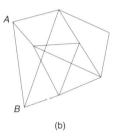

(a)   (b)

FIGURE 4.5: Determining a unicursal drawing

Solution

First, we check the degrees of the vertices in the diagram. All vertices have even degree, except for vertices A and B; see Figure 4.5(b). So it is possible to trace the figure using a single penstroke if you start at A (or at B). Exercise 7 asks you to trace out this figure starting from A; necessarily you must end up at the other odd-degree vertex B. ◁

(*Related Exercises: 4–7*)

Example 4.3 Determine Whether a Graph has an Eulerian Path

Does the graph in Figure 4.6 have an Eulerian path—a path that traverses each edge exactly once?

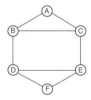

FIGURE 4.6: Determining an Eulerian path

Solution

We apply the criterion from Euler's theorem: a graph has an Eulerian path whenever it has either 0 or 2 vertices of odd degree. However, vertices B, C, D, E all have odd degree so we are assured that there cannot exist an Eulerian path. Equivalently, it is not possible to draw the associated diagram with a single penstroke. ◁

(Related Exercises: 8–9)

Various websites allow the user to trace out Eulerian circuits and Eulerian paths [OR3].

4.2   Routing Problems and Eulerian Circuits

Suppose a truck wants to leave the depot (at location A) and return to the depot after it has cleaned each street in the neighborhood described by Figure 4.7. For efficiency reasons, it doesn't want to travel down any street twice. Is such a route possible? Equivalently, we are looking for an Eulerian circuit starting and ending at A.

FIGURE 4.7: Street-cleaning graph

The street-cleaning graph shown in Figure 4.7 possesses an Eulerian circuit because each vertex has even degree. One possible routing of the truck is described by the sequence $ABEDBCFHIFEHGDA$. In this way, an Eulerian circuit provides an *optimal routing*, as each street is traversed exactly once and the truck returns to its original location.

DEFINITION **Routing Problems**
Situations in which we need to determine the most efficient way of routing objects, such as vehicles or messages, to serve different destinations. These optimization problems occur in transportation, communication, and delivery services (as well as other areas).

Example 4.4 Design a Route for Reading Utility Meters

The map of a housing development is shown in Figure 4.8, with its entrance designated by A. Every month the local utility company sends an employee to read the meters of all homes in the development. This can be done remotely by driving down each street. Can the utility company design a route for the meter reader that begins and ends at the entrance to the development and travels past every house just once?

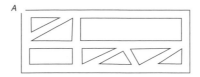

FIGURE 4.8: Housing map

Solution

To answer this question, we create a graph with vertices representing the street intersections and edges joining adjacent intersections. What we require then is an Eulerian circuit in the housing graph, shown in Figure 4.9, that starts at A and ends at A. Since all vertices have even degree (check this!), there exists an Eulerian circuit in this graph, such as $ABCGKJGFJIFEIHDEBDA$. This Eulerian circuit then translates into an acceptable route for the meter reader shown in Figure 4.10. ◁

(Related Exercises: 22–24)

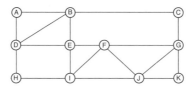

FIGURE 4.9: Housing graph

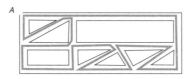

FIGURE 4.10: Route for the meter reader

Chinese Postman Problem

A postman must start and end at the post office, and needs to travel down each street of his assigned neighborhood. If the corresponding graph of the streets in the neighborhood has no Eulerian circuit, the postman will have to travel down some of the streets more than once. How then can the postman find a route with the *fewest* number of repeated streets? This is another example of an optimization problem in graphs. This problem was first introduced by a Chinese mathematician and was later renamed the *Chinese Postman Problem*.

HISTORICAL CONTEXT **Chinese Postman Problem**
The young Chinese mathematician Mei-Ko Kwan (Mei-Gu Guan) developed in 1960 the first algorithm to solve this routing problem for a rural postman.

Example 4.5 Design an Optimal Route for the Chinese Postman Problem

Figure 4.11(a) shows a street graph, with the post office at location A. Design a route that starts and ends at vertex A, traverses each street at least once, and involves the fewest number of repeated streets.

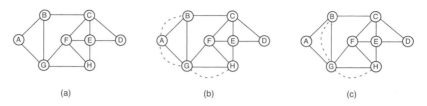

FIGURE 4.11: Street graph for postal delivery

Solution

Since there are two vertices (B, H) of odd degree, there is no Eulerian circuit. However, if we add the dashed edges BA, AG, GH (forming a path that connects the odd-degree vertices), then all vertices of the graph will now have even degree; see Figure 4.11(b). We can then trace out an Eulerian circuit, starting at A in this expanded graph. What this means is that by repeating three streets (edges BA, AG, GH) we can obtain a suitable routing for the postman that returns to location A.

However, we could connect the odd vertices (B, H) in a different way. We could duplicate edges BG, GH and create instead the graph in Figure 4.11(c); this expanded graph has an Eulerian circuit since all vertices now have even degree. Notice that this is a better solution because only two streets are repeated (instead of three). In fact, this is the best we can do: we now have an optimal postman route, for example, the sequence $ABGBCDECFHEFGHGA$. ◁

(*Related Exercises: 10–13*)

In real-world problems, optimal routes can also involve considerations of street lengths, traffic congestion, road work, etc. To reflect this, we might assign to each edge a number representing the time required to travel down that street. This now gives a more realistic optimization problem: find a minimum-time route that traverses each edge at least once.

Example 4.6 Find a Minimum-Time Route

The weighted graph in Figure 4.12(a) shows the time (in minutes) needed to traverse each edge. Determine a route that starts at vertex A, travels along each edge at least once, returns to vertex A, and takes the smallest total elapsed time.

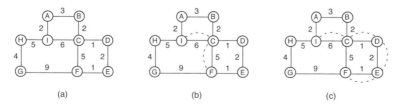

FIGURE 4.12: Optimal route with two vertices of odd degree

Solution

Notice that the graph contains two vertices F and I with odd degree, so the graph does not have an Eulerian circuit. Consequently, we will incur additional time by traversing some edges more than once. Suppose we identify the path with the fewest number of edges joining F and I, namely $F \to C \to I$. Adding the edges FC and CI to the graph results in the expanded graph of Figure 4.12(b); now all vertices have even degree and so we can find an Eulerian circuit. Since we added the extra edges FC and CI (meaning that we travel down these streets twice rather than once), this will increase the total time by $5 + 6 = 11$ minutes.

However, another way to connect F and I is through the path $F \to E \to D \to C \to I$. Adding edges FE, ED, DC, and CI to the original graph gives the expanded graph of Figure 4.12(c); since all vertices now have even degree, this graph possesses an Eulerian circuit. By adding these four edges, the total trip time will increase by $1 + 2 + 1 + 6 = 10$ minutes, which is shorter by 1 minute than the solution previously found. This is in fact the smallest total increase in time possible. As a result, tracing an Eulerian circuit in Figure 4.12(c) provides an optimal routing. One such routing is given by the Eulerian circuit $AIHGFCDEFEDCICBA$; the corresponding elapsed time is $2 + 5 + 4 + 9 + 5 + 1 + 2 + 1 + 1 + 2 + 1 + 6 + 6 + 2 + 3 = 50$ minutes. ◁

The lesson to be learned here is that we need to pair up the odd-degree vertices F and I in the best possible way to achieve the overall best (minimum-time) route. Aha! This is just a shortest path problem between vertices F and I; we know how to solve such a problem using, for example, Dijkstra's Algorithm from Section 3.2. Specifically, the shortest path between F and I consists of edges FE, ED, DC, CI with minimum cost $1 + 2 + 1 + 6 = 10$. This makes good sense: we know that we have to traverse each edge once (at least), so an optimal route will add the smallest incremental time (the length of a shortest path) to the sum of all the original edge traversal times.

(Related Exercises: 14–15)

How should we proceed if there are more than two vertices of odd degree? The following example illustrates this situation.

Example 4.7 Find a Minimum-Time Route with Several Pairs of Odd-Degree Vertices

The weighted graph in Figure 4.13 shows the time (in minutes) needed to traverse each edge. Determine a route that starts at vertex A, travels along each edge at least once, returns to vertex A, and takes the smallest total elapsed time.

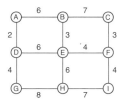

FIGURE 4.13: Graph with several odd-degree vertices

Solution

The graph shown in Figure 4.13 has four vertices B, D, F, H with odd degree. We can pair them up in three different ways: $(B\text{-}D, F\text{-}H)$, $(B\text{-}F, D\text{-}H)$, or $(B\text{-}H, D\text{-}F)$. When pairing two vertices, it makes sense to do so in the best possible way: that is, joining them by a shortest path. Figure 4.14 displays these three pairings; each individual pair of vertices is joined by a shortest path, indicated by dashed lines. For example, in the first pairing (*B-D, F-H*) shown in Figure 4.14(a), $B \to A \to D$ is the minimum-time path between vertices B and D, while $F \to E \to H$ is the minimum-time path between vertices F and H. Here is a summary of the three possible pairings and the additional time incurred by each pairing:

1. (*B-D, F-H*): path $B \to A \to D$ adds 8 minutes while $F \to E \to H$ adds 10 minutes.
2. (*B-F, D-H*): path $B \to E \to F$ adds 7 minutes while $D \to G \to H$ adds 12 minutes.
3. (*B-H, D-F*): path $B \to E \to H$ adds 9 minutes while $D \to E \to F$ adds 10 minutes.

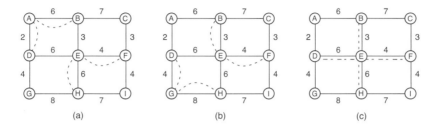

FIGURE 4.14: Three different pairings

The best choice is then Pairing 1, which increases the total route time by only 18 minutes. After adding the repeated edges BA, AD, FE, EH to the original graph, we can now find an Eulerian circuit. For example, $ADGHEFIHEFCBEDABA$ is a minimum-time route that traverses every edge at least once. ◁

Example 4.8 Find a Minimum-Time Route in Another Weighted Graph

In Figure 4.15 we wish to start and end at vertex A and traverse each edge at least once, at minimum total time. Use a table to list the different pairings of odd-degree vertices, the best edges to add for each pairing, and the extra time added to the route. Then identify a minimum-time solution for this situation.

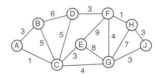

FIGURE 4.15: Another graph with several odd-degree vertices

Solution

The six odd-degree vertices are B, C, D, E, G, H so we consider various ways to arrange these vertices into three pairs. Table 4.1 shows the possible pairings. For example, the first row of the table pairs B with C, leaving the odd-degree vertices D, E, G, H to be paired; there are then three ways to match up vertex D: (a) D-E, (b)D-G, (c)D-H. After this second pair is selected, the remaining two vertices must be paired. The minimum time (using a shortest path) to join each pair of vertices is also shown in parentheses. For example, after B-C is selected (incurring the additional time 4) in the first row, the pairing (a) D-E and (d) G-H adds the additional time $8+5=13$, the pairing (b) D-G and (e) E-H adds $7+10=17$, and the pairing (c) D-H and (f) E-G adds $4+7=11$. Since the last pairing (c)-(f) is the best, we show in bold the respective additional times. The last entry of the row displays the sum of all three times for this pairing: $4 + 4 + 7 = 15$ minutes. Continuing, we see that the pairings in the second row achieve the smallest overall sum 14. As a result, the optimal pairing is *B-D, C-E, G-H*, which entails adding the corresponding shortest path edges BD, CE, GF, FH to the graph in Figure 4.15. Tracing an Eulerian circuit in this expanded graph gives an optimal routing: for example, $ABDFHJGHFGFEGCECDBCA$ with an overall route time of 79 minutes. ◁

(Related Exercises: 16–21)

TABLE 4.1: Pairings of odd-degree vertices

Pair 1	Pair 2			Pair 3			Time
	(a)	(b)	(c)	(d)	(e)	(f)	
$BC(\mathbf{4})$	$DE(8)$	$DG(7)$	$DH(\mathbf{4})$	$GH(5)$	$EH(10)$	$EG(\mathbf{7})$	15
$BD(\mathbf{6})$	$CE(\mathbf{3})$	$CG(4)$	$CH(9)$	$GH(5)$	$EH10)$	$EG(7)$	14
$BE(\mathbf{7})$	$CD(5)$	$CG(4)$	$CH(9)$	$GH(5)$	$DH(\mathbf{4})$	$DG(7)$	15
$BG(\mathbf{8})$	$CD(5)$	$CE(\mathbf{3})$	$CH(9)$	$EH(10)$	$DH(\mathbf{4})$	$DE(8)$	15
$BH(\mathbf{10})$	$CD(5)$	$CE(\mathbf{3})$	$CG(4)$	$EG(7)$	$DG(\mathbf{7})$	$DE(8)$	20

Here is a statement of the algorithm to solve the generalized Chinese Postman Problem in a graph with edge traversal times.

Algorithm for Optimal Routing in a Weighted Graph

1. Identify all vertices having odd degree.

2. List all pairings that group together in pairs the odd-degree vertices.

3. Find the minimum time to connect the two vertices of each pair by a shortest path.

4. Select a pairing having smallest total time for its constituent pairs.

5. Enlarge the original graph by adding shortest path edges between each pair of the selected pairing.

6. Construct an Eulerian circuit in the enlarged graph. This provides an optimal routing.

4.3   Applications of Eulerian Paths

In this section we present two applications that surprisingly involve the determination of Eulerian paths in an associated graph. The first is a recreational puzzle, and the second is a real-world application to DNA sequencing.

A Chip-Stacking Puzzle

In this game, there are 10 chips, which have certain symbols $\odot, \dagger, \triangle, \Diamond, \bigstar$ appearing on each side. The two sides of each chip contain different symbols and no two chips are the same. Now one chip is removed at random and hidden away. The players take turns selecting a chip from those remaining. The first person can choose any of the nine remaining chips. Each subsequent player must select a chip that matches either end of the growing stack of selected chips.

For example, the first player might choose the $(\dagger, \bigstar)$ chip, and then the next player could choose the $(\Diamond, \dagger)$ chip, giving the stack $(\Diamond, \dagger), (\dagger, \bigstar)$. Since the two ends of the stack of chips are now $\Diamond$ and $\bigstar$, the following player might select the $(\bigstar, \triangle)$ chip, giving the current stack $(\Diamond, \dagger), (\dagger, \bigstar), (\bigstar, \triangle)$. The game continues until all nine chips have been added to the stack (if both ends are the same, then a new stack can be started). Then by looking at the two ends of the final stack, we will know which chip was originally hidden.

To explain why this trick works, we will use the numbers $1, 2, \ldots, 5$ instead of the symbols $\odot, \dagger, \triangle, \Diamond, \bigstar$. We construct a graph G with five vertices, in which an edge joins each pair of distinct vertices (this is called the *complete graph* on five vertices; see Section 5.1). Each edge of the graph corresponds to precisely one of the 10 chips (that is how the set of chips was constructed). For example, the edge $(2, 5)$ represents the $(\dagger, \bigstar)$ chip.

Notice that each vertex of G has degree four; see Figure 4.16(a). Since each vertex has even degree, the graph G has an Eulerian circuit. Now suppose that the chip $(\odot, \triangle)$ was the one hidden. This means that the edge $(1, 3)$ was removed from G giving a new graph G' with exactly two vertices of odd degree—namely, vertex 1 and vertex 3. This is illustrated in Figure 4.16(b), in which edge $(1, 3)$ has been removed.

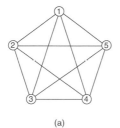 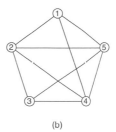

(a)   (b)

FIGURE 4.16: Graphs G and G'

Because we are growing a stack by adding chips to either end, we are actually creating a path in graph G' with nine edges (or possibly a path plus a circuit). Moreover, we do not repeat any edges (chips), so we are actually constructing an Eulerian path in G'. Since G' has exactly two vertices (1 and 3) of odd degree, the Eulerian path must start and end at these two vertices. Put another way, by growing our stack of chips through a series of choices (in which the participants have some latitude) we are actually constructing an Eulerian path, and one whose end vertices reveal the unknown symbols on the hidden chip.

DNA Hybridization

Suppose we are trying to decode an unknown DNA target sequence: for example, the sequence CATGTTCAGTATC. Through *hybridization experiments*, we can use a series of probes of small length to uncover the structure of this unknown DNA sequence.

DEFINITION **Hybridization**

An experiment to identify the composition of a strand of DNA by using several "probes" of nucleotides. For example, if the probe is CGGT, experiments can determine whether this length four sequence appears in the larger target sequence but not its exact location.

For example, lab results might show that the unknown target DNA contains the length three sequences CAT, ATG, TGT, GTT, TTC, TCA, CAG, AGT, GTA, TAT, ATC (listed in no particular order). If we knew the target DNA, then we could find all its different length three sequences. But how do we reverse this logic?

Example 4.9 Construct a Target DNA Sequence from its Length Three Sequences

Suppose you are given the following length three sequences appearing in some unknown target DNA: CAT, ATG, TGT, GTT, TTC, TCA, CAG, AGT, GTA, TAT, ATC. Use this information to construct a likely target DNA sequence containing just these length three probe sequences.

Solution

First, we construct the directed graph in Figure 4.17, whose vertices and edges are derived from the length three probes. For each such probe, a directed edge connects the two vertices that represent the first two letters and the last two letters of the probe. For example, the length three probe CAT produces the directed edge CA → AT, which is labeled with the "added" letter T. Note that there are 11 directed edges in this graph because there are 11 given probes with three letters.

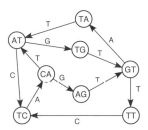

FIGURE 4.17: Directed graph for DNA sequencing

Next, we identify an Eulerian path in this directed graph. Notice that all but two of the vertices have even degree (an equal number of edges entering and leaving the vertex). For example, vertex AT has two entering edges and two leaving edges, while vertex TA has one entering edge and one leaving edge. The two vertices with odd degrees are CA (having one more edge leaving than entering) and TC (having one fewer edge leaving than entering).

The following directed Eulerian path traverses every edge exactly once, provided that we start at the "excess" vertex CA (with an extra leaving edge) and end at the "deficit" vertex TC (with an extra entering edge):

$$CA \to AT \to TG \to GT \to TT \to TC \to CA \to AG \to GT \to TA \to AT \to TC.$$

Because this path includes all 11 edges, it accounts for all 11 nucleotide probes of length three. So if we start with the initial vertex CA and add the letters that label each edge in the Eulerian path, we obtain the target DNA sequence CATGTTCAGTATC. Note that other Eulerian paths in Figure 4.17 might yield different target DNA sequences. Additional laboratory information can decide which candidate DNA sequence is the most likely. ◁

(*Related Exercises: 25–28*)

Here is a statement of the general procedure used above.

Hybridization Algorithm for DNA Sequencing

1. Construct a directed graph with vertices and edges defined by the initial and the final length $k-1$ segments of each given length k probe sequence.

2. Find an Eulerian path in this graph from the excess vertex to the deficit vertex.

3. From this Eulerian path, construct a target DNA sequence that is consistent with the evidence.

Chapter Summary

This chapter introduces the notions of Eulerian circuits and Eulerian paths, which arise in a variety of contexts, from recreational puzzles (figure tracing, chip stacking) to contemporary applications (street cleaning, postal delivery routes, DNA sequencing). Again we follow a consistent overall approach: the problem is modeled using a graph, a solution (Eulerian circuit, Eulerian path) is identified in the graph, and this graph structure is then translated into a solution in the context of the original problem. In addition, solution of the minimum-time routing problem illustrates how one can decompose the original problem into separately solvable parts: finding a best pairing of odd-degree vertices (using shortest paths) and then finding an Eulerian circuit in an expanded graph.

Applications, Representations, Strategies, and Algorithms

Applications		
Routing (4–8)		Biology (9)

Representations		
Graphs		
Weighted	Undirected	Directed
6–9	1–5	9

Strategies	
Rule-Based	Composition/Decomposition
1–9	6–8

Algorithms		
Exact	Inspection	Sequential
1–9	2–3	1, 4–9

Exercises

Eulerian Circuits and Eulerian Paths

1. Does the following graph have an Eulerian circuit? If so, list one; if not, explain why it does not exist.

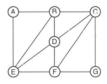

2. Does the following graph have an Eulerian circuit? If so, list one; if not, explain why it does not exist.

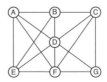

3. Does the following graph have an Eulerian circuit? If so, list one; if not, explain why it does not exist.

4. Can the following diagram be traced with a single penstroke? If so, list the sequence of points visited in such a drawing. Otherwise, explain why it is not possible.

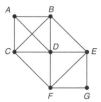

5. Can the following diagram be traced with a single penstroke? If so, list the sequence of points visited in such a drawing. Otherwise, explain why it is not possible.

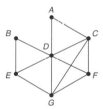

6. Can the following diagram be traced with a single penstroke? If so, list the sequence of points visited in such a drawing. Otherwise, explain why it is not possible.

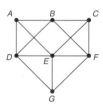

7. Add letters to the remaining vertices in Figure 4.5(b) and then trace out an Eulerian path, starting at A.

8. Determine whether each of the following graphs has an Eulerian circuit (if so, list one); if not, determine whether it has an Eulerian path (if so, list one). If it has neither, explain why this is the case.

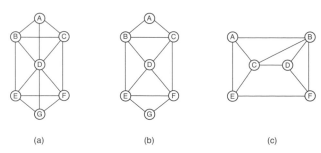

(a)   (b)   (c)

9. Determine whether each graph below has an Eulerian circuit (if so, list one); if not, determine whether it has an Eulerian path (if so, list one). If it has neither, explain why this is the case.

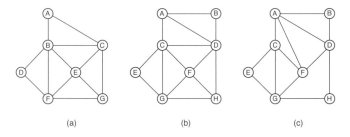

(a)   (b)   (c)

Optimal Routing Problems

10. A postman wishes to deliver mail along all streets of the neighborhood map below, starting and ending the route at location A.

    (a) Is it possible to do so without traversing any street more than once? Explain.

    (b) If not, determine a route that minimizes the number of repeated streets.

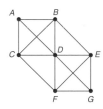

11. A postman wishes to deliver mail along all streets of the following neighborhood map, starting and ending the route at location A.

    (a) Is it possible to do so without traversing any street more than once? Explain.

    (b) If not, determine a route that minimizes the number of repeated streets.

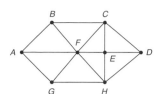

12. A waste collection vehicle needs to traverse all streets of the neighborhood map below, starting and ending the route at location *A*.

    (a) Is it possible to do so without traversing any street more than once? Explain.

    (b) If not, determine a route that minimizes the number of repeated streets.

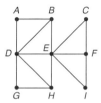

13. A waste collection vehicle needs to traverse all streets of the following neighborhood map, starting and ending the route at location *A*.

    (a) Is it possible to do so without traversing any street more than once? Explain.

    (b) If not, determine a route that minimizes the number of repeated streets.

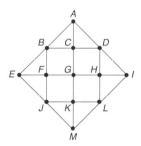

14. Identify the odd-degree vertices in the weighted graph below and then determine an optimal (minimum-time) pairing of these vertices.

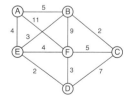

15. Identify the odd-degree vertices in the following weighted graph and then determine an optimal (minimum-time) pairing of these vertices.

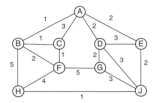

16. A snow removal truck needs to plow all streets of the neighborhood map shown in Exercise 14, starting the route at location A and returning to location A. The travel times along each street are as indicated. Determine a minimum-time route that travels along each street at least once.

17. A snow removal truck needs to plow all streets of the neighborhood map shown in Exercise 15, starting the route at location A and returning to location A. The travel times along each street are as indicated. Determine a minimum-time route that travels along each street at least once.

18. A meter reader needs to travel down all streets of the following neighborhood map, starting the route at location A and returning to location A. The travel times along each street are as indicated.

   (a) Find the best pairing of the odd-degree vertices.

   (b) Determine a minimum-time route that travels along each street at least once.

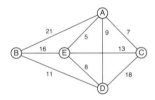

19. A tiny robot needs to examine the welds along each segment of the pipeline system shown below, starting at location A and returning to location A. The length of each pipeline segment is as indicated.

   (a) Find the best pairing of the odd-degree vertices.

   (b) Determine a minimum-length route for the robot so that each pipeline segment will be examined at least once.

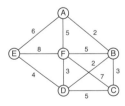

20. A garbage truck needs to travel down all streets of the following neighborhood map, starting the route at location A and returning to location A. The travel times along each street are as indicated.

   (a) Find the best pairing of the odd-degree vertices.

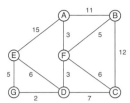

(b) Determine a minimum-time route that travels along each street at least once.

21. The following grid must be duplicated hundreds of times by a mechanical plotter (located at position A). In order to draw the grid as quickly as possible, we want to take advantage of the fact that a horizontal edge requires twice the time to draw as a vertical edge.

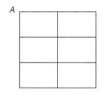

(a) Label the remaining vertices of this grid and find a best pairing of the odd-degree vertices.

(b) Determine an optimal route that minimizes the total time needed to produce the plot, returning at the end to the home location A.

Applications

22. Below is the floor plan of several office suites, certain of which are connected by doors. A security guard wants to determine a patrol that will take him through every office door (including the doors to the outside) so he can lock each door as he leaves.

   (a) Construct a graph representation of the floor plan, with vertices corresponding to offices (as well as the outside), and edges corresponding to doors.

   (b) Find an Eulerian path in this graph and use it to describe an efficient patrol that the security guard can use to lock every door.

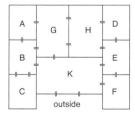

23. You have a set of 11 dominos (shown in the following table), each containing two different numbers from 0 to 6. You would like to arrange all of them in a line (called a *train*). The restriction is that any two consecutive dominos in the train must contain matching numbers where the two dominos are adjacent. For example, the fifth and sixth dominos can be placed in the order 2-1, 1-4.

Domino										
0	0	0	0	1	1	2	2	3	3	5
3	4	5	6	2	4	3	4	4	6	6

(a) Draw a graph to represent this problem, with vertices $0, 1, \ldots, 6$ and with an edge corresponding to each domino.

(b) Is it possible to construct a train using all the dominos? If so, exhibit such a train; if not, explain why it is not possible.

24. As in the previous exercise, you have a set of 11 dominos (shown in the following table) and would like to arrange all of them in a single train.

Domino										
0	0	1	1	1	1	2	2	3	4	5
3	5	2	3	4	6	3	4	4	5	6

(a) Draw a graph to represent this problem, with vertices $0, 1, \ldots, 6$ and with an edge corresponding to each domino.

(b) Is it possible to construct a train using all the dominos? If so, exhibit such a train; if not, explain why it is not possible.

25. In Example 4.9, an Eulerian path from CA to TC was found in the directed graph of Figure 4.17 and used to identify a target DNA sequence.

(a) Find three other Eulerian paths from CA to TC.

(b) List the target DNA sequences associated with these Eulerian paths.

26. We want to find target DNA sequences consistent with the following observed length three probe sequences: ACT, AGC, ATG, CAG, CAT, CTG, GAC, GCA, GCT, TGA, TGC.

(a) Draw the directed graph associated with this problem.

(b) Using the graph from part (a), find three different target DNA sequences consistent with the given length three probe sequences.

27. We want to find target DNA sequences consistent with the following observed length three probe sequences: ACT, ATC, ATG, CGG, CTG, GAC, GAT, GCT, GGA, TCG, TGA, TGC.

(a) Draw the directed graph associated with this problem.

(b) Using the graph from part (a), find three different target DNA sequences consistent with the given length three probe sequences.

28. We want to find target DNA sequences consistent with the following observed length four probe sequences: AATC, AGCA, ATAG, ATCC, ATGC, CAAT, CATA, CATC, CCAT, GCAA, GCAT, TAGC, TCCA, TGCA.

(a) Draw the directed graph associated with this problem.

(b) Using the graph from part (a), find three different target DNA sequences consistent with the given length four probe sequences.

Projects

29. Euler's theorem indicates that a circuit or path traversing each edge exactly once will exist if there are either 0 or 2 vertices of odd degree. Naturally, we would wonder what happens if there are (say) 1 or 3 vertices of odd degree. Explore this possibility.

    (a) Examine various examples of graphs (with and without multiple edges) and determine the sum of all vertex degrees. Can you relate this to the number of edges in the graph? Formulate a conjecture and argue why it must hold in general.

    (b) Based on this conjecture, conclude that it is *impossible* for there to be an odd number of vertices with odd degree.

30. Euler's theorem indicates that a figure can be drawn unicursally if it contains either 0 or 2 vertices of odd degree. We would like to know the fewest number of penstrokes needed to draw any figure.

    (a) Suppose that a figure contains four vertices A, B, C, D with odd degree. Start a path at vertex A and continue as far as you can before you are stuck. Notice that you must end at another vertex of odd degree; say it is B. Now remove from the original figure all edges of this $A \to B$ path. How does that affect the number of vertices of odd degree in the new figure? How many penstrokes will be needed to trace out this new figure?

    (b) Generalize your observations in part (a) to obtain a formula for the fewest number of penstrokes needed to draw a figure having $2k$ vertices with odd degree.

Bibliography

[Assad 2007] Assad, A. A. 2007. "Leonhard Euler: A brief appreciation." *Networks* 49: 190–198.

[Grötschel 2011] Grötschel, M., and Yuan, Y.-X. 2010. "Euler, Mei-Ko Kwan, Königsberg, and a Chinese Postman." *Documenta Mathematica* Extra Volume ISMP: 43–50.

[Higgins 2007] Higgins, P. M. 2007. *Nets, puzzles, and postmen*, Chapter 5. New York: Oxford University Press.

[Ore 1990] Ore, O., and Wilson, R. J. 1990. *Graphs and their uses*, Chapter 2. Washington, D.C.: Mathematical Association of America.

[Wilson 1990] Wilson, R. J., and Watkins, J. J. 1990. *Graphs: An introductory approach*, Chapter 6. Hoboken, NJ: John Wiley & Sons.

[OR1] http://designomnivore.blogspot.com/2012/01/fiona-ross-unicursal-painter.html (accessed November 23, 2013).

[OR2] http://www.pdmi.ras.ru/EIMI/EulerBio.html (accessed November 25, 2013).

[OR3] http://www.flashandmath.com/mathlets/discrete/graphtheory/euler.html (accessed November 23, 2013).

Chapter 5

Traveling Salesmen and Optimal Orderings

First things first, but not necessarily in that order.

—*Doctor Who*, British science fiction television program

Chapter Preview

The Traveling Salesman Problem (TSP) involves constructing a minimum distance tour that visits each city exactly once and returns to the starting point. As well as its applications to puzzles, the TSP underlies a variety of problems in which we need to determine an optimal ordering: for example, manufacturing a circuit board or determining a likely chronology for archaeological finds. Unlike the previously encountered shortest path problem and the Eulerian circuit problem, there are no known solution techniques that can solve every TSP efficiently. As a result, a number of heuristic (approximate) techniques have been developed to find good, though not necessarily optimal, solutions.

5.1 Hamiltonian Circuits and the TSP

Chapter 4 discussed the problem of finding an Eulerian circuit in a graph: a way of traversing the graph so that all edges are visited once. A related problem is that of traversing a graph so that all vertices are visited once. Such a traversal (or tour) is known as a *Hamiltonian circuit*, named after an illustrious Irish mathematician and physicist [OR1]. The following example illustrates this concept.

Example 5.1 Determine if a Graph Contains a Hamiltonian Circuit

Consider the two graphs shown in Figure 5.1, both having six vertices and nine edges. Which of these have a Hamiltonian circuit and which do not?

Solution

The sequence of vertices $ADECBFA$ in Figure 5.1(a) describes a circuit that visits every vertex exactly once and so provides a Hamiltonian circuit. However, it is not obvious how to find a Hamiltonian circuit in Figure 5.1(b). In fact, this is impossible! We can convince ourselves of this result in the following way. Since vertex C has degree 2, any Hamiltonian

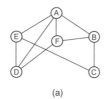 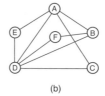

(a)   (b)

FIGURE 5.1: Determining Hamiltonian circuits in graphs

circuit must use the two edges (A, C) and (C, D). Similarly, vertex E has degree 2, so any Hamiltonian circuit must use the two edges (A, E) and (D, E). We are then forced to use these four edges. However, these edges themselves form a circuit on four vertices, precluding a Hamiltonian circuit on the full six vertices. ◁

(Related Exercises: 1–2)

DEFINITION **Hamiltonian Circuit**
A path that starts and ends at the same vertex and visits every vertex exactly once. (It need not traverse all the edges.)

HISTORICAL CONTEXT **William Rowan Hamilton (1805–1865)**
An eminent Irish mathematician and physicist, known for his significant contributions to algebra as well as to the study of optics and mechanics.

In 1857, Hamilton invented the *Icosian Game* puzzle in which the player seeks to visit each of 20 cities in the world exactly once and return home, played using the edges of a dodecahedron, a 12-sided solid figure with 20 corners (representing the cities). A two-dimensional version of the puzzle is shown as the graph in Figure 5.2(a). The question then is whether one can find a Hamiltonian circuit in this graph. One possible solution is the circuit displayed in bold in Figure 5.2(b), which starts and ends at the top vertex A: namely, $ABCDENOTPQRSMLKJIHGFA$. Each one of the 20 vertices is visited once before returning to the starting vertex. There are other solutions to this puzzle.

(Related Exercises: 3–4)

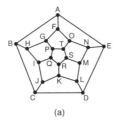 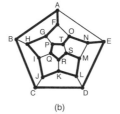

(a)   (b)

FIGURE 5.2: The Icosian game and a solution

Example 5.2 Design a Patrol for a Night Watchman

The floor plan of a museum is diagrammed in Figure 5.3(a), with doorways indicated between various rooms. A night watchman wants to start his hourly patrol at the museum entrance (room A) and visit each of the rooms once, returning back to his post at the museum entrance. Is it possible to design such a patrol of the museum rooms?

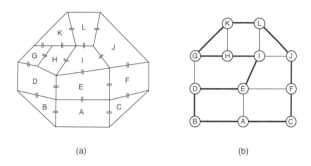

(a)   (b)

FIGURE 5.3: A Hamiltonian circuit in the museum graph

Solution

The museum layout can be represented by a graph in which each vertex corresponds to a room and an edge joins two vertices if their associated rooms are joined by a doorway. This graph is shown in Figure 5.3(b). Now the original question translates into finding a Hamiltonian circuit, starting at vertex A, in this graph. One possible solution is the tour $ABDEIHGKLJFCA$, shown in bold in Figure 5.3(b). This defines an efficient patrol for the night watchman. ◁

(Related Exercises: 5, 8–10)

Example 5.3 Solve the *Instant Insanity* Puzzle

Instant Insanity^{TM} is a puzzle that involves four plastic cubes, each of whose six faces is colored with one of the four colors R, B, G, Y. The object is to stack the four cubes on one another so that each of the four sides of the resulting tower displays all four different colors. Suppose that the four cubes contain the following colors on the pairs of opposite faces:

$$\text{Cube 1: } B\text{-}G, \ B\text{-}Y, \ Y\text{-}G$$
$$\text{Cube 2: } B\text{-}Y, \ B\text{-}G, \ R\text{-}G$$
$$\text{Cube 3: } B\text{-}G, \ R\text{-}R, \ R\text{-}Y$$
$$\text{Cube 4: } R\text{-}G, \ R\text{-}Y, \ Y\text{-}Y$$

For example, the colors on a flattened Cube 2 are shown in Figure 5.4(a). It turns out that we can solve this puzzle by getting one pair of opposite sides of the tower to display four different colors and then doing likewise for the other pair of opposite sides of the tower.

Solution

Draw a graph in which each vertex corresponds to a color; an edge is drawn between two vertices when a cube contains these two colors on opposite faces. Each such edge is labeled with the number of the cube it represents. For example, Cube 4 generates the edges

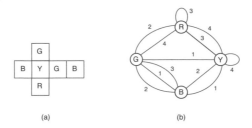

FIGURE 5.4: Graph for the *Instant Insanity*™ puzzle

(R, G) and (R, Y) as well as the "loop" edge (Y, Y). The graph so constructed is shown in Figure 5.4(b).

In this graph, we identify the Hamiltonian circuit $BYRGB$ in which the associated cube numbers $2, 3, 4, 1$ occur exactly once. This corresponds to a way of stacking up the four cubes so that the opposite sides of the tower contain each color once. This is shown in the second column of Table 5.1, arranged in order of the cubes. Next, we find a second such Hamiltonian circuit, edge disjoint from the first: for instance, $YBGRY$ with associated cube numbers $1, 3, 2, 4$, shown in the third column of Table 5.1. Together we obtain two stacks of cubes that can be merged to provide a solution to the *Instant Insanity*™ puzzle. ◁

(Related Exercises: 6–7)

TABLE 5.1: Two stacks of opposite faces

Cube	First Stack	Second Stack
1	G-B	Y-B
2	B-Y	G-R
3	Y-R	B-G
4	R-G	R-Y

In Example 5.3, we found a Hamiltonian circuit for each pair of opposite sides of the tower. More generally, we could use a collection of edge-disjoint circuits instead of a full Hamiltonian circuit: for example, the circuit BGB using Cubes 1 and 2, plus the disjoint circuit RYR using Cubes 3 and 4.

The concept of a Hamiltonian circuit also arises in transportation networks, in which there is a *weight* (time, distance, or cost) associated with each edge. Specifically, consider a salesman who needs to visit clients in a number of cities before returning home. As there may be many such *tours* (directed Hamiltonian circuits) in the network, the salesman would like to construct a tour that involves the minimum total weight. This famous problem is known as the *Traveling Salesman Problem*.

DEFINITION **Traveling Salesman Problem (TSP)**
This problem requires finding a minimum weight Hamiltonian circuit in a graph with edge weights.

Another version of this problem could be termed the "Traveling Politician Problem," which arose during campaigns leading up to the 2012 Presidential election [OR2]. Several candidates spent time at the end of 2011 making stops in all of the 99 counties of Iowa. If the criterion is to minimize the total distance traveled, an optimal Iowa political tour would require 2739 miles (and take 55.5 hours of driving time) to visit all 99 counties.

Is there an efficient way to find the best tour, to solve the TSP? A natural first approach is to use brute force to list all the possibilities and then choose the best.

Example 5.4 Use Brute Force to Solve the TSP

The network in Figure 5.5 displays travel distances between four cities A, B, C, D. Suppose a tour begins at city A, visits each city once, and returns to city A. Use brute force to list all distinct (undirected) Hamiltonian circuits and then select the minimum distance tour.

FIGURE 5.5: Travel distances between four cities

Solution

There are six possible directed tours: $ABCDA, ABDCA, ACBDA, ACDBA, ADBCA$, and $ADCBA$. However, the total distance for tour $ABCDA$ and its reversal $ADCBA$ are exactly the same, so Figure 5.6 only shows as bold the three (undirected) Hamiltonian circuits in the graph, as well as their total distance. The optimal TSP tour is seen to be $ACBDA$ with minimum distance 1895. ◁

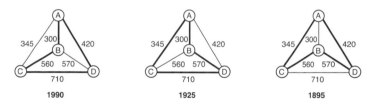

FIGURE 5.6: Three Hamiltonian circuits on four cities

Example 5.4 introduces a type of graph in which every pair of vertices (cities) are joined by an edge. Such a graph is called a *complete graph*, since all possible edges joining different vertices are present. These types of graphs are implicit in many transportation systems, where distances or shipping costs can be calculated between any two locations in the graph. Examples of complete graphs with 3, 4, and 5 vertices are displayed in Figure 5.7. There the notation K_n is used to denote a complete graph with n vertices.

FIGURE 5.7: Examples of complete graphs

> DEFINITION **Complete Graph**
> A graph in which any two distinct vertices are joined by an edge.

While it was easy in Example 5.4 to list all the possible tours involving four cities, the number of possible tours grows rapidly with the number of vertices. To illustrate this growth, the number of (undirected) Hamiltonian circuits in K_n for small values of n is shown in Table 5.2.

TABLE 5.2: Number of Hamiltonian circuits in complete graphs

n	3	4	5	6	7	8	9	10
# Hamiltonian Circuits	1	3	12	60	360	1260	10,080	90,720

The number of Hamiltonian circuits in a complete graph with n vertices can be expressed by the mathematical formula $\frac{(n-1)!}{2}$, where the *factorial* function $k!$ gives the product of the first k positive integers: $k! = k \times (k-1) \times (k-2) \times \cdots \times 2 \times 1$. (This function will be used again in Chapter 13 when probabilities are discussed.) For example, the entry for $n = 7$ in Table 5.2 can be directly computed using the formula $\frac{(7-1)!}{2} = \frac{6!}{2} = \frac{6 \times 5 \times 4 \times 3 \times 2 \times 1}{2} = \frac{720}{2} = 360$.

(Related Exercise: 11)

As demonstrated in Table 5.2, the number of possible Hamiltonian circuits grows quite rapidly. Suppose for example that we want to solve the TSP for $n = 21$ cities. Then the number of different Hamiltonian circuits is $\frac{(n-1)!}{2} = \frac{(21-1)!}{2} = \frac{20!}{2} = 1.216 \times 10^{18}$. In order to find a minimum distance tour by brute force, we would need to examine every one of these. Even if we could process a million tours per second, it would still take about 390 centuries of calculations. So there is a clear need for an efficient solution technique, in contrast to using brute force.

(Related Exercises: 12–13)

Unfortunately, no general procedure is known that will solve every TSP problem in a reasonable amount of time. The TSP has in fact been classified as an *NP-hard problem* [OR3] by computational mathematicians who attempt to devise efficient solution algorithms. A problem is *polynomially solvable* (or *tractable*) if an algorithm can be found whose running time grows no faster than some fixed power of the size of the input data. For example, sorting a list of numbers into ascending or descending order is a tractable problem; it can be solved efficiently, which is why commercial database programs are successful in manipulating vast collections of customer records. By contrast, the TSP is not believed to be a tractable problem because no algorithm has yet been found that runs in polynomial time. In Chapter 6 we will talk about certain graph coloring problems, which also turn out to be difficult NP-hard optimization problems. This is in stark contrast to the shortest path problem of Chapter 3 and the Chinese Postman Problem of Chapter 4, which can be efficiently solved.

Despite the lack of a generally applicable solution approach, a great deal of computational progress has been achieved on specific instances of the TSP. In 2001, the TSP was optimally solved for visiting 15,112 cities in Germany. In 2004, the TSP was optimally solved for 24,978 cities in Sweden. Then a TSP on 85,900 vertices was optimally solved using clusters of computers between February 2005 and April 2006 for a problem that arose at AT&T

Bell Laboratories in the design of customized computer chips. The webpage [OR4] shows portions of the optimal tours found for the cities in Germany and Sweden. The tours in these examples follow very intricate, non-obvious patterns that interconnect the cities.

5.2   Heuristic Algorithms

Since an efficient exact solution method for the TSP is not known, several greedy-type algorithms have been developed for the TSP. Most basic of these is the *Nearest Neighbor Algorithm* in which the salesman builds up a path of cities visited by greedily selecting the next city to add. Specifically, if the currently constructed path reaches city i, the salesman travels to the closest unvisited neighbor of city i. Once all cities have been visited along this path, the salesman needs to return home to complete a Hamiltonian circuit. This is a *heuristic* algorithm, as it may or may not lead to a minimum weight Hamiltonian circuit.

> DEFINITION **Heuristic Algorithm (Heuristic)**
> A computational procedure that is not guaranteed to find an optimal solution, but produces a solution that is typically close to optimal in a reasonable amount of time.

The Nearest Neighbor Algorithm (NNA) prevents you from choosing an edge that would prematurely create a circuit before all vertices have been visited. It is greedy in that it chooses the nearest neighbor (in terms of distance or cost) from the city that has just been reached. The Nearest Neighbor Algorithm will produce a genuine tour, which we hope will have near-minimum overall weight.

Example 5.5 Apply the Nearest Neighbor Heuristic to a Complete Graph

Apply the NNA to the example in Figure 5.5. (a) Starting at city A, determine the resulting tour and its overall distance. (b) Starting at city B, determine the resulting tour and its overall distance.

Solution

(a) Starting at city A, we first select the minimum weight edge (A, B); at city B, the minimum weight edge is (B, C); at city C, the only valid edge is (C, D) since the other two cities have already been visited. Our current path is then $A \to B \to C \to D$. Since we have visited all cities, we must now return home to A using the edge (D, A). The resulting circuit $ABCDA$ incurs a total distance of $300 + 560 + 710 + 420 = 1990$, compared to the minimum distance of 1895 obtained in Example 5.3.

(b) If we start instead at city B, the NNA produces the circuit $BACDB$, with a total distance of $300 + 345 + 710 + 570 = 1925$. This is an improvement over the distance of 1990, but is still not optimal. ◁

The results obtained in Example 5.5 illustrate why the NNA is a heuristic and does not necessarily produce an optimal solution. Here is a description of the steps of the NNA.

Nearest Neighbor Algorithm (NNA)

1. Start at an origin vertex.

2. Travel to a vertex that hasn't yet been visited, using an edge with the smallest weight. (If there is a tie, make the selection arbitrarily.)

3. Continue until all vertices have been visited.

4. Travel back to the origin vertex to complete the tour.

It is also possible to apply the NNA heuristic to graphs that are not complete, but care must be taken that we don't exhaust the available choices before completing the tour. The next example shows the application of the NNA heuristic to a graph that is not complete.

Example 5.6 Apply the Nearest Neighbor Heuristic to a Graph That is Not Complete

Apply the NNA to the weighted graph shown in Figure 5.8, with edge costs as specified. Start the heuristic at each of the different vertices and record the Hamiltonian circuit found. Which is the best solution identified using this heuristic?

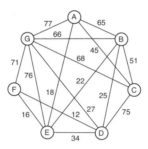

FIGURE 5.8: Applying the NNA to another graph

Solution

Table 5.3 records the NNA tour obtained when beginning at the specified vertex as well as its overall cost. It is seen that six different tours are produced, when starting the Nearest Neighbor heuristic at the seven different origin vertices. The best (smallest cost) tour found is that produced by starting at either vertex D or vertex G. ◁

(Related Exercises: 14–18)

The NNA follows a very nearsighted approach: it selects the next vertex to add to the current path in a greedy way, before closing the path to form a Hamiltonian circuit. This approach has two drawbacks: (1) it never changes the order of vertices on the current path, and (2) the final edge of the tour (from the last vertex back to the starting vertex) might incur a very large cost. To address these difficulties, improved TSP heuristics have been developed. Rather than extending a *path* at each stage, we can extend an existing *circuit*. Then a new vertex is inserted within the current circuit in such a way as to incur the least additional cost. This process is repeated until the circuit contains all vertices and thereby defines a genuine tour.

Specifically, the *Cheapest Insertion Algorithm* closely parallels Dijkstra's Algorithm from

TABLE 5.3: Hamiltonian circuits produced by the NNA

Origin Vertex	Hamiltonian Circuit	Total Cost
A	$AEFDBCGA$	267
B	$BEFDGCAB$	255
C	$CAEFDBGC$	250
D	$DFEACBGD$	**235**
E	$EFDBCAGE$	302
F	$FDBEACGF$	261
G	$GDFEACBG$	**235**

Chapter 3. At each step we have a set S of vertices already in the circuit and we determine a vertex j nearest the set S: that is, vertex j is chosen to minimize the cost c_{ij} where vertex i is in the set S and j is not in S. While Dijkstra's Algorithm calculates the sum of the permanent label on vertex i plus the cost $c_{i,j}$, here we only consider the simpler quantity $c_{i,j}$. Once this vertex j has been selected to enter S, we need to insert it in the best possible place, between two consecutive vertices v and w on the current circuit. Since this process will *remove* the edge (v, w) and *introduce* the new edges (v, j) and (j, w), the criterion used is to minimize the net increase in cost $c_{v,j} + c_{j,w} - c_{v,w}$. The next example shows how this procedure works in a small example.

Example 5.7 Apply the Cheapest Insertion Algorithm to a Complete Graph

Apply the Cheapest Insertion Algorithm to the graph of Example 5.4, shown again in Figure 5.9; costs are indicated for using each edge. Begin the algorithm with the trivial circuit that starts and ends at city A.

FIGURE 5.9: Example to illustrate the Cheapest Insertion Algorithm

Solution

The initial circuit can be written as AA, meaning that we start and end at A. The set S consists of vertex A alone. The cheapest edge leaving $S = \{A\}$ leads to vertex B, so that will be the next vertex added to S, giving the circuit ABA (cost 600). Among all edges leaving $S = \{A, B\}$, the minimum cost edge is (A, C) leading to C. So vertex C is added to S and is inserted between A and B on the circuit, giving the expanded circuit $ACBA$. Finally, only vertex D remains outside the circuit so it is added to S. Now we need to decide where to add vertex D within the circuit $ACBA$ (cost 1205).

- insert D between A and C: cost increase is $420 + 710 - 345 = 785$

- insert D between C and B: cost increase is $710 + 570 - 560 = 720$

- insert D between B and A: cost increase is $570 + 420 - 300 = 690$

The smallest net cost increase 690 occurs by inserting D between B and A, producing the tour $ACBDA$ with overall cost 1895. As it turns out, this is the optimal TSP tour (see Example 5.4). However, in general the Cheapest Insertion Algorithm is just a heuristic; it need not always produce an optimal tour. ◁

(Related Exercises: 19–21)

The steps of the Cheapest Insertion heuristic can be summarized as follows:

Cheapest Insertion Algorithm

1. Start with some initial circuit from s to s. Let $S = \{s\}$.
2. Select a vertex j not in S that is closest to S via edge (i, j), where i is in S.
3. Add vertex j to S and insert j between consecutive vertices v and w on the current circuit to minimize $c_{v,j} + c_{j,w} - c_{v,w}$.
4. Repeat Steps 2–3 until a tour on all vertices is obtained.

Further information on these and other heuristic approaches to the TSP, as well as animated illustrations, can be found at the sites [OR5, OR6].

(Related Exercise: 29)

5.3 Other Optimal Ordering Problems

What the Traveling Salesman Problem ultimately requires is an *optimal ordering*—a specific sequence of cities to be visited in order to minimize the total distance or cost. This section introduces some other optimal ordering problems that are closely related to the TSP.

Example 5.8 Formulate a Circuit Board Sequencing Problem as a TSP

A manufacturing plant uses an expensive machine that precisely drills holes in a printed circuit board. The geometric positions (coordinates) of the holes in the circuit board are shown in Figure 5.10(a). What is the best sequence for this machine to drill the holes, assuming that the apparatus needs to return to position A at the end of the processing?

Solution

It seems reasonable to select a sequence that minimizes the total time traveled by the drill head, starting at A. The time needed to travel between any two positions is directly proportional to the geometric (straight line) distance between these positions. So we require a Hamiltonian circuit (starting and ending at A) that visits all the positions of the required holes and incurs the smallest total distance. This is just a standard TSP instance on the set of hole locations plus location A.

One possible sequence of drilling the holes is given in Figure 5.10(b), and an optimal sequence is given in Figure 5.10(c). ◁

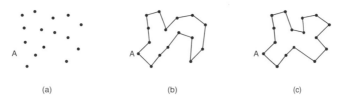

(a) (b) (c)

FIGURE 5.10: A circuit board sequencing problem

Example 5.9 Search for a Knight's Tour on a Chessboard

Recall that a knight moves in an L-shaped manner on a chessboard: two steps in one direction and one step in the other. (a) In the 3×4 chessboard shown in Figure 5.11, is it possible to move a knight from position to position, visiting each square and returning to the starting square? (b) If not, is there a way to move the knight from position to position so that each square is visited, but we do not return to the starting square?

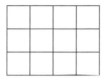

FIGURE 5.11: Knight's tour on a 3×4 chessboard

Solution

(a) We can represent this problem using a graph on 12 vertices (representing the 12 chessboard squares), with edges representing valid moves; see Figure 5.12(a). The Knight's Tour we are seeking is then a Hamiltonian circuit in this graph. Notice that the six circled vertices have degree two, meaning that any Hamiltonian circuit must use both of their incident edges. The portion of the graph, showing just these required edges, is displayed in Figure 5.12(b). We have highlighted these required edges (solid and blue), to show they form two disjoint circuits, containing six edges each. However, these 12 edges do not form a Hamiltonian circuit (on the 12 vertices), so there can be no Knight's Tour in a 3×4 chessboard.

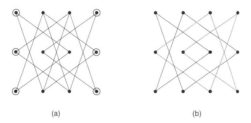

(a) (b)

FIGURE 5.12: Searching for a Hamiltonian circuit in the chessboard graph

(b) On the other hand, Figure 5.13(a) shows that we can remove an edge from each of these smaller circuits (forming two paths) and then reconnect them by an edge (shown dashed) to form a *Hamiltonian path*. This then provides a tour for the knight that visits all squares, but does not return to the starting position. This tour in the graph can be translated back into the 3×4 chessboard and results in the numbered sequence of squares visited, shown on the chessboard in Figure 5.13(b). ◁

(Related Exercises: 22–23, 30)

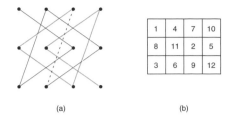

(a)   (b)

FIGURE 5.13: Finding a Hamiltonian path in the chessboard graph

---

DEFINITION **Hamiltonian Path**
A path in a graph that visits every vertex exactly once. It need not return to the starting vertex.

Example 5.10 Determine a Likely Chronology for Archaeological Finds

Archaeologists have uncovered artifacts at several sites. At each excavation site, the belief is that these artifacts were in use at more or less the same period of geological time. Figure 5.14 shows a schematic of which artifacts $(a, b, \ldots, f)$ were found at each of the sites $1, 2, \ldots, 7$. Determine a probable ordering of the times at which these artifacts were in use.

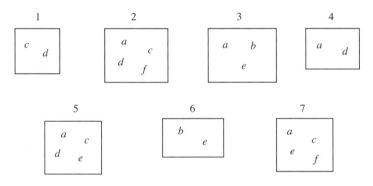

FIGURE 5.14: Artifacts appearing at seven archaeological sites

Solution

Figure 5.14 shows that artifact a is found at sites $2, 3, 4, 5, 7$; we represent this collection of sites by writing sites$(a) = \{2, 3, 4, 5, 7\}$. Similarly, we have

$$\text{sites}(b) = \{3, 6\}, \qquad \text{sites}(c) = \{1, 2, 5, 7\}, \qquad \text{sites}(d) = \{1, 2, 4, 5\},$$
$$\text{sites}(e) = \{3, 5, 6, 7\}, \qquad \text{sites}(f) = \{2, 7\}.$$

These data suggest, for example, that artifacts c and d might likely be contemporaneous since they are found together at three sites: 1, 2, and 5. We can represent this information as a graph in which the vertices are the artifacts and edges join pairs of artifacts that have at least one site in common. Moreover, we can associate a weight with this edge, representing the number of sites in common. Our example produces the weighted graph shown in Figure 5.15.

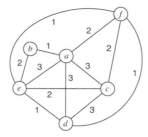

FIGURE 5.15: Graph associated with the archaeological sites

A path in this graph through all the vertices defines a possible chronology or sequencing of the artifacts, and a path that has a high total weight is considered more likely than a path with small weight. Using this criterion, we seek a Hamiltonian path of maximum weight in the graph. In this example, a maximum weight Hamiltonian path is $b \to e \to a \to d \to c \to f$ with weight 13. So a possible sequencing of the artifacts is b, e, a, d, c, f; yet another is the reverse sequencing f, c, d, a, e, b. (Additional information would help us decide between these two choices.) ◁

(Related Exercises: 24–25)

It turns out that this maximum weight Hamiltonian path problem can be transformed into a standard Traveling Salesman Problem by negating the edge weights; this changes the maximization objective into the minimization objective of the TSP. Also, by adding a "source" vertex s, joined to every other vertex by an edge of weight 0, we can change the Hamiltonian path requirement into a Hamiltonian circuit requirement. Consequently, the heuristic algorithms developed for the TSP can be applied to solve this sequencing problem in archaeology.

(Related Exercises: 26–28)

Chapter Summary

This chapter discusses another type of traversal problem arising in graphs. Unlike the case treated in the previous chapter (Eulerian circuits), here the objective is to visit all the vertices of the graph exactly once, returning to the starting vertex. A weighted version of this Hamiltonian circuit problem is the famous Traveling Salesman Problem. Since this problem is quite difficult to solve efficiently with an exact algorithm, we illustrate two greedy heuristics (the Nearest Neighbor Algorithm and the Cheapest Insertion Algorithm) that can be quickly applied. Several applications are discussed that require the optimal sequencing of choices, ranging from puzzles to manufacturing to seriation methods in archaeology. To solve such problems, we again follow a common modeling paradigm: convert the original problem into an equivalent problem in a suitable graph, identify an exact (or approximate) solution in the graph, and then translate this graph solution into the original problem context.

Applications, Representations, Strategies, and Algorithms

Applications			
Routing (2)	Business (4)	Manufacturing (8)	Archaeology (10)

Representations		
Graphs		Symbolization
Unweighted	Weighted	Geometric
1–2, 9	3, 4–7, 10	8

Strategies		
Brute-Force	Solution-Space Restrictions	Composition/Decomposition
1–2, 4, 8, 10	5–7	3, 9

Algorithms				
Exact	Approximate	Greedy	Inspection	Sequential
1–4, 8–10	5–7	5–7	1–4, 8–10	5–7

Exercises

Hamiltonian Circuits

1. Determine whether there is a Hamiltonian circuit in each of the graphs below. If so, display the circuit found. If not, justify why there cannot be one.

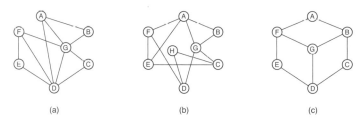

    (a)   (b)   (c)

2. Determine whether there is a Hamiltonian circuit in each of the graphs below. If so, display the circuit found. If not, justify why there cannot be one.

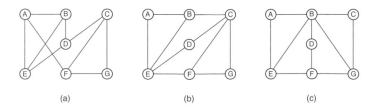

    (a)   (b)   (c)

3. Find a solution to the Icosian game in Figure 5.2 that uses edge AB and edge BH.

4. Find a solution to the Icosian game in Figure 5.2 that uses edge AE and edge EN.

5. The floor plan of a museum is shown below. A watchman would like to start at room A and visit all the rooms exactly once before returning to A. Model this as a Hamiltonian circuit problem in a graph and then determine a suitable patrol for the watchman.

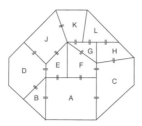

6. Consider the graph in Figure 5.4(b) associated with the *Instant Insanity*™ puzzle discussed in Example 5.3. A possibility for the first stack of cubes might use the three disjoint circuits BGB, RR, YY for Cubes 1–4, respectively. Find an appropriate Hamiltonian circuit disjoint from these that can be used for the second stack of cubes. Then deduce a different solution to the puzzle than that given in Table 5.1.

7. Construct the graph associated with an *Instant Insanity*™ puzzle having the following pairs of opposite faces on the cubes. Identify two disjoint Hamiltonian circuits, each using four different cube numbers, and then deduce a solution to the puzzle.

Cube 1: B-Y, R-G, R-Y

Cube 2: B-G, B-Y, R-G

Cube 3: B-R, R-Y, G-G

Cube 4: B-Y, R-R, G-Y

8. You have a set of 11 dominos (see the table below), each containing two different numbers from 0 to 6. You would like to arrange as many of them as possible in a circle so that each of the numbers 0 to 6 appears exactly twice. Moreover, any two consecutive dominos in the circle must contain matching numbers where the two dominos are adjacent. For example, the fourth and fifth dominos can be placed in the order 2-1, 1-5. Model and solve this as a graph problem on the vertices $0, 1, \ldots, 6$ with an edge corresponding to each domino.

Domino										
0	0	0	1	1	1	2	2	3	3	4
3	5	6	2	5	6	3	4	5	6	6

9. You have the set of 11 dominos shown in the table below. As in the previous exercise, you would like to arrange as many of them as possible in a circle so that each of the numbers 0 to 6 appears exactly twice. Model this as a graph problem on the vertices $0, 1, \ldots, 6$ with an edge corresponding to each domino. Using this graph representation, determine whether it is possible to arrange the dominos as described.

Domino										
0	0	0	1	1	1	2	2	2	3	4
2	4	5	2	4	6	3	4	5	4	6

10. You have the set of 11 dominos shown in the table below. As in the previous two exercises, you would like to arrange as many of them as possible in a circle so that each of the numbers 0 to 6 appears exactly twice. Model this as a graph problem on the vertices $0, 1, \ldots, 6$ with an edge corresponding to each domino. Using this graph representation, determine whether it is possible to arrange the dominos as described.

Domino										
0	0	0	0	1	1	2	3	3	4	5
1	2	3	5	2	4	6	4	6	5	6

11. Label the vertices of the complete graph K_5 with the letters $A, B, \ldots, E$. List all the (undirected) Hamiltonian circuits in this graph and verify that you obtain the number found in Table 5.2. Then compute this number using the mathematical formula that involves the factorial function.

12. List all (undirected) Hamiltonian circuits in the graph below, with edge costs indicated. Use the brute-force method to evaluate the cost of each such tour and thereby determine the Hamiltonian circuit with minimum total cost.

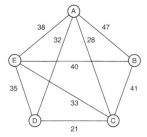

13. List all (undirected) Hamiltonian circuits in the graph below, with edge costs indicated. Use the brute-force method to evaluate the cost of each such tour and thereby determine the Hamiltonian circuit with minimum total cost.

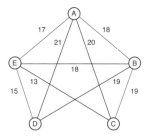

TSP Heuristics

14. Apply the NNA to the graph shown in Exercise 12, first starting at vertex A and then at vertex C. Display the resulting Hamiltonian circuits as well as their cost. Which of these tours is better? Is either of them optimal?

15. Apply the NNA to the graph shown in Exercise 13, first starting at vertex A and then at vertex C. Display the resulting Hamiltonian circuits as well as their cost. Which of these tours is better? Is either of them optimal?

16. Apply the NNA to the following graph, first starting at vertex A and then at vertex E. Display the resulting Hamiltonian circuits as well as their cost. Which of these tours is better?

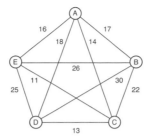

17. Apply the NNA to the following graph, first starting at vertex A and then at vertex D. Display the resulting Hamiltonian circuits as well as their cost. Which of these tours is better?

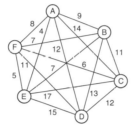

18. Apply the NNA to the following graph, first starting at vertex C and then at vertex F. Display the resulting Hamiltonian circuits as well as their cost. Which of these tours is better?

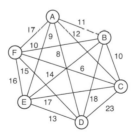

19. Apply the Cheapest Insertion Algorithm to the following graph, starting at vertex B. Display the Hamiltonian circuit that results as well as its cost.

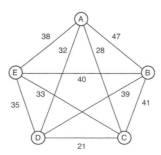

20. Apply the Cheapest Insertion Algorithm to the graph displayed in Exercise 16, starting at vertex C. Display the Hamiltonian circuit that results as well as its cost.

21. Apply the Cheapest Insertion Algorithm to the following graph, starting at vertex A. Display the Hamiltonian circuit that results as well as its cost.

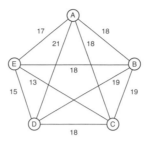

Ordering Problems

22. Consider the Knight's Tour on the truncated 4×4 chessboard below. We want to know if there is a way to start the knight at (say) square 1 and visit each of the other squares exactly once before returning home to square 1.

    (a) Construct a graph with 12 vertices, corresponding to the 12 squares, and with edges defined by valid knight moves. It may be useful to place vertices $4, 5, 9, 8$ on the outside of your diagram and cyclically arrange the other vertices $10, 12, 11, 7, 3, 1, 2, 6$ in the middle.

    (b) Determine if there is a Hamiltonian circuit in your graph, and thereby answer the original question.

		1	2	
3	4	5	6	
7	8	9	10	
	11	12		

23. We want to know if there is an "open" Knight's Tour on a 3×3 chessboard; namely, is there a way to start a knight at some square and visit each of the other squares exactly once? The knight need not however return to the starting square. Answer this problem by examining the associated graph, constructed as in Exercise 22.

24. The excavation of six sites has unearthed artifacts a, b, c, d, e as specified in the following table.

    (a) Construct a graph G with vertices representing artifacts and positive edge weights indicating the number of sites in which the two artifacts are present.

    (b) Using brute force, determine a Hamiltonian path in G with the largest weight.

    (c) Suppose it is also known that artifact d is definitely older than artifact a. Determine a probable chronology of the five artifacts.

Site	1	2	3	4	5	6
Artifacts	a, b	a, c, e	a, d, e	c, d	b, d, e	c, d, e

25. The excavation of seven sites has unearthed artifacts a, b, c, d, e as specified in the following table.

    (a) Construct a graph G with vertices representing artifacts and positive edge weights indicating the number of sites in which the two artifacts are present.

    (b) Using brute force, determine a Hamiltonian path in G with the largest weight.

    (c) Suppose it is also known that artifact e is definitely older than artifact b. Determine a probable chronology of the five artifacts.

Site	1	2	3	4	5	6	7
Artifacts	b, c, d	b, d, e	c, e	a, c, e	a, b, e	a, e	b, c

26. Suppose that a graph G on n vertices has positive edge weights. One approach to finding a largest weight path in G is to list all paths with $n - 1$ edges and select one with the largest weight. To see the difficulty of this approach, suppose that G is a complete graph K_n on n vertices.

    (a) List all the paths of length 2 in K_3. Since a path and its reversal have the same weight, we only need to list one of these. How many such paths are there?

    (b) Answer part (a) for paths of length 3 in K_4.

    (c) Conjecture a formula for the number of paths of length $n - 1$ in K_n.

27. Suppose we would like to find a largest weight path in the following graph G.

    (a) Transform this into a standard TSP by constructing an enlarged graph H from G by adding a new vertex s joined to each original vertex by a weight 0 edge. Also, to accommodate the maximization rather than the minimization objective, the original edge weights are negated.

    (b) Use the NNA starting from vertex A in graph H to obtain a Hamiltonian circuit. Then translate that into a Hamiltonian path for graph G; what is its weight?

    (c) Repeat part (b) when starting at vertex B, starting at vertex C, and starting at vertex D.

    (d) Which is the best solution found using these four starting vertices?

    (e) Determine the optimal solution by brute-force enumeration of all paths in G with three edges.

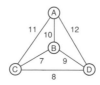

28. Suppose we would like to find a largest weight path in the following graph G.

    (a) Transform this into a standard TSP as described in Exercise 27.

(b) Use the NNA starting from vertex A in the enlarged graph to obtain a Hamiltonian circuit. Then translate that into a Hamiltonian path for graph G; what is its weight?

(c) Repeat part (b) when starting at vertex B, starting at vertex C, and starting at vertex F.

(d) Which is the best solution found using these four starting vertices?

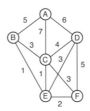

Projects

29. Another heuristic for finding an approximate solution to a TSP is the *Farthest Insertion Algorithm*. It follows the same steps as the Cheapest Insertion Algorithm, except that the vertex j chosen to enter the set S is one that maximizes the distance to the set S, rather than minimizes the distance. The rationale for this method is that it incorporates distant vertices into the set S early, avoiding having to insert them later on at high cost. Apply this heuristic to the graphs in Exercises 16–19.

30. Investigate whether there is an open Knight's tour on $3 \times n$ chessboards, where $n = 5, 6, 7, 8$.

Bibliography

[Cook 2012] Cook, W. J. 2012. *In pursuit of the traveling salesman problem: Mathematics at the limit of computation.* Princeton, NJ: Princeton University Press.

[Wilson 1990] Wilson, R. J., and Watkins, J. J. 1990. *Graphs: An introductory approach*, Chapter 7. Hoboken, NJ: John Wiley & Sons.

[OR1] http://www.britannica.com/EBchecked/topic/253431/Sir-William-Rowan-Hamilton (accessed November 29, 2013).

[OR2] http://eqn.princeton.edu/2011/12/ (accessed November 29, 2013).

[OR3] http://www.youtube.com/watch?v=7hRLOnTn7o4 (accessed November 29, 2013).

[OR4] http://www.math.uwaterloo.ca/tsp/optimal/index.html (accessed November 29, 2013).

[OR5] http://www-e.uni-magdeburg.de/mertens/TSP/node2.html (accessed November 29, 2013).

[OR6] http://riot.ieor.berkeley.edu/~cander/cs270/ (accessed November 29, 2013).

Chapter 6

Vertex Colorings and Edge Matchings

Illinois is green, Indiana is pink ... I've seen it on the map, and it's pink.

—*Tom Sawyer Abroad*, Mark Twain, American novelist, 1835–1910

Chapter Preview

Solutions to scheduling and assignment problems strive to avoid conflicts or incompatibilities while seeking to optimize various criteria. Such problems can be modeled using graphs with the focus either on vertex coloring or on edge matching. The concept of vertex coloring involves assigning different colors to adjacent vertices so that all vertices are colored and the fewest colors are used. The idea behind edge matching involves choosing edges that have no vertices in common and choosing as many such non-adjacent edges as possible. Similar to the Traveling Salesman Problem, there are no known efficient solution techniques that are guaranteed to find the fewest colors needed in a vertex coloring (the chromatic number of the graph). Several heuristic (approximate) techniques have been developed for coloring a graph, and we illustrate the Greedy Coloring Algorithm, perhaps the simplest technique for sequentially assigning colors to the vertices. By contrast, there are efficient techniques for determining the largest size of an edge matching in a graph.

6.1 Chromatic Numbers and Planar Graphs

When coloring a map, our goal is to color neighboring countries with different colors so that we can tell them apart. It would seem that as a map becomes more complex, with many countries, it would require more and more colors. But is this the case? In answering this question, we will introduce the idea of coloring the vertices of a graph, which has significant implications for finding solutions to a number of puzzles as well as real-world problems.

First, we discuss how to convert a map into a graph. Let each country be represented by a vertex and place an edge between any two adjacent countries (those sharing a common boundary). Figure 6.1(a) shows a map of a portion of Eastern Europe, with its associated graph displayed in Figure 6.1(b). In order to obtain a map coloring so that any two neighboring countries have different colors, we can equivalently color adjacent vertices of the associated graph with different colors.

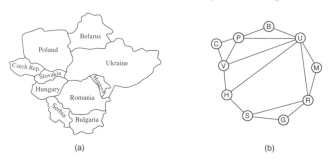

FIGURE 6.1: Map of Eastern Europe and its corresponding graph

In general, a *vertex coloring* is a coloring of the vertices such that any two vertices connected by an edge (adjacent vertices) have different colors. Figure 6.2 shows two vertex colorings of the graph in Figure 6.1(b); for convenience, the different colors are represented by different numbers. The vertex coloring in Figure 6.2(a) uses seven colors and one might think that seven colors are needed because Ukraine (U) is adjacent to six other countries. But the vertex coloring in Figure 6.2(b) reveals that the graph can be colored using only three colors. In fact, three is the fewest number of colors; Poland (P), Belarus (B), and Ukraine (U) all share common borders, so two colors would not work even for this part of the map.

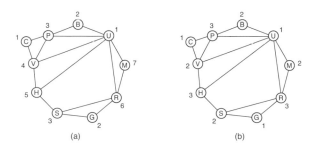

FIGURE 6.2: Two vertex colorings of a graph

DEFINITION **Vertex Coloring**
Each vertex of a graph is assigned a color (or a number representing the color) that is different from the colors assigned to any of its adjacent vertices.

DEFINITION **Chromatic Number**
The fewest number of colors required to color a graph so that adjacent vertices receive different colors.

Mapmakers are interested in using the fewest colors possible when printing maps, so determining the chromatic number of the associated graph is appropriate. Also, as will be seen in Section 6.3, other applications arise where it is advantageous to color an underlying graph with the minimum number of colors. An online site that invites the user to color graphs using the fewest colors is found at [OR1].

Example 6.1 Find a Map Coloring Using a Graph

A map of a portion of Western Europe is shown in Figure 6.3. (a) Draw a graph representation of this map. (b) Find a vertex coloring that uses the fewest number of colors and thereby determine the chromatic number of the graph. (c) Translate this vertex coloring back into a coloring of the original map.

FIGURE 6.3: Western Europe map

Solution

(a) There are eight countries shown in the map and therefore the graph representation has the eight vertices displayed in Figure 6.4(a). Vertices are considered adjacent if the corresponding countries share a common border.

(b) Two possible colorings are shown in Figure 6.4(b) and Figure 6.4(c). The latter coloring involves three colors and in fact uses the fewest number of colors. Since France (F), Germany (G), and Switzerland (W) are all mutually adjacent, three colors are needed just to color these three countries. So the chromatic number of the graph is 3.

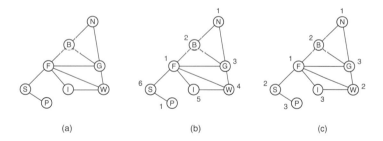

(a)   (b)   (c)

FIGURE 6.4: Graph of the Western Europe map and two colorings

(c) The vertex coloring in Figure 6.4(c) can now be used to color the original map. Specifically, the Netherlands and France can be colored red (1); Belgium, Switzerland, and Spain can be colored blue (2); while Portugal, Italy, and Germany can be colored green (3). ◁

(*Related Exercises: 1–3*)

Example 6.2 Find an Optimal Coloring for a Puzzle Diagram

The diagram in Figure 6.5 is taken from a popular iPhone app called *Doodle Fill* [OR2], in which you try to color (within a time limit) the indicated areas so that no two bordering areas are colored the same. (a) Draw a graph representation of the diagram shown in Figure 6.5. Find a vertex coloring of this graph using the fewest colors and thereby determine the chromatic number. (b) Translate this vertex coloring back into a coloring of the original diagram that uses the fewest colors.

FIGURE 6.5: Puzzle diagram to be colored

Solution

(a) The corresponding graph of the puzzle is shown in Figure 6.6(a); Figure 6.6(b) shows a vertex coloring using four colors. In fact, at least four colors are needed because vertices A, B, C, D are all mutually adjacent and so they require using four different colors. Since we have found a coloring using only four colors, the chromatic number is 4.

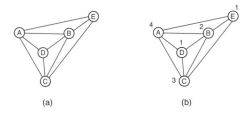

(a)　　　　(b)

FIGURE 6.6: Coloring of the puzzle graph

(b) By interpreting this vertex coloring in the original puzzle, areas D and E can be colored red, B can be colored blue, C can be colored green, and A can be colored yellow. ◁

(Related Exercise: 4)

Mathematicians have proven that any map can be colored with at most four colors. In addition, any map can be drawn as a *planar graph* where edges cross only at vertices (as in Figure 6.4 and Figure 6.6). On the other hand, the complete graph K_5 on five vertices (see Section 5.1) is nonplanar. It is not possible to redraw this graph, shown in Figure 6.7, without having some edges cross. Moreover, this graph requires the use of five colors since any two different vertices are adjacent, so each vertex must receive a separate color.

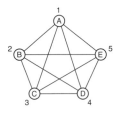

FIGURE 6.7: A coloring of K_5 with the fewest colors

DEFINITION **Planar Graph**
A graph that can be drawn on the plane without crossings—so that its edges intersect only at their endpoints.

> RESULT **Four-Color Theorem (1976)**
> At most four colors are needed to color any planar graph (and therefore any map).

Example 6.3 Find the Chromatic Number of a Graph

Find the chromatic number of the planar graph in Figure 6.8(a) and the nonplanar graph in Figure 6.8(b). Explain why you have correctly determined these chromatic numbers.

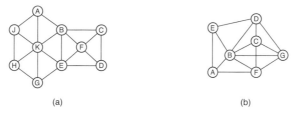

(a)   (b)

FIGURE 6.8: Planar and nonplanar graphs

Solution

The coloring shown in Figure 6.9(a) uses three colors; we cannot use fewer colors since A, B, K are all mutually adjacent and so require three colors among themselves. Thus, the chromatic number of the planar graph is 3. The coloring shown in Figure 6.9(b) uses four colors; we can't use fewer colors because B, C, F, G are all mutually adjacent. The chromatic number of the nonplanar graph is 4. ◁

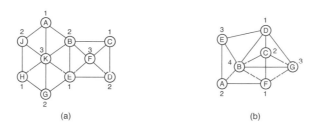

(a)   (b)

FIGURE 6.9: Optimal colorings of two graphs

Example 6.4 Find the Chromatic Number of Circuits

Figure 6.10 displays circuits on four and five vertices, respectively. Determine the chromatic number of each graph.

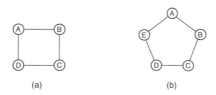

(a)   (b)

FIGURE 6.10: Circuits on four and five vertices

Solution

Since vertices A and B in Figure 6.10(a) are adjacent, they must be colored differently so at least two colors are needed to color that graph. The coloring shown in Figure 6.11(a) uses only two colors so it is an optimal coloring, meaning that the 4-circuit has chromatic number 2.

The coloring shown in Figure 6.11(b) uses three colors, so the chromatic number is at most 3. Could we use just two colors? A "forcing" argument shows that we cannot. Namely, the adjacent vertices A and B must have different colors, so let's call those colors 1 and 2. Then C must be colored 1 (we only have two colors, remember), and likewise D must have color 2. Now vertex E is adjacent to A and D, with respective colors 1 and 2, so there is no way to assign a different color to E if we are allowed just two colors. This verifies that the chromatic number of the 5-circuit is 3. ◁

(Related Exercises: 5–8)

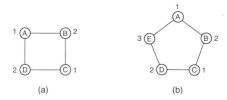

(a)   (b)

FIGURE 6.11: Optimal colorings of two circuits

In previous examples we have seen that when a graph contains three vertices that are all mutually adjacent (forming the complete graph K_3), the chromatic number cannot be smaller than 3. Or when the graph contains four vertices that are all mutually adjacent (forming the complete graph K_4), the chromatic number cannot be smaller than 4. On the other hand, if every vertex of a graph has degree at most 2 (as in Example 6.4), we can color the graph using at most $2 + 1 = 3$ colors. These observations are special cases of the following general principles about vertex colorings.

RESULT **Lower Bound on the Chromatic Number**

If the graph contains some K_n (in which the n vertices are mutually adjacent), any vertex coloring must use at least n colors. In other words, the chromatic number of the graph is at least n.

RESULT **Upper Bound on the Chromatic Number**

If the maximum degree of any vertex in the graph is d, then any vertex coloring will use at most $d + 1$ colors. In other words, the chromatic number of the graph is at most $d + 1$.

Figure 6.12(a) displays a graph that contains a K_3 (e.g., the triangle BCF), so its chromatic number is at least 3. Also, the maximum degree of a vertex in this graph is 5 (occurring at vertex F), so the graph will need at most $5 + 1 = 6$ colors; this means that the chromatic number is at most 6. As it turns out, the chromatic number of this graph is 4, which falls between the bounds of 3 and 6; Figure 6.12(b) provides an optimal coloring with four colors.

(Related Exercises: 9, 29)

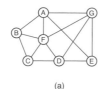 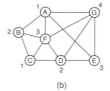

(a)   (b)

FIGURE 6.12: Application of general coloring principles

We can see how these general principles about vertex colorings may not ensure that we are able to pin down exactly the chromatic number of a graph. Moreover, a brute-force approach that tries all possible colorings requires just too much time, especially as the graphs become larger. In fact, vertex coloring has been shown to be an NP-hard problem, just as the Traveling Salesman Problem (see Section 5.1). This signifies that no efficient algorithm has yet been devised to determine the chromatic number of an arbitrary graph, though much effort has been expended to find one. Consequently, we will again content ourselves with using heuristic algorithms for vertex coloring, which need not necessarily produce an optimal coloring (having the fewest colors).

6.2   Greedy Coloring Algorithm

Let's examine a simple greedy approach to serve as a heuristic for vertex coloring. It involves sequentially processing the vertices of the graph in a fixed order; at each step, the current vertex is assigned the smallest number/color not already assigned to one of its (already colored) neighbors. We will see that the order chosen can affect the number of colors used.

Example 6.5 Color a Graph Using the Greedy Coloring Algorithm

The diagram in Figure 6.13 is also taken from the iPhone app called *Doodle Fill* [OR2], in which you try to color (within a time limit) the indicated areas so that no two adjacent areas are colored the same. (a) Create the graph corresponding to the diagram, where vertices $A, B, \ldots, K$ represent the eleven areas of the diagram. (b) Use the Greedy Coloring Algorithm, where vertices are processed in reverse alphabetical order, to produce a vertex coloring of the graph and thereby a coloring of the puzzle diagram.

FIGURE 6.13: Puzzle diagram

Solution

(a) The graph representation of the diagram in Figure 6.13 is provided in Figure 6.14(a).

(b) The Greedy Coloring Algorithm, using reverse alphabetical ordering, first assigns color 1 to K, then 2 to J (since J is adjacent to K), and then 3 to I (the smallest color available). We then assign the color 4 to H, since H is adjacent to vertices already colored 1, 2, 3. Next, color/number 1 is assigned to G, since that vertex is adjacent to vertices already colored 2 and 4. Continuing to apply the Greedy Coloring Algorithm in reverse alphabetical order results in a final coloring using four colors, shown in Figure 6.14(b). This coloring can be interpreted in the original diagram so that areas A, G, K are colored the same (say with red), areas D, E, J are colored blue, areas C, I are colored green, and areas B, F, H are colored yellow. Because areas H, I, J, K are all mutually adjacent, four is the fewest number of colors possible and so we have found an optimal coloring of the diagram. ◁

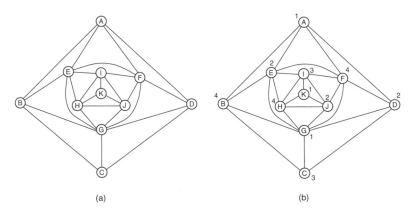

(a)   (b)

FIGURE 6.14: Graph and greedy coloring for the puzzle diagram

Example 6.6 Apply the Greedy Coloring Algorithm Using Different Vertex Orderings

Carry out the Greedy Coloring Algorithm twice on the graph in Figure 6.15. (a) Start at vertex A and proceed through the graph in alphabetical order, assigning the smallest color available for each vertex. (b) Start at vertex K and proceed through the graph in reverse alphabetical order, again assigning the smallest color available for each vertex. (c) Compare your results. Does order matter?

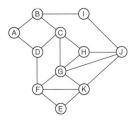

FIGURE 6.15: Graph to illustrate different vertex orderings

Solution

(a) Coloring in alphabetical order produces the coloring shown in Figure 6.16(a). Notice that vertex J requires color 4 because it is adjacent to G, H, and I, which have already been colored with 2, 3, and 1, respectively. Similarly, vertex K is assigned color 5 since it is adjacent to vertices already colored with 1, 2, 3, 4. Overall, this coloring uses five colors.

(b) Coloring in reverse alphabetical order produces Figure 6.16(b), which uses only three colors. This is an optimal coloring because of the existence of a K_3 structure (the triangle formed by E, F, K) within the graph.

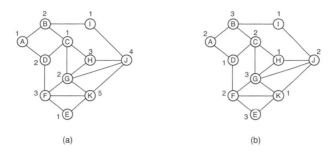

(a)   (b)

FIGURE 6.16: Greedy colorings with different vertex orderings

(c) Order affects not only the coloring, but also the number of colors used. ◁

Example 6.7 Apply the Greedy Coloring Algorithm and Verify an Optimal Coloring

Carry out the Greedy Coloring Algorithm on the graph shown in Figure 6.17 using alphabetical ordering of the vertices. How many colors are needed? Could we use fewer colors?

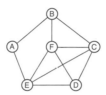

FIGURE 6.17: Another graph for greedy coloring

Solution

Applying the Greedy Algorithm with the alphabetical ordering $A, B, \ldots, F$ produces the coloring in Figure 6.18. Vertices receive the successive colors 1, 2, 1, 2, 3, 4, and so four colors are used. This is an optimal coloring since the graph contains a K_4 structure on the four vertices C, D, E, F. Consequently, the graph has chromatic number 4. ◁

(Related Exercises: 10–14, 30–31)

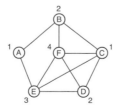

FIGURE 6.18: An optimal greedy coloring

We summarize the greedy approach to coloring as follows:

Greedy Coloring Algorithm

1. Choose an ordering of the vertices of the graph (e.g., alphabetical order)
2. Assign color 1 to the first vertex in the order.
3. Assign to the next vertex v in the order the smallest number (color) not appearing on a vertex adjacent to v.

Using this algorithm will produce a legitimate coloring of the graph, but it may not use the fewest number of colors. In other words, the Greedy Coloring Algorithm is a heuristic.

6.3   Applications of Vertex Coloring

This section presents several instances of assignment and scheduling problems that can be viewed as types of coloring problems.

Example 6.8 Use a Vertex Coloring to Assign Cell Phone Frequencies

Cell phone networks can be divided into different coverage regions. Two regions that are next to each other need to be assigned different frequencies in order to avoid interference problems. By using vertices to represent cell towers, and connecting two vertices with an edge if the two cell towers *cannot* use the same frequency, we can determine an optimal assignment of frequencies to cell towers through a vertex coloring.

Specifically, assume that any two cell towers within 150 miles of each other cannot use the same frequency because their signals will interfere. Distances between the cell towers $A, B, \ldots, F$ are shown in Table 6.1. (a) Draw the "interference" graph for this problem. (b) Determine an optimal coloring. (c) What is the minimum number of frequencies required? Show a corresponding assignment of frequencies to cell towers using this minimum number.

TABLE 6.1: Distances between cell towers

	A	B	C	D	E	F
A		108	175	165	95	120
B	108		125	180	155	85
C	175	125		105	160	130
D	165	180	105		120	115
E	95	155	160	120		145
F	120	85	130	115	145	

Solution

(a) The interference graph is shown in Figure 6.19. For instance, there is an edge between towers A and B because the distance between them is less than 150 miles. Towers A and B

must be assigned different colors (different frequencies) because if they both used the same frequency, there would be signal interference. There is an edge between towers B and F for the same reason. On the other hand, there is no edge between B and E because these towers are more than 150 miles apart and could be assigned the same frequency.

FIGURE 6.19: Interference graph for frequency assignment

(b) Figure 6.19 also displays a coloring that uses four colors. We can argue that four is the chromatic number using a forcing argument as follows. Suppose that only three colors are available. Vertices B, C, F form a K_3, so we must assign them three different colors (say, 1, 2, 3). Vertex A, which is adjacent to B and F with colors 1 and 3, must then receive color 2; vertex E, which is adjacent to vertices with colors 2 and 3, must receive color 1. Now vertex D will be adjacent to vertices C, E, F with colors 2, 1, 3 and so must receive a fourth color. This shows that three colors are not enough. As Figure 6.19 shows a coloring with four colors, it must be optimal.

(c) Since the chromatic number is 4, only four frequencies need to be assigned. Using the coloring in Figure 6.19 produces the assignment shown in Table 6.2. Namely, vertices colored the same can be assigned the same frequency. ◁

(Related Exercises: 15–17)

TABLE 6.2: An optimal frequency assignment

Frequency 1	B, E
Frequency 2	A, C
Frequency 3	F
Frequency 4	D

Example 6.9 Use a Vertex Coloring to Schedule Exams

A university needs to schedule final exams and it wishes to avoid having a student take more than one exam on the same day. The courses are labeled $A, B, \ldots, F$. In Table 6.3, a star in any cell, such as cell (A, B), indicates that courses A and B have at least one student in common so they should not be scheduled on the same day. (a) Construct a "conflict" graph from the table, where vertices represent exams; an edge between two vertices means that these two exams conflict (have students in common) and so cannot be scheduled on the same day. (b) Find an optimal coloring of the conflict graph. (c) Interpret the resulting coloring as an exam schedule that uses the fewest number of days.

Solution

(a) Figure 6.20 shows the conflict graph, in which the vertices represent courses and two courses are adjacent if they have at least one student in common. For example, edge (A, B) appears since these courses conflict, whereas edge (B, E) does not since courses B and E have no students in common.

TABLE 6.3: Courses with students in common

	A	B	C	D	E	F
A		*	*	*	*	*
B	*		*			*
C	*	*		*		
D	*		*		*	
E	*			*		*
F	*	*			*	

(b) Applying the Greedy Algorithm using alphabetical ordering produces the coloring in Figure 6.20 that uses four colors. This is in fact the smallest number of colors possible (see Exercise 18); the chromatic number of the graph is 4.

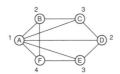

FIGURE 6.20: Graph of exam conflicts

(c) In a coloring of the conflict graph, any two courses that are adjacent will receive different colors and so they will be scheduled at different times. Coloring this graph with the fewest number of colors produces the fewest number of days to accommodate the exams without conflicts. As a result, the university only needs four days for the exams. A corresponding schedule (Table 6.4) can be found by placing on each day those vertices/exams that are all colored the same. The coloring shown in Figure 6.20 is not the only possible coloring with four colors. Exercise 18 asks you to find another coloring and exam schedule. ◁

(Related Exercises: 18–19)

TABLE 6.4: An optimal exam schedule

Day	Exam
1	A
2	B, D
3	C, E
4	F

Notice that the graph constructed in Example 6.9 is a planar graph. It can be redrawn as shown in Figure 6.21 without crossings and so we are assured that it requires no more than four colors (by the Four-Color theorem). Interestingly enough, the scheduling graph in Example 6.9 needs four colors, even though it does not contain a K_4 structure.

Example 6.10 Use a Vertex Coloring to Schedule Experiments

Seven solar experiments are to be conducted at various observatories around the world. Each experiment begins on a specified day and ends on a specified day. Because of equipment limitations, an observatory can perform only one solar experiment at a time. The problem

FIGURE 6.21: Planar representation of the exam scheduling graph

is to determine the minimum number of observatories required to perform all experiments. Table 6.5 lists the time periods required by the seven experiments $A, B, \ldots, G$. For example, Experiment A starts on March 10 and continues through June 15.

TABLE 6.5: Durations of experiments

	A	B	C	D	E	F	G
Start	3/10	4/21	5/11	6/3	7/10	7/25	8/28
End	6/15	7/18	7/8	8/12	9/6	10/1	10/5

(a) Draw an "overlap" graph representing this problem, where vertices correspond to experiments and edges join two vertices whenever the corresponding experiments overlap in time. (b) Find an optimal coloring using the Greedy Coloring Algorithm with vertices processed in alphabetical order. (c) What is the fewest number of observatories needed in order to conduct all seven experiments? Construct an assignment of experiments to observatories that uses the fewest number of observatories.

Solution

(a) The overlap graph is shown in Figure 6.22(a). An edge joins two vertices whenever the experiments they represent overlap in time. For example, A and D are adjacent because experiments A and D both need to be conducted in early June. Similarly, there is an edge between D and F because experiments D and F overlap in late July. However, no edge is placed between A and E because these experiments do not overlap in time.

(b) Using the Greedy Coloring Algorithm with vertices processed in alphabetical order results in a coloring with four colors; see Figure 6.22(b). This greedy coloring is in fact optimal, since the graph contains a K_4 structure on the vertices A, B, C, D and so it cannot be colored using fewer than four colors.

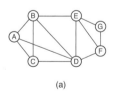
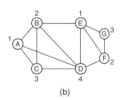

(a)   (b)

FIGURE 6.22: Overlap graph and an optimal coloring

(c) Since vertices assigned the same color do not overlap and can be carried out by the same observatory, only four observatories (colors) are needed. The experiments can be assigned to the four observatories as specified in Table 6.6. ◁

(*Related Exercises: 20–21*)

TABLE 6.6: Observatory experiments

Observatory 1	A, E
Observatory 2	B, F
Observatory 3	C, G
Observatory 4	D

Example 6.11 Solve a Word Puzzle Using a Vertex Coloring

You are given the following 19 three-letter words:

> BOX, COT, DEW, DRY, FIX, FOE, FOX, JAR, LED, LEG,
> PAW, PIT, RAY, SAY, TIC, TOE, WAR, WAS, WET

These words are composed of exactly 18 distinct letters. The puzzle is to find a way to put the 18 letters on three cubes (each is six-sided) so that you can spell out each of the above 19 words using one letter from each of the three cubes. There is no order to the cubes. The only requirement is that for each of the 19 words, we can rearrange the cubes so that the letters appearing on the top will spell out the given word.

(a) Create a graph where the vertices are the 18 distinct letters appearing in the set of 19 words. Put an edge between two letters if they appear together in any of the 19 words; this edge signifies a "conflict" in the sense that those two letters can't appear on the same cube. (b) Find a coloring of your graph using the Greedy Coloring Algorithm, with vertices processed in alphabetical order. (c) Can you identify a better coloring of the graph? Explain how an optimal coloring identifies a solution to the puzzle.

Solution

(a) The graph that results is shown in Figure 6.23(a). Each of the 19 words forms a triangle in this graph. For example, the word BOX generates the edges (B,O), (O,X), (B,X).

(b) Applying the Greedy Coloring Algorithm in alphabetical order to this graph results in a coloring that uses five colors; see Figure 6.23(b).

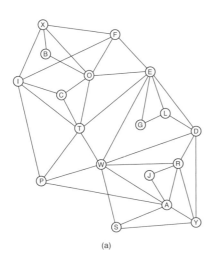

(a)

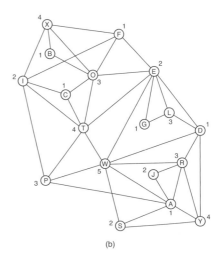

(b)

FIGURE 6.23: Graph and greedy coloring for the word puzzle

(c) The coloring obtained in part (b) uses five colors, which is too many colors since we want to put all letters with color 1 on the first cube, all letters with color 2 on the second cube, and all letters with color 3 on the third cube. We want to use six of each color. How can we find such a coloring?

We start with triangle FOE and color the vertices with $1, 2, 3$, respectively. Then we must use color 1 for T (looking at triangle TOE). Continuing in this manner to look for "forced" colors, we are able to successfully color the graph shown in Figure 6.23(a) with three colors:

> Color 1: A, B, D, F, G, T (all go on cube 1)
>
> Color 2: I, J, L, O, W, Y (all go on cube 2)
>
> Color 3: C, E, P, R, S, X (all go on cube 3)

You can check that all 19 words can be spelled out using distinct cubes: for example, BOX (cubes 1-2-3), COT (cubes 3-2-1), and DEW (cubes 1-3-2). ◁

(Related Exercise: 22)

6.4 Matching Problems

In a vertex coloring, no two vertices having the same color can be adjacent to each other. We can extend this idea to the coloring of edges, where we now require that no two edges having the same color can be adjacent to each other. When we try to find a set of edges that can be colored the same because they are not adjacent (do not share a vertex), we are in fact looking for a *matching*.

Figure 6.24 shows matchings in three different graphs; edges of the matching are shown as bold. Figure 6.24(a) shows a matching of size two. Even though no additional edges can be added to maintain a matching, it is not a *maximum matching* since a matching of larger size can be found in the graph: namely, edges $(A, E), (C, F), (D, G)$. Figure 6.24(b) shows a *maximum matching* of size three, as it is not possible to find four (non-adjacent) matching edges in this graph. Notice that not all vertices are included in this matching. Figure 6.24(c) shows a *perfect matching*: a maximum matching that includes all vertices.

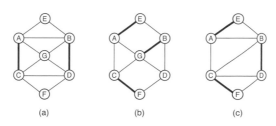

(a)   (b)   (c)

FIGURE 6.24: Matchings in graphs

Several types of puzzles as well as important practical problems involve the search for matchings in a graph.

(Related Exercises: 23–25)

DEFINITION **Matching**
A set of edges in a graph, no two of which share a common vertex.

DEFINITION **Maximum Matching**
A matching in which as many edges as possible are included in the matching.

DEFINITION **Perfect Matching**
A matching in which the edges of the matching include all vertices.

Example 6.12 Solve the Rook's Problem on a Chessboard

In the chessboard of Figure 6.25, place as many non-taking rook pieces as possible, using only the vacant (unshaded) squares. A rook can move any number of squares in the horizontal direction or any number of squares in the vertical direction. So for rooks to be non-taking, no two can appear in the same row or the same column. (a) Construct a *bipartite graph G*, with one set of vertices representing the rows $R_1, \ldots, R_5$ and another set of vertices $C_1, \ldots, C_4$ representing the columns of the chessboard. An edge is placed between a row vertex and a column vertex only when the corresponding square in the chessboard is vacant. (b) Find a maximum size matching in the graph G. (c) Use this matching to identify a placement of the maximum number of non-taking rooks on the chessboard.

FIGURE 6.25: Placing rooks on a 5×4 chessboard

Solution

(a) Figure 6.26(a) contains nine edges placed between row vertices and column vertices, each signifying a vacant square in the chessboard.

(b) Rooks placed at squares $(1, 3)$ and $(2, 2)$ are *non-taking*; these rooks correspond to the *non-adjacent* edges (R_1, C_3) and (R_2, C_2). In other words, these two edges form a matching. The original problem of finding the largest number of non-taking rooks is equivalent to finding a matching of maximum size in this graph. In our particular example, the edges $(R_1, C_3), (R_2, C_2), (R_4, C_4)$ form a matching of size 3; these are shown in bold in the graph of Figure 6.26(b). This is in fact a maximum matching.

(c) We have identified $(R_1, C_3), (R_2, C_2), (R_4, C_4)$ as a maximum matching in the graph. Translating this into the context of the original problem, we can assert that rooks placed at squares $(1, 3), (2, 2), (4, 4)$ give a largest number of non-taking rooks on the chessboard. ◁

In Example 6.12, we claimed that three is the maximum size of any matching in the graph of Figure 6.26(a). This claim can be solidly verified. Notice that since there are four column vertices, we can select at most four edges to form a matching: no two edges can meet at the same vertex. Moreover, vertices R_4, C_2, C_3 *cover* all of the edges in G. So this limits the size of any matching to at most three: any collection of more than three edges would contain two edges that meet at one of these three vertices and so would not constitute a matching.

(Related Exercises: 26–28)

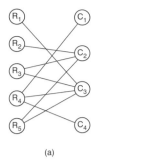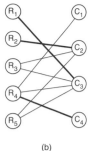

(a)   (b)

FIGURE 6.26: Maximum matching in a bipartite graph

DEFINITION **Bipartite Graph**
A graph in which the vertices can be separated into two parts; edges only are permitted to join vertices in different parts.

DEFINITION **Vertex Cover**
A set of vertices such that every edge in the graph touches at least one of these vertices.

More generally, there is a relation between matchings of edges and covers of vertices. We implicitly used the following fact to verify the size of a maximum matching in Example 6.12. It will be helpful again in later examples.

RESULT **The Match-Cover Theorem**
The number of edges in any matching is less than or equal to the number of vertices in any vertex cover.

Example 6.13 The Perfect Square Dance Puzzle

Vanessa invited 17 guests to her estate for an evening of dinner and dance. Each guest was assigned a number from 2 to 18; Vanessa kept the number 1 for herself. When all nine pairs of individuals were dancing, the butler observed that the sum of the two numbers for each pair of dancers was a perfect square (i.e., the square of an integer). What was the number of the guest with whom Vanessa was dancing? (a) To answer this question, draw a graph G in which the vertices correspond to the numbers assigned to the 18 individuals. Draw an edge between vertices i and j when $i + j$ is a perfect square. (b) Find a perfect matching of all vertices of G so that every vertex is matched with another vertex. (c) Use this information to deduce the number of Vanessa's partner.

Solution

(a) The "perfect square" graph G is shown in Figure 6.27(a). For example, G contains edges $(3, 6)$ and $(3, 13)$ since $3 + 6 = 9 = 3^2$ and $3 + 13 = 16 = 4^2$.

(b) Starting with vertex 16 we see that it must be matched with vertex 9; similarly, 18 must be matched with 7; 2 must be matched with 14; and so on. Continuing this process, we obtain the unique perfect matching shown in Figure 6.27(b).

(c) Since vertex 1 is matched uniquely with vertex 15, it must be the case that Vanessa is dancing with guest 15. ◁

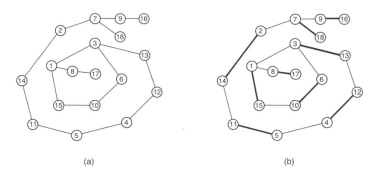

(a)   (b)

FIGURE 6.27: Square dance graph and its unique perfect matching

Paired Kidney Exchanges

In the U.S. today there is a severe shortage of organs for use in kidney transplants. The waiting list for kidney transplants contains over 100,000 individuals and continues to grow [OR3]. To partially alleviate this situation, some hospitals have organized a paired kidney donation program. It is often the case that a relative would gladly donate a kidney to a particular patient, but this is not possible because of incompatibilities in blood type or because there is a "positive cross-match" (patients may have antibodies that react negatively with the donor's blood).

Consider two recipients R_1, R_2 and their willing but incompatible donors D_1, D_2. Suppose that instead D_1 could donate to R_2 and D_2 could donate to R_1. Such a swap is able to successfully find compatible donors for both patients. This illustrates the idea of a paired kidney exchange.

Example 6.14 Maximize the Number of Kidney Exchanges

We can model a paired kidney donation problem using a graph. Each of the 12 vertices in Figure 6.28(a) represents a pair of incompatible donor-recipients. Two vertices are connected by an edge if the donors of each pair are compatible matches for the recipients of the other pair. Find the largest number of compatible kidney exchanges by identifying a matching of maximum size in this graph.

Solution

Figure 6.28(b) shows a matching of size five, using edges $(A, H), (B, L), (C, G), (E, J), (I, K)$. Since there are 12 vertices, perhaps there could be an even better matching—one with six edges. How can we demonstrate that this selection is in fact a maximum matching?

Recall the relationship between matchings and vertex covers discussed after Example 6.12. Namely, the size of any vertex cover provides an upper bound on the size of any matching. In Figure 6.29, we have circled the five vertices A, C, E, K, L. We can see that every edge meets at least one of the circled vertices, so these five vertices form a vertex cover of the graph. By the Match-Cover theorem, this means that any matching can have at most five edges. Since our particular matching does have five edges, we have demonstrated that no better solution can exist! ◁

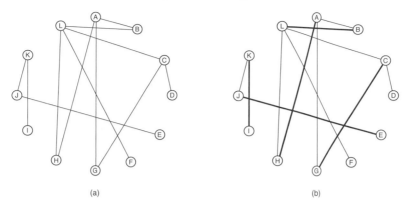

FIGURE 6.28: Feasible kidney exchanges

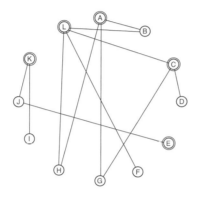

FIGURE 6.29: Vertex cover in the kidney exchange graph

HISTORICAL CONTEXT **Matchings and the Nobel Prize**
Lloyd S. Shapley (1923–) and Alvin E. Roth (1951–) shared the 2012 Nobel Prize in Economics for their work on theoretical and practical aspects of matchings. Recent applications of this work include not only paired kidney exchanges, but also allocation of doctors to hospitals for their first year of residency, as well as the assignment of students to high schools in an equitable way.

(Related Exercise: 32)

Chapter Summary

The examples in this chapter demonstrate how vertex colorings and matchings arise in a variety of contexts, spanning the range from puzzles to practical applications (map making, frequency assignment, exam scheduling, kidney donations). Since vertex coloring is classified as a notoriously hard problem to solve in general, a greedy approach is followed to obtain a coloring with relatively few colors. Interestingly enough, we are sometimes able to verify the optimality of a proposed coloring by exhibiting a certain graph structure (a complete graph) or by invoking a forcing type of argument. In a similar vein, we are able to verify the optimality of a proposed maximum matching by exhibiting an appropriate vertex cover.

Applications, Representations, Strategies, and Algorithms

Applications			
Geography (1)	Communication (8)		
Scheduling (9–10)	Medicine (14)		

Representations			
Graphs	Tables		
Unweighted	Data		
1–14	8–10		

Strategies			
Solution-Space Restrictions	Rule-Based		
1–14	5–7, 9–11		

Algorithms			
Exact	Greedy	Inspection	Sequential
1–4, 8, 11–14	5–7, 9–11	1–4, 8, 12, 14	5–7, 9–11, 13

Exercises

Maps, Graphs, and the Chromatic Number

1. The map below shows six states and territories of Australia.

   (a) Create a graph representation of this map.

   (b) Find a vertex coloring of the graph using the fewest number of colors (and thereby determine the chromatic number). Justify your answer for the chromatic number.

   (c) Translate this vertex coloring back into a coloring of the original map.

2. The following map shows eight southern provinces of Sweden.

   (a) Create a graph representation of this map.

(b) Find a vertex coloring of the graph using the fewest number of colors (and thereby determine the chromatic number). Justify your answer for the chromatic number.

(c) Translate this vertex coloring back into a coloring of the original map.

3. The map below shows nine countries in Central and Eastern Europe.

   (a) Create a graph representation of this map.

   (b) Find a vertex coloring of the graph using the fewest number of colors (and thereby determine the chromatic number). Use a forcing argument to justify your answer for the chromatic number.

   (c) Translate this vertex coloring back into a coloring of the original map.

A = Austria
C = Czech Republic
G = Germany
H = Hungary
I = Italy
P = Poland
V = Slovakia
L = Slovenia
W = Switzerland

4. Consider the diagram below, which consists of regions $A, B, \ldots, H$.

   (a) Create a graph representation of this diagram.

   (b) Find a vertex coloring of the graph using the fewest number of colors (and thereby determine the chromatic number). Use a forcing argument to justify your answer for the chromatic number.

   (c) Translate this vertex coloring back into a coloring of the original diagram.

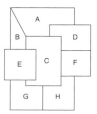

5. Find a vertex coloring of the graph below that uses the fewest number of colors. Verify that your coloring is optimal.

6. Find a vertex coloring of the graph below that uses the fewest number of colors. Verify that your coloring is optimal.

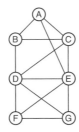

7. Example 6.4 determined the chromatic number of circuits with four and five vertices. Generalize these observations to determine the chromatic number of a circuit with n vertices.

8. The graph below is a *wheel graph* on six vertices, consisting of five vertices arranged in a circuit all of which are joined to a central vertex.

   (a) What is the chromatic number of this wheel graph? Justify your answer.

   (b) What is the chromatic number of a wheel graph on seven vertices? Justify your answer.

   (c) Generalize your observations to wheel graphs on n vertices.

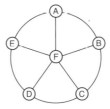

9. Figure 6.12(b) shows a coloring of the graph in Figure 6.12(a) that uses four colors. Use a forcing argument to show that this graph cannot be colored using three colors.

Greedy Coloring Algorithm

10. Apply the Greedy Coloring Algorithm to color the following graph, processing the vertices in alphabetical order. Is the resulting coloring optimal? Explain.

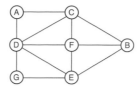

11. Apply the Greedy Coloring Algorithm to color the graph below, processing the vertices in alphabetical order. Is the resulting coloring optimal? Explain.

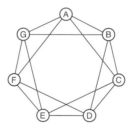

12. Apply the Greedy Coloring Algorithm to color the graph below, processing the vertices in alphabetical order. Is the resulting coloring optimal? Explain.

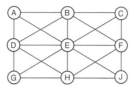

13. Apply the Greedy Coloring Algorithm to color the graph below, processing the vertices in alphabetical order.

   (a) Is the resulting coloring optimal?

   (b) Can you find another ordering of the vertices so that the Greedy Coloring Algorithm, applied with this ordering, produces an optimal coloring?

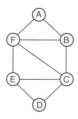

14. Apply the Greedy Coloring Algorithm to color the following graph, processing the vertices in alphabetical order.

   (a) Is the resulting coloring optimal?

   (b) Can you find another ordering of the vertices so that the Greedy Coloring Algorithm, applied with this ordering, produces an optimal coloring?

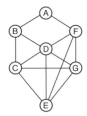

Applications of Coloring

15. Suppose that the signal strength of cell towers has increased in Example 6.8 so that any two towers within 170 miles must be assigned different frequencies. The same distances provided in Table 6.1 remain valid.

    (a) Construct the frequency interference graph for this situation.

    (b) Determine an optimal coloring of this graph and then translate this coloring into a best assignment of frequencies to cell towers.

16. A cell phone area is divided into eight regions $A, B, \ldots, H$ each served by a cell tower. Distances between cell towers are given in the table below. Towers that are within 125 miles of one another cannot be assigned the same frequency.

    (a) Construct the frequency interference graph for this situation.

    (b) Determine an optimal coloring of this graph and then translate this coloring into a best assignment of frequencies to cell towers.

	A	B	C	D	E	F	G	H
A		110	135	129	95	121	131	132
B	110		98	142	133	108	120	149
C	135	98		122	128	141	117	128
D	129	142	122		118	137	98	123
E	95	133	128	118		114	132	118
F	121	108	141	137	114		107	119
G	131	120	117	98	132	107		121
H	132	149	128	123	118	119	121	

17. Suppose that the signal strength of cell towers has increased in Exercise 16 so that any two towers within 130 miles must be assigned different frequencies. The same distances provided in Exercise 16 remain valid.

    (a) Construct the frequency interference graph for this situation.

    (b) Determine an optimal coloring of this graph and then translate this coloring into a best assignment of frequencies to cell towers.

18. The exam scheduling problem in Example 6.9 produced the graph in Figure 6.20, which was colored with four colors.

    (a) Verify that this graph has chromatic number 4 by using a forcing argument.

(b) Find a different coloring that also uses four colors and then produce a new exam schedule using four days.

19. A university needs to schedule exams for courses $A, B, \ldots, G$ so that no student has more than one exam on any day. Below we list for each course X any other course that also enrolls a student who is in course X.

$A : B, E, F$ $B : A, C, D, E$ $C : B, D, E, F, G$ $D : B, C, G$
$E : A, B, C, F, G$ $F : A, C, E, G$ $G : C, D, E, F$

(a) Construct the course interference graph for this situation.

(b) Determine an optimal coloring of this graph and thus determine the fewest days needed to schedule the exams. Translate your coloring into an allowable schedule that uses the fewest days.

20. Some species of fish will attack each other and cannot be stored in the same tank. We have 10 types of fish, denoted $A, B, \ldots, K$. A star in the table below indicates that the species in that row and column are incompatible. We want to use the fewest number of tanks to contain all our fish.

	A	B	C	D	E	F	G	H	J	K
A		*	*	*						
B	*		*	*						
C	*	*		*	*			*	*	
D	*	*	*		*					
E		*	*			*				
F				*			*	*	*	
G					*			*	*	
H					*	*			*	*
J			*			*	*	*		*
K			*					*	*	

(a) Construct the associated conflict graph, where each vertex represents a species of fish. An edge between two species means they cannot be placed in the same tank.

(b) Color the conflict graph using the Greedy Coloring Algorithm; process the vertices in alphabetical order.

(c) Is this an optimal coloring? Justify your answer.

(d) Show an optimal assignment of fish to tanks.

21. Certain chemicals need to be kept at separate locations to avoid explosions or other reactions. The following table indicates by a star which pairs of chemicals $(A, B, \ldots, G)$ cannot be stored together.

(a) Construct the conflict graph for this situation.

(b) Determine an optimal coloring of this graph and thus determine the fewest number of locations needed. Translate your coloring into a corresponding assignment that uses the fewest locations.

	A	B	C	D	E	F	G
A		*				*	*
B	*		*	*			
C		*		*	*		
D		*	*		*	*	
E			*	*		*	*
F	*			*	*		*
G	*				*	*	

22. Use a forcing argument to color Figure 6.23(a), starting with the assignment of colors 1, 2, 3 to vertices F, O, E, respectively. Verify that you obtain a coloring in which six vertices are colored with each of the three colors.

Matchings

23. In the graph shown below, determine a matching with the largest number of edges. Is it also a perfect matching?

24. In the graph shown below, determine a matching with the largest number of edges. Is it also a perfect matching?

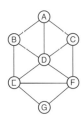

25. The following graph has 10 vertices so that a perfect matching would contain five edges. Is this possible? If not, find a maximum matching in this graph.

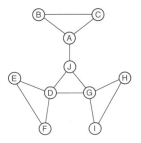

26. In order to put as many planes in the air as possible at one time, an airline with international routes wants to match up pilots and copilots who are compatible in terms of language and navigational skills. The following list shows which of the copilots (E, F, G, H) are compatible with each of the pilots (A, B, C, D). Find a maximum matching between pilots and copilots (in an appropriate bipartite graph), so the largest set of planes can be deployed. Verify that you have found a maximum matching by exhibiting an appropriate vertex cover.

$$A : F, H \qquad B : G, H \qquad C : F, G \qquad D : E, H$$

27. Suppose there are four subjects (French, German, Math, and Physics) that must be taught at a particular high school and there are four instructors (A, B, C, D) who are qualified to teach one or more of these subjects, as specified below. Find a matching in an appropriate bipartite graph that ensures that all four subjects will be taught and no instructor will teach more than one course.

$$A : \text{Math, Physics} \quad B : \text{French, German} \quad C : \text{French, Physics} \quad D : \text{German, Math}$$

28. A pharmaceutical company has solicited five volunteers $(A, B, \ldots, E)$ to test five new antibiotics $(a, b, \ldots, e)$. Each volunteer has known allergic reactions to certain drugs, as specified below. For example, volunteer A is allergic to antibiotics a and c.

$$A : a, c \qquad B : d, e \qquad C : a, c, e \qquad D : a, c \qquad E : a, c, d$$

   (a) Construct a bipartite graph whose parts correspond to volunteers and antibiotics; an edge indicates *compatibility* between the volunteer and the antibiotic.

   (b) Determine a maximum matching in this graph. Verify that you have found a maximum matching by exhibiting an appropriate vertex cover.

   (c) Is it possible to assign the five different antibiotics to five different volunteers? If so, specify such an assignment.

Projects

29. Recall the upper bound result from Section 6.1: if the maximum degree of a vertex in graph G is d, then no more than $d + 1$ colors are needed to color G. This can be verified as follows.

    Suppose that you have $d + 1$ colors available when applying the Greedy Coloring Algorithm. You are about to color the next vertex v (say in alphabetical order). What is the maximum number of colors that can appear on the (already colored) vertices adjacent to v? Then argue that v can now be colored using one of the $d + 1$ available colors. Conclude that G can be legitimately colored using at most $d + 1$ colors.

30. Previous examples have shown that the Greedy Coloring Algorithm, using a particular vertex ordering, may not produce an optimal coloring. Will there always be some vertex ordering for the Greedy Coloring Algorithm that produces an optimal ordering? Investigate this possibility by starting with an optimal coloring of the graph and then seeing how you might define a vertex ordering that will yield this optimal coloring.

31. We have illustrated the simplest type of greedy approach for coloring a graph, which uses an alphabetical ordering to process the vertices. However, a more intelligent

ordering of vertices is to process them by decreasing vertex degree. In other words, we first select a vertex of highest degree and assign it color 1. Then we successively select a vertex with the next highest degree to process, breaking any ties arbitrarily. The reasoning behind this ordering is to avoid being confronted in the last stages of the Greedy Coloring Algorithm with processing a vertex of high degree whose adjacent vertices have many different colors; this may cause the current vertex to use a new color.

Investigate the efficacy of this *Largest Degree First* (LDF) heuristic on a number of sample graphs, comparing its performance with that of the Greedy Coloring Algorithm used in this chapter. Also, construct examples for which the LDF heuristic uses fewer colors than the Greedy Coloring Algorithm.

32. Example 6.14 illustrates the utility of identifying a maximum matching in the graph of donor-recipient pairs. Each matching edge represents a two-way exchange. It is also logistically feasible to have a three-way exchange, which corresponds to a circuit of length three in a related directed graph. Investigate this extension by consulting references such as [Saidman 2006].

Bibliography

[Higgins 2007] Higgins, P. M. 2007. *Nets, puzzles, and postmen*, Chapter 4. New York: Oxford University Press.

[Jensen 1995] Jensen, T., and Toft, B. 1995. *Graph coloring problems*. New York: John Wiley & Sons.

[Saaty 1986] Saaty, T., and Kainen, P. 1986. *The four-color problem: Assaults and conquest*. New York: Dover Publications.

[Saidman 2006] Saidman, S., Roth, A., Sönmez, T., Ünver, M., and Delmonico, F. "Increasing the opportunity of live kidney donation by matching for two- and three-way exchanges." *Transplantation* 81(5): 773-782.

[Wilson 2002] Wilson, R. 2002. *Four colors suffice: How the map problem was solved*. Princeton, NJ: Princeton University Press.

[OR1] http://www.cut-the-knot.org/Curriculum/Combinatorics/ColorGraph.shtml (accessed December 10, 2013).

[OR2] https://itunes.apple.com/us/app/doodle-fill-most-addictive/id664553211?mt=8 (accessed December 8, 2013).

[OR3] http://www.unos.org (accessed December 8, 2013).

Unit II

Logic: Rational Inference and Computer Circuits

Chapter 7

Inductive and Deductive Arguments

It is useless to attempt to reason a man out of a thing he was never reasoned into.

—Jonathan Swift, Anglo-Irish essayist, 1667–1745

Chapter Preview

Determining the logical correctness of an inductive or deductive argument is central to problem solving. Inductive and deductive logic comprise the standards and techniques for evaluating both types of arguments. Inductive logic allows us to make inferences about the likelihood of a conclusion based on accumulated evidence. On the other hand, deductive logic allows us to analyze the internal consistency of an argument. A set representation is used to abstract the essential features of a deductive argument and a Venn diagram provides a visual method of determining the validity or invalidity of such arguments. Tables can also assist in solving certain deductive puzzles. Other arguments can be rejected quickly and without the aid of such representations because they are based on fallacious reasoning and designed only to persuade, not justify. When errors are more subtle and arguments are more complex, evaluation becomes more difficult and, as we will see in subsequent chapters, more detailed representations and techniques are required to judge the validity of arguments.

7.1 Rational Arguments

Persuasive arguments are made every day by speakers on radio and TV, by participants on blogs and twitter, by writers in books, magazines, newspapers and scholarly journals, by teachers and colleagues at work, and by family members in order to convince us to follow their reasoning. Good reasons must be presented to justify political positions, scientific hypotheses, medical claims, and other facts. Rational arguments are distinct from emotional appeals, eloquent persuasion, indoctrination, or coercion.

Rational arguments always provide reasons or evidence in the form of *premises* for the purpose of supporting certain *conclusions*. Logic is concerned with two types of arguments: (a) *inductive arguments*, where the goal is to establish that the premises probably support the conclusion, and (b) *deductive arguments*, where the goal is to establish that the premises necessarily support the conclusion.

Below are two proposed arguments in which the reasoning proposed is insufficient to justify the conclusion.

A. Inductive argument:

> For hundreds of years, Europeans observed that every swan they saw was white.
> So, they inferred that all swans are white.

Even though this was a reasonable inductive generalization, the conclusion was proven false when black swans were found in Australia.

B. Deductive argument:

> If Michael buys bonds, then Michael is investing his money.
> Michael is investing his money.
> Therefore, Michael buys bonds.

Assuming the truth of both premises, one cannot infer the truth of the stated conclusion because Michael could buy stocks, gold, annuities, etc.

DEFINITION **Premise**
A statement in an argument (inductive or deductive) that is claimed to justify a particular conclusion.

DEFINITION **Conclusion**
A statement in an argument (inductive or deductive) that is claimed to be justified by the given premises.

To evaluate the rationality of arguments, we need to understand what constitutes sufficient justification to support conclusions in both inductive and deductive arguments. We now study these types of arguments in more detail.

7.2 Inductive Arguments and Inductive Logic

The goal of a good inductive argument is to provide evidence for the probable truth of the conclusion. Figure 7.1 depicts an inductive argument as a list of steps, beginning with the given premises, followed by a variety of legitimate inferences from these premises, and ending with the desired conclusion.

The truth of each premise is either established in advance (on the basis of other arguments) or is viewed as unassailable through knowledge common to the audience at which the argument is directed. If the premises are not accurate, relevant, or sufficient to support the conclusion, then a new argument based on additional evidence or methodology will be needed to increase the *strength* of the inference.

> **Premise** (Given)
> **Premise** (Given)
> $\vdots$
> **Premise** (Given)
>
> intermediate steps based on the rules of
> statistics and probability
>
> Therefore, it is *probable* that the **Conclusion** is true.

FIGURE 7.1: Structure of an inductive argument

DEFINITION **Strong Inductive Argument**
An argument in which the truth of the conclusion is highly probable based on statistical analysis of reliable evidence.

DEFINITION **Weak Inductive Argument**
An argument in which either the premises are false or the truth of the premises is insufficient to establish the claim that the conclusion is probable.

Inductive logic teaches us how to evaluate arguments by using important tools of scientific procedure: namely, *statistics* and *probability*.

DEFINITION **Inductive Logic**
The study of proper strategies and methods for choosing premises (gathering evidence, conducting experiments, etc.) and drawing legitimate probabilistic conclusions.

DEFINITION **Statistics**
Sets standards for how to collect reliable evidence (e.g., unbiased surveys and careful, reproducible experiments) and how to interpret the evidence based on knowledge of uncertain events.

DEFINITION **Probability**
Used to calculate the likelihood of an event occurring. It provides a quantitative measure of how well the evidence supports the stated conclusion.

Example 7.1 Evaluate an Inductive Argument

In the inductive argument below, the premises are denoted P1, P2, P3 and the conclusion is denoted C. Assess its strength or weakness.

P1: Numerous well-designed, large-sample studies have been conducted by prominent researchers on the negative impact of smoking on human health.

P2: These studies have been repeated over time.

P3: Results have been published in highly respected, peer-reviewed journals.

C: It is probable that smoking is bad for your health.

Solution

Using information about statistical data collection, and statistical inference and probability, we can determine whether the evidence cited is sufficient to support the conclusion.

We should ask which experts sanction the conclusions and why—are they independent, respected authorities? Have the experiments and outcomes been reproduced by others? Have the data been analyzed and evaluated through peer review? Assuming that the studies and journal articles in the above example meet these criteria, the argument in Example 7.1 is a strong inductive argument. ◁

Example 7.2 Evaluate an Inductive Argument

Evaluate the strength or weakness of the following inductive argument:

P1: A lab in San Jose claims to have verified the effectiveness of magnets on arthritis.
P2: Many people have testified that magnets helped their arthritis.
C: It is probable that magnets help arthritis.

Solution

Assuming that no well-respected lab has been able to reproduce the results of the San Jose lab, the first premise may be true, but this does not provide convincing evidence. Also, it is well known that human beings are susceptible to believing that they being helped by any intervention claiming to ameliorate their condition (*placebo effect*). Moreover, those who have testified may represent a self-selected group (only those experiencing improvement). As a result, the second premise is not supportive. The inductive argument in Example 7.2 is judged to be weak. ◁

Example 7.3 Evaluate an Inductive Argument

Evaluate the strength or weakness of the following inductive argument:

P1: There was money missing at Freedom Bank on Wednesday.
P2: Mary is a bank teller.
P3: Mary needed money for her mortgage payment.
P4: Mary was working at Freedom Bank on Wednesday.
P5: Mary made her mortgage payment on Thursday.
C: Mary stole money from Freedom Bank on Wednesday.

Solution

The above argument is an example of inductive reasoning that seems logical but is frequently found to generate false conclusions. For example, in detective fiction and TV crime series, inductive evidence will be gathered and point to one suspect, only to be overturned later in the book or show. The above argument about Mary is weak because we need more evidence, such as Mary's bank statements, information about her character, other possible suspects, and video surveillance footage. Suppose that Mary received a loan from a relative so she could pay her mortgage on Thursday, and further analysis reveals that similar unsolved shortages occurred at Freedom Bank before Mary was hired. All these facts, taken together, strongly suggest that a different employee should be investigated. ◁

(*Related Exercises: 1–7*)

Example 7.4 Solve an Inductive Mystery

In the short story *The Silver Blaze* by Arthur Conan Doyle, Sherlock Holmes investigates the theft of a racehorse from an estate that was guarded by a watchdog. Holmes comments that the dog did not bark on the night of the theft because the people sleeping in the loft of the stables did not wake up. He concludes that the person stealing the horse must have been someone that the dog knew. Is this a strong inductive argument?

Solution

It is true that dogs usually bark at strangers, but there are many other possibilities: the dog could have been drugged, the intruder could have provided food to the dog, the people may have been sleeping exceptionally soundly, etc. Therefore, Holmes did not present us with the strong argument that he asserts he has. ◁

Example 7.5 Solve an Inductive Puzzle

In this inductive reasoning problem, we are presented with the following pattern of successive integers. What would probably be the next number in the sequence?

$$4, 3, 3, 5, 4, 4, 3, 5, 5, 4, \ldots$$

Solution

One might argue that the next number should be 6 because ascending identical pairs of numbers are seen to follow each single digit: 4, **3**, **3**, 5, **4**, **4**, 3, **5**, **5**, 4, **6**, **6**. But this isn't a strong inductive argument since it doesn't explain the pattern for the single digits.

Indeed, one might argue with higher probability that these 10 numbers represent the number of letters in each word in the following sequence: zero, one, two, three, four, five, six, seven, eight, nine. Therefore, the next number is much more likely to represent the number of letters in ten, which is 3. ◁

(Related Exercises: 8–12)

In the case of the earlier "white swan" argument, the existence of black swans showed that the argument was weak because the previous evidence was insufficient. The existence of black swans presented a *counterexample* to the stated conclusion.

> DEFINITION **Counterexample to an Inductive Argument**
> A single exception showing that the stated conclusion need not hold in the context presented.

7.3   Deductive Arguments and Deductive Logic

The goal of a good deductive argument is to provide evidence for the claim that the conclusion is necessarily true. Figure 7.2 depicts an argument as a list of steps, beginning with the given premises, followed by a variety of legitimate inferences from these premises, and ending with the desired conclusion. We use certain well-established logical rules to guide us from the premises to the conclusion.

Premise (Given)

Premise (Given)

      ⋮

Premise (Given)

   intermediate steps based on certain rules of
inference and rules of replacement

Therefore, it is *certain* that the **Conclusion** is true.

FIGURE 7.2: Structure of a deductive argument

The truth of a premise may either be established in advance (on the basis of other reasoning) or is viewed as unassailable through knowledge common to the audience at which the argument is directed. If a premise is not true, this will affect the usefulness of a deductive argument, but as we will show later, false premises can logically imply the conclusion as long as the form of the argument meets certain conditions.

Deductive logic enables us to evaluate arguments by revealing the *structure* of an argument; that is, by making explicit the connections that exist between the set of premises and the conclusion. If the chain of reasoning from the premises to the conclusion follows certain rules, then the conclusion necessarily follows from the premises and the argument is said to be *valid*.

DEFINITION **Deductive Logic**
The study of techniques of symbolization enabling the analysis of argument forms and the study of rules of inference explaining how statements are structured in valid arguments.

DEFINITION **Valid Deductive Argument**
The premises and the conclusion are connected in such a way that if the premises are true, then the conclusion is necessarily true.

DEFINITION **Sound Deductive Argument**
A valid argument whose premises are in fact true.

In this section we will show how one representation, using sets and a visual representation of sets, can be useful in determining the validity of arguments. We begin by describing how to translate verbal statements into the language of sets.

Example 7.6 Translate a Statement Using Sets and Venn Diagrams

Express the following statement using the language of sets: "Snakes are reptiles."

Solution

The statement involves two sets: the set S of snakes and the set R of reptiles. Moreover, it indicates a *subset* relation between the sets: namely, every snake is also a reptile. Another common way of saying this is the assertion that all snakes are reptiles. Because this statement claims a particular relation between sets, it is termed a *categorical proposition*.

We can also show this subset relation visually, using the *Venn diagram* illustrated in Figure 7.3. We have drawn a circle and an oval to represent the two sets S and R. Also, we have shown the relationship between S and R by placing set S entirely within set R. This indicates that every snake is a reptile, or S is a *subset* of R. In fact, S is a *proper subset* of R, meaning that there are reptiles (such as iguanas) that are not snakes. We have also drawn a large rectangle to indicate the *universe* of discourse—the set of all creatures, which itself contains both sets S and R. ◁

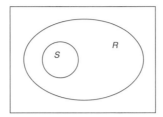

FIGURE 7.3: Venn diagram to represent the relationship between sets

---

DEFINITION **Set**
A collection of objects (or members).

DEFINITION **Subset**
Set A is a subset of set B if every member of A is a member of B. Any set is a subset of itself.

DEFINITION **Proper Subset**
If set A is a subset of set B and if A is not equal to B, then A is a proper subset of B.

DEFINITION **Categorical Proposition**
A proposition that indicates a relation between the members of certain sets.

DEFINITION **Venn Diagram**
A diagram that shows the logical relations among a number of sets.

HISTORICAL CONTEXT **John Venn (1834–1923)**
A British logician and philosopher, who is best known for his creation (in 1881) of diagrams that represent categorical propositions.

Example 7.7 Determine if a Deductive Argument is Valid by Using Venn Diagrams

Use a Venn diagram to determine if the following argument is valid:

P1: All smokers put their health at risk.
P2: Walter smokes cigarettes.
C: Walter puts his health at risk.

Solution

The premises in the argument above involve two sets of objects: the set A of people who smoke and the set B of people who put their health at risk. According to the first premise, set A (smokers) is a subset of set B (people who risk their health). The Venn diagram showing this subset relation is depicted in Figure 7.4(a), with the large rectangle indicating the universe of all people. The second premise states that Walter (w) smokes, so Figure 7.4(b) shows that w is a member of set A.

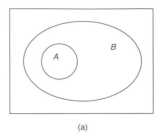 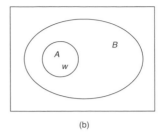

(a)   (b)

FIGURE 7.4: Venn diagrams to represent a valid deductive argument

From the Venn diagram in Figure 7.4(b), it follows that w must also be a member of set B. In other words, Walter puts his health at risk, and so the stated conclusion logically follows. The given argument is valid. It is also a sound argument if both the premises really are true. ◁

The following argument has the same form as the argument in Example 7.7 and so it is also valid. However, now one of the premises is clearly false.

P1: All birds can fly.

P2: A penguin is a bird.

C: Penguins can fly.

Because a penguin is a bird that cannot fly, the argument is unsound. The argument though is valid—the same Venn diagrams in Figure 7.4 apply, where now A is the set of birds and B is the set of flying creatures. Validity simply means that *if* the premises were both true, *then* the conclusion would necessarily follow.

Example 7.8 Show a Deductive Argument is Invalid by Using Venn Diagrams

Use a Venn diagram to determine if the following argument is valid:

P1: All smokers put their health at risk.

P2: Yolanda does not smoke.

C: Yolanda does not put her health at risk.

Solution

The first premise asserts that the set A of smokers is a subset of the set B of people who risk their health. The second premise states that Yolanda (y) is outside the set of smokers. If we try to place Yolanda in the Venn diagrams of Figure 7.5, we see that y can be placed either inside set B or outside set B. This means we cannot logically conclude that Yolanda is *necessarily* outside the set of risk-takers.

Therefore, the conclusion C does not logically follow from the given premises P1 and P2. This deductive argument is invalid; a defensible link does not exist between the premises and the conclusion. ◁

(*Related Exercises: 13–15*)

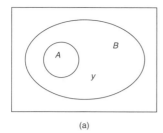
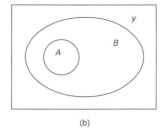

(a)   (b)

FIGURE 7.5: Venn diagrams to illustrate an invalid deductive argument

If we accept the truth of the two premises in Example 7.8, do we also have to accept the conclusion as true—or is there an example that will make the conclusion false, yet allow for the truth of the premises? Such an example is called a *counterexample*.

DEFINITION **Counterexample to a Deductive Argument**
An example that shows it is possible for the premises to be true and the conclusion to be false.

Example 7.9 Show a Deductive Argument is Invalid by Finding a Counterexample

Find a counterexample to the argument put forth in Example 7.8:

P1: All smokers put their health at risk.
P2: Yolanda does not smoke.
C: Yolanda does not put her health at risk.

Solution

In this case, obesity could provide a counterexample to the above argument. Even if Yolanda does not smoke, she could still put her health at risk by being obese. For this situation the premises would be true and yet the conclusion would be false. This one counterexample shows that the argument is invalid. ◁

Example 7.10 Show a Deductive Argument is Invalid by Finding a Counterexample

Find a counterexample to the following argument:

P1: If the switch is off, then the room is dark.
P2: The switch is not off.
C: Therefore, the room is not dark.

Solution

The intention of any deductive argument is to show that the conclusion is necessarily true if the premises are true. But in this case a counterexample can be exhibited: the switch could be on, and yet the room is still dark because the light bulb has burned out. In other words, there is a scenario in which the premises are true and the conclusion is false, so the argument is invalid. ◁

What happens if we now change the form of the argument in Example 7.10 by rearranging premise P1? The argument then turns out to be valid. This can be demonstrated by the use of Venn diagrams.

Example 7.11 Show a Deductive Argument is Valid by Using Venn Diagrams

Use a Venn diagram to show that the following argument is valid:

P1: If the room is dark, then the switch is off.
P2: The switch is not off.
C: Therefore, the room is not dark.

Solution

The Venn diagram of Figure 7.6(a) shows the two sets A (rooms that are dark) and B (rooms that have the light switch off), both subsets of the universe of all rooms (the large rectangle). Premise P1 asserts that set A is a subset of set B. Premise P2 asserts that a specific room (r) is not in set B, as shown in Figure 7.6(b). It is now seen that this room r must also lie outside set A, meaning that the room is not dark. This establishes the validity of the argument. ◁

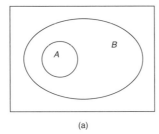
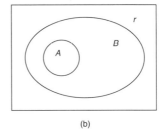

(a)   (b)

FIGURE 7.6: Venn diagrams to illustrate a valid deductive argument

Example 7.12 Solve a Game Show Puzzle Using Deductive Reasoning

You are a participant on a game show and need to find the grand prize hidden either in Box 1 or Box 2. The game show host tells you that the two assistants A and B standing on stage both know where the prize is hidden and that you may ask *one* of them a *single* question. The host adds that one of the assistants always tells the truth and the other one always lies. You decide to ask B the following question: "What would A say if I asked A which box contains the prize?" Assistant B now replies that A would say Box 2. Will you now be able to choose the correct box and take home the grand prize?

Solution

We could set up a valid deductive argument as follows. If B lies, then (a) A would say the prize is in Box 1, and (b) A is the truth-teller. So you should pick Box 1. On the other hand, if B tells the truth, then (a) A would say the prize is in Box 2, and (b) A is the liar. So again you should pick Box 1. By the way, the same logic applies if you had instead asked A the analogous question "What would B say if I asked B which box contains the prize?" So by asking this question, you are able to take home the prize even though you do not know which assistant is telling the truth and which is lying. ◁

Example 7.13 Solve a Deductive Puzzle

Adams, Beckett, Castle, and Davis make their living as carpenter, mechanic, painter, and plumber, though not necessarily in that order. Beckett has never met the mechanic or the painter. The painter is a brother of the carpenter, who has done remodeling work for both Adams and Beckett. Davis is an only child. Using logical deductions, determine which people have the respective occupations.

Solution

It is convenient to represent the information using a table, with rows corresponding to the individuals and columns corresponding to the professions. We place an X in an entry of the table if that combination is not possible. In particular, the first clue shows that Beckett is neither the mechanic nor the painter, so those entries are filled with an X, as shown in Table 7.1. Also, since Davis is an only child, Davis cannot be either the painter or the carpenter. The other clues are used to complete the table.

TABLE 7.1: Tabular solution of a logic puzzle

	Carpenter	Mechanic	Painter	Plumber
Adams	X			
Beckett	X	X	X	
Castle				
Davis	X		X	

This table now shows that the only possible profession for Beckett is as plumber, and that the carpenter must be Castle. Once this information is deduced, we can remove the rows for Beckett and Castle as well as the columns for plumber and carpenter, to obtain the reduced Table 7.2. Consequently, Davis must be the mechanic and then Adams is the painter. ◁

(Related Exercises: 16–17)

TABLE 7.2: The reduced logic table

	Mechanic	Painter
Adams		
Davis		X

Similar types of logic puzzles can be found online, for example at the site [OR1].

7.4   Irrational Arguments

A "serious" argument (inductive or deductive) is one that has been put forward in the sincere belief that the evidence supports the conclusion. Such arguments can contain *logical flaws* that may or may not be easy to identify. In this section, we will look at errors in reasoning so common throughout history that speakers (e.g., politicians and advertisers) have deliberately used them to persuade their audience irrationally, in lieu of strong inductive or valid deductive arguments to support their positions. These common logical flaws are referred to as *fallacies*. We will also look at an example of a *paradox*, where the argument put forward is so defective that it seems to yield an inherent contradiction.

DEFINITION **Logical Flaw**
An error that results even though a serious attempt was made to apply the rules of inductive or deductive logic.

DEFINITION **Fallacy**
A common pattern of poor reasoning (yielding weak or invalid arguments) that is potentially deceptive and, to an unpracticed audience, appears to be good reasoning.

DEFINITION **Logical Paradox**
A chain of reasoning that appears to use meaningful and consistent premises, yet it yields contradictory statements. Somehow an initial mistake has allowed a contradiction.

We discuss a number of fallacies in this section to illustrate how blatant errors in bad arguments are often exploited for the purpose of persuasion rather than to inform. It is difficult to categorize fallacies because many fallacies cross over any division we might try to impose. However, for ease of explanation, these fallacies are grouped according to the type of error they are exploiting: emotional appeal, misrepresenting evidence, false presumption, and distorted language.

Emotional Appeal

Below are four fallacies in which the argument is based on irrelevant side issues, with the goal of provoking emotional responses from the listener.

1. Personal-Attack Fallacy

When a lawyer attacks the credibility of a witness in the courtroom by introducing negative characteristics or qualities of the witness, this is not fallacious. It is relevant to evaluating the testimony of the witness. When someone points out that a celebrity Jennifer is being paid to endorse a product, so you cannot assume that the product is everything that Jennifer claims it is, this fact about Jennifer is relevant. But when arguments are based on qualities or circumstances of an individual that the audience would find distasteful and are not relevant to the topic under discussion, then the arguments involve the personal-attack fallacy.

A too-frequent version of the personal-attack fallacy involves comparing your opponent in an argument to Hitler or to the Nazis. Suppose some argues: "Wilson has explained why he is in favor of abortion. But remember, Hitler was also in favor of abortion. Therefore, you should reject Wilson's argument." Here is an attempt to demonize Wilson; while this historical fact is true, it does not constitute a reason to reject Wilson's abortion argument. Moreover, the "Hitler Card" is too often used inappropriately when comparing any behavior of a contemporary political figure with the behavior of Hitler and the horrors of Nazism. This simply offends people on all sides of an argument and ends any meaningful discussion.

Example 7.14 Identify the Personal-Attack Fallacy

"You can't trust Jones' theory of black holes because he is an atheist."

Solution

This is not a reasoned argument; rather, it is a personal attack. The speaker knows the audience and intends to demean the opposition's character and thereby demean the opposition's theory. ◁

2. Empathy Fallacy

Empathy is the ability to understand what others are feeling and to experience emotions that match theirs. Appeals to empathy are always fallacious when intended to influence our judgment, because emotional effect is not a substitute for reasons in support of a position. If the goal of an argument is also to motivate people to act, then it could be reasonable to include evidence that appeals to empathy, but only if it is accompanied by evidence demonstrating the validity of the argument.

Example 7.15 Identify the Empathy Fallacy

"These children are starving. Therefore, you should contribute to our charity."

Solution

This is an appeal to sympathy or pity for a situation, but it is not sufficient evidence for the conclusion. Will my check to this charity really go to help these children? Is there a better way to help them? Does the evidence demonstrate a causal connection between a check to this organization and help for these children? ◁

3. Popularity Fallacy

The popularity fallacy is used in arguments where instead of defending a proposal on its own merits, the premises express how the product is popular. Examples include: "This car is worth the price because it is driven by attractive people with style and money," "This book is worth reading because it has been on the best-seller list for nine weeks," or "You should see this film because it made $94 million in the first week it was released."

Example 7.16 Identify the Popularity Fallacy

"Armed forces should be sent to North Korea. Polls show that 75% of Americans are in favor of intervention."

Solution

The opinion of most Americans (even if this number were accurate) isn't relevant to the argument why there should be such a significant intervention in the foreign affairs of another country. ◁

4. Red-Herring Fallacy

When a red herring is dragged across a fox trail in the sport of fox hunting, it throws the hounds off the scent of the fox. When topics of greater interest to the audience than the topic under debate are introduced into an argument, these distract and redirect the focus, yet they do not contribute to weakening the argument under discussion.

A red herring fallacy introduces a second topic B that is unrelated to the first topic A; but since topic B is more important or of more value to the audience, it is distracting. Issues with higher priority for someone always carry more commitment and invoke greater emotional response. Topic B may be an issue that generates audience support, but the argument concerns topic A, and topic B is currently irrelevant.

Example 7.17 Identify the Red-Herring Fallacy

"Why worry about whether Siberian tigers are becoming extinct when we haven't solved the plight of the homeless? Therefore, let's reject these conservation efforts."

Solution

Perhaps something should be done about the homeless, but this is not the topic under discussion; it is irrelevant to the current argument about whether Siberian tigers are becoming extinct and what we ought to do about it. The plight of the homeless, though an emotionally more pressing topic for the audience, is irrelevant to the current discussion. (If we were discussing government budgetary priorities for funding new housing initiatives, it would however be relevant.) ◁

Misrepresenting Evidence

Below are four fallacies based on a flawed characterization of the evidence supporting a conclusion. Evidence is misconstrued or manufactured, while other relevant evidence may be ignored.

1. Slippery-Slope Fallacy

Arguments based on the slippery-slope fallacy are constructed in the following way. If A happens, then B will happen, then C will happen, ..., and finally Z will happen; since Z should not happen, then A should not happen. This form of argument is not necessarily fallacious. For example, one could make a strong argument that you ought to prepare for Exam 1 in Calculus because failing Exam 1 will contribute to failing Exam 2, and then failing Exam 3, then failing the final exam, and so failing the course. Such an argument only becomes fallacious when the intermediate steps lack the necessary causal strength and the so-called evidence doesn't really exist.

Example 7.18 Identify the Slippery-Slope Fallacy

"Legalizing abortion on demand will weaken respect for human life, which would eventually lead to euthanasia of the sick, the elderly, and the mentally and physically challenged."

Solution

There is insufficient evidence to show that this chain of reasoning is probable or even likely. The speaker deliberately claims that there is evidence (when there is not) to support an implication that it is believed that audience will want to reject. ◁

2. Hasty-Generalization Fallacy

When a statistician presents the results of a study and the study is criticized for generalizing to the entire population from a sample that is too small, the statistician is not accused of the hasty-generalization fallacy. There is a legitimate debate about sample size in complex situations that requires discussion among professional statisticians. The hasty-generalization fallacy occurs in arguments where the sample size is obviously too small, and it becomes quickly clear that the argument was poorly formed or that the goal of the argument was to persuade an audience rather than to justify the conclusion.

We should realize that if a population being studied is completely homogeneous or composed of members that were all the same, we could sample just one member and make a reasonable generalization. But most populations are variable, and increasing variability (heterogeneity) requires samples of increasing size.

Example 7.19 Identify the Hasty-Generalization Fallacy

"My roommate said that her philosophy class was hard, and the one I'm in is hard, too. All philosophy classes must be hard."

Solution

This generalization about all philosophy classes is not supported by sufficient evidence. It is unjustified because we cannot make generalizations about a (varied) population based upon a sample that is too small or not known to be representative. ◁

3. Correlation/Causation Fallacy

The Latin phrase *non causa pro causa*, meaning "non-cause for cause," captures what happens when committing this fallacy. If an argument implies a causal relationship simply because two events occur at the same time (are correlated) and does so without examining or presenting any further evidence (perhaps done intentionally to deceive), then we observe the fallacy of confusing correlation with causation.

This fallacy is encountered when we observe that Y occurs soon after X happens; therefore, the claim is made that Y occurs because of X. Presented with correlation, this argument in fact claims causation. To illustrate, suppose that after some children play a video game for four hours, we observe disruptive behavior by these children. The correlation/causation fallacy occurs if we immediately conclude that playing four hours of this video game causes the disruptive behavior. The mere occurrence of one situation following the other is not evidence that the former causes the latter.

Example 7.20 Identify the Correlation/Causation Fallacy

"Governor Williams raised state taxes, and then the jobless rate increased. Williams is responsible for the rise in unemployment."

Solution

Coincidence and correlation do not constitute sufficient evidence of causation. The speaker chooses to hide evidence that there might be factors other than the increase in taxes that could cause unemployment to rise, such as a growing population in the state, companies going out of business, or companies moving out of the country for cheaper labor. ◁

4. Pseudo-Authority Fallacy

We all rely on expert opinion in scientific or technical matters where we lack time or training to make informed decisions. But we have to be alert to the pseudo-authority fallacy which is committed in the following ways: when the alleged authority is some entity that is not an expert or reliable source, when the expert is biased on the topic under discussion, or when the expert's claims are out-of-sync with the evidence produced by other experts in the field.

Example 7.21 Identify the Pseudo-Authority Fallacy

A. "The Excel Lab has done careful double-blind experiments on the power of crystals and we have extensive data to prove that crystals have healing powers. Therefore, you should believe in the healing power of crystals."

B. "Dr. Jones, a professor of Religious Studies at Icarus University, agrees with us that creationism should be taught in the biology department as a legitimate scientific theory. Therefore, you should believe this too."

Solution

Here we have two arguments claiming to be based on solid evidence. Argument A appeals to the name of a lab and then uses scientific language to describe the experiments, perhaps in order to hide the lack of objective research. There is no peer-reviewed scientific evidence presented that crystals can heal. If other labs cannot reproduce the results claimed by the Excel Lab, then Excel Lab is not a reliable authority and the power of crystals is simply a *pseudo-fact*.

Argument B appeals to an alleged authority who is not a disinterested authority. Moreover, the expertise of Dr. Jones is not in biology; rather it is in religious studies. Even if he were an expert in biology, he would not be representative of the majority of experts in this field with respect to the argument's conclusion. ◁

DEFINITION **Pseudo-fact**
A statement based on deceptive or weak inductive arguments, involving observation or experimentation that cannot be reproduced and verified by others, including experts in relevant fields.

False Presumption

Below are three fallacies based on false presumptions by the person putting forward the argument. Such arguments fail because they (a) ascribe to the opposition a position that the opposition does not actually hold, yet proceed to attack this position anyway, (b) claim incorrectly that there are only limited options available, or (c) propose a seemingly similar situation to redirect the discussion.

1. Straw-Man Fallacy

Arguments using the straw-man fallacy often attack a position or organization by focusing on their extreme positions, which are the easiest to attack. But this may not be the position under discussion. For example, a straw-man argument might presume that anti-abortion positions are making the claim that all abortions are wrong without exception (when there are many variants of this position) or it might presume that abortion-rights positions are making the claim that abortions should never be restricted (when there are many variants of this position).

Example 7.22 Identify the Straw-Man Fallacy

"Opposition to the Euro amounts to nothing but anti-European prejudice."

Solution

Someone can hold no prejudice toward Europe and yet still oppose the adoption of the Euro. This is an attempt to win an argument by attributing to the opposing side a position that it does not hold. There is no evidence provided to support this claim. ◁

2. False-Dichotomy Fallacy

An argument employing the false-dichotomy fallacy presents the audience with a choice, A or B, and presumes that they accept this choice. Since the audience doesn't want A, the conclusion is that they must accept B. The fallacy occurs because the argument omits other legitimate choices (C, D, ...) available to the audience. In other words, the premise "A or B" is false. Normally, a false premise is not considered a fallacy because its truth or falsity is not decided by deductive logic, but rather by science or common sense. But the falsity of this *or* statement is the result of presumptions being made in order to validate the logical form of the argument.

Example 7.23 Identify the False-Dichotomy Fallacy

"City Hall is in bad shape. Either we tear it down and put up a new building, or we continue to risk the public's safety. Obviously we shouldn't risk anyone's safety, so we must tear the building down."

Solution

The argument is based on the fallacy that the opposition will accept that they only have two choices when they really have more. The speaker has dismissed any evidence that supports other choices, such as remodeling city hall or only tearing down part of the building. ◁

3. False-Analogy Fallacy

No analogy is perfect, but when arguing using an analogy, the extent of similarity or dissimilarity possible between the objects being compared is important. Presumptions are made that the audience will accept the appropriateness of the analogy. If an analogy is intended to provide all the evidence needed for a particular conclusion, it may be too weak for this purpose even though there are several (superficial) similarities. The false-analogy fallacy is often used deceptively with the intention to garner support even though the argument itself is too weak to succeed.

Example 7.24 Identify the False-Analogy Fallacy

"We use seat belts in cars in order to protect ourselves, so why shouldn't we be able to use offshore tax havens to protect our financial resources?"

Solution

The above fallacy is from *The Daily Show* (April 2, 2014). But the analogy that seat belts are like tax havens is too absurd to even begin to support the conclusion of this argument. There is no evidence presented for the conclusion that tax havens are as necessary in the same way that seatbelts are. ◁

Distorted Language

Distorted and convoluted language can mask or lead to devastating errors in reasoning. In some cases the errors are so fundamental that, once recognized, there is no real argument to analyze because either the conclusion is also a premise or the proposed argument contains an internal contradiction. Below are two fallacies where such arguments just can't get off the ground!

1. Circular-Reasoning Fallacy

An argument involving this fallacy assumes what it is trying to prove: the conclusion is repeated as a (disguised) premise. Most circular arguments are sufficiently complex that it may be difficult to identify the specific premise that is the same as the conclusion, especially when different words or loaded phrases are used to say the same thing. But the point remains: there is no real argument being presented. The speaker is deceptively (or ignorantly) creating the illusion of support for a conclusion. Even though an argument employing the circular-reasoning fallacy is a valid argument (given premise P, we are assured of the conclusion P), the argument does not show us anything new or advance our understanding in any way.

Example 7.25 Identify the Circular-Reasoning Fallacy

"Euthanasia of the terminally ill is morally defensible. It is justifiable and morally responsible to assist someone to avoid prolonged and excruciating pain through death."

Solution

The argument assumes what it is trying to prove; the reasoning is circular. The conclusion is contained in the premise, but the language used has (temporarily) camouflaged this circularity. Note that the phrase "morally defensible" in the premise is the same as "justifiable and morally responsible" in the conclusion. ◁

2. Paradoxical-Reasoning Fallacy

Somewhere in the premises of an argument with paradoxical reasoning lies a basic mistake that leads to an impossible situation: an internal contradiction. It is irrational and inconsistent to be committed to the truth of a statement and also to the truth of its negation. When this happens, the deductive argument is in serious trouble since all premises cannot be true and so they are now capable of implying anything. There is really no legitimate argument to evaluate.

Example 7.26 Identify the Paradoxical-Reasoning Fallacy

"A fair coin is tossed. The first 10 flips produce 10 heads. Sam argues that the next flip will produce a tail because a tail is due. But Ginny argues that the next flip will produce a head, because a head will probably occur again. It seems that both are right. How can this be?"

Solution

The premises contain a contradiction: that the next flip will be a head and the next flip will not be a head. This paradox arises because we have made an initial mistake in assuming that the "universe" somehow carries a memory of past results that tend to favor or disfavor future outcomes. With a fair coin, the outcomes in successive flips are statistically independent and the probability of getting a head (or a tail) on a single flip is exactly $\frac{1}{2}$ (one in two). ◁

(Related Exercises: 18–32)

7.5 How Rational Arguments Fail

Inductive arguments can be weak and deductive arguments can be invalid. We have seen arguments where it is obvious when these situations occur because counterexamples to the conclusion are easy to find. We have also encountered weak or invalid arguments in which the 13 fallacies summarized in Table 7.3 can be used to camouflage the lack of real evidence to logically support a conclusion. Other fallacies can be found at the site [OR2].

TABLE 7.3: Four categories of fallacies

Emotional Appeal	Misrepresenting Evidence	False Presumption	Distorted Language
Personal Attack	Slippery-Slope	Straw-Man	Circular Reasoning
Empathy	Hasty Generalization	False Dichotomy	Paradoxical Reasoning
Popularity	Correlation/Causation	False Analogy	
Red-Herring	Pseudo-Authority		

There are other factors beyond these fallacies that also need to be considered. Finding weaknesses in complex arguments requires precision of language, careful analysis of the evidence by experts, and experience in assessing logical inferences. An inductive argument fails when experts reveal the following difficulties:

- The evidence is biased. For example, the data might not have been collected properly, the experiments have been poorly designed, incorrect sampling techniques have been used, or the results have not been verified by independent and respected researchers.
- Contradictory evidence has been overlooked.
- The evidence, though not biased, is insufficient to justify the conclusions or inferences. For example, extrapolation beyond the range of the data occurs, probabilities have not been calculated correctly, or inappropriate methods of statistical analysis have been used.

Our discussion of inductive logic will continue in Unit 3 with discussions of the proper and improper use of probability.

A deductive argument fails when it is invalid—when the form of the argument allows the premises to be true and yet the conclusion is false. This can occur when

- Premises have not been made explicit.
- Premises are explicit but not sufficient to justify the conclusion.
- Premises would justify the conclusion, but the argument fails to show the logical connection.

To determine whether a complex deductive argument succeeds or fails, the next chapters will introduce symbolization (to reveal the form of a valid argument), derivations, and formal proof techniques.

Chapter Summary

When constructing inductive arguments, we need to provide empirical evidence that strongly supports the stated conclusion. This is done by analyzing the components of the argument and determining the likelihood of its supporting premises. A single exception, a counterexample, can cast doubt on an inductive argument by providing an instance where the premises hold but the conclusion does not. On the other hand, deductive arguments are assessed based on their validity, the internal consistency of the form of the argument. To this end, we provide a set representation to model relations between various objects and use visual methods (Venn diagrams) to test for validity and invalidity. We also discuss various fallacies that can arise in the statement of both deductive and inductive arguments.

Applications, Representations, Strategies, and Algorithms

Applications			
Medicine (1, 2, 7–9)	Law Enforcement (3–4)		Biology (6)

Representations			
Tables	Sets		Symbolization
Data	Lists	Venn Diagrams	Numerical
13	1–4, 9–10	6–8, 11	5

Strategies			
Brute-Force	Solution-Space Restrictions	Rule-Based	Composition/ Decomposition
10	5, 13	1–26	1–5, 12

Algorithms					
Exact	Approximate	Inspection	Logical	Geometric	Classification
1–13	14–26	9–10	12–13	6–8, 11	14–26

Exercises

Inductive Arguments and Counterexamples

1. Evaluate the following inductive argument and assess its strength:

> There is a cooler with 50 cans of soft drinks.
> I've pulled out three cans and they were all Diet Pepsi.
> Therefore, all the cans are Diet Pepsi.

2. Evaluate the following inductive argument and assess its strength:

There is no proof that elephants can't fly.
Therefore, elephants can fly.

3. Evaluate the following inductive argument and assess its strength:

> The moon has been visible from Earth in a clear sky every night since people
> began recording their experiences.
> Therefore, the moon will be visible from Earth in a clear sky tomorrow.

4. Evaluate the following inductive argument and assess its strength:

> Yesterday you bought 34 losing lottery tickets.
> Therefore, the next ticket you buy will probably be a winner.

5. Evaluate the following inductive argument and assess its strength:

> The sun is shining and the weather is hot.
> The ocean is calm.
> The seagulls and pelicans are flying.
> Therefore, it is a good day to walk along the beach.

6. Evaluate the following inductive argument and assess its strength:

> A sample of 15,000 eligible voters from around the state have been polled,
> and 70% support the governor's plan to increase funding for education.
> This sample of eligible voters is highly correlated with the state population
> of eligible voters in terms of income, political affiliation, and age.
> These demographic features are known to affect opinions on public expen-
> ditures.
> Therefore, 70% of eligible voters support the governor's plan to increase
> funding for education.

7. Evaluate the following inductive argument and assess its strength:

> The door is open and the door frame is broken.
> There are muddy footprints outside and also in the house.
> Some jewelry and computers are missing.
> Therefore, the house was burglarized.

Inductive Puzzles

8. Reason inductively to find the next number in the sequence 2, 5, 10, 17, 26,... Suppose
   you observe that each number seems to be 1 more than a squared number (2 is one
   more than 1^2; 5 is one more than 2^2; 10 is one more than 3^2, and so on). Is this a
   strong inductive argument? What would you predict to be the next two numbers in
   the sequence?

9. A *composition* of the integer n is a way of expressing it as a sum of positive parts,
   where the order of the parts matters. For example, we can express 2 as 2 or as $1+1$, so
   2 has a total of two compositions. Likewise, 3 can be expressed as $3, 2+1, 1+2, 1+1+1$,
   so 3 has a total of four compositions.

   (a) Determine the number of compositions of 4.

(b) Determine the number of compositions of 5.

(c) Can you conjecture a formula for the number of compositions of n?

10. The previous exercise defined a composition of the integer n. Now suppose that we want to express n as a sum of exactly three nonnegative parts (before we only allowed positive parts). For example, 1 can be expressed as $1 + 0 + 0, 0 + 1 + 0, 0 + 0 + 1$, so that 1 has three of these (tripartite) compositions. Likewise, 2 can be expressed as $2 + 0 + 0, 1 + 1 + 0, 1 + 0 + 1, 0 + 2 + 0, 0 + 1 + 1, 0 + 0 + 2$, so that 2 has six of these (tripartite) compositions.

(a) Determine the number of tripartite compositions of 3 into nonnegative parts.

(b) Determine the number of tripartite compositions of 4 into nonnegative parts.

(c) What would you predict to be the number of tripartite compositions of 6 into nonnegative parts? Can you conjecture a formula for the number of tripartite compositions for the integer n?

Raymond Smullyan is known for his collections of logic puzzles. Several of these puzzles concern the anthropologist Edgar Abercrombie and his trip to an island inhabited by knights (who always tell the truth) and by knaves (who always lie).

11. Abercrombie encounters a native of the island and remembers that his name was either Paul or Saul. When he asks the native his name, the native replies "Saul." From this reply, he cannot tell for sure whether this native is a knight or knave, but he can tell with a high degree of probability. How?

12. Again, Abercrombie meets a native on an island inhabited by knights (who always tell the truth) and by knaves (who always lie). The native claims "I am a knave and there is no gold on this island." What can Abercrombie deduce about whether there is gold on the island?

Deductive Arguments and Venn Diagrams

13. Use a Venn diagram to determine if the following argument is valid or invalid:

> If you live on the Bayou, you are familiar with hurricanes.
> Marisa is familiar with hurricanes.
> Therefore, Marisa lives on the Bayou.

14. Use a Venn diagram to determine if the following argument is valid or invalid:

> If you like James Patterson detective fiction, you will like Lee Child detective fiction.
> Omar likes James Patterson detective fiction.
> Therefore, Omar likes Lee Child detective fiction.

15. Use a Venn diagram to establish that the following argument is invalid and give a counterexample:

> If a student receives an A on the calculus exam, then he passes the calculus course.
>
> Jones did not receive an A on the calculus exam.
>
> Therefore, Jones did not pass the calculus course.

Deductive Puzzles

16. Reason deductively to solve the following 4×4 Sudoku puzzle. Each of the numbers 1–4 must be used only once in each of the four 2×2 blocks, only once in each row, and only once in each column.

4		2	
1			
			3
		4	

17. Miguel was in charge of delivering mail every morning to different floors of his company's building. He had a regular daily schedule starting at 8 am, and he visits a new floor every 20 minutes. Identify his schedule from the following clues.

    Offices on the first floor received mail before the offices on the second floor but after the basement offices. The offices on the third floor received mail 20 minutes after the basement offices. The fourth floor received mail after the second floor offices.

Fallacies

Classify the following fallacies.

18. If we don't reduce public spending, our economy will collapse like that of Greece.

19. Stephen Colbert, American political satirist and comedian, when arguing and disagreeing with a guest on his show (February 8, 2007), stated "You can say no and I can say yes and my word has three letters."

20. You should not vote for Senate Bill #454 because the senator who brought it to the floor is a racist.

21. Global warming has the most terrible consequences. As a result of global warming, polar bears are threatened with extinction and this is morally unacceptable. Consequently, slowing global warming should be our first priority.

22. Either we cut government programs, such as education funding, or we live with huge deficits. Therefore, we must cut education funding.

23. If the proposed health care bill passes, then senior citizens will not receive the care they need, they will become a burden on their families, and they will fall into depression and think that ending their lives is the only solution. Therefore, we should not vote for this health care bill.

24. Driving a car is like riding a bicycle. Many people teach themselves how to ride a bicycle. It is a matter of coordination and watching where you are going. Consequently, people don't need driving lessons; they should be able to teach themselves how to drive a car.

25. People who are against capital punishment believe that the lives of murderers are more important than the lives of the police and prison guards who protect us. This is not the case. We should support capital punishment.

26. The last few times that I wore my old torn baseball cap, our team won the game. Therefore, I should keep wearing this hat.

27. Sixty percent of students surveyed on their way out of church said that they believe in God. Therefore, it is quite likely that sixty percent of all students believe in God.

28. Children who play violent video games are more likely to show violent behavior. This happens because they copy the behavior displayed in the games. Therefore, such games should be prohibited.

29. The levels of dioxins and PCBs in seafood may be unsafe, but what will fishermen do to support their families?

30. Minds are like rivers—the broader they are, the shallower they are. Therefore, students need to specialize; they should not be expected to take the required liberal arts courses.

31. The senator from Indiana is an effective speaker. Therefore, he is a good communicator.

32. In the argument given in the following paragraph, the first sentence is the conclusion and the subsequent assertions (1)–(6) represent six premises, each of which commits a different fallacy. Identify these six fallacies.

> The argument in favor of universal health care by the federal government has no merit. (1) Universal health care advocates obviously have never been helped by their families, since being for universal health care by Big Brother and having strong families are mutually exclusive. (2) Many important people agree that families or charitable organizations should help those who can't afford health care, instead of the government. (3) Federal universal health care advocates should take a lesson from my parents—they both work two jobs to pay for their health insurance and they refuse to take handouts from the government. (4) If universal health care advocates get their way and use federal funds for this purpose, federal spending will accelerate, our foreign debt will increase, our economic future will be dismal, and the very existence of our country will be at stake. (5) In light of these consequences, universal health care advocates are enemies of the state and their ideas should be rejected. (6) Truly, the universal health care argument should be put aside, because the majority of people don't want it.

Project

33. Watch the video [OR3] of the first 2012 presidential debate between Barack Obama and Mitt Romney. Identify and categorize various fallacies that are committed by the participants.

Bibliography

[Fahnestock 2004] Fahnestock, J., and Secor, M. 2004. *A rhetoric of argument: A text and reader.* New York: McGraw-Hill.

[Gardner 1957] Gardner, M. 1957. *Fads and fallacies in the name of science.* New York: Dover.

[Schiller 2009] Schiller, A. 2009. *Stephen Colbert and philosophy: I am philosophy (and so can you!).* Chicago: Open Court.

[Smullyan 2009] Smullyan, R. 2009. *The lady or the tiger? & other logic puzzles.* Mineola, NY: Dover Recreational Math.

[Smullyan 2011] Smullyan, R. 2011. *What is the name of this book?: The riddle of Dracula and other logical puzzles.* Mineola, NY: Dover Recreational Math.

[OR1] `http://www.puzzlersparadise.com` (accessed April 9, 2014).

[OR2] `http://www.fallacyfiles.org` (accessed April 10, 2014).

[OR3] `http://www.cbsnews.com/videos/the-entire-first-2012-presidential-debate` (accessed April 9, 2014).

Chapter 8

Deductive Arguments and Truth-Tables

The hard but just rule is that if the ideas don't work, you must throw them away. Don't waste any neurons on what doesn't work. Devote those neurons to new ideas that better explain the data. Valid criticism is doing you a favor.

—Carl Sagan, American astrophysicist and author, 1934–1996

Chapter Preview

This chapter introduces a way to formulate deductive arguments using a representation by logical expressions (symbolic logic). This representation reveals the general form of deductive arguments, which can then be analyzed for their validity. Translating simple statements from ordinary language into symbols is the first step. Then more complex statements can be represented by using logical operations such as *and, or, not,* and *if-then.* These logical connectives are defined for all the possible truth-values of their component statements. Truth-tables provide a useful tool for organizing this information; they also support an iron-clad method of determining whether a proposed argument is valid, based on a truth-function analysis of its premises and conclusion. The following chapter explores alternative ways to evaluate deductive arguments, especially useful when the truth-table method becomes too tedious.

8.1 Symbolization of Deductive Arguments

While the use of Venn diagrams in Chapter 7 provided a visual way of deciding the validity of deductive arguments, this method can be unwieldy when there are more complex premises and therefore more than a few sets are involved. We now turn our attention to a more flexible way of determining validity. It involves the use of symbols to reveal the formal structure of a complex argument.

We will focus on *propositional logic*, where the smallest unit of symbolization is a simple statement (or proposition) that is either true or false. For example, "The grass is green" and "2 + 4 = 8" are simple statements. The first statement has a truth-value of "true" whereas the second has a truth-value of "false." Larger and more complex statements can be constructed from simple statements by using *connectives* such as *and, or, not, if-then*. Statements and connectives are the basic building blocks of propositional logic.

DEFINITION **Propositional Logic**
A branch of symbolic logic that deals with the relationships formed between propositions (which are the smallest units of symbolization) and truth-functional connectives.

DEFINITION **Proposition**
A declarative statement that is either true or false, as distinguished from a question, exclamation, or command.

DEFINITION **Connectives**
Logical operations—*and, or, not, if-then*—that link simple statements together into compound statements. The truth of any compound statement is determined by the truth-values of its component statements and the definitions of its connectives.

DEFINITION **Compound Statement**
A statement that contains truth-functional connectives and other statements as its components.

HISTORICAL CONTEXT **George Boole (1815–1864)**
An English mathematician and philosopher, who first defined an algebraic system of propositional logic, sometimes referred to as Boolean logic or Boolean algebra. It is based on propositions and their connectives.

The use of symbols in propositional logic can be illustrated by looking at a type of argument found in analytical reasoning sections of standardized tests, such as the LSAT (Law School Admission Test):

P1: If Cindy goes to the dance, then Barbara will not go.

P2: If Allison does not go to the dance, then Cindy will go.

P3: If Barbara goes to the dance, then either Cindy will not go or Allison will not go.

P4: Allison went to the dance.

C: Barbara did not go to the dance.

Does the conclusion C necessarily follow from the premises P1, P2, P3, and P4? We will show that the answer is no, and so this is an invalid argument. To do this we first dissect the given argument, using lowercase letters to represent positive simple statements and using the symbol "$\sim$" to represent the connective *not*:

a = "Allison goes to the dance" $\sim a$ = "Allison does not go to the dance"

b = "Barbara goes to the dance" $\sim b$ = "Barbara does not to the dance"

c = "Cindy goes to the dance" $\sim c$ = "Cindy does not to the dance"

We assign other symbols to the statement connectives: *and* ($\wedge$), *or* ($\vee$), *if-then* ($\rightarrow$). In the conditional statement "If Cindy goes to the dance, then Barbara will not go to the dance," we refer to "Cindy goes to the dance" as the *antecedent* of the *if-then* statement, and we refer to "Barbara will not go to the dance" as the *consequent* of the *if-then* statement.

> DEFINITION **Antecedent and Consequent**
> In the conditional construction *if p then q*, the statement *p* is referred to as the antecedent of the conditional, and the statement *q* is referred to as the consequent of the conditional.

Example 8.1 Assign Statement Variables and Connective Symbols

Translate the following LSAT logic problem into symbolic form:

P1: If Cindy goes to the dance, then Barbara will not go.
P2: If Allison does not go to the dance, then Cindy will go.
P3: If Barbara goes to the dance, then either Cindy will not go or Allison will not go.
P4: Allison went to the dance.
C: Barbara did not go to the dance.

Solution

We substitute lowercase letters for the respective statements and use the notation for the logical connectives to replace the verbal argument by the following symbolic argument:

P1: $c \rightarrow \sim b$
P2: $\sim a \rightarrow c$
P3: $b \rightarrow \sim c \vee \sim a$
P4: a
C: $\sim b$

We now have obtained a symbolic form of the original argument. ◁

Once the argument has been symbolized, we reveal its *argument form*. The statement variables can represent any statements. Arguments having the same form will all be valid or will all be invalid. Validity or invalidity of an argument form can be established using truth-tables.

8.2 Truth-Tables

A compound statement can contain several statement variables. For example, the compound statement "*a and b*" contains the statement variables *a* and *b*, each of which represents a simple statement. Each simple statement variable can take on different truth-values. We are going to introduce a table that will help us organize the truth-values of a complex statement for all possible values of its constituent simple statements.

In Example 8.1, the proposition *a* ("Allison goes to the dance") will be either true (T) or false (F). Similarly, *b* ("Barbara goes to the dance") will be either true (T) or false (F). Suppose we combine these simple statements into the larger compound statement $a \wedge b$: that is, "Allison goes to the dance and Barbara goes to the dance." We now have four possible scenarios: both *a* and *b* are true, *a* is true and *b* is false, *a* is false and *b* is true, or both are false. It is helpful to arrange these four different possibilities into a table, called a *truth-table*. Each row of Table 8.1 indicates one of the possible combinations of the truth-values of the two simple statements *a* and *b*.

TABLE 8.1: Possible truth-values for two simple statements

a	b
T	T
T	F
F	T
F	F

DEFINITION **Truth-Table**
This table systematically lists all possible combinations of the truth-values of its component simple statements.

Increasing the number of simple statements in a proposition increases the number of possible true and false combinations. Let's see how the number of rows in the truth-table grows with the number of component statements by using a tree representation. As shown in Figure 8.1, with one simple statement a, then there are just 2 rows in the truth-table (corresponding to T and F). If we now add statement b, there are two possibilities for each truth-value of a, giving $2 \times 2 = 4$ rows. If we now add statement c, there are two possibilities for each specification of truth-values for a and b, giving $2 \times 4 = 8$ rows.

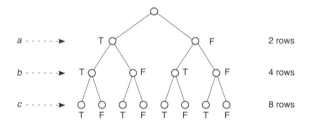

FIGURE 8.1: Number of rows in a truth-table

In general the number of rows doubles each time, since a truth-table with n simple statements contains two copies of the truth-table with $n-1$ statements: one copy associated with the first statement being T and a second copy associated with the first statement being F. We illustrate this recursive construction for the case of $n = 3$ statements in Table 8.2. The first block is just the truth-table for two statements, associated with a being T; the second block is the very same truth-table for two statements, associated with a being F.

The truth-table for $n = 4$ statements would then contain two copies of Table 8.2, for a total of $2 \times 8 = 16 = 2^4$ rows. In general, the truth-table for n statements will require 2^n rows.

Defining Connectives Using Truth-Tables

We are now ready to define four connectives (*and, or, not, if-then*) using the truth-table representation. Throughout, we will use the symbols p, q, r to indicate simple statements.

The *and* Connective (Conjunction)

Consider the compound statement "I will have coffee with you and I will go to the library." In ordinary language, you would say that I am telling the truth only when I really do both of these things—that is, only when both of the simple statements are true.

TABLE 8.2: Recursive nature of a truth-table

a	b	c
T	T	T
T	T	F
T	F	T
T	F	F
F	T	T
F	T	F
F	F	T
F	F	F

DEFINITION *and* (**Conjunction**)
A conjunction is true only when each of its component statements is true.

Table 8.3 shows the truth-table for the connective *and* ($\wedge$). The first row of this truth-table indicates that if p is true and q is true, then "*p and q*" is defined as true. No other combination of truth-values assigned to p and q will make a conjunction true.

TABLE 8.3: Definition of *and*

p	q	$p \wedge q$
T	T	T
T	F	F
F	T	F
F	F	F

The *or* Connective (Disjunction)

Consider the compound statement "I will have coffee with you or I will go to the library." In most common English usage, you would say that I am telling the truth as long as I do at least one of these things. That is, you would say that I have made a false statement only if I fail to do both of these things.

DEFINITION *or* (**Disjunction**)
A disjunction is false only when both of its component statements are false.

Table 8.4 shows the truth-table for the connective *or* ($\vee$). The last row of this truth-table asserts that if p is false and q is false, then "*p or q*" is false. No other combination of truth-values assigned to p and q will make a disjunction false.

Remark. There is a second type of disjunction called the "exclusive-or" in which the compound statement is considered to be false when both simple statements are true. As an example, when someone states that "I am going to work at McDonald's or I am going to work at Pizza Hut," the meaning is that the person works at one or the other, but not both. We will not be discussing further the *exclusive-or* here, although it is an extremely useful

TABLE 8.4: Definition of *or*

p	q	$p \vee q$
T	T	T
T	F	T
F	T	T
F	F	F

operation in the design of electronic circuits (Section 10.3) and in cryptographic applications (Section 23.2).

(Related Exercise: 32)

Example 8.2 Compound *and* Statements and Compound *or* Statements

(a) What would you have to demonstrate in order to show that the statement $p \wedge q \wedge r$ is true? (b) What would you have to demonstrate in order to show that the statement $p \vee q \vee r$ is true?

Solution

(a) You would have to show that *each* simple statement p, q, r is true.

(b) You would have to show that at least *one* of the simple statements p, q, r is true. ◁

Example 8.3 Nuances of Ordinary Language with *and/or*

In English we express conjunction with many words other than *and*; we also express disjunction with words other than *or*. Sometimes an *and* or an *or* may connect two nouns and we need to rephrase the compound statement so that both simple statements are explicit. Symbolize the following statements using the connectives *and/or*. (a) Both theater and ballet tickets are expensive. (b) The airplane arrived on time, but baggage retrieval took 45 minutes. (c) Either the book or the movie will be worth the money. (d) I'd like to spend my vacation by relaxing or reading.

Solution

(a) $p =$ "Theater tickets are expensive"; $q =$ "Ballet tickets are expensive": $p \wedge q$

(b) $p =$ "The airplane arrived on time"; $q =$ "Baggage retrieval took 45 minutes": $p \wedge q$

(c) $p =$ "The book is worth the money"; $q =$ "The movie is worth the money": $p \vee q$

(d) $p =$ "I'd like to spend my vacation by relaxing"; $q =$ "I'd like to spend my vacation by reading": $p \vee q$ ◁

The *not* Connective (Negation)

Consider the negation of a simple statement, such as "It is raining." The negation simply changes the truth-value of the statement. If "It is raining" is true, then "It is not raining" is false, and vice versa.

> DEFINITION *not* (**Negation**)
> A negation is false only when the statement it precedes is true, and is true only when the statement it precedes is false.

Table 8.5 shows the truth-table for *not* ($\sim$). Unlike the previous two connectives, this logical operation only applies to a single statement. As seen in Table 8.5, negation switches the truth-value of its input.

TABLE 8.5: Definition of *not*

p	$\sim p$
T	F
F	T

There are some compound statements involving negations that can never be false and some that can never be true.

> DEFINITION **Tautology**
> A statement that can never be false, such as "it is raining or it is not raining": $p \vee \sim p$.
>
> DEFINITION **Contradiction**
> A statement that can never be true, such as "it is raining and it is not raining": $p \wedge \sim p$.

Example 8.4 Nuances of Ordinary Language with *not*

In English we sometimes express negation by using phrases other than *not*. Also, a single negation may apply to more than one simple statement. Symbolize the following statements using negation. (a) Offshore bank accounts in general are legal, but offshore bank accounts used to avoid paying taxes are illegal. (b) It is false that I am going to visit my sister. (c) Jessie is not a freshman, and neither is Isabella. (d) It is not the case that I read the book and saw the movie.

Solution

(a) p = "Offshore bank accounts in general are legal"; q = "Offshore bank accounts used to avoid paying taxes are legal": $p \wedge \sim q$

(b) p = "I am going to visit my sister": $\sim p$

(c) p = "Jessie is a freshman"; q = "Isabella is a freshman": $\sim p \wedge \sim q$

(d) p = "I read the book"; q = "I saw the movie": $\sim(p \wedge q)$ ◁

The *if-then* Connective (Conditional)

Consider the compound statement "If I win the lottery, then I will buy you a new car." In ordinary language, you would claim that I am lying in the case that I win the lottery, yet I don't buy you a new car. Similarly, consider the statement "If you take the medicine, then you will feel better." There is a suggested causal connection. When the antecedent of the conditional is true (I take the medicine), but the consequent is false (I don't feel better), we label the *if-then* claim as false.

DEFINITION *if-then* (**Conditional**)
A conditional statement is false only when its antecedent is true and its consequent is false.

Table 8.6 shows the truth-table for the *if-then* connective ($\rightarrow$), indicating that a conditional is false only when a false consequent q is obtained from a true antecedent p. As emphasized earlier, you would rightly challenge the truth of a conditional when the antecedent is true, and the consequent is not; such would be a clear instance of "false advertising."

TABLE 8.6: Definition of *if-then*

p	q	$p \rightarrow q$
T	T	T
T	F	F
F	T	T
F	F	T

The first row of Table 8.6 specifies that the conditional is true when both the antecedent and the consequent are true; and this seems quite reasonable. But it is not obvious why the *if-then* should be labeled true in the last two rows where the antecedent is false. To address this concern, consider the conditional statement "If I receive the raise, then I will buy you a new computer." When the antecedent is false, we do not have a case of false advertising. Regardless of whether the consequent is true or not, all bets are off: no promise has been made, no causal claim is being tested.

The only way a conditional is false is when the antecedent is true and the consequent is false. This is often misunderstood. To clarify this, consider the following situation (known as the *Wason selection task*). There are four two-sided cards containing letters on one side and numbers on the other. On the table you see appearing face-up A, D, 4, 7. You are given the following conditional as a rule: "If a card has a vowel on one side, then it has an even number on the other side." The question is then asked "Which card(s) do you need to turn over in order to determine if the rule is violated"? The answer is to turn over the cards in which A and 7 appear, because the only way to make the rule false is having a card with a vowel and an odd number. No other cards need to be examined.

Remark. It is simple to remember the definitions of the *and, or,* and *if-then* connectives by noting the single case in which the truth-value is different from that in the other three cases. Table 8.7 summarizes these special cases: *and* is only true when both simple statements are true; *or* is false only when both simple statements are false; *if-then* is false only when the antecedent is true and the consequent is false.

TABLE 8.7: Special cases for *and, or, if-then*

p	q	$p \wedge q$	$p \vee q$	$p \rightarrow q$
T	T	T		
F	F		F	
T	F			F

Example 8.5 Nuances of Ordinary Language with *if-then*

In English we sometimes express the conditional relation by using words other than *if-then*. In such cases, we have to think about which simple statement is the antecedent and which is the consequent. Symbolize the following statements using the conditional. (a) The Peso is strong if the Euro is not strong. (b) All whales are mammals. (c) Sunlight is a necessary condition for plant growth. (d) Having ten dollars is sufficient for buying dinner at a fast food restaurant. (e) Don't start anything that you can't finish.

Solution

(a) p = "The Peso is strong"; q = "The Euro is strong": $\sim q \rightarrow p$

(b) p = "A creature is a whale"; q = "A creature is a mammal": $p \rightarrow q$

(c) p = "Sunlight is present"; q = "Plant growth occurs": $q \rightarrow p$

(d) p = "You have ten dollars"; q = "You can buy dinner at a fast food restaurant": $p \rightarrow q$

(e) p = "You start something"; q = "You finish the task": $\sim q \rightarrow \sim p$

Generally, a *necessary* condition for some occurrence is the consequent of the conditional (because it always occurs when the antecedent occurs). A *sufficient* condition for some occurrence is the antecedent of the conditional (because when it occurs, the consequent can occur, though the consequent can also occur without it). ◁

(*Related Exercises: 1–12, 33*)

Determining the Truth-Value of Compound Statements

In propositional logic we can always determine the truth-value of any compound statement using the definitions of the connectives *and, or, not, if-then* in addition to knowing the truth-value of the simple statements.

Example 8.6 Evaluate the Truth-Value of a Compound Statement

Symbolize the following compound statement, assign truth-values to each simple statement, and calculate the truth-value of the compound statement:

"If squares have four sides and triangles have three sides, then pentagons have six sides."

Solution

Let p represent the statement "Squares have four sides," let q represent "Triangles have three sides," and let r represent "Pentagons have six sides." The correct truth-values are p = T, q = T, and r = F. The truth-value of the compound statement is obtained using the definitions of the connectives and using the proper order of operations (evaluating the expression inside parentheses first):

$$(p \wedge q) \rightarrow r$$
$$(T \wedge T) \rightarrow F$$
$$T \rightarrow F$$
$$F$$

We see that the compound conditional statement is false. ◁

(*Related Exercises: 13–17*)

Example 8.7 Use a Truth-Table to Find All Truth-Values of a Compound Statement

(a) Symbolize the following compound statement: "If Mary goes shopping and doesn't stop at the library, then she will miss the movie." (b) Construct a truth-table to list the various combinations of truth-values that can be assigned to the component statements and then evaluate the final column of truth-values.

Solution

(a) We first symbolize the simple (positive) statements comprising the compound statement. Let p = "Mary goes shopping," q = "Mary stops at the library," and r = "Mary will miss the movie." Then the compound statement can be represented by the logical expression $(p \wedge \sim q) \to r$.

(b) The set of input values p, q, r (called the *reference values*) are listed in the truth-table displayed in Table 8.8. We first evaluate the negation in front of q, and then the conjunction within the parentheses $(p \wedge \sim q)$, listing this as the fifth column of the truth-table. For clarity, we next repeat the column for r. Finally, the seventh column contains the truth-values for the conditional. Note that this compound statement is false in just one situation: when p is true, q is false, and r is false. ◁

(*Related Exercises: 18–22*)

TABLE 8.8: Truth-table evaluation of a compound statement

p	q	r	$\sim q$	$p \wedge \sim q$	r	$(p \wedge \sim q) \to r$
T	T	T	F	F	T	T
T	T	F	F	F	F	T
T	F	T	T	T	T	T
T	F	F	T	T	F	F
F	T	T	F	F	T	T
F	T	F	F	F	F	T
F	F	T	T	F	T	T
F	F	F	T	F	F	T

Remark. As in arithmetic or algebra, there is an order in which various operations are performed. Here each row of truth-values is manipulated in the following order: (1) calculate the negation in front of any simple statement, (2) calculate the connective inside any parentheses, (3) calculate any negation in front of parentheses, and (4) calculate the final connective outside of any parentheses.

In the next section we will see that truth-tables can be valuable in assessing the validity of deductive arguments.

8.3　Truth-Tables, Implications, and Validity of Arguments

In a valid argument, the conclusion follows necessarily from the premises. The premises are said to imply the conclusion; the conclusion is said to be *entailed* by the premises.

Implication or *entailment* is a relationship between the premises and the conclusion that holds when an argument is valid. It indicates that there is no row in the truth-table that makes all the premises true and the conclusion false.

DEFINITION **Logical Implication (Entailment)**
The relationship holding between premises and conclusion when an argument is valid, meaning that the truth of the conclusion follows from the truth of all the premises.

The validity of an argument can be demonstrated in the following way. First, we construct a compound statement by placing conjunctions between all the premises and by placing a conditional between these conjoined premises and the conclusion. This compound statement is then evaluated using a truth-table. Here is a description of the procedure.

Determining Argument Validity Using a Truth-Table

1. Conjoin all the premises: $P1 \land P2 \land \cdots \land Pn$.
2. Connect these linked premises with the desired conclusion C by forming the conditional statement $(P1 \land P2 \land \cdots \land Pn) \to C$.
3. Calculate the column of truth-values for this conditional.
4. By consulting this column, determine the validity of the argument.

How do we draw conclusions about the validity or invalidity of the argument? If there is even one F in the column for the conditional (as illustrated in Table 8.9), the argument is invalid because it is possible that all the premises are true and yet the conclusion is false. On the other hand, when the column for the conditional contains all true values (as illustrated in Table 8.10), the conditional is a tautology (always true) and the argument is valid. In this case, the truth of the conclusion logically follows from the truth of the premises. We use a double-arrow to express that logical implication holds: $(P1 \land P2 \land \cdots \land Pn) \Rightarrow C$.

TABLE 8.9: Detecting an invalid argument

All Premises	Conclusion	Conditional
$P1 \land P2 \land \cdots \land Pn$	C	$(P1 \land P2 \land \cdots \land Pn) \to C$
T	T	T
T	F	F
F	T	T
F	F	T

TABLE 8.10: Detecting a valid argument

All Premises	Conclusion	Conditional
P1 ∧ P2 ∧ ··· ∧ Pn	C	(P1 ∧ P2 ∧ ··· ∧ Pn) → C
T	T	T
F	T	T
F	F	T

Example 8.8 Show an Argument is Valid Using a Truth-Table

(a) Symbolize the following argument as a set of premises and a conclusion. (b) Connect the premises and conclusion using a conditional, and then construct a truth-table to show that the argument is valid.

> If you receive an A in this course, you will graduate.
> You receive an A in this course.
> Therefore, you will graduate.

Solution

(a) Let p = "You receive an A in this course" and q = "You will graduate." Then the argument is represented symbolically as

P1: $p \to q$
P2: p
C: q

(b) Conjoin the premises and connect them with the conclusion using a conditional, forming the expression $((p \to q) \land p) \to q$. Because the entire argument involves only two simple statements p and q, the truth-table will have $2^2 = 4$ rows. The first two columns of Table 8.11 are the reference columns, showing all possible combinations of truth-values for p and q. Notice that P1 is calculated using the truth-values of p and q; the conjunction P1 ∧ P2 is calculated using the truth-values from the first and third columns; and the conditional (P1 ∧ P2) → C is calculated using the truth-values from columns two and four. Since the final (conditional) column contains all true values, the given argument is valid. ◁

TABLE 8.11: Establishing validity using a truth-table

P2	C	P1	P1 ∧ P2	(P1 ∧ P2) → C
p	q	$p \to q$	$((p \to q) \land p)$	→
T	T	T	T	T
T	F	F	F	T
F	T	T	F	T
F	F	T	F	T

The abstract form of the argument in Example 8.8 can be described as follows: P1 is a conditional, P2 affirms the antecedent, and C affirms the consequent. This particular form arises frequently and is known by the Latin name *modus ponens* (translated as "mode that affirms"). The truth-table constructed in Table 8.11 demonstrates that this is a valid argument form.

HISTORICAL CONTEXT **Aristotle and Argument Forms**
The philosopher Aristotle (384–322 BC) in his logical treatises (*The Organon*) gave us insights into deductive reasoning by studying various valid forms of arguments, such as *modus ponens*.

Example 8.9 Show an Argument is Invalid Using a Truth-Table

(a) Symbolize the following argument as a set of premises and a conclusion. (b) Connect the premises and conclusion using a conditional, and then construct a truth-table to show that the argument is invalid.

If you receive an A in this course, you will graduate.

You graduated.

Therefore, you received an A in this course.

Solution

(a) Let $p =$ "You receive an A in this course" and $q =$ "You will graduate." Then the argument is represented symbolically as

P1: $p \to q$
P2: q
C: p

Intuitively, we doubt the validity of this argument since it is possible to graduate yet receive a grade of B in the course.

(b) We will more formally establish the invalidity of the argument using the truth-table constructed in Table 8.12. To begin, we conjoin the premises and connect them with the conclusion using a conditional, forming the expression $((p \to q) \land q) \to p$. Again we build up the final conditional column by creating intermediate columns for $p \to q$ and then $(p \to q) \land q$. Notice that the final column $((p \to q) \land q) \to p$ in Table 8.12 contains a false value, so the argument is demonstrated to be invalid. ◁

TABLE 8.12: Establishing invalidity using a truth-table

C	P2	P1	P1 ∧ P2	(P1 ∧ P2)→ C
p	q	$p \to q$	$(p \to q) \land q$	$\to$
T	T	T	T	T
T	F	F	F	T
F	T	T	T	F
F	F	T	F	T

By looking at the highlighted row in Table 8.12 containing this false value, the truth-table reveals a situation in which the premises are all true, but the conclusion is not. Namely, when p is false and q is true (row 3), the premises $p \to q$ and q are both true yet the conclusion p is false. By exhibiting just this one *counterexample*, we have definitively shown the invalidity of the proposed argument.

Example 8.10 Determine Whether an Argument is Valid or Invalid Using a Truth-Table

(a) Symbolize the following argument as a set of premises and a conclusion. (b) Connect the premises and conclusion using a conditional, and then construct a truth-table to determine whether the argument is valid or invalid.

John will take Psychology or he will take Economics.
John will not take Economics.
Therefore, John will take Psychology.

Solution

(a) Let p = "John will take Psychology" and q = "John will take Economics." The argument has the following form:

P1: $p \lor q$
P2: $\sim q$
C: p

The given argument is represented by the compound statement $((p \lor q) \land \sim q) \to p$.

(b) We then construct a truth-table to evaluate this expression. Since the final (conditional) column in Table 8.13 contains all true values, the argument is established as valid. ◁

(*Related Exercises: 23–31*)

TABLE 8.13: Investigating validity using a truth-table

C		P1	P2	P1 ∧ P2	(P1 ∧ P2) → C
p	q	$p \lor q$	$\sim q$	$((p \lor q) \land \sim q)$	$\to$
T	T	T	F	F	T
T	F	T	T	T	T
F	T	T	F	F	T
F	F	F	T	F	T

The abstract form of the argument in Example 8.10 can be described as follows: P1 is a disjunction, P2 negates one of the statements in the disjunction, and C affirms the other. This particular form is called "disjunctive syllogism," and we will encounter it again in Chapter 9.

8.4   Limitations of Truth-Tables

We can use truth-tables to demonstrate the validity or invalidity of an argument, but this method has significant limitations—namely, the size of the associated truth-table grows rapidly with the number of statements. The following two examples involve three component statements.

Example 8.11 Use a Truth-Table to Determine Validity of an Argument

Consider the following argument:

> If the ground is icy or the crew is not available, then the flight will be delayed. The flight was not delayed. So, the ground is not icy and the crew is available.

Symbolize this argument and use a truth-table to assess its validity.

Solution

Let p = "The ground is icy," q = "The crew is available," and r = "The flight is delayed." So the argument takes the following form:

P1: $(p \lor \sim q) \to r$
P2: $\sim r$
C: $\sim p \land q$

Since the final column of the truth-table in Table 8.14 displays a tautology, this argument is in fact valid. We have used 8 rows and 11 columns to carry out this verification. ◁

TABLE 8.14: Establishing validity using a truth-table

| | | | | | | P1 | P2 | | C | (P1 ∧ P2) → C |
p	q	r	$\sim q$	$p \lor \sim q$	$\sim p$	$(p \lor \sim q) \to r$	$\sim r$	P1 ∧ P2	$\sim p \land q$	$\to$
T	T	T	F	T	F	T	F	F	F	T
T	T	F	F	T	F	F	T	F	F	T
T	F	T	T	T	F	T	F	F	F	T
T	F	F	T	T	F	F	T	F	F	T
F	T	T	F	F	T	T	F	F	T	T
F	T	F	F	F	T	T	T	T	T	T
F	F	T	T	T	T	T	F	F	F	T
F	F	F	T	T	T	F	T	F	F	T

Example 8.12 Use a Truth-Table to Determine Validity of the LSAT Argument

Using the symbolization of the LSAT argument presented in Example 8.1, develop a truth-table to assess its validity.

Solution

The symbolization of the argument is

P1: $c \to \sim b$
P2: $\sim a \to c$
P3: $b \to (\sim c \lor \sim a)$
P4: a
C: $\sim b$

We construct the truth-table in Table 8.15 to represent the four premises, the conclusion, and the conditional (P1 ∧ P2 ∧ P3 ∧ P4) → C. It is convenient to use columns A and B to store intermediate results. In particular, B represents the conjoined premises P1 ∧ P2 ∧ P3 ∧ P4. The second row of the truth-table makes all the premises true and the conclusion false, so the argument is seen to be invalid. ◁

TABLE 8.15: Exploring invalidity using a truth-table

P4			C	P1		P2		A	P3	B	B → C
a	b	c	$\sim b$	$c \to \sim b$	$\sim a$	$\sim a \to c$	$\sim c$	$\sim c \vee \sim a$	$b \to A$	P1 $\wedge \cdots \wedge$ P4	$\to$
T	T	T	F	F	F	T	F	F	F	F	T
T	T	F	F	T	F	T	T	T	T	T	F
T	F	T	T	T	F	T	F	F	T	T	T
T	F	F	T	T	F	T	T	T	T	T	T
F	T	T	F	F	T	T	F	T	T	F	T
F	T	F	F	T	T	F	T	T	T	F	T
F	F	T	T	T	T	T	F	T	T	F	T
F	F	F	T	T	T	F	T	T	T	F	T

If two more component statements were added to the arguments described in Examples 8.11 and 8.12, then the truth-table would contain $2^5 = 32$ rows, and the calculations become even more laborious. Recall that as the number n of statements $p, q, r, \ldots$ increases, the size of the truth-table grows exponentially as 2^n. So we can see that it is only feasible to use a truth-table when we have relatively few component statements. Automated evaluation of truth-tables can be found at the sites [OR1, OR2].

In Chapter 9, we will discuss a better method of establishing validity—using *derivations*. It will build upon some fundamental mathematical properties of logical expressions, and these mathematical properties are generally concise enough to be established using truth-tables. In summary, our ability to establish validity using truth-tables can be parlayed into a more flexible and applicable approach for testing the validity of arguments.

Chapter Summary

When evaluating deductive arguments, we need appropriate tools to aid rational thought. We need to strip away the ambiguities of ordinary language and make precise the underlying argument. Using propositional logic, we are able to symbolize statements and determine their truth-values. Arguments are made up of statements (premises) offered in support of another statement (the conclusion). A deductive argument only succeeds if the truth of its premises necessarily implies the truth of its conclusion. To determine validity, we isolate the form of the argument through symbolization, calculate the truth-value of each premise and the conclusion, and then determine if the conditions for validity are met. Using truth-tables is just one way to establish validity, and we will encounter additional tools for accomplishing this in the next chapter.

Applications, Representations, Strategies, and Algorithms

Applications		
Language (2–7)		

Representations		
Tables		Symbolization
Decision		Logical
7–12		1–12

Strategies		
Rule-Based		
1–12		

Algorithms		
Exact	Sequential	Logical
1–12	7–12	1–12

Exercises

Symbolization

Symbolize the following statements using *and, or, not, if-then*:

1. If the candidate has wealthy supporters and campaigns hard, he will be elected.

2. It requires skill or courage to climb the mountain.

3. If Smith is sick, then he needs a doctor and if Smith is in an accident, then he needs a lawyer.

4. Canada and the U.S. are not in the Eastern Hemisphere.

5. Emily Dickinson was a poet, and Walt Whitman was too.

6. Being able to run a distance of over 5 kilometers is a sufficient condition for being physically fit.

7. You can take math or statistics to satisfy the quantitative reasoning requirement.

8. Everyone here is either an Independent or a Republican, but not both an Independent and a Republican.

9. Louis can't drive and neither can Rosa.

10. Improvements in health information technology are necessary for us to achieve low-cost, high-quality health care.

11. If you are the president, then you are not the provost, and if you are the provost, then you are not the president.

12. It is not the case that you will eat Thai or Mexican food for dinner.

Truth-Values of Compound Statements

Symbolize the following compound statements, assign the known truth-values to each simple statement, and using the definitions of the connectives, calculate the truth-value of the compound statement.

13. If the Pope is Catholic, then Mars is a planet and the sun revolves around Mars.

14. Clinton is not a Republican and Bush is not a Democrat.

15. If red and blue are primary colors, then orange is not a primary color.

16. If Chicago is in Illinois or Chicago is in Indiana, then Chicago is not in Iowa.

17. Pears and carrots are vegetables, or apples are fruits.

Truth-Tables for Compound Statements

In the following, use a truth-table to list the various combinations of truth-values that can be assigned to the component statements, and then evaluate the resulting truth-values.

18. $(p \land q) \to \sim q$

19. $\sim(p \to q) \land p$

20. $(\sim p \land q) \lor (\sim p \to q)$

21. $(p \lor q) \to (\sim r \land p)$

22. $(\sim p \lor \sim q) \land (\sim r \to q)$

Symbolization of Arguments

Symbolize the following arguments, labeling the premises (P1, P2, P3, ...) and the conclusion (C). Note that the conclusion as presented may not necessarily be the last statement in an argument.

23. The contract is legal. If it was drawn up using appropriate terminology and Smith signed it in front of witnesses, then the contract is legal. Smith signed it in front of witnesses and the contract was written using appropriate terminology.

24. Robbery was the motive for the crime or revenge was motive for the crime. If the victim had his wallet in his pocket, then robbery wasn't the motive for the crime. The victim did not have his wallet in his pocket. Therefore, revenge was not the motive for the crime.

25. Epidemic X can be prevented if we provide an adequate number of inoculations. If we provide an adequate number of inoculations, we do not have to quarantine infected persons. We did quarantine infected persons. So, epidemic X was not prevented.

26. Definitely, San Diego is south of Los Angeles. If San Diego is not south of Los Angeles, then San Diego is not south of San Francisco, but San Diego is south of San Francisco.

27. If you are interested in volcanos, then you visited Mt. Vesuvius. If you have visited Mt. Vesuvius, then you have visited Mt. Etna. If you have visited Mt. Etna, then you also have visited Mauna Loa and Kilauea. You did not visit Kilauea. So, you are not interested in volcanos.

Validity and Invalidity of Arguments

In the following, (a) symbolize the premises and conclusion, (b) conjoin the premises, (c) place a conditional between the premises and conclusion, (d) calculate the truth-values of the conditional for each row in the truth-table, and (e) determine if the argument is valid.

28. The tax code will be simplified if Congress is willing to implement reform. Congress is not willing to implement reform. Therefore, the tax code will not be simplified.

29. The stock market will crash if inflation is not under control. Inflation is not under control. So, the stock market will crash.

30. If K-12 education improves in the U.S., then standards for student performance at each level must be implemented and parents must become more involved. Standards for student performance are not implemented. So, K-12 education will not improve in the U.S.

31. If unemployment is high, then kids join gangs. If kids join gangs, then prisons are overcrowded. Unemployment is high. As a result, prisons are overcrowded.

Projects

32. Define the following two new connectives: *exclusive-or* (where p and q cannot both be true for the compound statement to be true) and *equivalence* (where p and q must both be true or both be false for the compound statement to be true). Construct truth-tables for these connectives and give examples where you might use these new connectives. Also, show that $(p \rightarrow q) \wedge (q \rightarrow p)$ has the same meaning as the equivalence connective.

33. There are four ways in which two statements p and q may be related to each other as necessary or sufficient conditions: (1) p is necessary but not sufficient for q; (2) p is sufficient but not necessary for q; (3) p is both necessary and sufficient for q; and (4) p is neither necessary nor sufficient for q. Give specific examples of these relationships, drawing from a variety of real-world contexts.

Bibliography

[Copi 2010] Copi, I., Cohen, C., and McMahon, K. *Introduction to logic*, Chapters 1 and 6. New York: Prentice-Hall.

[Hurley 2012] Hurley, P. *A concise introduction to logic*, Chapters 5 and 6. Boston: Cengage.

[OR1] http://staff.science.uva.nl/~jaspars/AUC/apps/javascript/proptab/index.html (accessed April 20, 2014).

[OR2] http://demonstrations.wolfram.com/TruthTables/ (accessed April 20, 2014).

Chapter 9

Deductive Arguments and Derivations

When you have eliminated the impossible, whatever remains, however improbable, must be the truth.

—Sherlock Holmes, fictional detective conceived by Scottish author, Sir Arthur Conan Doyle, 1859–1930

Chapter Preview

We can establish the validity of an argument by showing explicitly how the premises can be interwoven to obtain the stated conclusion. The premises can be manipulated (taken apart or conjoined) through a series of intermediate steps, each of which is justified by citing a specific Rule of Inference. Each such rule provides a legitimate inference or rephrases a statement in a logically equivalent form. Such a formal demonstration is termed a derivation. We will look at three particular techniques used in constructing derivations: direct proofs, indirect proofs, and the method of mathematical induction. Also, we introduce the derivation graph, which provides a visual representation of the internal structure of an argument.

9.1 Counterexamples and Derivations

In the previous chapter, we used truth-tables to demonstrate validity or invalidity of an argument and noted the impracticality of truth-tables due to size limitations. To establish validity, we needed to verify that the final column represented a tautology (all truth-values being T). On the other hand, to show invalidity we only needed to identify one row where the premises were true and the conclusion false. This occurred in Example 8.12, where we were able to find an assignment of truth-values that made all premises true, yet the conclusion was false. If we can find a more direct way to make all the premises true and the conclusion false, then we can produce a counterexample and therefore establish invalidity. The next example shows how this might be done.

Example 9.1 Use the Method of Assigning Truth-Values to Prove Invalidity

Find an assignment of truth-values that produces a counterexample to the argument presented in Example 8.12: that is, an assignment that makes all premises true and the conclusion false. Begin by assigning F to the conclusion. Then deduce a sequence of assignments needed in order to make each premise T.

P1: $c \to \sim b$

P2: $\sim a \to c$

P3: $b \to (\sim c \lor \sim a)$

P4: a

C: $\sim b$

Solution

The conclusion must be false, so $\sim b$ is assigned F which means we assign T to b. Since $\sim b$ is F, for P1 to be true, c must be F. Since c is F and P2 is to be made true, $\sim a$ must be F and so a is given the truth-value T. Since a is T, P4 is clearly true. Finally, given the assignment $a = \text{T}$, $b = \text{T}$, $c = \text{F}$, we now substitute into the last remaining premise P3, giving $\text{T} \to (\sim\text{F} \lor \sim\text{T})$ or $\text{T} \to \text{T}$, which is true. To summarize, we have found an assignment that makes each premise true and the conclusion C false. Now look at the truth-table (Table 8.15) developed in Chapter 8 for this example. The row of the truth-table that produced F for the conditional contains exactly this same assignment of truth-values for a, b, c. ◁

(Related Exercises: 1–3)

This method of assigning truth-values to prove invalidity may not always succeed as easily. A clear chain of necessary assignments may not always be as obvious as occurred in Example 9.1. Note that *not* finding an assignment that makes the premises true and the conclusion false doesn't mean that argument is valid; it just means that we haven't produced a counterexample, and thus a definitive answer still remains.

Now we turn to a method of establishing validity without truth-tables, called *proof by derivation*. The objective is to demonstrate that the conclusion is implied by the premises through a series of legitimate steps. This can provide a more succinct demonstration of validity than by using a truth-table. However, not finding a derivation does not mean that an argument is invalid; it just means that someone else might be able to find it, and currently we don't know whether the argument is valid or invalid.

DEFINITION **Proof by Derivation**
The premises of an argument are followed by intermediate steps, each of which is justified by citing a specific Rule of Inference. The last step is the conclusion, which is now justified by the preceding chain of steps.

In developing a framework for logical derivations, we identify a small number of fundamental *Rules of Inference*. The validity of each such Rule of Inference can be shown by constructing a truth-table and showing that the final column is a tautology, containing all true values. From these few building blocks, we can develop chains of "proof steps" that can lead to surprising results.

DEFINITION **Rule of Inference**
A "mini-valid argument" or "interchangeable equivalent statement" identified by logicians and given names (such as *modus ponens*, *modus tollens*).

The seven rules listed in Table 9.1 are among the most common, and they can be used to verify the truth of each step inside longer, more complicated arguments.

TABLE 9.1: Seven rules of inference

Rule of Inference	Terminology
$p \rightarrow q$ p $\therefore q$	Modus Ponens (MP)
$p \rightarrow q$ $\sim q$ $\therefore \sim p$	Modus Tollens (MT)
$p \wedge q \qquad p \wedge q$ $\therefore p \qquad \therefore q$	Simplification (SMP)
p q $\therefore p \wedge q$	Conjunction (CN)
p is equivalent to $\sim\sim p$	Double Negation (DN)
$p \vee q \qquad p \vee q$ $\sim p \qquad \sim q$ $\therefore q \qquad \therefore p$	Disjunctive Syllogism (DS)
$p \qquad q$ $\therefore p \vee q \quad \therefore p \vee q$	Addition (ADD)

For example, the first rule shown in Table 9.1 is the valid argument form discovered in Example 8.8. For brevity, the two premises ($p \rightarrow q$, p) are listed first and the conclusion (q) is identified using the symbol $\therefore$ (therefore).

The derivation process carries out the following steps:

Determining Argument Validity Using a Derivation

1. Symbolize the argument.
2. Create a list by numbering the premises and each new step leading up to the desired conclusion.
3. Justify each new step by citing previous steps and a specific Rule of Inference.

Example 9.2 Use the Rules of Inference in a Derivation

In the following argument, (a) symbolize the premises and the conclusion, and (b) derive the conclusion. Be sure to cite the abbreviation for any Rule of Inference used in the derivation, along with the number of any previous step(s) being referenced by this rule.

> If Charles receives the appropriate vaccinations, then it is not the case that he will contract smallpox or chickenpox.
> Charles did contract chickenpox.
> Therefore, Charles did not receive the appropriate vaccinations.

Solution

(a) Let p = "Charles receives the appropriate vaccinations," q = "Charles contracts small-pox," and r = "Charles contracts chickenpox." The argument then takes the form

P1: $p \rightarrow \sim(q \vee r)$
P2: r
C: $\sim p$

(b) A possible derivation is the following:

1. $p \rightarrow \sim(q \vee r)$ Given
2. r Given
3. $q \vee r$ 2, ADD
4. $\sim\sim(q \vee r)$ 3, DN
5. $\therefore \sim p$ 1, 4, MT

Step 3 follows from Step 2 by the Rule of Inference called Addition. Since r is true, then $q \vee r$ is also true because a disjunction is true whenever one of its component statements is true.

Step 4 applies "double negation" to the expression in Step 3. Since the expression in Step 3 is true, a double negation doesn't change its truth-value.

Step 5 is the result of applying *modus tollens* to Steps 1 and 4. Since Step 1 is an implication and Step 4 negates its consequent, *modus tollens* guarantees that the negation of its antecedent is necessarily true. So we have now justified the stated conclusion. ◁

Example 9.3 Symbolize an Argument and Derive its Conclusion

Below is a variation of the LSAT argument appearing in Example 8.1. (a) Symbolize the argument, and then (b) prove its validity by constructing a formal derivation.

If Cindy goes to the dance, then Barbara will not go.
If Allison does not go to the dance, then Barbara will go.
Barbara went to the dance, but Allison did not.
Therefore, Cindy did not go to the dance.

Solution

(a) First, we symbolize the argument, by letting a = "Allison goes the dance," b = "Barbara goes the dance," and c = "Cindy goes the dance." The above argument becomes

P1: $c \rightarrow \sim b$
P2: $\sim a \rightarrow b$
P3: $b \wedge \sim a$
C: $\sim c$

(b) We create a logical derivation to show the validity of this argument. (There is more than one way to create a derivation to show the validity of this argument.)

1. $c \rightarrow \sim b$ Given
2. $\sim a \rightarrow b$ Given
3. $b \wedge \sim a$ Given
4. b 3, SMP
5. $\sim\sim b$ 4, DN
6. $\therefore \sim c$ 1, 5, MT

Notice that our derivation did not use the second premise. ◁

(*Related Exercises: 4–8*)

Derivation Graphs

A directed graph can be used to represent logical relationships between the premises and the intermediate steps in a derivation, leading to the conclusion. This graphical representation helps us quickly visualize the connections between each premise and each intermediate step, and it shows how the conclusion logically flows from the truth of these earlier steps. To illustrate, we return to the derivation found in Example 9.2:

1. $p \to \sim(q \vee r)$ Given
2. r Given
3. $q \vee r$ 2, ADD
4. $\sim\sim(q \vee r)$ 3, DN
5. $\therefore \sim p$ 1, 4, MT

In Figure 9.1, we show the corresponding *derivation graph*, in which the vertices correspond to premises, intermediate results, or the conclusion. The edges (implicitly directed from top to bottom) indicate how the truth of certain statements implies the truth of other statements by applying an appropriate Rule of Inference (which labels the edge). For example, the truth of vertex $q \vee r$ follows (by ADD) from the truth of vertex r, whereas the truth of the conclusion $\sim p$ follows from the premise $p \to \sim(q \vee r)$ and the intermediate result $\sim\sim(q \vee r)$.

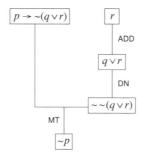

FIGURE 9.1: Derivation graph with two premises

Example 9.4 Construct a Derivation Graph

Construct a derivation and the corresponding derivation graph for the following argument:

P1: $\sim p \vee \sim q$
P2: $r \to q$
P3: $t \to p$
P4: $\sim r \to \sim s$
P5: s
C: $q \wedge \sim t$

Solution

One derivation for this argument is given below:

1. $\sim p \lor \sim q$ Given
2. $r \to q$ Given
3. $t \to p$ Given
4. $\sim r \to \sim s$ Given
5. s Given
6. $\sim\sim s$ 5, DN
7. $\sim\sim r$ 4, 6, MT
8. r 7, DN
9. q 2, 8, MP
10. $\sim\sim q$ 9, DN
11. $\sim p$ 1, 10, DS
12. $\sim t$ 3, 11, MT
13. $\therefore q \land \sim t$ 9, 12, CN

The corresponding derivation graph is shown in Figure 9.2, with the top five vertices corresponding to premises and the bottom vertex corresponding to the conclusion. ◁

(Related Exercises: 9–11)

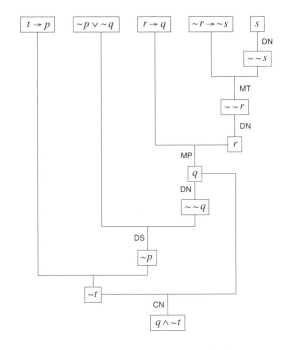

FIGURE 9.2: Derivation graph with five premises

The derivations provided in Examples 9.2–9.4 represent direct proofs of the claimed conclusion. We now explore an alternative method of establishing the validity of an argument, but now proceeding in an indirect manner.

9.2 Indirect Proof

The derivations in Section 9.1 moved directly from premises to conclusion via intermediate steps. A different strategy, called *indirect proof* (or *proof by contradiction*), assumes that the negation of the conclusion is true. With this new "premise" we then try to obtain a contradiction. The underlying idea is to assume the *opposite* of what you are trying to prove; if this leads to an absurdity, then this assumption must be in error and so we must accept the truth of the conclusion.

DEFINITION **Indirect Proof**
A demonstration that proceeds by assuming the opposite of the desired conclusion. If a contradiction is obtained, we then can assert that the desired conclusion must hold.

DEFINITION **Contradiction (or Contradictory Statements)**
Two statements that cannot both be true or both be false at the same time.

The indirect method of proof is similar to the procedure followed in crime scene investigations. Suppose that suspect A has been detained in a murder investigation. Person A is known to have made threats to the victim B and would have financially profited from the death of B. Assuming that A was in fact guilty, then A would need to have committed the crime between 1 am and 2 am, based on the coroner's estimate of the time of death. However, surveillance video at an ATM shows that A was on the other side of town during that time period. This evidence provides a contradiction to the assumed guilt of A, so A could not in fact have committed the crime.

Below are the steps used in an Indirect Proof (IP):

Determining Argument Validity Using an Indirect Proof

1. Assume the negation of the conclusion you are trying to establish.
2. Derive a contradiction, such as p and $\sim p$, which is something that cannot hold.
3. Conclude that the original conclusion must be true—since its negation has led to an absurdity.

Example 9.5 Prove Validity of an Argument Using an Indirect Proof

Use an indirect proof to establish the conclusion in the argument from Example 9.3:

P1: $c \to \sim b$
P2: $\sim a \to b$
P3: $b \wedge \sim a$
C: $\sim c$

Solution

We add the negation of the conclusion C to the set of premises and try to obtain a contradiction.

1. $c \rightarrow \sim b$ Given
2. $\sim a \rightarrow b$ Given
3. $b \wedge \sim a$ Given
4. $\sim\sim c$ Assume (IP)
5. c 4, DN
6. $\sim b$ 1, 5, MP
7. b 3, SMP
8. $\therefore \sim c$ 4, 6, 7, IP

The very last line of the proof above indicates that a contradiction has been reached (Steps 6 and 7) and concludes that the assumption made in Step 4 must be erroneous, so $\sim c$ in fact holds. ◁

(Related Exercises: 12–16)

We can also diagram the indirect argument in Example 9.5 using a derivation graph, shown in Figure 9.3. The initial assumption $\sim\sim c$ is shown highlighted as an added premise. This derivation graph clearly shows the contradiction $\sim b$ and b obtained; this means that the initial assumption $\sim\sim c$ must be false, so we conclude at the very last step (labeled IP) that $\sim c$ must indeed hold. The graph also reveals that the second premise is not needed in this derivation.

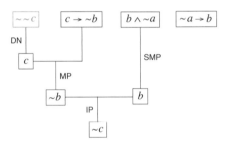

FIGURE 9.3: Derivation graph for an indirect proof

Example 9.6 Prove Validity of Another Argument Using an Indirect Proof

(a) Symbolize the following argument, (b) show a derivation using an indirect proof, citing the appropriate justification for each step, and (c) draw the corresponding derivation graph.

> If the ground is icy or the crew is not available, then the flight will be delayed and Sarah will miss her meeting. If the flight is delayed, then Tobias will get a hotel room. However, Tobias did not stay at a hotel. So, the ground is not icy.

Solution

(a) First, we symbolize the argument using the propositions p = "The ground is icy," q = "The crew is available," r = "The flight will be delayed," s = "Sarah will miss her meeting," and t = "Tobias will get a hotel room." The argument then has the following form:

P1: $(p \lor \sim q) \to (r \land s)$
P2: $r \to t$
P3: $\sim t$
C: $\sim p$

(b) In the indirect proof shown below, we assume the negation of the desired conclusion as a step in the derivation.

1. $(p \lor \sim q) \to (r \land s)$	Given
2. $r \to t$	Given
3. $\sim t$	Given
4. $\sim\sim p$	Assume (IP)
5. p	4, DN
6. $p \lor \sim q$	5, ADD
7. $r \land s$	1, 6, MP
8. r	7, SMP
9. $\sim r$	2, 3, MT
10. $\therefore \sim p$	4, 8, 9, IP

(c) Figure 9.4 displays the corresponding derivation graph, which clearly indicates the contradiction r and $\sim r$ obtained. As a result, the initial assumption $\sim\sim p$ cannot hold, so $\sim p$ is established by this indirect proof. The graph also reveals that all premises are needed in this derivation. ◁

(Related Exercises: 17–19)

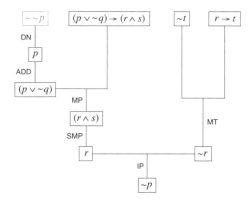

FIGURE 9.4: Derivation graph for an indirect proof with three premises

Can the kind of logical reasoning we describe above be performed by a computer? Can a computer give us an answer about logical consequences—whether one or more logical expressions imply another? The answer is yes! Under the broad area of artificial intelligence or expert systems is the area of automated reasoning. This discipline deals with programming computers in order to obtain logical consequences from all their stored knowledge (stored premises).

For example, logical programming languages such as *Prolog* apply the method of indirect proof and establish that a conclusion follows from a set of premises by assuming that the conclusion is false and then generating a contradiction from its knowledge base.

Logic systems such as *Prolog*, and proofs in mathematics or computer science, often use notation from the propositional logic that we have been studying, as well as notation from *predicate* (or *quantificational*) logic.

HISTORICAL CONTEXT **Predicate (Quantificational) Logic**
Predicate logic is usually credited to Gottlob Frege (1848–1925), whereas propositional logic was developed earlier and credited to Gottfried Leibniz, George Boole, and Augustus De Morgan. Predicate logic breaks down the internal structure of statements, relating subject and predicate. It introduces $\forall$ (a universal quantifier, meaning all x) and $\exists$ (an existential quantifier, meaning for some x) to indicate whether the predicate applies to all or some members of a set.

9.3   Mathematical Induction

This section will introduce a third method of presenting valid arguments, called *mathematical induction*, a technique that can be used to establish mathematical patterns. First, however, let's look at a direct proof of a result concerning positive integers. Note that each integer can be classified as either *even* (a multiple of 2) or as *odd* (not a multiple of 2). Consider the following argument:

   P1: n is an odd integer
   C: n^2 is an odd integer

First, let's see if the proposed conclusion seems reasonable. If $n = 1$, then $n^2 = 1^2 = 1$ is odd; when $n = 3$, then $n^2 = 3^2 = 9$ is odd; when $n = 5$, then $n^2 = 5^2 = 25$ is odd. So far, the conjectured result appears to hold. To establish this conclusively, we can carry out a direct (algebraic) proof. In the derivation that follows, the rules of inference referenced are standard definitions and rules of algebra. These provide the justifications needed for each intermediate step in the direct proof.

1. n is an odd integer	Given
2. $n = 2k + 1$, where k is an integer	1, Definition of odd integer
3. $n^2 = (2k + 1)^2$	2, Square both sides
4. $n^2 = 4k^2 + 4k + 1$	3, Algebraic expansion
5. $n^2 = 2(2k^2 + 2k) + 1$	4, Factoring
6. $r = 2k^2 + 2k$	5, Definition of a new integer
7. $n^2 = 2r + 1$	5, 6, Substitution
8. $\therefore n^2$ is odd	7, Definition of odd integer

Mathematical Induction is a second technique for directly establishing a result. The term "mathematical induction" is a bit misleading because it is a technique of deductive logic, not inductive logic: we are interested in validity, not probability. This technique was designed primarily for proving statements about the *natural numbers* (positive integers) 1, 2, 3, ... or the *whole numbers* (nonnegative integers) 0, 1, 2, 3, ...; however, we will later see other applications of this technique.

Suppose someone asserts that a particular property holds for all integers. One way to show this argument is invalid is to produce a single counterexample. Consider, for instance, the following argument. Is it valid—does the conclusion necessarily hold?

If n is any nonnegative integer, then $n^2 + n + 41$ is a prime.

Recall that *prime numbers* are integers greater than 1 that have no divisors other than themselves and 1. For example, the first few primes are 2, 3, 5, 7, 11, 13, 17, 19, 23, 29. However, 15 is not a prime because it can be factored as $15 = 5 \times 3$. Large prime numbers are the key to modern cryptographic systems and the security of financial transactions (see Chapter 24).

Substituting $n = 0, 1, \ldots, 30$ into the formula $n^2 + n + 41$ gives the following list of integers, each of which is prime:

41, 43, 47, 53, 61, 71, 83, 97, 113, 131, 151, 173, 197, 223, 251, 281, 313, 347, 383, 421, 461, 503, 547, 593, 641, 691, 743, 797, 853, 911, 971.

So the proposition looks quite promising. However, it turns out that the formula fails for $n = 40$, because it gives $40^2 + 40 + 41 = 1681 = 41 \times 41$, which is not prime. This then provides a *counterexample* to the stated conclusion, making the argument invalid.

Example 9.7 Find a Counterexample to a Conjecture

Explore the following assertion and see if you can find a counterexample:

If n is any positive integer, then $1^5 + 2^5 + 3^5 + \cdots + n^5 \leq (n + 1)^5$ holds.

Solution

Trying a few values shows that this inequality holds for $n = 1, 2, \ldots, 7$; for example, when $n = 3$ this assertion states that

$$1^5 + 2^5 + 3^5 \leq (3 + 1)^5 \text{ or } 276 \leq 1024$$

which is indeed true. However, the assertion fails for $n = 8$ since it claims

$$1^5 + 2^5 + 3^5 + \cdots + 8^5 \leq (8 + 1)^5 \text{ or } 61{,}776 \leq 59{,}049$$

which is false. So the argument is invalid. We have found a case in which the premise (n is a positive integer) is true but the conclusion (an algebraic inequality) is false. ◁

(Related Exercises: 20–24)

On the other hand, if we can't find a counterexample to the conclusion of some proposed argument, this does not constitute a proof that the conclusion is true and that the argument is valid. (Recall the hasty-generalization fallacy from Chapter 7.) Unfortunately, if the conclusion references an infinite number of integers n, we can't possibly check each one individually. It is useful then to have a technique that allows us to establish that an argument is valid for an infinite number of cases. Mathematical induction is such a technique.

Proving Validity Using Mathematical Induction

Mathematical Induction is simply a way to climb a ladder with an infinite number of rungs. First, we establish that we can get on the first rung. Second, we need to show that we can always get from one rung to the very next rung. Proofs that make use of mathematical induction typically take the following form:

Structure of a Proof Using Mathematical Induction

1. *Initial Step*: Prove the proposition is true for $n = 0$.
2. *Inductive Hypothesis*: Assume that the proposition holds for $n = k$.
3. *Inductive Step*: Prove that if the proposition is true for $n = k$, then it necessarily holds for $n = k + 1$.
4. *Conclusion*: The proposition is true for all integers $n \geq 0$.

Remark. Step 1 above specifies what is termed the *base case*, indicating the starting point for the asserted proposition. Depending on the application, the base case could be $n = 0$, $n = 1$, or some other specified starting point.

We can relate this technique to application of our Rules of Inference, specifically the repeated use of *modus ponens*. Let P(0) stand for the statement that the proposition holds for $n = 0$, and let P(1) stand for the statement that the proposition holds for $n = 1$. *Modus ponens* shows the validity of the following argument:

$$P(0)$$
$$P(0) \rightarrow P(1)$$
$$\therefore P(1)$$

We interpret this as saying that (a) if we can verify the statement P(0), and (b) if we can establish the conditional P(0) $\rightarrow$ P(1), then we will have established the truth of P(1). Taking this one step further, suppose we can also establish the conditional P(1) $\rightarrow$ P(2). Using *modus ponens* again shows the validity of

$$P(1)$$
$$P(1) \rightarrow P(2)$$
$$\therefore P(2)$$

which then establishes the truth of P(2). Continuing in this way to apply *modus ponens* will eventually establish the truth of P(3), P(4), P(5), and so on. That is, we will have shown the truth of the proposition for an *infinite* number of cases by carrying out the process of mathematical induction.

Example 9.8 Establish a Change-Making Claim

Prove by mathematical induction that we can make change for any (integer) amount greater than or equal to 4 cents using only 2-cent and 5-cent coins. In other words, the claim is that any integer $n \geq 4$ can be expressed using a certain number a of 2-cent coins and a certain number b of 5-cent coins: that is, $n = 2a + 5b$, where a and b are nonnegative integers.

Solution

Initial Step:

Verify the claim for the base case (namely $n = 4$). When $n = 4$, we can make change for 4 cents by using two 2-cent coins: $4 = 2(2) + 5(0)$. Thus, P(4) is true.

Inductive Hypothesis:

We assume P(n) holds for some $n = k \geq 4$ (k is an integer).

Inductive Step:

We need to consider how to extend the truth of P(k) so that the claim continues to hold for the next larger value $n = k + 1$. We reason that if the change for k cents uses a 5-cent coin, we can replace it with three 2-cent coins to form $k + 1$ cents. Otherwise, the change for k cents uses only 2-cent coins; moreover, since $k \geq 4$ we must have at least two 2-cent coins. Then we can replace these two 2-cent coins with a single 5-cent coin and thereby make change for $k + 1$ cents. So in any case we can make change for $k + 1$ cents using only 2-cent and 5-cent coins, based on assuming that it is possible to make change for k cents.

Conclusion:

P(n) holds for all $n \geq 4$. ◁

Example 9.9 Prove a Bowling-Pin Formula

In the standard game of bowling there are 10 pins, arranged in a triangle with four rows having 1, 2, 3, and 4 pins, respectively. Let's generalize this situation to the case when we have instead $n \geq 1$ rows of pins. The claim is that the total number of pins $S(n) = 1 + 2 + \cdots + n$ is given by the formula

$$S(n) = \tfrac{1}{2}n(n + 1).$$

This formula seems to work. When $n = 3$, it gives $S(3) = \tfrac{1}{2}(3)(4) = 6$ pins and when $n = 4$, it gives $S(4) = \tfrac{1}{2}(4)(5) = 10$ pins.

(a) Prove by mathematical induction that the stated formula works for all $n \geq 1$. (b) What would be the largest triangle of pins that could be constructed if we had 50 pins available?

Solution

(a) An inductive proof follows.

Initial Step:

Verify the formula for the initial case (namely $n = 1$). When $n = 1$, there is just one pin and the formula gives $S(1) = \frac{1}{2}(1)(2) = 1$, as desired. Thus, P(1) is true.

Inductive Hypothesis:

We assume P(n) holds for some $n = k \geq 1$ (k is an integer). That is,

$$S(k) = \tfrac{1}{2}k(k+1).$$

Inductive Step:

We need to show that the formula continues to hold for the next larger value $n = k + 1$. The sum of the first $k + 1$ positive integers would therefore need to be

$$S(k+1) = \tfrac{1}{2}(k+1)(k+1+1) = \tfrac{1}{2}(k+1)(k+2).$$

We manipulate the formula for $S(k)$ using algebra to show that if the formula holds for $S(k)$, then it also holds for $S(k+1)$:

1. $S(k) = 1 + 2 + \cdots + k = \frac{1}{2}k(k+1)$ Inductive hypothesis
2. $S(k+1) = 1 + 2 + \cdots + k + (k+1) = \frac{1}{2}k(k+1) + (k+1)$ 1, Add $(k+1)$
3. $S(k+1) = (k+1)(\frac{k}{2} + 1)$ 2, Factor out $(k+1)$
4. $S(k+1) = \frac{1}{2}(k+1)(k+2)$ 3, common denominator

Conclusion:

P(n) holds for all $n \geq 1$. So the stated formula for $S(n)$ holds for all positive integers. ◁

Example 9.10 Conjecture and Establish a Formula Concerning Trees

In Unit I we encountered trees in several contexts, from ways to search a given maze to its role in finding shortest paths from a given origin vertex s. Figure 9.5 displays examples of undirected trees on $n = 2, 3, 4$ vertices. We have also shown the degrees of the vertices. Notice that the sum of the vertex degrees seems to follow a pattern: 2, 4, 6. In particular, even though there are two types of trees on $n = 4$ vertices, the sum of the vertex degrees is the same value 6.

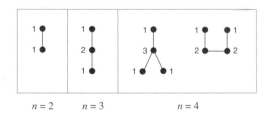

$n = 2$ $n = 3$ $n = 4$

FIGURE 9.5: Trees on $n = 2, 3, 4$ vertices

(a) Conjecture a formula for the sum of the vertex degrees for any tree on n vertices. (b) prove that this formula holds for all $n \geq 2$ using mathematical induction.

Solution

(a) First, we notice a pattern: the sum of vertex degrees is always an even integer (2, 4, 6) and it increases by 2 each time with the number of vertices n. This suggests the formula $2(n - 1)$ as the sum of the vertex degrees of a tree on n vertices.

(b) To establish that this formula is valid for all trees with at least two vertices, we use mathematical induction.

Initial Step:

Verify the claim for the initial case (namely, $n = 2$). When $n = 2$, each vertex has degree 1 so the sum of the vertex degrees is 2. The formula gives $2(n - 1) = 2(2 - 1) = 2$ and thus $P(2)$ is true.

Inductive Hypothesis:

We assume $P(n)$ holds for $n = k$ vertices, where $k \geq 2$. That is, any tree on k vertices has degree sum of $2(k - 1)$.

Inductive Step:

We want to extend the truth of the formula for $n = k$ vertices so that the formula continues to hold for $n = k+1$ vertices. Namely, we need to show that the degree sum is $2[(k+1)-1] = 2k$ when there are $k + 1$ vertices. The key here is to notice that any tree T_{k+1} on $k + 1$ vertices can be obtained by adding a new vertex $k + 1$ to a tree T_k on vertices $1, 2, \ldots, k$ and joining it to one of the vertices of T_k, say vertex r. So in T_{k+1} the degree of r increases by 1 and the new vertex $k + 1$ has degree 1. Overall the sum of vertex degrees increases by 2. However, by the inductive hypothesis, the degree sum for T_k is $2(k - 1)$, so the degree sum for T_{k+1} is $2(k - 1) + 2 = 2k$.

Conclusion:

$P(n)$ holds for all $n \geq 2$. That is, any tree on n vertices has degree sum of $2(n - 1)$. ◁

(Related Exercises: 25–30)

Proving Validity Using Strong Induction

There is a version of mathematical induction called *strong induction*, which allows us to assume more in the inductive step. In regular inductive proofs, we only assume $P(k)$ and then try to prove $P(k + 1)$; in this stronger version, we are able to assume the truth of all the previous statements $P(0)$, $P(1)$, $P(2)$, ..., $P(k)$ on the way to establishing $P(k + 1)$. These extra assumptions can make proving $P(k + 1)$ easier, as seen in the next example.

Example 9.11 Use Strong Induction to Solve a Change-Making Problem

Use strong induction to prove that in a currency with coins having the two denominations 3 cents and 7 cents, we can make change for any amount $n \geq 12$.

Solution

Initial Step:

It is desirable to first establish the claim for several base cases. Here we show that it holds for $n = 12, 13, 14$. Namely, $12 = 3(4)$; $13 = 3(2) + 7(1)$; $14 = 7(2)$. So $P(12)$, $P(13)$, $P(14)$ hold.

Inductive Hypothesis:

We assume that $P(12)$, $P(13)$, ..., $P(k)$ all hold for some $k \geq 14$, and then we want to establish that $P(k + 1)$ holds.

Inductive Step:

We need to consider how to extend the truth of P(12), P(13), ..., P(k) so that the claim continues to hold for the next larger value $n = k+1$. Since $k \geq 14$, then $k+1 \geq 15$. To make change for $k+1$ cents we first use a 3-cent coin, leaving $(k+1) - 3 = k - 2$ cents. However, by the inductive hypothesis we can make change for this smaller remaining amount since $(k+1) - 3 \geq 15 - 3 = 12$. Together with the first 3-cent coin, we now have made change for the given $k+1$ cents. So P($k+1$) is true.

Conclusion:

P(n) holds for all $n \geq 12$. ◁

Remark. It should now be less mysterious why we first needed to establish three base cases in Example 9.11. That is, the verification of P($k+1$) extended back three units since we removed a 3-cent coin and we needed to know that we could make change for the remaining amount $(k+1) - 3$. For standard induction, the "history" just extends back one unit.

(Related Exercises: 31–34)

Chapter Summary

This chapter presents several techniques for establishing the validity of deductive arguments. Both direct proofs and indirect proofs are based on derivations that begin with the premises and apply certain Rules of Inference to obtain the stated conclusion. In contrast to a direct proof, an indirect proof assumes the opposite of the desired conclusion in the hope of reaching a contradiction. The derivation graph provides a useful way to visualize how such proofs flow from the premises to the desired conclusion. In practice, we may accumulate evidence that suggests the truth of some conjecture. However, a rigorous proof is needed to be sure that we have not overlooked a counterexample to the conjecture. The technique of mathematical induction can be applied to establish the truth of such conjectures, based on applying a particular Rule of Inference (*modus ponens*) to build an infinite ladder of propositions indexed by the integers.

Applications, Representations, Strategies, and Algorithms

Applications					
Finance (8, 11)					

Representations					
Graphs		Sets	Symbolization		
Directed	Tree	Lists	Logical	Algebraic	
4–6	10	1–6	1–6	7–11	

Strategies			
Brute-Force	Solution-Space Restrictions	Rule-Based	Composition/Decomposition
7	1	1–6	8–11

Algorithms					
Exact	Inspection	Sequential	Numerical	Logical	Recursive
1–11	7	1–6	7–10	1–6	8–11

Exercises

Assigning Truth-Values to Prove Invalidity

In the following problems, find an assignment of truth-values that will make each of the following arguments invalid. That is, we want to make each premise T and the conclusion F. Based on this objective, see if you can successively deduce truth-values that must hold for the constituent statements.

1. P1: $p \to q$
    P2: $r \to q$
    C: $p \to r$

2. P1: $p \to (q \vee r)$
    P2: $r \to (s \wedge t)$
    P3: $\sim s$
    C: $p \to t$

3. P1: $p \to q$
    P2: $r \to s$
    P3: $q \vee r$
    C: $p \vee s$

Derivations and Valid Arguments

Provide a direct proof of the validity of the following arguments. Number your steps and justify each step citing previously numbered steps and the Rule of Inference used.

4. P1: $p \lor (q \land r)$
   P2: $p \to r$
   P3: $\sim r$
   C: q

5. P1: $(p \lor q) \to (r \land s)$
   P2: $(s \lor t) \to u$
   P3: p
   C: u

6. P1: $p \to r$
   P2: $\sim p \to q$
   P3: $q \to s$
   P4: $\sim r$
   C: s

7. P1: $p \to (q \land r)$
   P2: $(q \lor u) \to t$
   P3: $s \lor p$
   P4: $\sim s$
   C: t

8. P1: $(p \to \sim q) \land (r \to s)$
   P2: $(\sim q \to t) \land (s \to \sim m)$
   P3: $(\sim t \to \sim n) \land (\sim m \to u)$
   P4: $p \land r$
   C: $\sim n \lor u$

Derivations and Derivation Graphs

9. (a) Symbolize the following argument; (b) provide a derivation (specifying the rules used and citing the previously numbered steps) to show its validity; (c) construct the associated derivation graph.

   If we visit Norway, then we don't stay in Stockholm. If we don't stay in Stockholm, then we will stay in Oslo. We didn't stay in Oslo. So, we didn't visit Norway.

10. (a) Symbolize the following argument; (b) provide a derivation (specifying the rules used and citing the previously numbered steps) to show its validity; (c) construct the associated derivation graph.

    If Brazil doesn't join the alliance, then both Portugal and Argentina will boycott it. If Brazil joins the alliance, then Chile will not boycott it. Chile will boycott it. Therefore, Portugal will boycott it.

11. (a) Symbolize the following argument; (b) provide a derivation (specifying the rules used and citing the previously numbered steps) to show its validity; (c) construct the associated derivation graph.

If Maria goes to the party, then Caitlin will not go to the party. If Tamika goes to the party, then Rhonda will not go to the party. Either Maria or Tamika went to the party. Rhonda went to the party. Therefore, Caitlin did not go to the party.

Indirect Proofs and Valid Arguments

12. Provide an indirect proof to establish validity of the following argument. Number your steps and justify each step citing the previously numbered steps and the rule used.

    P1: $p \to q$
    P2: $p \vee q$
    C: q

13. Provide an indirect proof to establish validity of the following argument. Number your steps and justify each step citing the previously numbered steps and the rule used.

    P1: $\sim p \wedge \sim s$
    P2: $r \to q$
    P3: $\sim p \to r$
    C: r

14. Provide an indirect proof to establish validity of the following argument. Number your steps and justify each step citing the previously numbered steps and the rule used.

    P1: $p \to r$
    P2: $\sim p \to q$
    P3: $q \to s$
    P3: $\sim r$
    C: s

15. Provide an indirect proof to establish validity of the following argument. Number your steps and justify each step citing the previously numbered steps and the rule used.

    P1: $(s \vee p) \to \sim q$
    P2: $r \to q$
    P3: $\sim q \to r$
    C: $\sim(s \vee p)$

16. (a) Provide an indirect proof to establish validity of the following argument. Number your steps and justify each step citing the previously numbered steps and the rule used. (b) Draw the associated derivation graph.

    P1: $p \to \sim q$
    P2: $\sim q \to r$
    P3: $\sim r$
    C: $\sim p$

Indirect Proofs and Derivation Graphs

17. (a) Symbolize the following argument; (b) provide an indirect proof to establish its validity; (c) construct the associated derivation graph.

    Steve does not have his parking permit or his car is out of gas. Steve has his parking permit or his car doesn't start. Steve's car is not out of gas or his car doesn't start. Therefore, his car does not start.

18. (a) Symbolize the following argument; (b) provide an indirect proof to establish its validity; (c) construct the associated derivation graph.

    If Melissa doesn't visit her aunt, she will go to traffic court. If Melissa does visit her aunt, she will not have dinner with Lauren. Either Melissa will not visit her aunt or she will have dinner with Lauren. Therefore, Melissa will go to traffic court.

19. (a) Symbolize the following argument; (b) provide an indirect proof to establish its validity; (c) construct the associated derivation graph.

    If I go to Chicago or Cleveland, then if I go to Chicago, I will not go to Minneapolis. If I do not go to Madison or I go to Ann Arbor, then I will go to Chicago and I will also go to Minneapolis. So, I will go to Madison.

Counterexamples to Conjectures

20. Find a counterexample to the following conjecture: For all real numbers x and y, if $x^2 = y^2$, then $x = y$.

21. Find a counterexample to the following conjecture: If x and y are any two positive real numbers, then $(x + y)^2 > 4xy$.

22. Find a counterexample to the following conjecture: If x is a positive real number, then $x^3 > x^2$.

23. Find a counterexample to the following conjecture: If n is any integer and n^2 is evenly divisible by 4, then n is evenly divisible by 4.

24. Find a counterexample to the following conjecture: Adding 1 to the product of the first $n \geq 1$ primes results in a prime. (Recall that a *prime number* is an integer greater than 1 that is evenly divisible only by itself and 1.) For example, the first two primes are 2 and 3 so adding 1 to their product gives $(2 \times 3) + 1 = 7$, a prime. Adding 1 to the product of the first three primes gives $(2 \times 3 \times 5) + 1 = 31$, a prime.

Mathematical Induction

25. Prove by mathematical induction that we can make change for any (integer) amount greater than or equal to 8 cents using only 3-cent and 5-cent coins. In other words, the claim is that any integer $n \geq 8$ can be expressed using a certain number a of 3-cent coins and a certain number b of 5-cent coins: that is, $n = 3a + 5b$, where a and b are nonnegative integers.

26. Prove by mathematical induction that a tree with $n \geq 1$ vertices necessarily has $n-1$ edges.

27. Prove by mathematical induction that $n^2 - 5n - 1 > 0$ holds for all integers $n \geq 6$.

28. Let the sequence $a_n, n \geq 1$, be defined by the following: $a_1 = 2$; $a_{n+1} = 2a_n - n$. For example, $a_2 = 2a_1 - 1 = 2(2) - 1 = 3$. Also, $a_3 = 2a_2 - 2 = 2(3) - 2 = 4$. Use mathematical induction to prove that $a_n = n + 1$ for $n \geq 1$.

29. Let the sequence $b_n, n \geq 1$, be defined by the following: $b_1 = 3$; $b_{n+1} = b_n + 2n - 1$. For example, $b_2 = b_1 + 2(1) - 1 = 3 + 1 = 4$. Also, $b_3 = b_2 + 2(2) - 1 = 4 + 3 = 7$. Use mathematical induction to prove that $b_n = (n-1)^2 + 3$ for $n \geq 1$.

30. Figure 8.1 shows an example of a full, complete binary tree: the vertices are arranged on successive levels, and every vertex (except for those at the last level) has two immediate descendants at the next level. Use mathematical induction to prove that such a binary tree with $n \geq 1$ levels has a total of $2^n - 1$ vertices. [*Hint*: Note that the root vertex is connected to two full, complete binary trees, each having $n - 1$ levels.]

Strong Induction

31. Use strong induction to prove that in a currency with coins having the two denominations 2 cents and 5 cents, we can make change for any amount $n \geq 4$.

32. Use strong induction to prove that in a currency with coins having the two denominations 4 cents and 5 cents, we can make change for any amount $n \geq 12$.

33. Use strong induction to prove that in a currency with coins having the three denominations 3 cents, 8 cents, and 10 cents, we can make change for any amount $n \geq 8$.

34. Use strong induction to prove that every integer $n \geq 1$ can be expressed as a sum of distinct powers of two. For example, $11 = 1 + 2 + 8 = 2^0 + 2^1 + 2^3$. [*Hint*: Consider two cases, depending on whether n is even or n is odd.]

Bibliography

[Velleman 2006] Velleman, D. *How to prove it: A structured approach.* New York: Cambridge University Press.

[OR1] `http://www.aaai.org/Magazine/Watson/watson.php` (accessed April 22, 2014).

[OR2] `http://www.wired.co.uk/news/archive/2013-02/11/ibm-watson-medical-doctor` (accessed May 21, 2014).

Chapter 10

Deductive Logic and Equivalence

I am the literary equivalent of a Big Mac and fries.

—Stephen King, American author, 1947–present

Chapter Preview

It is important to know when two logical statements are equivalent in the context of various applications, whether it be understanding a convoluted statement, specifying a computer search, or designing an efficient computer circuit. We develop certain Rules of Replacement that aid in verifying equivalences; these rules also assist us in carrying out deductive proofs. Replacing the truth-values T and F with the binary numbers 1 and 0, respectively, we can transform compound logical statements into diagrams of digital circuits constructed from certain basic logic gates (*and-*, *or-*, *not-*gates). The Rules of Replacement, together with a graphical representation, can be used to produce efficient and economical designs for digital circuits.

10.1 Equivalences and Conditionals

In querying search engines, such as those available online or in libraries, we often invoke logical expressions to specify those items being sought. For example, to search for web pages containing images of kiwis (birds endemic to New Zealand), one might specify "kiwis but not fruit or people." That is, "I want images of kiwis, but I do not want images of fruit and I do not want images of people." We can use p, q, and r to represent these three simple statements, but it is unclear whether to symbolize the compound statement as $p \wedge (\sim q \wedge \sim r)$ or as $p \wedge \sim(q \vee r)$. Will either one work?

We need to determine if these expressions produce exactly the same results. The answer hinges on whether $\sim q \wedge \sim r$ and $\sim(q \vee r)$ are *logically equivalent*. In Chapter 9, we saw an earlier example of the logical equivalence of $\sim\sim p$ and p.

DEFINITION **Logical Equivalence**

Two statements are logically equivalent if they always have the same truth-value in the same context.

We can use a truth-table to determine logical equivalence by checking whether the columns of values for each statement are identical. In evaluating each statement (which may involve a complex logical expression), there is an order of operations similar to that used in algebra and arithmetic. Start within the innermost parentheses of the statement and successively work your way to the outside. A negation in front of enclosing parentheses reverses the truth-value of the expression inside those parentheses.

Example 10.1 Use a Truth-Table to Determine if Two Search Queries are Equivalent

Decide if the expressions $\sim q \wedge \sim r$ and $\sim(q \vee r)$ are logically equivalent by constructing a truth-table.

Solution

The truth-table shown in Table 10.1 begins with the $2^2 = 4$ rows of possible truth-values for q and r. We then proceed to determine the corresponding truth-values for the two logical expressions $\sim q \wedge \sim r$ and $\sim(q \vee r)$. Since columns five and seven of the truth-table are identical, we conclude that the expressions $\sim q \wedge \sim r$ and $\sim(q \vee r)$ are logically equivalent. Expressed in words, the conjunction of two negations is the same as the negation of their disjunction. In particular, the statement "I do not want images of fruit and I do not want images of people" is the same as "It is not the case that I want images of fruit or images of people." ◁

(Related Exercises: 1–3)

TABLE 10.1: Determining the logical equivalence of two expressions

q	r	$\sim q$	$\sim r$	$\sim q \wedge \sim r$	$q \vee r$	$\sim(q \vee r)$
T	T	F	F	F	T	F
T	F	F	T	F	T	F
F	T	T	F	F	T	F
F	F	T	T	T	F	T

The equivalence established in Example 10.1 is referred to as one of the *De Morgan laws*. There is a counterpart to this equivalence, in which we negate the conjunction of two statements.

Example 10.2 Use a Truth-Table to Verify Another of the De Morgan Laws

Use a truth-table to verify that $\sim q \vee \sim r$ and $\sim(q \wedge r)$ are logically equivalent.

Solution

We construct the truth-table shown in Table 10.2 that has four rows for the possible truth-values for q and r. We then determine the columns of truth-values for both $\sim q \vee \sim r$ and $\sim(q \wedge r)$. Since these columns are identical, $\sim q \vee \sim r$ and $\sim(q \wedge r)$ are logically equivalent. As a concrete illustration, the statement "I do not want both fame and fortune" is the same as saying "Either I do not want fame or I do not want fortune." ◁

TABLE 10.2: Verifying one of the De Morgan laws

q	r	$\sim q$	$\sim r$	$\sim q \vee \sim r$	$q \wedge r$	$\sim(q \wedge r)$
T	T	F	F	F	T	F
T	F	F	T	T	F	T
F	T	T	F	T	F	T
F	F	T	T	T	F	T

HISTORICAL CONTEXT **Augustus De Morgan (1806–1871)**
An English mathematician and logician. He is credited with formally stating the Four-Color Conjecture (see Section 6.1) in a 1852 letter to William Rowan Hamilton.

RESULT **De Morgan's Laws**
$\sim(q \vee r) \Leftrightarrow \sim q \wedge \sim r$
$\sim(q \wedge r) \Leftrightarrow \sim q \vee \sim r$

In the above statement of De Morgan's laws, we have used a double-arrow $\Leftrightarrow$ between two statements to indicate that they are logically equivalent. The next example shows that it is possible to use the notion of equivalence to simplify an otherwise complex expression.

Example 10.3 Use a Truth-Table to Verify Another Logical Equivalence

Use a truth-table to establish that $\sim(p \wedge \sim q) \wedge p$ is logically equivalent to $p \wedge q$.

Solution

We use the truth-table shown in Table 10.3 to evaluate the two indicated expressions. Since the last two columns are identical, we have $\sim(p \wedge \sim q) \wedge p \Leftrightarrow p \wedge q$. ◁

TABLE 10.3: Simplifying a logical expression

p	q	$\sim q$	$p \wedge \sim q$	$\sim(p \wedge \sim q)$	$\sim(p \wedge \sim q) \wedge p$	$p \wedge q$
T	T	F	F	T	T	T
T	F	T	T	F	F	F
F	T	F	F	T	F	F
F	F	T	F	T	F	F

Example 10.4 Use a Truth-Table to Simplify a Form of Negated Conditional

Use a truth-table to verify that $\sim(p \rightarrow q)$ is logically equivalent to $p \wedge \sim q$.

Solution

We use the truth-table shown in Table 10.4 to evaluate the two indicated expressions. Since the last two columns are identical, we have $\sim(p \rightarrow q) \Leftrightarrow p \wedge \sim q$. ◁

(Related Exercises: 4–9)

TABLE 10.4: Simplifying a logical expression

p	q	$\sim q$	$p \to q$	$\sim(p \to q)$	$p \wedge \sim q$
T	T	F	T	F	F
T	F	T	F	T	T
F	T	F	T	F	F
F	F	T	T	F	F

Example 10.5   Use a Truth-Table to Determine Logical Equivalence

Use a truth-table to determine whether $\sim p \to q$ is equivalent to $p \vee \sim q$.

Solution

We construct the truth-table shown in Table 10.5 and then proceed to evaluate the two indicated expressions. Since the last two columns are *not* identical, we conclude that $\sim p \to q$ is not logically equivalent to $p \vee \sim q$. ◁

(*Related Exercises: 10–13, 39*)

TABLE 10.5: Testing a possible logical equivalence

p	q	$\sim p$	$\sim q$	$\sim p \to q$	$p \vee \sim q$
T	T	F	F	T	T
T	F	F	T	T	T
F	T	T	F	T	F
F	F	T	T	F	T

Contrapositives, Opposites, Inverses, and Converses

In our everyday speech, we use conditional statements (those involving *if-then* constructs) in a variety of forms. Some of these forms are equivalent to one another, yet others are not. It is important to study these different forms and their relation to one another. For example, statement B below is called the *contrapositive* of statement A:

    A. If Bill has the password, then Steve does not: $b \to \sim s$

    B. If Steve has the password, then Bill does not: $s \to \sim b$

DEFINITION **Contrapositive Statements**
Two conditional statements are contrapositives of each other if one statement exchanges and negates the antecedent and consequent of the other: for example, the statements $p \to q$ and $\sim q \to \sim p$ are contrapositives.

Example 10.6   Use a Truth-Table to Verify that Contrapositives are Logically Equivalent

Demonstrate that the contrapositive statements $p \to q$ and $\sim q \to \sim p$ are logically equivalent by using a truth-table.

Solution

We use the truth-table shown in Table 10.6 to construct columns of truth-values for the two indicated expressions. Since the last two columns are identical, the statements $p \to q$ and $\sim q \to \sim p$ are logically equivalent. $\lhd$

TABLE 10.6: Verifying that contrapositives are logically equivalent

p	q	$\sim p$	$\sim q$	$p \to q$	$\sim q \to \sim p$
T	T	F	F	T	T
T	F	F	T	F	F
F	T	T	F	T	T
F	F	T	T	T	T

There are three other ways of transforming a conditional statement into another logical statement. These are summarized below.

DEFINITION **Opposite Statements (Negations)**
Two statements are opposites or negations of each other if they always have opposite truth-values: for example, the statements $p \to q$ and $\sim(p \to q)$ are opposites of one another.

DEFINITION **Inverse Statements**
Two conditional statements are inverses of each other if one statement negates both the antecedent and the consequent of the other: for example, the statements $p \to q$ and $\sim p \to \sim q$ are inverses of one another.

DEFINITION **Converse Statements**
Two conditional statements are converses of each other if one statement exchanges the antecedent and consequent of the other: for example, the statements $p \to q$ and $q \to p$ are converses of one another.

(Related Exercises: 14–17)

Recall the statement A: If Bill has the password, then Steve does not have the password. The inverse would be the statement C: If Bill does not have the password, then Steve has the password. The converse would be the statement D: If Steve does not have the password, then Bill has the password. How do these statements relate to one another? To answer this question, we can use truth-tables to demonstrate relationships among these various types of conditional statements.

Example 10.7 Use a Truth-Table to Study the Relationships Between Conditionals

Explore which of the variants of conditional statements are equivalent to one another.

Solution

In the truth-table shown in Table 10.7, we develop columns of truth-values for the contrapositive, opposite, inverse, and converse of the original conditional $p \to q$. We see that the columns for Conditional and Contrapositive are identical; also, the columns for Inverse and Converse are identical. Therefore, we have the logical equivalences

$$p \to q \Leftrightarrow \sim q \to \sim p, \qquad \sim p \to \sim q \Leftrightarrow q \to p.$$

Notice that the second equivalence is just a thinly disguised version of the first equivalence: namely, $\sim p \to \sim q$ is obtained by simply forming the contrapositive of $q \to p$. ◁

TABLE 10.7: Relationships between various conditional statements

		Conditional	Contrapositive	Opposite	Inverse	Converse
p	q	$p \to q$	$\sim q \to \sim p$	$\sim(p \to q)$	$\sim p \to \sim q$	$q \to p$
T	T	T	T	F	T	T
T	F	F	F	T	T	T
F	T	T	T	F	F	F
F	F	T	T	F	T	T

As a result of Example 10.7, we are assured that statements C and D have the same meaning, just as statements A and B have the same meaning. However, statements A and C have quite different meanings.

Venn Diagrams

So far we have seen how to use truth-tables to demonstrate when two logical statements are equivalent. Alternatively, we can also use Venn diagrams (Chapter 7) to demonstrate certain equivalences. Specifically, the Venn diagrams of two types of conditionals will be the same precisely when the conditionals are logically equivalent.

To illustrate, consider the two statements $p =$ "A person is a farmer" and $q =$ "A person works on the land." These statements involve the set F consisting of all farmers and the set L consisting of all people working on the land. Both of these are sets contained within the larger set of All People, indicated by the large rectangle in Figure 10.1. The conditional $p \to q$ indicates that if a person is a farmer, then he or she works on the land; in other words, all farmers work on the land. Pictorially, this means that the set F is a subset of the set L, as shown in Figure 10.1(a). On the other hand, the conditional $\sim q \to \sim p$ indicates that if a person does not work on the land (is outside set L), then he or she is not a farmer (is outside set F). This relationship is shown by the same exact diagram in Figure 10.1(a). This geometric representation then demonstrates the equivalence of a conditional $p \to q$ and its contrapositive $\sim q \to \sim p$.

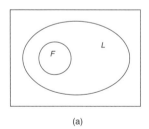 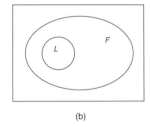

         (a)   (b)

FIGURE 10.1: Venn diagram representations of conditionals

We can also interpret geometrically the equivalence of the inverse statement $\sim p \to \sim q$ and the converse $q \to p$. The former states that if a person is not a farmer (is outside F), then he or she does not work on the land (is outside set L); the latter states that if a person works on the land (is inside set L), then he or she is a farmer (is inside set F). Both of these conditionals correspond to the exact same diagram shown in Figure 10.1(b).

Example 10.8 Use Venn Diagrams and Truth-Tables to Determine Equivalence

Symbolize the following statements. Then use both truth-tables and Venn diagrams to identify which statements are logically equivalent.

A. If a person is hired by FedEx, then he or she works weekends.
B. If a person works weekends, then he or she is hired by FedEx.
C. If a person does not work weekends, then he or she is not hired by FedEx.
D. If a person is not hired by FedEx, then he or she does not work weekends.

Solution

Let p = "A person is hired by FedEx" and let q = "A person works weekends." Then the statements above can be symbolized as follows:

A. $p \rightarrow q$
B. $q \rightarrow p$
C. $\sim q \rightarrow \sim p$
D. $\sim p \rightarrow \sim q$

We now construct the truth-table in Table 10.8 and examine the four columns labeled A, B, C, D. We see that statements A and C are logically equivalent, as are statements B and D. There are no other equivalences displayed in this table.

TABLE 10.8: Truth-table verification of equivalences

		A	B			C	D
p	q	$p \rightarrow q$	$q \rightarrow p$	$\sim p$	$\sim q$	$\sim q \rightarrow \sim p$	$\sim p \rightarrow \sim q$
T	T	T	T	F	F	T	T
T	F	F	T	F	T	F	T
F	T	T	F	T	F	T	F
F	F	T	T	T	T	T	T

Alternatively, let sets F and W indicate those persons hired by FedEx and those persons working weekends, respectively. The Venn diagram in Figure 10.2(a) expresses the relationship in statements A and C, while the Venn diagram in Figure 10.2(b) expresses the relationship in statements B and D. ◁

(Related Exercises: 18–19)

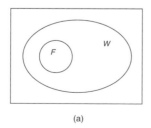
(a)

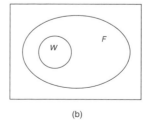
(b)

FIGURE 10.2: Two Venn diagram representations

When we seek to *disprove* a conditional $p \to q$, we are in essence establishing the truth of its opposite $\sim(p \to q)$. As Example 10.4 shows, $\sim(p \to q)$ is logically equivalent to $p \wedge \sim q$: that is, we try to establish that the hypothesis p is true yet the conclusion q is false. This is precisely the strategy adopted in Section 9.1 for finding a *counterexample* to the assertion $p \to q$: we search for an instance in which p holds but q does not. More generally, when there are several hypotheses to an argument, a counterexample is an instance that makes all the hypotheses true but the conclusion false.

10.2   Rules of Replacement

In Example 10.4, we discovered the equivalence $\sim(p \to q) \Leftrightarrow p \wedge \sim q$. Using this known equivalence, we can establish a new equivalence $p \to q \Leftrightarrow \sim p \vee q$ without resorting to construction of a truth-table. Namely, we can make use of De Morgan's laws by taking the negation of both sides of the first equivalence.

1. $\sim(p \to q) \Leftrightarrow p \wedge \sim q$ Example 10.4
2. $\sim\sim(p \to q) \Leftrightarrow \sim(p \wedge \sim q)$ 1, take negations
3. $\sim\sim(p \to q) \Leftrightarrow \sim p \vee \sim\sim q$ 2, De Morgan's laws
4. $p \to q \Leftrightarrow \sim p \vee q$ 3, Double Negation

This shows that we can derive other equivalences from known equivalences. In turn, these equivalences can assist us in analyzing the validity of deductive arguments. We designate certain well-known equivalences as our *Rules of Replacement*. These can be established either by setting up a truth-table or by transforming other equivalences (as done in the argument above).

DEFINITION **Rule of Replacement**
An "interchangeable equivalent statement" that can be used as part of a logical chain in establishing the validity of a given argument.

We list various Rules of Replacement in Table 10.9 together with their common designations: the Absorption, Associative, Commutative, Double Negation, Distributive, and De Morgan laws. Some convenient abbreviations are also provided. Some of these rules are already familiar, and others can be established by constructing truth-tables. These Rules of Replacement serve as fundamental building blocks in establishing other equivalences as well as other logical inferences.

We now illustrate how these Rules of Replacement, together with the Rules of Inference from Chapter 9, can be used to carry out logical derivations or to simplify logical expressions.

TABLE 10.9: Rules of replacement

Rule of Replacement	Terminology
$p \vee (p \wedge q) \Leftrightarrow p$ $p \wedge (p \vee q) \Leftrightarrow p$	Absorption (ABS)
$p \vee (q \vee r) \Leftrightarrow (p \vee q) \vee r$ $p \wedge (q \wedge r) \Leftrightarrow (p \wedge q) \wedge r$	Associative (ASSOC)
$p \vee q \Leftrightarrow q \vee p$ $p \wedge q \Leftrightarrow q \wedge p$	Commutative (COM)
$\sim\sim p \Leftrightarrow p$	Double Negation (DN)
$p \wedge (q \vee r) \Leftrightarrow (p \wedge q) \vee (p \wedge r)$ $p \vee (q \wedge r) \Leftrightarrow (p \vee q) \wedge (p \vee r)$	Distributive (DIST)
$\sim(p \vee q) \Leftrightarrow \sim p \wedge \sim q$ $\sim(p \wedge q) \Leftrightarrow \sim p \vee \sim q$	De Morgan (DM)

Example 10.9 Use the Rules of Inference and the Rules of Replacement to Derive a Conclusion from Given Premises

Beginning with the premises P1, P2, P3 stated below, show a derivation of the conclusion C, numbering each intermediate step and indicating the rule used.

P1. $\sim(r \wedge q) \to (p \wedge q)$
P2. $\sim p \vee \sim q$
P3. $p \wedge s$
C. $(p \wedge q) \wedge r$

Solution

The following steps establish the truth of the conclusion based on the given premises. There may be alternative ways to organize a logical argument leading to this same conclusion.

1. $\sim(r \wedge q) \to (p \wedge q)$ Given
2. $\sim p \vee \sim q$ Given
3. $p \wedge s$ Given
4. $\sim(p \wedge q)$ 2, DM
5. $\sim\sim(r \wedge q)$ 1, 4, MT
6. $r \wedge q$ 5, DN
7. p 3, SMP
8. $q \wedge r$ 6, COM
9. $p \wedge (q \wedge r)$ 7, 8, CN
10. $(p \wedge q) \wedge r$ 9, ASSOC

◁

Example 10.10 Use the Rules of Replacement to Simplify an Expression

Beginning with the following expression, transform it into a simpler expression, numbering each intermediate step and indicating the Rule of Replacement used.

$$[(p \wedge q) \vee (p \wedge r)] \wedge (q \wedge r).$$

Solution

1. $[(p \wedge q) \vee (p \wedge r)] \wedge (q \wedge r)$ Given
2. $[p \wedge (q \vee r)] \wedge (q \wedge r)$ 1, DIST
3. $p \wedge [(q \vee r) \wedge (q \wedge r)]$ 2, ASSOC
4. $p \wedge [((q \vee r) \wedge q) \wedge r]$ 3, ASSOC
5. $p \wedge [(q \wedge (q \vee r)) \wedge r]$ 4, COM
6. $p \wedge (q \wedge r)$ 5, ABS

◁

(Related Exercises: 20–22)

We will see an application of this kind of simplification in Chapter 11, where the Rules of Replacement can be used to efficiently locate a limited number of fire stations so that they can serve all zones of a city.

10.3   Computer Circuit Design

Everything in the digital world is based on the binary number system—which uses only the two symbols 0, 1. This is appropriate because electrical impulses can be characterized as having two activation states: *off* (0) or *on* (1). We can apply truth-tables and logical connectives to circuit design by making the following conversions from truth-values to *bits* and from connectives to *gates*:

F (false) ↠ 0 (off)

T (true) ↠ 1 (on)

and, or, not connectives ↠ logic gates

Most digital circuits are built up from the three gates *and, or, not.* Sometimes circuit designers also use *nand* ("not-*and*") gates as well as *nor* ("not-*or*") gates. These gates just reverse the output of *and*-gates and *or*-gates, respectively.

DEFINITION **Logic Gate**
A simple circuit with one or two inputs and one output. Electrical impulses enter the gate, which then converts the impulses according to certain rules.

DEFINITION **Binary Digit (Bit)**
A variable having only two possible values, the binary digits 0 and 1. These values are often interpreted as activation states (off/on).

HISTORICAL CONTEXT **Deductive Logic and Digital Circuits**
In his 1937 MIT master's thesis *A Symbolic Analysis of Relay and Switching Circuits,* Claude Shannon connected binary numbers and logic to the design of digital circuits. This work became the foundation underlying all modern electronic digital computers.

The translations between truth-values and binary digits, as well as between logical connectives and logic gates, are illustrated in Tables 10.10 and 10.11. The first table identifies p and q as truth-values and lists the operations of negation, conjunction, and disjunction. The second table shows the equivalent representation in which p and q are binary digits and provides outputs for the corresponding logic gates *not, and, or* as well as the derived logic gates *nand, nor*.

TABLE 10.10: Truth-values and logical connectives

p	q	$\sim p$	$p \wedge q$	$p \vee q$
T	T	F	T	T
T	F	F	F	T
F	T	T	F	T
F	F	T	F	F

TABLE 10.11: Operations on binary digits

		not	*and*	*or*	*nand*	*nor*
p	q	$\sim p$	$p \wedge q$	$p \vee q$	$\sim(p \wedge q)$	$\sim(p \vee q)$
1	1	0	1	1	0	0
1	0	0	0	1	1	0
0	1	1	0	1	1	0
0	0	1	0	0	1	1

Basic Logic Gates

Every logic circuit corresponds to a logical expression, whose symbols (such as p, q, r) represent the inputs to the circuit. A logic gate takes the values from its inputs and combines them according to the definitions of the connectives (*not, and, or*) to produce an output. We describe now three basic logic gates that can be used to design more complicated circuits.

Not-Gate (indicated in Figure 10.3 by a triangle with a circle at its output) has a single input p. This gate reverses or inverts the value of the input bit p. For example, with the specific input $p = 0$, the *not*-gate will produce the output $\sim p = \sim 0 = 1$.

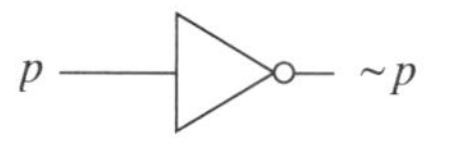

FIGURE 10.3: Representation of the *not*-gate

And-Gate (indicated in Figure 10.4 by the horseshoe shape) has the two inputs p and q, and one output. The output will be 0 if any of the inputs is 0. For example, with the specific inputs $p = 0$ and $q = 1$, the *and*-gate will produce the output $p \wedge q = 0 \wedge 1 = 0$.

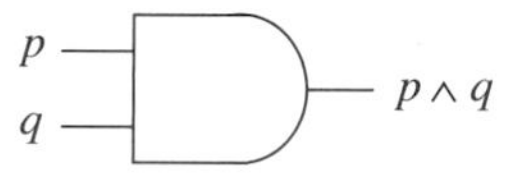

FIGURE 10.4: Representation of the *and*-gate

***Or*-Gate** (indicated in Figure 10.5 by the arrow shape) has the two inputs p and q, and one output. The output will be 1 if any of the inputs is 1. For example, with the specific inputs $p = 0$ and $q = 1$, the *or*-gate will produce the output $p \vee q = 0 \vee 1 = 1$.

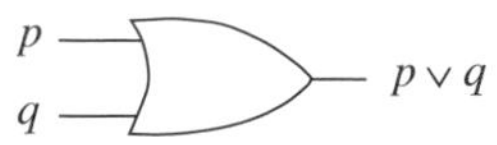

FIGURE 10.5: Representation of the *or*-gate

Each gate performs a very simple computation. A more complex circuit can be built from these simple logic gates and is represented by a *circuit diagram*: the inputs enter from the left, progress through the circuit from left to right, and produce an output on the right. Such circuits form the core of computations carried out by computers; they are found in the portion of the central processing unit called the arithmetic logic unit (ALU).

DEFINITION **Circuit Diagram**
A representation of an electrical circuit in which inputs appear on the left, and are successively converted to outputs as we proceed from left to right in the diagram.

For example, a *not*-gate can be combined with an *or*-gate or combined with an *and*-gate. The location of the *not*-gate affects the calculation. In Figure 10.6, the *not*-gate is applied to the output of the *or*-gate, and so corresponds to the logical expression $\sim(p \vee q)$. If the inputs to this circuit are $p = 0$ and $q = 1$, then the output will be $\sim(p \vee q) = \sim(0 \vee 1) = \sim 1 = 0$.

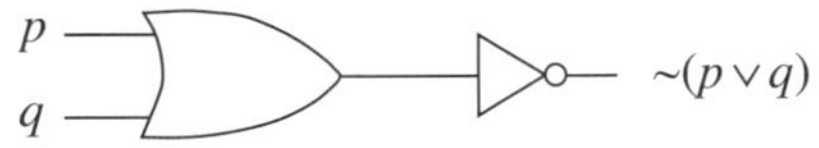

FIGURE 10.6: Applying a *not*-gate to the output of an *or*-gate

As another example, the *not*-gate might appear on one of the inputs to an *and*-gate and so change this input from p to $\sim p$. The circuit shown in Figure 10.7 shows this situation; it represents the logical expression $\sim p \wedge q$. If the inputs to this circuit are $p = 0$ and $q = 1$, then the output will be $\sim p \wedge q = \sim 0 \wedge 1 = 1 \wedge 1 = 1$.

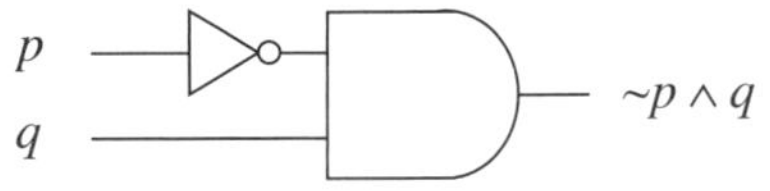

FIGURE 10.7: Applying a *not*-gate to the input of an *and*-gate

Example 10.11 Find the Symbolic Representation of a Circuit and its Output

Find the logical expression corresponding to the circuit shown in Figure 10.8. Then determine its output for the input values $p = 0$, $q = 0$, and $r = 0$.

FIGURE 10.8: A circuit with three inputs p, q, r

Solution

The output from the first *or*-gate is $\sim q \vee r$ and this becomes an input to the second *or*-gate, giving the final expression $p \vee (\sim q \vee r)$. The output produced for input values $p = 0, q = 0, r = 0$ is then $p \vee (\sim q \vee r) = 0 \vee (\sim 0 \vee 0) = 0 \vee (1 \vee 0) = 0 \vee 1 = 1$. ◁

(Related Exercises: 23–26)

Example 10.12 Draw the Circuit for a Logical Expression and Calculate its Output

Given the logical expression $\sim(\sim(p \vee \sim q) \wedge p)$, construct the associated circuit and evaluate it for the input values $p = 0$ and $q = 1$.

Solution

When drawing the circuit, it is helpful to start with the innermost connective of the expression and its component statements, and then work outward. In this case, we begin with $\sim q$ and then apply the logical connective $\vee$ to produce an *or*-gate with inputs $\sim q$ and p. Since the input p appears later as input to the *and*-gate, the signal p is split so that it can be used as input to a second gate. Figure 10.9 shows the resulting circuit. The output produced for input values $p = 0$ and $q = 1$ is then $\sim(\sim(p \vee \sim q) \wedge p) = \sim(\sim(0 \vee \sim 1) \wedge 0) = \sim(\sim 0 \wedge 0) = \sim(1 \wedge 0) = 1$. ◁

(Related Exercises: 27–32)

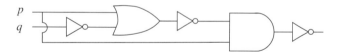

FIGURE 10.9: A circuit with two inputs p, q

Given the correspondence between circuits and logical expressions, we can use our previous tools to easily determine if two different circuits do indeed perform the same function.

Example 10.13 Determine if Two Circuits are Equivalent

Determine if the two circuits A and B shown in Figure 10.10 produce the same output given the same inputs. (a) First, represent the circuits as logical expressions. (b) Then use a truth-table to test for equivalence of the circuits.

Solution

(a) Circuit A represents the expression $p \vee \sim q$ while circuit B represents the expression $\sim(\sim p \wedge q)$.

(b) The truth-table shown in Table 10.12 shows that these two circuits are equivalent: that is, they produce the same output for the same inputs. Notice that the first circuit only

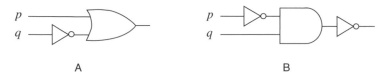

A   B

FIGURE 10.10: Two circuits with inputs p, q

TABLE 10.12: Truth-table representation of two circuits

				A		B
p	q	$\sim p$	$\sim q$	$p \vee \sim q$	$\sim p \wedge q$	$\sim(\sim p \wedge q)$
1	1	0	0	1	0	1
1	0	0	1	1	0	1
0	1	1	0	0	1	0
0	0	1	1	1	0	1

requires two gates, whereas the second requires three gates, so the first circuit might be considered more economical and reliable. ◁

We can now use logical representations and truth-tables to design circuits that perform specific tasks. In particular, we can build *recognizer circuits*.

DEFINITION **Recognizer Circuit**
A circuit that recognizes a certain input pattern.

Example 10.14 Design a Simple Recognizer Circuit

Design a circuit that recognizes the equality of two binary digits p and q. That is, the circuit should output 1 (true) when bits p and q are identical and should output 0 for every other combination.

Solution

We first construct the truth-table of Table 10.13 in which the third column records the required output for each pair of inputs. We then add a fourth column (*term*) that represents the corresponding logical expression: namely, this expression produces the specified output only when the corresponding inputs have their stated value. For example, we place $p \wedge q$ in the first row of column four, since this conjunction equals 1 only when $p = 1$ and $q = 1$. Similarly we place $p \wedge \sim q$ in the second row since this conjunction equals 1 only when $p = 1$ and $\sim q = 1$: that is, when $p = 1$ and $q = 0$.

Next, we focus on the rows in which the output is required to be 1 and form the disjunction of the terms associated with these rows. In our example, we obtain the expression

$$(p \wedge q) \vee (\sim p \wedge \sim q).$$

This disjunction will evaluate to 1 only when one of the component terms is 1: $p \wedge q$ will be 1 when p and q are both 1, whereas $\sim p \wedge \sim q$ will be 1 when p and q are both 0. So whenever p and q are the same, the disjunction will equal 1, as required by this recognizer circuit.

TABLE 10.13: Specifications for a recognizer circuit

p	q	output	term
1	1	1	$p \wedge q$
1	0	0	$p \wedge \sim q$
0	1	0	$\sim p \wedge q$
0	0	1	$\sim p \wedge \sim q$

Finally, we draw the circuit corresponding to $(p \wedge q) \vee (\sim p \wedge \sim q)$. Notice that this expression has five connectives, so the associated circuit will use five gates: two *and*-gates connected by an *or*-gate, with *not*-gates operating on p and q as they enter the second *and*-gate (see Figure 10.11). ◁

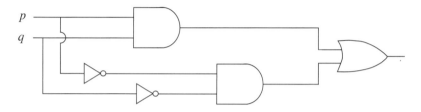

FIGURE 10.11: Circuit to recognize equality of inputs p, q

NOTATION **Logical Operations**

In designing more complicated circuits, it is convenient to abbreviate conjunctions by juxtaposition, disjunctions by +, and negations by an overbar. For example, $(p \wedge q) \vee (\sim p \wedge \sim q)$ is represented more succinctly as $pq + \bar{p}\bar{q}$.

10.4 Optimizing Circuits Using Graphs

In complicated circuit designs, the number of gates can be large and the circuit might be expensive to build. It becomes economically important to find equivalent circuits that use as few gates as possible. This can be a challenging task; when a circuit design involves millions of gates, such as the core processor in a desktop computer, specialized software is needed.

It turns out that we can use a graph representation to help reduce the number of gates used. The following examples illustrate how to design a circuit with a specified function and then make it more economical to fabricate.

Example 10.15 Design a More Complex Circuit

Design a circuit involving four inputs, labeled w, x, y, and z. The output of the circuit will depend on the specific values (0 or 1) given to each of the four inputs. Rather than listing all $2^4 = 16$ possible rows of input values or *bit strings* along with the corresponding output for each, let's focus on only those sets of input values that produce an output of 1 (as done in Example 10.14). All other sets of input values will necessarily give an output of 0 and can be omitted. Table 10.14 lists each of the seven input bit strings (labeled A–G) that should produce the output 1.

TABLE 10.14: Specifications for a more complex circuit

	w	x	y	z	output	term	# of gates
A	1	1	1	0	1	$wxy\bar{z}$	4
B	1	0	1	1	1	$w\bar{x}yz$	4
C	0	1	1	1	1	$\bar{w}xyz$	4
D	1	0	1	0	1	$w\bar{x}y\bar{z}$	5
E	0	1	0	1	1	$\bar{w}x\bar{y}z$	5
F	0	0	1	1	1	$\bar{w}\bar{x}yz$	5
G	0	0	0	1	1	$\bar{w}\bar{x}\bar{y}z$	6

Solution

We begin by completing the next two columns of Table 10.14. For example, when $w = 1$, $x = 1$, $y = 1$, and $z = 0$ (i.e., the input bit string is 1110) in row A, the circuit should output 1. As done in Example 10.14, we associate the term $w \wedge x \wedge y \wedge \sim z$ with this bit string; this is abbreviated using our shorthand notation as $wxy\bar{z}$. Notice that this expression evaluates to 1 *only* when $w = 1$, $x = 1$, $y = 1$, and $z = 0$: that is, this expression is a recognizer for the bit string 1110. To implement this term as a circuit would involve three *and*-gates and one *not*-gate, or four gates in all. In a similar way, we get the terms shown in rows B–G, as well as the number of gates required for each term.

Finally, a logical expression f that evaluates to 1 precisely under the conditions specified by A–G is obtained by forming the disjunction of the seven terms in Table 10.14:

$$f = wxy\bar{z} + w\bar{x}yz + \bar{w}xyz + w\bar{x}y\bar{z} + \bar{w}x\bar{y}z + \bar{w}\bar{x}yz + \bar{w}\bar{x}\bar{y}z.$$

This expression evaluates to 1 when one of its seven terms evaluates to 1 and so recognizes exactly the seven bit strings specified in A–G. However, such a circuit requires 33 gates for the individual terms plus 6 *or*-gates, for a total of 39 gates. ◁

(Related Exercises: 33–35)

We want to find a more economical expression (one using fewer gates) for f. To do so, we construct a graph according to the following rules:

1. Each vertex represents a row of Table 10.14 and is labeled with the row letter (A–G).
2. Two vertices are connected by an edge if their bit strings differ in exactly one position.
3. Each edge is labeled with the number of the position where the bit strings differ.

For example, we join A and D by an edge since their bit strings 1110 and 1010 differ in exactly one position (namely, position 2), and so the edge is labeled 2. The entire graph for this problem is shown in Figure 10.12. We have also indicated the term associated with each vertex.

Because each term in the expression for f must be accounted for, we say that each vertex in the graph must be *covered*. To obtain as few terms as possible, consider how we might combine individual terms into groups and obtain a covering of all vertices using a smaller number of groups.

For example, instead of using two terms to cover A and D individually, we can cover them

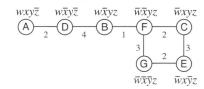

FIGURE 10.12: Graph associated with a complex circuit

both with a single term $wy\bar{z}$. This is justified because $wy\bar{z}$ represents what the two terms have in common. More precisely, these two logical expressions can be combined using the distributive property of Table 10.9:

$$wxy\bar{z} + w\bar{x}y\bar{z} = wy\bar{z}(x + \bar{x}) = wy\bar{z}.$$

We have also used here the fact that $x + \bar{x} = x \vee \sim x = 1$. Notice that the bit strings for A and D differ only in position 2, and so have in common $wy\bar{z}$. In a similar way, any two vertices joined by an edge can be combined to form a new term that covers both of the vertices. Clearly, the combined term involves fewer gates.

Now look at the square in Figure 10.12 defined by vertices F, C, E, G. These four terms can be combined into a single term $\bar{w}z$. Namely, all four vertices have $\bar{w}z$ in common:

$$\bar{w}\bar{x}yz + \bar{w}xyz + \bar{w}x\bar{y}z + \bar{w}\bar{x}\bar{y}z = \bar{w}z(\bar{x}y + xy + x\bar{y} + \bar{x}\bar{y}) = \bar{w}z(\bar{x} + x)(y + \bar{y}) = \bar{w}z.$$

Notice that the square $FCEG$ has edges labeled with 2 and 3 (these are the differing positions), so the four vertices should indeed have in common those inputs in positions 1 ($\bar{w}$) and 4 (z). The resulting term $\bar{w}z$ can be implemented using just one *and*-gate and one *not*-gate; this is a much simpler representation than using individual terms for F, C, E, G or using two terms that correspond to edges FC and EG.

So, we are led to the following graph covering problem.

DEFINITION **Graph Covering Problem**
Cover all the vertices of a graph using the fewest number of squares, edges, and vertices.

In order to obtain the best covering (fewest number of grouped terms), we are guided by first covering those vertices for which there is only one good choice for the grouped term. In the current example, vertex A can be covered by the single term $wxy\bar{z}$, but clearly it is better to cover it with the edge AD (giving the grouped term $wy\bar{z}$). As a result, vertex D is covered also. Now look at vertex E. It can be covered by vertex E itself, by either edge CE or EG, or by the square $FCEG$. The best choice is to use the square $FCEG$, giving the covering term $\bar{w}z$; this grouped term covers F, C, E, and G. In general using squares is better than using edges, while using edges is better than using single vertices. The current situation is depicted in Figure 10.13, which shows a partial covering of the graph.

Now we have covered all vertices except B. It is best to use either edge DB (term $w\bar{x}y$) or edge BF (term $\bar{x}yz$), rather than just vertex B. This gives two possible coverings of the graph, corresponding to the two expressions

$$wy\bar{z} + w\bar{x}y + \bar{w}z, \qquad wy\bar{z} + \bar{x}yz + \bar{w}z.$$

Each of these uses 5 *and*-gates, 2 *or*-gates, and 3 *not*-gates. We can further simplify the above expressions by making use of the distributive property in Table 10.9, giving

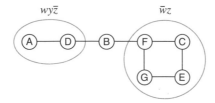

FIGURE 10.13: A partial covering of the graph

$$wy(\bar{z} + \bar{x}) + \bar{w}z, \qquad wy\bar{z} + (\bar{x}y + \bar{w})z.$$

These expressions yield equivalent circuits that use only 8 and 9 gates, respectively, instead of the original 39 gates. There is no guarantee that the procedure used here guarantees the fewest number of gates. However, this *heuristic* procedure was quite effective, by reducing the size of the circuit from 39 gates to 8 gates!

Exercise 34 asks you to place the first of these reduced expressions in a truth-table in order to establish that it produces outputs of 1 for exactly the seven specified bit strings (and 0 otherwise), so we indeed have found a simpler, equivalent circuit.

Example 10.16 Use the Graphical Method to Design a Simple Voting Machine

A voting machine indicates that a measure passes whenever a majority of three voters approves the measure. A circuit for this simple machine should have three inputs and should produce the output 1 (the measure passes) whenever at least two of the three inputs are 1 (yes votes). (a) Use a truth-table to design a circuit that accomplishes this task. (b) Simplify this circuit using the graphical method.

Solution

(a) Table 10.15 shows the eight possible voting results, depending on the inputs x, y, z from the three voters. A corresponding logical term is placed in the four rows (A, B, C, E) where the measure passes.

TABLE 10.15: Specifications for a simple voting machine

	x	y	z	output	term	# of gates
A	1	1	1	1	xyz	2
B	1	1	0	1	$xy\bar{z}$	3
C	1	0	1	1	$x\bar{y}z$	3
D	1	0	0	0		
E	0	1	1	1	$\bar{x}yz$	3
F	0	1	0	0		
G	0	0	1	0		
H	0	0	0	0		

The following logical expression f (a disjunction of the terms in rows A, B, C, E) will then produce a 1 exactly when a majority of voters is in favor of the measure:

$$f = xyz + xy\bar{z} + x\bar{y}z + \bar{x}yz.$$

Of course, this does not lead to a very efficient circuit representation, as it requires a total of 14 gates. (Verify this!)

(b) To obtain a more economical circuit, we first construct a graph with vertices A, B, C, E. An edge is placed between two vertices whose bit strings differ in only one position, with that position used to label the edge. See Figure 10.14.

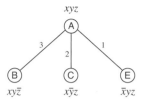

FIGURE 10.14: Graph for the simple voting machine

Note that this graph contains no squares, so any covering must use either vertices alone or edges. In order to cover B, it is best to use edge AB (xy); to cover C we use AC (xz); and to cover E we use AE (yz). Notice that A is then automatically covered, in fact multiple times. The covering using the three edges AB, AC, AE is shown in Figure 10.15. The resulting logical expression is $xy + xz + yz = x(y + z) + yz$. So, we need only 4 gates instead of the 14 gates required in (a). ◁

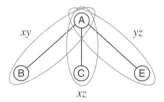

FIGURE 10.15: A covering of the graph

Example 10.17 Use the Graphical Method to Design a Car Alarm System

Design a circuit for a car alarm system that produces a buzzing sound only when certain conditions apply. Engineers have decided that the important variables are as follows: k indicates whether or not ($k = 1$ or $k = 0$) the key is in the ignition; d indicates whether or not ($d = 1$ or $d = 0$) the driver's door is shut; and b indicates whether or not ($b = 1$ or $b = 0$) the seat belt is engaged. The alarm system should buzz only under the following conditions: the key is in the ignition, the door is shut, but the seat belt is not engaged; the key is in the ignition, the seat belt is engaged, but the door is not shut; the key is not in the ignition and the seat belt is engaged. (a) Use a truth-table to design a circuit that accomplishes this task. How many gates will this circuit require? (b) Simplify this circuit using the graphical method. How many gates will this improved circuit require?

Solution

(a) Table 10.16 lists the four conditions A, B, C, D in which the buzzer should activate (output the value 1). It also shows the corresponding terms and the number of gates required for each term.

A logical expression f that will produce the output 1 only in rows A–D is then

$$f = kd\bar{b} + k\bar{d}b + \bar{k}db + \bar{k}\bar{d}b.$$

TABLE 10.16: Specifications for a car alarm system

	k	d	b	output	term	# of gates
A	1	1	0	1	$kd\bar{b}$	3
B	1	0	1	1	$k\bar{d}b$	3
C	0	1	1	1	$\bar{k}db$	3
D	0	0	1	1	$\bar{k}\bar{d}b$	4

Implementing this expression would require 16 gates: $3 + 3 + 3 + 4 = 13$ for the individual terms plus 3 gates to implement the disjunction of the four terms.

(b) The graph constructed in Figure 10.16 has vertices A, B, C, D and edges join two vertices whenever their bit strings differ in exactly one position (with that position used to label the edge).

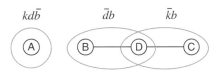

FIGURE 10.16: Graph for the car alarm system

Note that the graph contains no squares. Because vertex A is isolated, the only way to cover it is by using vertex A itself $(kd\bar{b})$. Now consider vertex B; it is best to cover it by using the edge BD $(\bar{d}b)$. Likewise, vertex C should be covered using the edge CD $(\bar{k}b)$. Automatically, vertex D is covered in the process. This produces the covering shown in Figure 10.17.

$$kd\bar{b} \qquad\qquad \bar{d}b \qquad\qquad \bar{k}b$$

$$(A) \qquad (B)\!-\!(D)\!-\!(C)$$

FIGURE 10.17: A covering of the graph

The corresponding logical expression is then

$$kd\bar{b} + \bar{d}b + \bar{k}b = kd\bar{b} + (\bar{d} + \bar{k})b,$$

which has been further simplified by using the distributive property. The final expression enables us to build a circuit that outputs the desired results, using 8 gates instead of the original 16 gates. In summary, this heuristic (graph covering) approach has enabled us to design a much more efficient circuit to implement the proposed car alarm system. ◁

(*Related Exercises: 36–38*)

Another example of the effectiveness of circuit minimization, and an associated applet, can be found at the site [OR1].

Chapter Summary

Conducting an accurate search of the Internet or designing economical computer circuits involves knowing when logical expressions are equivalent. To test equivalence, we employ a variety of representations such as symbolization, truth-tables, and Venn diagrams. Certain Rules of Replacement are valuable in simplifying logical expressions and in providing crucial steps in carrying out a deductive argument. To represent and manipulate digital circuits, we take advantage of a correspondence between logical variables and binary digits as well as between logical operations and gates of a circuit. The design of a complicated circuit can be accomplished by using truth-tables to form a logical expression that carries out the required specifications of the given circuit. Moreover, the simplification of such circuits can be viewed as a problem of simplifying logical expressions. This task is greatly aided by using a graph representation and then solving an intriguing type of graph covering problem.

Applications, Representations, Strategies, and Algorithms

Applications				
Technology (1, 11–15)		Society/Politics (16)		Transportation (17)

Representations					
Graphs		Tables		Sets	Symbolization
Weighted	Directed	Decision	Lists	Venn Diagrams	Logical
15–17	11–14	1–8, 13–17	9–10	8	1–17

Strategies		
Solution-Space Restrictions	Rule-Based	Composition/Decomposition
15–17	1–17	11–12, 14

Algorithms				
Exact	Approximate	Sequential	Logical	Geometric
1–14	15–17	1–10, 13–17	1–17	8

Exercises

Internet Searches

1. Suppose you want to find information on the use of the software language Python outside of Silicon Valley. Determine whether the following expressions are equivalent: "Python and not Silicon Valley," "not (not Python or Silicon Valley)."

   (a) Symbolize the two expressions. (Let p = Python, q = Silicon Valley.)

(b) Using a truth-table, determine whether they are equivalent.

2. Suppose you want to find information on Toyotas that are not hybrids. Determine whether the following expressions are equivalent: "Toyotas and not Hybrids," "not (Toyotas or Hybrids)."

   (a) Symbolize the two expressions. (Let p = Toyotas, q = Hybrids.)

   (b) Using a truth-table, determine whether they are equivalent.

3. Suppose you want to find information on the budget negotiations being conducted separately by the President and by the House Republicans. You decide to type in the query: "(President and not House Republicans) or (Budget and House Republicans)". Would the following be an equivalent way of obtaining the same results: "(President or (Budget and House Republicans)) and (Budget or not House Republicans)"?

   (a) Symbolize the two expressions. (Let p = President, q = Budget, r = House Republicans.)

   (b) Using a truth-table, determine whether they are equivalent.

Equivalences

4. Use a truth-table to verify the equivalence of $(p \vee q) \wedge p$ and p.

5. Use a truth-table to verify the equivalence of $(p \vee \sim q)$ and $\sim(\sim p \wedge q)$.

6. Use a truth-table to verify the equivalence of $\sim p \rightarrow (\sim q \rightarrow p)$ and $p \vee q$.

7. Use a truth-table to verify the equivalence of $(p \vee q) \wedge (p \vee r)$ and $p \vee (q \wedge r)$.

8. Use a truth-table to verify the equivalence of $(p \vee q) \wedge (r \vee \sim p)$ and $(p \wedge r) \vee (q \wedge (r \vee \sim p))$.

9. Use a truth-table to evaluate $(p \wedge q) \vee (p \wedge \sim q)$. What does this expression simplify to?

10. Determine if the following are equivalent: $p \rightarrow (q \vee r)$, $(p \rightarrow q) \vee (p \rightarrow r)$.

11. Determine if the following are equivalent: $p \rightarrow (q \wedge r)$, $(p \rightarrow q) \wedge (p \rightarrow r)$.

12. Determine if the following are equivalent: $p \rightarrow (q \rightarrow r)$, $(p \wedge q) \rightarrow r$.

13. Determine if the following are equivalent: $p \rightarrow (q \rightarrow r)$, $(p \rightarrow q) \rightarrow r$.

Conditional Statements

14. Find the contrapositive, inverse, and converse of the conditional statement "If Rita gets an A on the midterm, then she will be exempt from the final exam."

15. Find the contrapositive, inverse, and converse of the conditional statement "If Smith is elected mayor, then he will resign from the county council."

16. Find the contrapositive, inverse, and converse of the conditional statement "If it does not snow, then the hike will take place." Simplify your answers as appropriate.

17. Find the contrapositive, inverse, and converse of the conditional statement "If Brandon does not pass the English course, then he will not graduate." Simplify your answers as appropriate.

Venn Diagrams and Truth-Tables

18. James asserts that "If you have the password, you can access the website." His friend Seymour interprets this as meaning "If you can't access the website, you don't have the password," while another friend Felix asserts that it means "If you don't have the password, you can't access the website." Which friend is in complete agreement with James?

    (a) Symbolize all three statements.

    (b) Draw a Venn diagram to represent each of the assertions, and then answer the question posed.

    (c) Use a truth-table to represent all three statements, and then answer the question posed.

19. Monica asserts that "If the Braves don't win this game, they will not be in the play-offs." Her friend Dana interprets this as "If the Braves win this game, then they will be in the playoffs," while another friend Katelyn asserts that it means "If the Braves are in the playoffs, then they won this game." Which friend is in complete agreement with Monica?

    (a) Symbolize all three statements.

    (b) Draw a Venn diagram to represent each of the assertions, and then answer the question posed.

    (c) Use a truth-table to represent all three statements, and then answer the question posed.

Rules of Replacement

20. Use the Rules of Replacement to show that the logical expression $\sim(\sim p \vee q) \vee \sim(p \vee q)$ can be simplified to $\sim q$.

21. Use the Rules of Replacement to show that the logical expression $\sim(p \wedge \sim q) \wedge (p \vee \sim q)$ can be simplified to $(p \wedge q) \vee (\sim p \wedge \sim q)$.

22. Use the Rules of Replacement to show that the logical expression $\sim(\sim p \wedge q) \wedge (p \vee \sim r)$ can be simplified to $p \vee \sim(q \vee r)$.

Logical Expressions from Circuits

23. Represent the following circuit as a logical expression with inputs p, q and then evaluate the circuit using $p = 1$, $q = 0$.

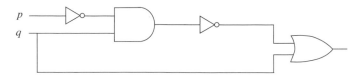

24. Represent the circuit below as a logical expression with inputs p, q and then evaluate the circuit using $p = 0$, $q = 1$.

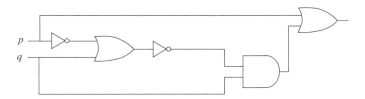

25. Represent the circuit below as a logical expression with inputs p, q, r and then evaluate the circuit using $p = 0$, $q = 1$, $r = 1$.

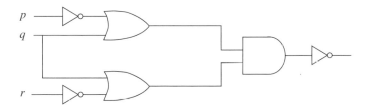

26. Represent the circuit below as a logical expression with inputs p, q, r and then evaluate the circuit using $p = 1$, $q = 0$, $r = 0$.

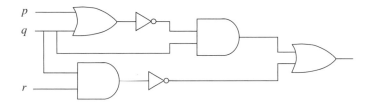

Circuits from Logical Expressions

27. Design a circuit corresponding to the logical expression $\sim(p \wedge q) \vee q$.

28. Design a circuit corresponding to the logical expression $(p \vee \sim q) \wedge (\sim p \vee q)$.

29. Design a circuit corresponding to the logical expression $(\sim p \wedge q) \vee (p \wedge \sim q)$.

30. Design a circuit corresponding to the logical expression $\sim(p \wedge \sim q) \wedge \sim(q \vee r)$.

31. Design a circuit corresponding to the logical expression $(\sim p \wedge q) \vee (\sim q \vee r)$.

32. Design a circuit corresponding to the logical expression $((\sim p \wedge q) \vee r) \wedge (q \vee \sim r)$.

Circuit Design and Simplification

33. The reduced logical expression $(x \wedge (y \vee z)) \vee (y \wedge z)$ was obtained in Example 10.16. Use a truth-table to verify that this expression satisfies the requirements of the circuit design: namely, that it evaluates to 1 only for the four input bit strings A, B, C, E.

34. The reduced logical expression $(w \wedge y \wedge (\sim z \vee \sim x)) \vee (\sim w \wedge z)$ was obtained in Example 10.15. Use a truth-table to verify that this expression satisfies the requirements of the circuit design: namely, that it evaluates to 1 only for the seven input bit strings A–G.

35. We would like to design a circuit that unlocks a mobile device (outputs a 1) when one or both of its two inputs x, y is 1 (on). That is, the specifications ask that the circuit output 1 when either x or y or both are 1, and output 0 when they are not.

    (a) Use a truth-table to find a logical expression that satisfies these specifications.

    (b) Draw the corresponding circuit and count the number of gates used.

    (c) Construct the associated graph and produce a simplified logical expression for the circuit. How many gates does your simplified design require?

36. We want to design a circuit that produces the output 1 for exactly the inputs A–F shown in the following table.

	w	x	y	z	output	term	# of gates
A	0	0	0	1	1		
B	0	0	1	0	1		
C	0	0	1	1	1		
D	1	0	0	1	1		
E	1	0	1	1	1		
F	1	1	0	1	1		

    (a) Fill in the last two columns of this table and then construct the associated logical expression. How many gates are required for a circuit that implements this logical expression?

    (b) Draw the graph representation and produce a simplified logical expression for the circuit. How many gates does your simplified design require?

37. We wish to design a voting machine that will accept four inputs w, x, y, z representing the votes of four members of a city council. A measure will pass if it receives at least three affirmative (yes) votes.

    (a) Using a truth-table, construct a logical expression that will output 1 (yes) only under the desired conditions. How many gates does the corresponding circuit require?

    (b) Draw the graph representation and produce a simplified logical expression for the circuit. How many gates does your simplified design require?

38. Design a circuit for a home alarm system that enables a siren to go off (output is 1) only in the following six situations.

    A. The front door is closed ($w = 1$), the back door is closed ($x = 1$), the front and back windows are open ($y = 0$), and the side windows are closed ($z = 1$).

B. The front door is open, the back door is open, the front and back windows are open, and the side windows are closed.

C. The front door is open, the back door is closed, the front and back windows are closed, and the side windows are closed.

D. The front door is closed, the back door is open, the front and back windows are open, and the side windows are closed.

E. The front door is open, the back door is open, the front and back windows are closed, and the side windows are closed.

F. The front door is open, the back door is closed, the front and back windows are open, and the side windows are closed.

(a) Use a truth-table to find a logical expression that satisfies these specifications.

(b) Determine the total number of gates required for the associated circuit.

(c) Draw the graph representation and produce a simplified logical expression for the circuit. How many gates does your simplified design require?

Project

39. Develop a spreadsheet that will evaluate a given logical expression and that can then be used to verify whether two expressions are equivalent. For this purpose, represent the truth-values T and F by the binary digits 1 and 0, respectively. Notice that $p \vee q$ can be implemented as $\text{MAX}(p, q)$, while $p \wedge q$ can be implemented as $\text{MIN}(p, q)$. Also, the negation of p is found as $1 - p$, and the conditional $p \to q$ can be implemented using the logically equivalent representation $\sim p \vee q$.

Bibliography

[Kohavi 2010] Kohavi, Z., and Jha, N. 2010. *Switching and finite automata theory*, Chapters 3 and 4. Cambridge: Cambridge University Press.

[Parks 2014] Parks, W. 2014. *An introduction to boolean algebra and switching circuits*. Hershey, PA: William R. Parks.

[OR1] http://babbage.cs.qc.edu/courses/Minimize/ (accessed April 22, 2014).

Chapter 11

Modeling Using Deductive Logic

Before I proceed further I will make some experiments, because it is my intention
to cite the experiment first and then to demonstrate by reasoning how such an
experiment must necessarily take effect in such a manner.

—Leonardo Da Vinci, Italian artist and inventor, 1452–1519

Chapter Preview

Deductive logic provides tools for modeling a variety of applications and for finding solutions
to problems involving optimal decisions. Specifically, this chapter will illustrate how logical
expressions can be used to model decisions involved in city planning and to model the
evolution of a biological system over time. In addition, an extension of deductive logic,
called game theory, can be used to model situations of competition or conflict in which we
seek a rationally defensible strategy in the face of uncertain actions by an adversary. Certain
paradoxes that seem to defy rational behavior are discussed, and we see how reliance on
chance mechanisms can actually aid in making optimal decisions in adversarial situations.

11.1 A City Planning Problem

In Chapter 10, we discussed some well-known equivalences called the Rules of Replacement.
These are listed for easy reference in Table 11.1. Such equivalences can be used to simplify
logical expressions that arise in a variety of applications and may thereby result in savings
of time or money. We will next discuss how logical expressions can be used to model a
particular application that arises in the optimal location of emergency facilities.

Locating Fire Stations

The city of Centerville consists of five different residential zones. Each of these zones is a
potential location for one of several planned fire stations; together these new fire stations
will serve the entire city. It would be too expensive to build five new fire stations, one in
each zone. So the city planner would like to consider all possible ways of efficiently placing
fire stations in order to serve all the zones of the city. We will see that this is yet another
type of *covering* problem, somewhat different from the covering problem encountered in the

TABLE 11.1: Rules of replacement

Rule of Replacement	Terminology
$p \vee (p \wedge q) \Leftrightarrow p$ $p \wedge (p \vee q) \Leftrightarrow p$	Absorption (ABS)
$p \vee (q \vee r) \Leftrightarrow (p \vee q) \vee r$ $p \wedge (q \wedge r) \Leftrightarrow (p \wedge q) \wedge r$	Associative (ASSOC)
$p \vee q \Leftrightarrow q \vee p$ $p \wedge q \Leftrightarrow q \wedge p$	Commutative (COM)
$\sim\sim p \Leftrightarrow p$	Double Negation (DN)
$p \wedge (q \vee r) \Leftrightarrow (p \wedge q) \vee (p \wedge r)$ $p \vee (q \wedge r) \Leftrightarrow (p \vee q) \wedge (p \vee r)$	Distributive (DIST)
$\sim(p \vee q) \Leftrightarrow \sim p \wedge \sim q$ $\sim(p \wedge q) \Leftrightarrow \sim p \vee \sim q$	De Morgan (DM)

efficient design of computer circuits (Section 10.4). The next three examples show how we can model, analyze, and solve the city planner's problem.

Example 11.1 Model the Location of Fire Stations to Serve All Zones of a City

(a) Construct a graph representation to model the zones of the city as well as the possible placements of fire stations. (b) Then use logical variables to design an expression that evaluates to 1 whenever a placement of fire stations is acceptable (all zones are covered) and 0 otherwise.

Solution

(a) We construct an undirected graph in which each of the zones A, B, C, D, and E is represented by a vertex. An edge is placed between two zones whenever they are close enough to one another that a fire station located in one zone can serve the other zone. The *proximity graph* corresponding to the five zones of Centerville is displayed in Figure 11.1.

FIGURE 11.1: Graph indicating proximities of the five zones

(b) Corresponding to each vertex will be a logical variable (0/1) that indicates the absence or presence of a fire station in that zone. Note that in order to cover zone A, a fire station needs to be placed at A or at one of the adjacent vertices: that is, a fire station needs to be placed at A or at B or at C, which we represent by the logical expression $a \vee b \vee c$. A similar disjunction can be constructed for the remaining zones. To cover zone B we require a station at A, B, C, or D; to cover zone C we require a station at A, B, C, D, or E; to cover zone D we require a station at B, C, D, or E; and to cover zone E we require a station at C, D, or E. This gives the five disjunctive terms $(a \vee b \vee c)$, $(a \vee b \vee c \vee d)$, $(a \vee b \vee c \vee d \vee e)$, $(b \vee c \vee d \vee e)$, and $(c \vee d \vee e)$.

Since we need all zones to be covered, we form the conjunction of the five individual terms to form the single logical expression

$$f = (a \vee b \vee c) \wedge (a \vee b \vee c \vee d) \wedge (a \vee b \vee c \vee d \vee e) \wedge (b \vee c \vee d \vee e) \wedge (c \vee d \vee e).$$

This will evaluate to 1 only under the prescribed conditions. For example, the placement of fire stations at A and E is acceptable (all zones will be covered), and we likewise see that f evaluates to 1 since

$$(1 \vee 0 \vee 0) \wedge (1 \vee 0 \vee 0 \vee 0) \wedge (1 \vee 0 \vee 0 \vee 0 \vee 1) \wedge (0 \vee 0 \vee 0 \vee 1) \wedge (0 \vee 0 \vee 1) = 1 \wedge 1 \wedge 1 \wedge 1 \wedge 1 = 1.$$

On the other hand, the placement of fire stations at A and B is not acceptable since the fifth term in f will be $0 \vee 0 \vee 0 = 0$, meaning that the conjunction f will evaluate to 0. ◁

Example 11.2 Use the Absorption Rule to Simplify a Conjunction

By applying the Absorption Rule from Table 11.1, simplify the conjunction of the first two terms in the expression f obtained in Example 11.1.

Solution

The Absorption Rule enables us to replace $p \wedge (p \vee q)$ by the simpler expression p. Applying this to the first two terms in the expression f means that $(a \vee b \vee c) \wedge (a \vee b \vee c \vee d)$ can be replaced by the equivalent but simpler $(a \vee b \vee c)$. It is convenient here to highlight in color those common terms that are being combined. ◁

Example 11.3 Use the Rules of Replacement to Simplify a Logical Expression

Use the Rules of Replacement to transform the expression f into a simpler equivalent expression and then use this simplified expression to obtain efficient solutions to the city planner's problem.

Solution

Below we justify each step by using one of the Rules of Replacement applied to the previous step. Again we highlight in color the terms combined using the Absorption Rule.

1. $(a \vee b \vee c) \wedge (a \vee b \vee c \vee d) \wedge (a \vee b \vee c \vee d \vee e) \wedge (b \vee c \vee d \vee e) \wedge (c \vee d \vee e)$ Given
2. $(a \vee b \vee c) \wedge (a \vee b \vee c \vee d \vee e) \wedge (b \vee c \vee d \vee e) \wedge (c \vee d \vee e)$ 1, ABS
3. $(a \vee b \vee c) \wedge (a \vee b \vee c \vee d \vee e) \wedge (c \vee d \vee e) \wedge (c \vee d \vee e \vee b)$ 2, COM
4. $(a \vee b \vee c) \wedge (a \vee b \vee c \vee d \vee e) \wedge (c \vee d \vee e)$ 3, ABS
5. $(a \vee b \vee c) \wedge (a \vee b \vee c \vee (d \vee e)) \wedge (c \vee d \vee e)$ 4, ASSOC
6. $(a \vee b \vee c) \wedge (c \vee d \vee e)$ 5, ABS
7. $(c \vee a \vee b) \wedge (c \vee d \vee e)$ 6, COM
8. $c \vee [(a \vee b) \wedge (d \vee e)]$ 7, DIST
9. $c \vee [((a \vee b) \wedge d) \vee ((a \vee b) \wedge e)]$ 8, DIST
10. $c \vee [(a \wedge d) \vee (b \wedge d) \vee (a \wedge e) \vee (b \wedge e)]$ 9, DIST
11. $c \vee (a \wedge d) \vee (b \wedge d) \vee (a \wedge e) \vee (b \wedge e)$ 10, ASSOC

Step 11 shows a simplified expression that is equivalent to the original expression. It provides an enumeration of five different ways to place a minimal number of fire stations covering all five zones: namely, {C}, {A, D}, {B, D}, {A, E}, {B, E}. For example, placing fire stations at locations A and D will cover all five zones, but no smaller subset of {A, D} will do the trick. Even though placing one fire station at location C will cover all zones, it might be very costly to acquire land or do construction in that zone. So it might be better to place fire stations at several locations. Consequently, we provide the decision maker (the city planner) with five efficient options, which can then be evaluated based on additional financial and political criteria. ◁

(Related Exercises: 1–7)

11.2   Gene Regulatory Models

We discuss here a very simple model of a chemical reaction that has been used to study the transport and metabolism of lactose (found in dairy products) in certain bacteria. This simple gene regulatory mechanism is fundamental to cellular regulation for all organisms. First, however, we introduce some background information on biochemical processes.

All living things must have an unceasing supply of energy and matter. Transformation of this energy and matter within the body is called metabolism. Various chemical reactions are the basis of metabolism and these involve the production of proteins. The production of essential proteins can be turned on and off by genetic mechanisms as follows: If lactose (L) is present, then a certain metabolite called allolactose (A) binds to a repressor site and activates a certain genetic region (M) which produces protein (P) for transport into the cell and protein (B) for the processing of the lactose into sugars (such as allolactose).

Example 11.4 Model a Gene Regulatory Process Using Logical Expressions

The sequence of actions in this regulatory process can be modeled using $(0/1)$ variables, where 1 indicates whether L, A, M, P, or B is present and active, while 0 indicates otherwise. Translate the following biological constraints into appropriate logical expressions.

1. M is activated only when A is present; thus, the value of M at a time step is the value of A at the previous time step.
2. Once M is activated, the proteins B and P can be produced, so the value of B as well as that of P is the value of M at the previous time step.
3. Allolactose (A) will be present at a time step if it was already present at the previous time step, or if lactose (L) was available and the protein B was present to convert it to A.
4. Lactose will be present if protein P is available for transport or if lactose is already present in the cell and has not been converted into sugars by protein B.

Solution

We can write equations that relate the values of logical variables at one step to the values of logical variables at the next time step.

1. $M = A$
2. $B = M$, $P = M$
3. $A = A \vee (L \wedge B)$
4. $L = P \vee (L \wedge \sim B)$

In the above equations, the quantity determined on the right-hand side is used to update the variable on the left-hand side at the next time step. ◁

Example 11.5 Predict How a Gene Regulatory System Evolves Over Time

The state of the system at any time is represented by the values taken on by M, B, A, L, and P, respectively. Suppose that the initial values are given by the state $(1, 1, 0, 0, 1)$. (a) Track the state of the system at successive time steps by applying the equations developed in Example 11.4. (b) Does the system reach a "steady state" in which the state does not further change with time?

Solution

(a) Table 11.2 lists the given logical equations. The initial state of the system is represented by the values $(1, 1, 0, 0, 1)$ placed in the first column. These values will evolve at the next time step by applying the logical equations and replacing the current state values with the updated state values. For example, the first row of the table corresponds to the evolution of variable M through time. Initially, M = 1. Since M = A, the value of M at the second step (the entry in Row 1, Column 2) will be 0 because the value of A at the first step (Row 3, Column 1) is 0. On the other hand, the value of A at step 2 is given by inserting the values for A, L, and B at the first step into $A \vee (L \wedge B)$, giving $0 \vee (0 \wedge 1) = 0 \vee 0 = 0$. The remaining entries in the second column are determined in a similar way, giving $(0, 1, 0, 1, 1)$ as the state of the system at step 2. The next states of the system are found to be $(0, 0, 1, 1, 0)$, $(1, 0, 1, 1, 0)$, $(1, 1, 1, 1, 1)$, and $(1, 1, 1, 1, 1)$.

TABLE 11.2: Evolution of state variables for a biological system

	First State	2nd	3rd	4th	5th	6th
M = A	M = 1	0	0	1	1	1
B = M	B = 1	1	0	0	1	1
A = A $\vee$ (L $\wedge$ B)	A = 0	0	1	1	1	1
L = P $\vee$ (L $\wedge$ ~B)	L = 0	1	1	1	1	1
P = M	P = 1	1	0	0	1	1

(b) Notice that starting with the sixth time step, the states keep repeating forever and so this represents an *equilibrium* state reached by the system, when started in state $(1, 1, 0, 0, 1)$. In other words, when the system reaches the state $(1, 1, 1, 1, 1)$ it remains there forever. In the next sections of this chapter, we will again encounter the concept of an equilibrium solution. ◁

(Related Exercises: 8–10, 33)

11.3 Rational Decision Making in Competitive Situations

An extension of deductive logic, called *Game Theory*, studies rational decision making in situations involving competition or conflict. The objective is to develop rational and optimal strategies for each person involved in such situations. More specifically, game theory examines the abstract properties of decision-making scenarios and quantifies the various outcomes. It has applications to games like poker and baseball, as well as to the social sciences, such as psychology, economics (market competition), and political science (arms races).

DEFINITION **Game Theory**

Game theory investigates situations involving competition or conflict and it provides insight into strategies available to the various players in order to make rational and optimal decisions.

HISTORICAL CONTEXT **John von Neumann (1903–1957)**

Considered the founder of game theory, he wrote the influential book *Theory of Games and Economic Behavior* (1944) in collaboration with the mathematical economist Oskar Morgenstern.

Games can be distinguished as being *deterministic* or *strategic*.

DEFINITION **Deterministic Game**

Players take turns in a specified manner and choose among legal moves in order to try to win the game. There is no secrecy involved, as players know each other's moves. (Chess, Checkers, Go, and Tic-Tac-Toe are examples of deterministic games.)

DEFINITION **Strategic Game**

Each player privately selects a move from a defined set of options. Outcomes are determined by the moves selected by these players, often taken simultaneously, without knowledge of the opponent's move.

In Section 2.3 we discussed the deterministic game *Thai 21*, in which players alternate in selecting a certain number of flags. It turned out to be advantageous to go first in this game, so you can always leave your opponent in a losing state (a multiple of four remaining flags). Here is another game that can be analyzed to determine a winning strategy.

Example 11.6 Analyze a *Nim* Game With Two Piles

Nim is a two-person game in which there are several piles of beads. The players alternate by removing one or more beads from a single pile. The player who removes the last bead is declared the winner. In the simplified version considered here, there are just two piles of beads with different sizes.

Let's designate the two players as FP (the player going first) and SP (the player going second). Consider the following "even-the-piles" strategy: namely, when it is your turn, remove enough beads from the larger pile so that its new size matches that of the smaller pile. Will either FP or SP be guaranteed a win using this strategy?

Solution

Consider the following sequence of moves:

> FP: Remove beads from the larger pile to make it the same size as the smaller pile.
> SP: Create piles of different sizes by removing some (or possibly all) beads from one pile.
> FP: Remove beads from the larger pile to make it the same size as the smaller pile.
> SP: Create piles of different sizes by removing some (or possibly all) beads from one pile.
>
> . . .

Notice that the game will eventually end since the total number of beads strictly decreases at each step. If the smaller pile ever contains no beads, FP uses the "even-the-piles" strategy to remove all beads in the remaining pile, and so wins. In other words, FP is guaranteed to win using this strategy in this simplified game of *Nim*, assuming we begin with piles of unequal size. Alternatively, if we begin with piles of equal size, then SP can adopt the "even-the-piles" strategy to assure an eventual win. ◁

There are more challenging versions of *Nim*, involving multiple piles and possibly restrictions on the number of beads that can be selected from each pile. Interestingly enough, it is possible to establish mathematically that for any *Nim* game, there is always a winning player and a winning strategy. We next illustrate another *Nim* game, now played with three piles of beads.

Example 11.7 Analyze a *Nim* Game With Three Piles

A *Nim* game is played with three piles, containing $2, 2, 1$ beads, respectively. Players take turn removing one or more beads from a single pile. The player removing the last bead is declared the winner. As in Example 11.6, the player going first is denoted FP and the one going second is denoted SP. (a) Construct a directed graph showing transitions between the states of the system: the number of beads remaining in each nonempty pile (without regard to any order of the piles). (b) Use this graph representation to design an optimal strategy for one of the players. Which player is guaranteed to win by using this strategy?

Solution

(a) Figure 11.2 shows the possible transitions from the initial state 221 after the players alternate making moves. In general, the state of the system is the configuration presented to a player who is about to make a move. For example, FP is presented with state 221 and can (1) remove one bead from a pile with two beads, producing state 211; (2) remove a bead from a pile with one bead, producing state 22; or (3) remove two beads from a pile with two beads, producing state 21. The graph is arranged by levels, depending on whether FP or SP is to make the next move.

(b) The shaded ovals in Figure 11.2 represent no beads remaining, so these are clearly losing (*L*) states if one is presented with this state—your opponent has just removed the last bead and won the game! Similar to the *Thai 21* game in Example 2.5, we can classify each state by looking at all the states adjacent from that state by a directed edge. If there is at least one state labeled *L* adjacent to the current state, then the current state is labeled *W*; this is justified since we can make a move from the current state to send the opponent into a losing state. For example, state 21 is classified as *W* since its adjacent states 11, 2, 1 are labeled *L, W, W*, respectively. On the other hand, state 22 is classified as *L* since its adjacent states 21 and 2 are labeled *W* and *W*, respectively. By working upward from the shaded ovals, the initial state 221 is eventually labeled *W* and so FP has a winning strategy. Namely, FP should remove one bead from the pile with one bead, producing state 22. If SP moves to state 2 (by removing two beads from a pile), then FP should remove two beads from the remaining pile with two beads, thereby winning. Exercise 11 asks you to describe the optimal strategy for FP if alternatively SP removes one bead from one of the two piles with two beads, producing state 21. ◁

(Related Exercises: 11–14)

The remainder of this chapter will introduce the basic principles of game theory with regard to strategic games, where one player does not know in advance the choice made by the

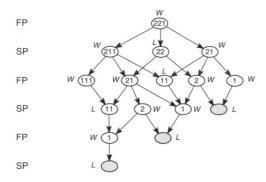

FIGURE 11.2: Graphical representation of a *Nim* game with three piles

other player. To begin, we present an example that challenges our intuition about rational behavior.

Prisoner's Dilemma and Nash Equilibria

Consider the following situation:

> Two suspects are captured by the police near the scene of a burglary and each has to choose whether to confess and implicate the other. Neither suspect is able to communicate with the other. If neither man confesses, then both will serve one year on a charge of carrying a concealed weapon. If both confess and implicate each other, they will each go to prison for eight years. However, if one confesses and implicates the other, while the other does not confess, the one who collaborates with the police will go free, while the other burglar will go to prison for 15 years on the maximum charge.

What is the best course of action for the suspects? We model this situation as a two-person (or two-player) game. Each player here has two *strategies*: either confess or don't confess. Depending on which strategies are selected by the two players, there will be certain *payoffs* to each player. The possible strategies and corresponding payoffs can be assembled in a *payoff matrix*, as illustrated in the following example.

DEFINITION **Strategies**
The courses of action available to a player.

DEFINITION **Payoffs**
Numbers attached to the outcomes of a game. Payoffs may represent profit, loss, utility, or may simply rank the desirability of outcomes to the players.

Example 11.8 Construct the Payoff Matrix for the Prisoner's Dilemma

Construct the payoff matrix for the two suspects, where each suspect has two strategies. The resulting payoffs (penalties, actually) to each player will be the sentences received, in years to be served.

Solution

In the payoff matrix shown in Table 11.3, one person is termed the Row Player (RP) and the other is the Column Player (CP). Each player has two strategies: either confess or not. Each entry in the matrix contains two numbers; the first number gives the payoff to RP while the second number gives the payoff to CP. For example, if RP confesses but CP does not, then RP receives 0 years while CP receives 15 years. ◁

TABLE 11.3: Payoff matrix for the Prisoner's Dilemma

		CP	
		confess	don't
RP	confess	$(8, 8)$	$(0, 15)$
	don't	$(15, 0)$	$(1, 1)$

Example 11.9 Analyze the Payoff Matrix for the Prisoner's Dilemma

Given the payoff matrix shown in Table 11.3, what should the suspects do, assuming they are acting rationally?

Solution

Rationality dictates that one should minimize the time spent in prison. Let's first analyze the thinking process of RP, based on the highlighted (first) payoffs in each entry of Table 11.3.

> If CP confesses, then it is better for me to confess (getting 8 years) than not to confess (getting 15 years).
> On the other hand, if CP does not confess, then it is better for me to confess (getting 0 years) than not to confess (getting 1 year).
> So regardless of what CP does, it is best if I confess.
> Therefore, I'll confess.

Of course CP can reason in the exactly same way (using the second payoff in each entry) and will therefore reach the same "rational" conclusion—to confess. ◁

Rationality then leads both suspects to confess and so both will go to prison for eight years. Yet, if they had acted "irrationally" and neither confessed, each would spend just one year in prison, a clearly better outcome for both. The source of this seeming paradox is that the pair of strategies (confess, confess) represents an *equilibrium position*—there is no incentive for either player to deviate unilaterally from this position. For example, if RP deviates from (confess, confess) to (don't confess, confess) then his payoff increases from eight years to 15 years. The same reasoning applies to CP. In other words, both players are stuck at the payoff entry $(8, 8)$ in the matrix. Notice that even if the players agreed beforehand to both use the strategy "don't confess", resulting in the payoff entry $(1, 1)$, then there would be an incentive for either player to defect from this position: for example, RP could reduce his sentence from 1 year to 0 years. We see that the strategy pair (don't confess, don't confess) is an unstable position.

In summary, rationality leads both suspects to a *Nash equilibrium* position in the payoff matrix. Namely, both confess and receive an eight-year sentence, even though an enforceable agreement for both not to confess would be a much better outcome for both.

DEFINITION **Nash Equilibrium**
A pair of strategies in which there is no incentive for either of the players to change his strategy unilaterally without hurting himself. Games can possess several Nash equilibria.

HISTORICAL CONTEXT **John F. Nash, Jr. (1928–)**
American mathematician who won the 1994 Nobel Prize in Economics for his work in game theory. (S. Nasar's biography of Nash, *A Beautiful Mind*, was made into a film that won the 2002 Academy Award for Best Picture.)

The result of the Prisoner's Dilemma—that individual rational decisions can result in both players being worse off in terms of their own self-interest—has had a large impact in the social sciences as there are many interactions (arms races, pollution reduction, price wars, congressional stalemates) that are similar in structure to the Prisoner's Dilemma. Cooperation would be preferable, but one needs external enforceable agreements to achieve this.

Example 11.10 Analyze a Congressional Vote

House Democrats and Republicans are about to vote on passing a continuing budget resolution. The strategists for both parties have determined that if both parties vote to pass the bill, then each will gain 10 points in their popularity; however, if both vote to defeat the bill, then they will only gain 5 points in popularity. On the other hand, if one party votes for the bill and the other votes against, then the party voting in favor will gain 3 points, while that voting against will gain 12 points (since they will be viewed as being more fiscally responsible). (a) Construct the payoff matrix for this situation. (b) Analyze the resulting two-person game.

Solution

(a) Each player (R = Republican, D = Democrat) has two strategies: vote for the bill or vote against it. Table 11.4 summarizes the payoffs to the respective players, where the first entry in each cell is the payoff to the Row Player R and the second is the payoff to the Column Player D.

TABLE 11.4: Payoff matrix for a Congressional vote

		D for	D against
R	for	$(10, 10)$	$(3, 12)$
	against	$(12, 3)$	$(5, 5)$

(b) Let's first consider the situation from the point of view of R. If D votes for the bill, then R receives, respectively, either 10 or 12 points, so strategy 2 (vote against) is preferred in this case. If D votes against the bill, then R receives, respectively, either 3 or 5 points, so again strategy 2 (vote against) is preferred. In either case then, it is rational for R to vote against the continuing resolution. By a symmetric argument, it is rational for D to vote against the bill. So rationality leads both players to the payoff entry $(5, 5)$, which represents a Nash equilibrium. However, they would both be better off at the entry $(10, 10)$, which unfortunately is not a stable solution. ◁

(*Related Exercises: 15–16, 34*)

Example 11.11 Determine All Nash Equilibria in a Game

In the payoff matrix shown in Table 11.5, RP has two strategies R1 and R2, while CP has three strategies C1, C2, and C3. Both players must make their decision in isolation and they want to maximize their resulting payoff (they prefer larger numbers). The first number in each pair is the payoff (profit) for RP while the second number is the payoff (profit) for CP. Determine any Nash equilibria for this game.

TABLE 11.5: Another payoff matrix

		CP		
		C1	C2	C3
RP	R1	$(10, 8)$	$(11, 7)$	$(3, 5)$
	R2	$(8, 3)$	$(7, 5)$	$(4, 6)$

Solution

In this example, Nash equilibria occur at both (R1, C1) and at (R2, C3). If a player deviates unilaterally from either of these entries, then that player will be worse off. For example, consider (R1, C1). If RP defects to R2, then RP's payoff decreases from 10 to 8. Likewise if CP defects from C1 to either C2 or C3, then CP's payoff decreases from 8 to either 7 or 5. A similar analysis applies to (R2, C3). Notice that (R1, C2) is not a Nash equilibrium: even though there is no incentive for RP to defect from this position, CP can increase the payoff from 7 to 8 by defecting from C2 to C1. ◁

(Related Exercises: 17–22)

11.4 Zero-Sum Games with Saddlepoints

In the Prisoner's Dilemma, there was an opportunity for both players to gain (if both choose "don't confess"). By contrast, we now study situations in which the players' interests are completely opposed—whatever one player wins, the other loses and vice versa. Such games are called *zero-sum* because the sum of the payoffs for each outcome is always zero: if I receive \$10, then my opponent receives −\$10 (loses \$10), and $10 + (-10) = 0$. Examples of zero-sum games involving two players occur in chess, poker, parlor games, and battles for TV viewership.

DEFINITION **Zero-Sum Game**
A two-person competitive game in which for each pair of strategies the amount that one player wins is the amount that the other player loses.

In zero-sum games, our convention is that larger payoffs are more desirable; they can be viewed as amounts gained. Moreover, since what player RP wins the other player will lose, we simply record the payoffs to RP in the payoff matrix; each payoff to CP will be the negative of that stated for RP. From the point of view of RP, we can formulate a conservative type of strategy, called a *maximin strategy*, that guarantees a certain payoff regardless of what CP does. If CP does not act rationally, RP might even do better (receive a higher payoff). This is termed a maximin strategy since RP will *maximize* the worst of the outcomes (*smallest* RP payoff) that might occur from CP's choice of strategy. Because payoffs are given in terms of what RP gains, a rational strategy for CP will be a *minimax strategy*: CP seeks to *minimize* the worst outcome (*maximum* payoff to RP) that might occur.

If these maximin and minimax strategies result in the same numerical outcome for both players, the game is said to have a *saddlepoint*, which corresponds to a Nash equilibrium in nonzero-sum games.

DEFINITION **Maximin Strategy**
Choose the best (largest) of the worst (smallest) options. The Row Player is guaranteed to receive *at least* the maximin value.

DEFINITION **Minimax Strategy**
Choose the best (smallest) of the worst (largest) options. The Column Player is guaranteed to pay out *at most* the minimax value.

DEFINITION **Saddlepoint**
A pair of strategies for the two players that represents a Nash equilibrium: neither player can improve by unilaterally departing from his or her strategy. A saddlepoint exists if the maximin value equals the minimax value, which is then termed the *value* of the game.

Example 11.12 Determine Maximin and Minimax Strategies for a Marketing Game

Consider the following scenario:

> Two soft-drink companies, Pepsi and Coke, can fund only one of three different advertising campaigns. Each decides which campaign to fund based on an analysis of their market share relative to their competitor. Pepsi's three alternative campaigns involve Pepsi Vanilla, Pepsi Lime, and Pepsi Wild Cherry. Coke's three alternative campaigns involve Lemon Coke, Raspberry Coke, and Cherry Coke. In the payoff matrix given in Table 11.6, Pepsi is designated the Row Player and Coke is the Column Player. Payoffs (percent increase or decrease in market share) are listed for Pepsi, which desires larger payoff values.

(a) Find the maximin and minimax strategies for Pepsi and Coke, respectively. (b) Then determine if this game possesses a saddlepoint.

TABLE 11.6: Payoff matrix for a marketing game

		Coke		
		LC	RC	CC
	PV	6	1	−3
Pepsi	PL	3	2	5
	PWC	−1	0	4

Solution

(a) Let's first focus on Pepsi (RP). If Pepsi chooses PV, then the worst outcome occurs if Coke chooses CC, giving Pepsi −3. However, if Pepsi chooses PL, then the worst outcome for Pepsi is 2, while if Pepsi chooses PWC, then the worst outcome for Pepsi is −1. Among these three worst outcomes (−3, 2, −1), Pepsi should rationally choose the largest, namely 2. In other words, using this "best of the worst" approach Pepsi can guarantee a payoff of 2, regardless of what Coke does, by adopting the maximin strategy PL.

Now we take the point of view of Coke (CP). If Coke chooses LC, then the worst outcome occurs if Pepsi chooses PV, giving Pepsi 6 (and thus Coke loses 6 to Pepsi). However, if Coke chooses RC, then the worst outcome for Coke gives Pepsi a gain of 2, while if Coke chooses CC, then the worst outcome for Coke gives Pepsi a gain of 5. Among these three worst outcomes (6, 2, 5), Coke chooses the smallest, namely a loss of 2 to Pepsi. In other words, using this "best of the worst" approach Coke can guarantee losing at most 2, regardless of what Pepsi does, by adopting the minimax strategy RC.

(b) Since the *maximin value* 2 is the same as the *minimax value* 2, the game has a saddlepoint at (PL, RC) and we say that the *value* of the game is 2. Notice that this saddlepoint is an equilibrium solution, since neither Pepsi nor Coke has any incentive to depart unilaterally from this solution. ◁

In games with saddlepoints, each player can calculate his or her (maximin/minimax) strategy before playing the game. Even if the selected *pure strategy* for one player is learned by the other player, there is no advantage gained. The other player should still stick with the predetermined (maximin/minimax) strategy. Games with saddlepoints thus do not require secrecy!

DEFINITION **Pure Strategy**
A specific move or action that a player will follow in every possible play of a game.

Example 11.13 Find the Saddlepoint in Another Payoff Matrix

Table 11.7 shows the payoffs to the Row Player (RP), who seeks higher payoffs, whereas the Column Player (CP) desires making smaller payoffs to RP. (a) Calculate the maximin value, the minimax value, and the corresponding strategies for both players. (b) Determine if this game has a saddlepoint.

TABLE 11.7: Payoff matrix for another game

		CP		
		C1	C2	C3
	R1	1	2	−3
RP	R2	4	2	−2
	R3	−5	3	−4

Solution

(a) It is convenient to compute all the row minimum values and display them on the right-hand side of the payoff matrix reproduced in Table 11.8. For example, the row minimum value for R1 is the smallest value among $1, 2, -3$; namely, -3. The maximum of the row minimum values (the maximin value) is then -2, which is marked with an asterisk. Also, we add the column maximum values at the bottom of Table 11.8. For example, the column maximum value for C1 is the largest value among $1, 4, -5$: namely, 4. The minimum of the column maximum values (the minimax value) is -2, and this is marked with an asterisk also. Thus, R2 is the maximin strategy for RP while C3 is the minimax strategy for CP.

(b) Because the maximin and minimax values are equal, this game has a saddlepoint at $(R2, C3)$. This game has the value -2, and so it inherently favors CP (by two units) since all payoffs are expressed as payments to RP. ◁

(Related Exercises: 23–27)

TABLE 11.8: Payoff matrix with row minima and column maxima

		CP			
		C1	C2	C3	Row Minimum
	R1	1	2	-3	-3
RP	R2	4	2	-2	-2*
	R3	-5	3	-4	-5
Column Maximum		4	3	-2*	

11.5   Zero-Sum Games without Saddlepoints

You might wonder whether all zero-sum games have saddlepoints, in which case each player can use a pure strategy given by the maximin/minimax calculation. To investigate further, consider the payoff matrix in Table 11.9 in which there are two strategies for RP and three strategies for CP. By computing the row minimum values, we determine that the maximin value is 2, guaranteeing that RP will gain at least 2. The minimax value for CP is 4, guaranteeing that CP will pay out at most 4. Since these values are not the same, this game has no saddlepoint.

TABLE 11.9: Payoff matrix with no saddlepoint

		CP			
		C1	C2	C3	Row Minimum
RP	R1	5	4	1	1
	R2	2	3	5	2*
Column Maximum		5	4*	5	

Moreover, we can now envision an endless loop with each player trying to outguess the other:

  RP: If CP sticks with the conservative minimax strategy C2, then I should choose R1 for a payoff of 4.

CP: I know what RP is thinking. RP will choose R1, so I should choose C3, reducing the payoff to 1.

RP: I know what CP thinks I am thinking. CP is going to choose C3, so I should choose R2 for a payoff of 5.

...

It is unclear how to break this vicious cycle. All we can say is that the true "value" of this game lies somewhere between a payoff of 2 and 4 (to RP), the respective maximin and minimax guaranteed values.

The brilliant insight of John von Neumann was that when there is no saddlepoint we can still achieve optimal play in a zero-sum game. The key is not to commit in advance to a fixed (pure) strategy, as tempting as that might be. Instead, we relinquish our choice to a random device that weights the different pure strategies in an optimal way. For example, if a player has three strategies, it might be best to weight these 1:3:2, giving most weight to the second strategy and least weight to the first. In other words, we select the strategies randomly with probabilities $\frac{1}{6}, \frac{3}{6}, \frac{2}{6}$; this is easily achieved by rolling a six-sided die and selecting strategy one if a 1 appears, strategy three if a 2 or 3 appears, and strategy two otherwise.

We will have more to say about probabilities in the next unit. For now, we illustrate how such *mixed strategies* (that use probabilities to weight the pure strategies) can be applied to break the vicious cycle and resolve the dilemma of how to outsmart a smart opponent.

DEFINITION **Mixed Strategy**
This consists of a set of pure strategies and associated probabilities that indicate how frequently each pure strategy is to be played.

We use a mixed strategy when there is no saddlepoint and thus no single optimal pure strategy. In this way, we continue to keep the opponent guessing as to our selected strategy. If the probabilities are chosen wisely, we can in fact achieve an optimal decision.

Example 11.14 Apply Mixed Strategies to a Game With no Saddlepoint

In the payoff matrix given in Table 11.10, the payoffs shown are to RP, who desires large values. (a) Verify that this game has no saddlepoint. (b) Use the mixed strategy $(\frac{4}{5}, \frac{1}{5})$ for RP and the mixed strategy $(\frac{3}{5}, \frac{2}{5})$ for CP to determine the value of the game.

TABLE 11.10: Payoff matrix with no saddlepoint

		CP	
		C1	C2
RP	R1	300	200
	R2	100	500

Solution

(a) As seen in Table 11.11, the maximin value is 200 and the minimax value is 300. As these

TABLE 11.11: Payoff matrix with row minima and column maxima

		CP C1	CP C2	Row Minimum
RP	R1	300	200	200*
	R2	100	500	100
Column Maximum		300*	500	

values are not the same, this game has no saddlepoint. We somehow believe that the actual value of the game should lie somewhere between 200 and 300.

(b) Now let's apply a mixed strategy for RP: namely, playing R1 80% ($\frac{4}{5}$) of the time and R2 20% ($\frac{1}{5}$) of the time. Imagine what would happen if RP used this mixed strategy for (say) 100 plays. We would expect RP to use R1 on 80 plays and R2 on 20 plays.

If CP plays C1, then RP should earn $(80 \times 300) + (20 \times 100) = 26{,}000$ over the 100 plays, or 260 per play. On the other hand, if CP plays C2, then RP should earn $(80 \times 200) + (20 \times 500) = 26{,}000$ over the 100 plays, or 260 per play. So RP can expect to earn 260 per play, regardless of which strategy CP chooses! This is better than the 200 that RP could guarantee by using a pure strategy alone.

Let's now see the result of CP playing the proposed mixed strategy ($\frac{3}{5}, \frac{2}{5}$). In 100 plays, we would expect CP to use C1 on 60 plays and C2 on 40 plays.

If RP plays R1, then CP should expect to pay $(60 \times 300) + (40 \times 200) = 26{,}000$ over the 100 plays, or 260 per play. On the other hand, if RP plays R2, then CP should expect to pay $(60 \times 100) + (40 \times 500) = 26{,}000$ over the 100 plays, also 260 per play. Using this mixed strategy, CP should pay 260 per play, regardless of which strategy RP chooses! This is better than the 300 that CP could guarantee by using a pure strategy alone.

Since now we have achieved equality of expected payoffs in the calculations for both players, it seems here that the value of the game should be 260 and that the proposed mixed strategies are in fact *optimal* for each player. Using mixed strategies has the added benefit that (in general) one player can announce his grand strategy (e.g., play strategy one $\frac{4}{5}$ of the time and strategy two $\frac{1}{5}$ of the time), and not suffer any disadvantage. The other player cannot capitalize on this added information—it is still optimal for that other player to play his or her own mixed strategy (which could be publicized as well).

Notice that the value of the game (260) lies between 200 (a guarantee for RP using only pure strategies) and 300 (a guarantee for CP using only pure strategies). ◁

Example 11.15 Find the Value of a Coin-Matching Game

Consider a coin-matching game in which two players, Rod and Carrie, decide to show each other either a head or a tail. Rod wins $5 if both coins are heads (HH) or wins $1 if both are tails ($TT$). If they don't match, then Carrie wins $3. On the surface, this game seems "fair": Rod wins $5 + 1 = 6$ if the coins match, while Carrie wins $3 + 3 = 6$ if the coins don't match (HT or TH). (a) Model this situation as a zero-sum game, with payoffs listed to Rod, and show there is no saddlepoint. (b) Using the mixed strategy ($\frac{1}{3}, \frac{2}{3}$) for each player, determine the value of the game and determine if the game is in fact fair.

Solution

(a) In this game Row Player Rod has two strategies (show a head or show a tail); Column Player Carrie also has the same two strategies. The stated payoffs are shown in Table 11.12, which also indicates the row minimum and column maximum values. We see that the maximin value is -3 and the minimax value is 1. This game has no saddlepoint, and we infer that the value of the game should lie between -3 and 1. A value of 0 would indicate a fair game.

TABLE 11.12: Payoff matrix with row minima and column maxima

		Carrie		
		H	T	Row Minimum
Rod	H	5	-3	-3^*
	T	-3	1	-3^*
Column Maximum		5	1^*	

(b) Suppose Rod uses the mixed strategy $(\frac{1}{3}, \frac{2}{3})$. If Carrie chooses H, then Rod expects to earn the weighted average of the payoffs 5 and -3: $(\frac{1}{3})(5) + (\frac{2}{3})(-3) = -\frac{1}{3}$. However, if Carrie chooses T, then Rod expects to earn the weighted average of the payoffs -3 and 1: $(\frac{1}{3})(-3) + (\frac{2}{3})(1) = -\frac{1}{3}$.

This means that regardless of what Carrie does, Rod can guarantee losing at most $\frac{1}{3}$ (about 33 cents per play) by using this mixed strategy. Symmetrically, if Carrie adopts the mixed strategy $(\frac{1}{3}, \frac{2}{3})$, she wins on average at least $\frac{1}{3}$ each time. Using these mixed strategies ensures equality between Rod's expected loss and Carrie's expected gain. Thus, $-\frac{1}{3}$ is the value of the game; since it is negative, this means that the game is not fair: it is slightly biased (by $\frac{1}{3}$) in favor of Carrie. ◁

(Related Exercises: 28–32, 35)

Justification of the calculations in part (b) of Example 11.15 leads us into the concepts of probability and expected value, which are covered more thoroughly in the next unit. Also, you may wonder how we arrived at the mixed strategy $(\frac{1}{3}, \frac{2}{3})$. We will see in Chapter 16 how to explicitly compute the optimal mixed strategy for each player in a two-person zero-sum game.

Chapter Summary

In deductive logic we use symbolization to clarify relationships between statements and to exhibit how these relationships necessarily yield other statements. In this chapter, logical expressions are used to model practical situations. In the case of locating fire stations, the simplification of a logical expression produces efficient candidate solutions (which can then be evaluated using additional criteria). Using a set of logical relations enables us to track the evolution of a biological system and determine its eventual equilibrium state. We also model competition or conflict between two players and apply principles of valid reasoning to derive optimal strategies for the players. Interestingly enough, we can also identify here an equilibrium position for the players, either using the concept of a saddlepoint or more generally a mixed strategy. In the latter case, we allow probabilities to guide our optimal decision making. The next unit will discuss in further detail the use of probabilities to model situations of uncertainty and to unravel apparent paradoxes.

Applications, Representations, Strategies, and Algorithms

Applications					
City Planning (1–3)		Biology (4–5)		Law Enforcement (8–9)	
	Society/Politics (10)		Marketing (12)		

Representations					
Graphs		Tables		Sets	Symbolization
Undirected	Directed	Data	Decision	Lists	Logical
1	7	8–15	5	4	1–5

Strategies		
Brute-Force	Rule-Based	Composition/ Decomposition
11	1–7, 11–15	9–10, 12–15

Algorithms					
Exact	Inspection	Sequential	Numerical	Logical	Recursive
1–15	11	1, 3, 5–6, 9–10, 12–15	6–15	1–5	7

Exercises

City Planning

1. The distances (in miles) between zones A, B, C, D are shown in the following table.

    (a) Suppose a fire station located in one zone can serve another zone that is no more than 75 miles away. Construct the associated proximity graph for this situation.

    (b) Determine all efficient placements of fire stations to serve all zones. This can be done by visual examination of the proximity graph constructed in part (a).

	A	B	C	D
A		60	63	78
B	60		55	73
C	63	55		49
D	78	73	49	

2. The distances (in miles) between zones A, B, C, D are those shown in Exercise 1.

   (a) Suppose a fire station located in one zone can serve another zone that is no more than 70 miles away. Construct the associated proximity graph for this situation.

   (b) Determine all efficient placements of fire stations to serve all zones. This can be done by visual examination of the proximity graph constructed in part (a).

3. Using the Rules of Replacement, simplify the logical expression $(a \lor b) \land (a \lor c) \land (a \lor c \lor d)$.

4. Using the Rules of Replacement, simplify the logical expression $(a \lor c \lor d) \land (a \lor d) \land (a \lor b \lor c)$.

5. Four zones A, B, C, D are to be served by fire stations. The logical expression obtained from the corresponding proximity graph is $f = (a \lor c) \land (b \lor d) \land (a \lor c \lor d) \land (b \lor c \lor d)$. Use the Rules of Replacement to simplify this expression and thereby obtain all efficient ways of locating fire stations to serve all four zones.

6. Four zones A, B, C, D are to be served by fire stations. The logical expression obtained from the corresponding proximity graph is $f = (a \lor b \lor c) \land (a \lor b \lor d) \land (a \lor c \lor d) \land (b \lor c \lor d)$. Use the Rules of Replacement to simplify this expression and thereby obtain all efficient ways of locating fire stations to serve all four zones.

7. Five zones A, B, C, D, E are to be served by fire stations. The logical expression obtained from the corresponding proximity graph is $f = (a \lor c) \land (b \lor c \lor e) \land (a \lor b \lor c \lor d) \land (c \lor d \lor e) \land (b \lor d \lor e)$. Use the Rules of Replacement to simplify this expression and thereby obtain all efficient ways of locating fire stations to serve all five zones.

Biological Models

8. Trace the evolution of the lactose metabolism model described in Example 11.4, when started in the initial state $(0, 1, 0, 1, 0)$ and also when started in the initial state $(1, 1, 0, 0, 0)$. Compare your results to those found in Example 11.5.

9. A biological system involves the entities A, B, C that affect one another according to the following equations: $A = B \lor C$; $B = \sim(A \land C)$; $C = A \lor (B \land C)$.

   (a) Investigate the evolution of this system, when started in the initial state $(1, 0, 0)$.

   (b) Investigate the evolution of this system, when started in the initial state $(0, 0, 0)$.

   (c) More generally, map out the possible transitions between states by defining a *state transition graph*, in which each vertex represents one of the eight possible states and a directed edge leads from a state to the next state obtained from that state by applying the above equations. [For example, there is an edge from $(0, 0, 1)$ to $(1, 1, 0)$ since substituting $A = 0$, $B = 0$, $C = 1$ into the equations produces $A = 0 \lor 1 = 1$, $B = \sim(0 \land 1) = 1$, $C = 0 \lor (0 \land 1) = 0$.] What does this graph show you about the equilibrium states reached by this system?

10. A biological system involves the entities A, B, C, D, E that affect one another according to the following equations: $A = B \lor \sim C$; $B = A \land D$; $C = E \land (B \lor D)$; $D = E \land (A \lor B)$; $E = B \land C$.

   (a) Investigate the evolution of this system, when started in the initial state $(0, 1, 0, 1, 0)$.

(b) Investigate the evolution of this system, when started in the initial state $(0, 0, 0, 1, 0)$, which differs from that given in (a) only in the initial value for entity B. What do you conclude?

(c) Investigate the evolution of this system, when started in the initial state $(0, 1, 0, 1, 1)$, which differs from that given in (a) only in the initial value for entity E. What do you conclude?

Games and Nash Equilibria

11. In Example 11.7, the best first move for FP is to remove one bead from the pile with a single bead. Suppose that SP next decides to remove one bead from one of the piles with two beads. Describe the optimal strategy that FP should then adopt for the remaining moves.

12. A *Nim* game is played with three piles, containing $2, 1, 1$ beads, respectively. As in Example 11.7, players take turns removing at least one bead from a single pile and the player taking the last bead wins. Is there a guaranteed win for one of the players? If so, which player? [*Hint*: Reason from the diagram in Figure 11.2.]

13. A *Nim* game is played with four piles, containing $2, 1, 1, 1$ beads, respectively. As in Example 11.7, players take turns removing at least one bead from a single pile and the player taking the last bead wins.

    (a) Construct the directed graph for this situation, rooted at the initial vertex 2111.

    (b) By working from the bottom level to the top level, assign W or L to each vertex in the graph.

    (c) Is there a guaranteed win for one of the players? If so, which player?

14. A *Nim* game is played with three piles, containing $2, 2, 2$ beads, respectively. As in Example 11.7, players take turns removing at least one bead from a single pile and the player taking the last bead wins.

    (a) Construct the directed graph for this situation, rooted at the initial vertex 222.

    (b) By working from the bottom level to the top level, assign W or L to each vertex in the graph.

    (c) What is the best first move for the first player FP?

    (d) Is there a guaranteed win for one of the players? If so, which player?

15. Two nearby sandwich shops A and B are in competition for customers. If both set a high price for their footlong sub, they both earn $9000 profit per week; if they both set a low price, they only earn $7000 profit per week. However, if one offers a low price and the other a high price, then the lower-priced shop attracts more customers and earns $14,000 profit per week, while the higher-priced shop earns only $5000 profit per week. Model this situation as a two-person game and then explain why it illustrates the Prisoner's Dilemma.

16. China and the U.S. are economic powers whose manufacturing sectors release significant amounts of CO_2 into the environment. Let us associate with the status quo (no reduction of emissions) the numerical utility of 0 for each country. As it is considered desirable by both countries (as well as the rest of the world) to reduce CO_2 emissions,

let us associate with that joint action a utility of 10 for each country. However, if one country reduces emissions and the other does not, the former country receives a utility of −5, while the other benefits greatly without expending its own funds and so receives a utility of 15. Model this situation as a two-person game and then explain why it illustrates the Prisoner's Dilemma.

17. Find all Nash equilibria (if any) of a game having the following payoff matrix. Assume that players desire larger payoffs.

		CP	
		C1	C2
RP	R1	(4,3)	(2,4)
	R2	(3,3)	(1,2)

18. Find all Nash equilibria (if any) of a game having the following payoff matrix. Assume that players desire larger payoffs.

		CP	
		C1	C2
RP	R1	(8,6)	(5,7)
	R2	(9,5)	(3,4)

19. Find all Nash equilibria (if any) of a game having the following payoff matrix. Assume that players desire larger payoffs.

		CP		
		C1	C2	C3
RP	R1	(3,2)	(4,1)	(2,1)
	R2	(2,3)	(5,3)	(3,4)

20. Find all Nash equilibria (if any) of a game having the following payoff matrix. Assume that players desire larger payoffs.

		CP		
		C1	C2	C3
RP	R1	(15,7)	(9,8)	(11,9)
	R2	(13,6)	(12,9)	(10,7)

21. Find all Nash equilibria (if any) of a game having the following payoff matrix. Assume that players desire larger payoffs.

		CP		
		C1	C2	C3
RP	R1	$(8,5)$	$(5,7)$	$(6,6)$
	R2	$(7,12)$	$(8,8)$	$(8,10)$
	R3	$(9,10)$	$(6,9)$	$(5,11)$

22. Two competing pharmaceutical representatives Rob and Chris have to choose which eye medications to promote in their overlapping territories. The new eye medications are denoted Anti-Bacterial (AB), Anti-Allergy (AA), and Anti-Inflammatory (AI). Each representative hopes to increase his company's share of the market. If both representatives emphasize medicines from the same category they will split the market gain equally, while if they emphasize different categories, one company will gain at the expense of the other. Using the following payoff matrix, determine any Nash equilibria.

		Chris		
		AB	AA	AI
Rob	AB	$(6,6)$	$(0,8)$	$(8,6)$
	AA	$(8,0)$	$(4,4)$	$(12,2)$
	AI	$(10,0)$	$(2,8)$	$(5,5)$

Saddlepoints in Zero-Sum Games

23. Determine maximin and minimax strategies in the zero-sum game with the following payoff matrix. All payoffs are expressed in terms of payments to the Row Player RP. Does this game possess any saddlepoints? If so, identify them and determine the value of the game.

		CP	
		C1	C2
RP	R1	7	5
	R2	4	3

24. Determine maximin and minimax strategies in the zero-sum game with the following payoff matrix. All payoffs are expressed in terms of payments to the Row Player RP. Does this game possess any saddlepoints? If so, identify them and determine the value of the game.

		CP	
		C1	C2
	R1	-2	4
RP	R2	0	3
	R3	1	2

25. Determine maximin and minimax strategies in the zero-sum game with the following payoff matrix. All payoffs are expressed in terms of payments to the Row Player RP. Does this game possess any saddlepoints? If so, identify them and determine the value of the game.

		\multicolumn{3}{c}{CP}		
		C1	C2	C3
RP	R1	4	−1	1
	R2	2	−3	0

26. Determine maximin and minimax strategies in the zero-sum game with the following payoff matrix. All payoffs are expressed in terms of payments to the Row Player RP. Does this game possess any saddlepoints? If so, identify them and determine the value of the game.

		\multicolumn{3}{c}{CP}		
		C1	C2	C3
RP	R1	−2	3	2
	R2	1	−2	0

27. Determine maximin and minimax strategies in the zero-sum game with the following payoff matrix. All payoffs are expressed in terms of payments to the Row Player RP. Does this game possess any saddlepoints? If so, identify them and determine the value of the game.

		\multicolumn{5}{c}{CP}				
		C1	C2	C3	C4	C5
	R1	5	0	−3	4	−2
	R2	4	1	2	3	1
RP	R3	6	−1	1	−2	0
	R4	2	1	4	4	1
	R5	3	0	3	5	−1

Mixed Strategies for Zero-Sum Games

28. Consider a zero-sum game with the following payoff matrix.

   (a) Determine the maximin and minimax strategies for both players and verify that there is no saddlepoint.

   (b) Verify that the mixed strategy $(\frac{3}{7}, \frac{4}{7})$ is optimal for RP and that the mixed strategy $(\frac{2}{7}, \frac{5}{7})$ is optimal for CP.

   (c) Determine the value of the game. Is it fair? If not, towards which player is it biased, and by how much?

		CP	
		C1	C2
RP	R1	3	−1
	R2	−2	1

29. Consider a zero-sum game with the following payoff matrix.

    (a) Determine the maximin and minimax strategies for both players and verify that there is no saddlepoint.

    (b) Verify that the mixed strategy $(\frac{5}{13}, \frac{8}{13})$ is optimal for RP and that the mixed strategy $(\frac{6}{13}, \frac{7}{13})$ is optimal for CP.

    (c) Determine the value of the game. Is it fair? If not, towards which player is it biased, and by how much?

		CP	
		C1	C2
RP	R1	4	−4
	R2	−3	2

30. Consider a zero-sum game with the following payoff matrix.

    (a) Determine the maximin and minimax strategies for both players and verify that there is no saddlepoint.

    (b) Verify that the mixed strategy $(\frac{5}{12}, \frac{7}{12})$ is optimal for RP and that the mixed strategy $(\frac{3}{5}, \frac{2}{5})$ is optimal for CP.

    (c) Determine the value of the game. Is it fair? If not, towards which player is it biased, and by how much?

		CP	
		C1	C2
RP	R1	14	−21
	R2	−10	15

31. In the following payoff matrix, the Row Player represents a batter who can anticipate either a fastball FB or a curveball CB from the Column Player, the pitcher. The values are the resulting batting averages achieved by the Row Player when confronted with that particular pitch.

    (a) Verify that there is no saddlepoint for this game.

    (b) Verify that the mixed strategy $(\frac{1}{5}, \frac{4}{5})$ is optimal for the batter and that the mixed strategy $(\frac{1}{4}, \frac{3}{4})$ is optimal for the pitcher.

    (c) Determine the value of the game and interpret what this means.

		Pitcher	
		FB	CB
Batter	FB	.340	.260
	CB	.265	.285

32. The situation is the same as in Exercise 31, where the batting averages achieved against the particular pitches are found in the following payoff matrix.

    (a) Verify that there is no saddlepoint for this game.

    (b) Verify that the mixed strategy $(\frac{2}{5}, \frac{3}{5})$ is optimal for the batter and that the mixed strategy $(\frac{1}{3}, \frac{2}{3})$ is optimal for the pitcher.

    (c) Determine the value of the game and interpret what this means.

		Pitcher	
		FB	CB
Batter	FB	.450	.270
	CB	.250	.370

Projects

33. Develop a spreadsheet that will track the states reached from a given starting state in a biological system defined by given logical equations. Notice that $A \vee B$ can be implemented as $MAX(A, B)$, while $A \wedge B$ can be implemented as $MIN(A, B)$. The negation of A can be calculated as $1 - A$. Then apply your routine to investigate the different equilibrium states that can be reached by the biological systems described in Example 11.4 and Exercises 8–10.

34. Play the iterated Prisoner's Dilemma found at site [OR1]. Here you play against an opponent for a number of rounds (which you do not know in advance, except that it does not exceed 100). Playing the game for many rounds allows for the possibility of cooperation to emerge, as well as retaliation. You can test out different strategies in order to optimize the total number of points received by the end of the game.

35. Develop a computer routine to simulate the behavior of a player who uses the mixed strategy $(p, 1 - p)$ in a two-person game. First, a random number x between 0 and 1 is randomly generated; if $x < p$, then strategy one is chosen, and otherwise strategy two is chosen.

    (a) Use this routine to play the optimal mixed strategies for both players using some of the examples in Section 11.5. Collect the resulting payoffs to the Row Player over a large number of repetitions (say 1000) and compare the average payoff received to the calculated value of the game.

    (b) Use this routine to play the optimal mixed strategy for the Row Player using these same examples, but now assuming that the Column Player selects each strategy with equal likelihood (i.e., 50% of the time). Compare your results, collected over a large number of repetitions, to those found in part (a).

Bibliography

[Guy 1989] Guy, R. K. 1989. *Fair game*. Arlington, MA: COMAP.

[Poundstone 1993] Poundstone, W. 1993. *Prisoner's dilemma*. New York: First Anchor Books.

[Schelling 1989] Schelling, T. C. 1989. *The strategy of conflict*. Cambridge, MA: Harvard University Press.

[Straffin 2002] Straffin, P. 2002. *Game theory and strategy*. Washington, D.C.: Mathematical Association of America.

[Williams 1986] Williams, J. D. 1986. *The compleat strategyst: Being a primer on the theory of games of strategy*. Mineola, NY: Dover.

[OR1] `http://www.iterated-prisoners-dilemma.net` (accessed April 9, 2014).

Unit III

Probability: Predictions and Expectations

Chapter 12

Probability and Counting

When one admits that nothing is certain one must, I think, also add that some things are more nearly certain than others.

—Bertrand Russell, British philosopher and mathematician, 1872–1970

Chapter Preview

Our behavior is based on our past experiences and outcomes. We infer something about the future from what we know at present because we believe the future will be related to the past. However, problems arise when our intuition misleads us. Learning how to draw legitimate probabilistic conclusions is aided by a careful study of probability and can be facilitated by using graph representations. We will use trees with weighted edges to enumerate the possible outcomes and their likelihood, enabling us to systematically calculate the probability of various events. Again, the strategy of decomposing a problem into its smaller components allows us to construct the solution from more easily understood building blocks.

12.1   Uncertainty and Probability

Whether investing in the stock market or organizing a concert, we make decisions in the face of uncertainty. Careful planning and preparation can be overridden by even one anomalous (unexpected) random event. In order to guide our decision making in these and other instances, we require a quantitative yardstick to measure the likelihood of certain events. This leads us to the study of probabilities.

In Chapter 7 we contrasted the certainty of valid deductive reasoning with the probable nature of inductive inference. Inductive logic provides a rational basis for decision making when there are uncertain outcomes in the future. While we can never be absolutely certain of the outcome of any decision, we can be comforted that our logic was correct and has guided us to the best decision possible under the uncertain circumstances.

To illustrate inductive logic using probabilities, consider the following argument:

400 patients took a new drug to treat macular degeneration.
300 of the patients showed improved vision with the new drug.

Jane has decided to use the new drug.
Therefore, Jane's vision will improve.

Clearly the above argument is not valid. However, we can qualify the conclusion and say: "There is a 75% chance that Jane's vision will improve." Even though we cannot be absolutely sure of the outcome (whether Jane's vision will improve), we are able to quantify the likelihood of that outcome. As a result of this quantification, Jane can make a more informed decision as to whether to take the new drug.

Not all uncertain situations are as simple as the one described above, where our intuition can be an accurate guide to behavior. As we will see in later examples, our intuition can at times lead us astray.

Example 12.1 Three Examples in Probability

Test your intuition in assessing the following uncertain situations.

A. Change of Luck

In the game of roulette, a player bets on the outcome of a spinning wheel. The wheel has 18 red positions, 18 black positions, and 2 green (neutral) positions. A player wins if he bets on red and the ball stops on red (or he bets on black and the ball stops there). The house always wins if the ball lands on green. Suppose a gambler notices that there have been eight consecutive spins in which the ball has stopped on red, and therefore decides to bet on black for the next spin. Do you agree with this logic?

B. License Plates

In the state of Virginia, the typical license plate consists of three letters followed by four digits, for example, UVA 9049. What is the chance that you observe a Virginia license plate having all four of the digits different? Is it likely or is it unlikely? Specifically, is it approximately

    (a) one in eight
    (b) one in four
    (c) one in three
    (d) one in two (equally likely)
    (e) five in eight?

C. Birthday Coincidences

In a class of 30 students, how likely is it that today is someone's birthday? Here is a second related question: how likely is it that two people (or more) in the room have the same birthday?

Solution

A. The gambler reasons that a black is "due" since in the long run there should be equal numbers of stops on red as on black. However, as we will see later, there is no logical reason to favor black, or to favor red for that matter.

B. It turns out that the correct answer is (d): it is just about as likely or not for all four digits to be different.

C. As will be verified in Section 12.4, there is only about a 1 in 13 chance that today is someone's birthday in a class of 30. On the other hand, there is about a 7 in 10 chance of two people (or more) sharing the same birthday, which represents fairly good odds. ◁

The preceding examples show that we need to quantify the idea of "chances" and "odds" before proceeding down the road to understanding uncertainty. Answers to the question "what are the chances?" often are given as percentages (there is a 10% chance), or as proportions (there is a 1 out of 10 chance, which is shown as $\frac{1}{10}$ or 0.10), or as odds (the odds are 9 to 1 against: nine chances of losing for every one chance of winning).

Suppose that in a sociology class of 30 students there were six grades of A on an exam. What is the probability that a randomly selected student in the class received an A? We would then say that there is a probability of $\frac{6}{30}$ or $\frac{1}{5}$ (or a 20% chance) that a randomly selected student in the class received an A.

Likewise, in the **Change of Luck** scenario in Example 12.1, the probability of the ball landing on red would be $\frac{18}{38}$ or approximately 0.474—since there are 18 red positions out of $18 + 18 + 2 = 38$ possible positions for the ball to land.

We now introduce an informal definition of the concept of *probability*. It requires the ability to count the number of times some outcome of interest occurs (denoted n) and the total number of possible outcomes (denoted m).

DEFINITION **Probability**
If all outcomes of an experiment are equally likely, then

$$\Pr[\text{outcome of interest}] = \frac{\text{\# of times outcome of interest occurs}}{\text{\# of possible outcomes}} = \frac{n}{m}.$$

Empirical and Theoretical Probabilities

Political surveys are based on taking samples drawn from a large community, rather than surveying the entire community—which might be too expensive or too time-consuming. Quality control in a factory is also based on examining only a sample of the items produced instead of all items in a production run. Suppose a manufacturer of computer hard drives observes two defective hard drives out of 500 drives tested. He concludes that the *empirical probability* of failure is $\frac{2}{500}$ or 0.004. If he had observed seven failures out of 2000 drives tested, the empirical probability would change to 0.0035. Because it is not economical to test all the hard drives, the empirical probability obtained from a sample is only an approximation to the true or *theoretical probability*.

DEFINITION **Empirical Probability**
A probability based on counting the outcomes in a sample taken from the population.

DEFINITION **Theoretical Probability**
A probability based on counting all outcomes in the entire population.

A theoretical probability results when we know all the outcomes of a roulette wheel or all the outcomes of the roll of a die. Suppose a six-sided die is rolled once and we ask for the probability that a 4 will land face up. Since the entire set of possible outcomes is known (namely, 1 through 6) and each of the outcomes is just as likely as any other (with a fair die), the probability of 4 landing face up is $\frac{1}{6}$. This is a theoretical probability.

If we toss the die 100 times, and if a 4 appears face up 15 times, then the empirical probability would be $\frac{15}{100}$ or 0.15, which is different from the theoretical probability of $\frac{1}{6}$. However, if we continue to roll the die many more times, the observed empirical probability will get closer to the theoretical probability of $\frac{1}{6}$. This behavior is called the *Law of Large Numbers*, and we will discuss it further in Chapter 16.

Using a demonstration available at the web site [OR1], we can simulate the rolling of a die. After rolling the die 100 times, we observed 0.21 as the empirical probability of a 4 appearing. However, after rolling the die 1000 times, we observed 0.158 as the empirical probability. After 10,000 rolls, the empirical probability was 0.164—this is getting closer to the theoretical probability of $\frac{1}{6}$, which is approximately 0.167.

Decision Trees and Outcomes

Theoretical probabilities can often be determined by representing the underlying experiment or process using a *decision tree*.

DEFINITION **Decision Tree**
A graphical representation (as a tree) of a process that unfolds in stages and has a finite number of options at each stage.

Consider tossing a fair coin twice. Figure 12.1 shows this two-stage decision process as a tree, in which s represents the start of the process. The first toss, or first stage, either produces a head (sending us to the state labeled H) or produces a tail (sending us to the state labeled T). If we arrive at state H at the end of the first toss, then the second toss either leads to state HH or to state HT. On the other hand, if we arrive at state T at the end of the first toss, then the second toss either leads to TH or to TT. There are altogether four paths from s to the four possible *outcomes* on the right-hand side of Figure 12.1.

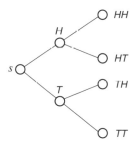

FIGURE 12.1: A decision tree

Using this decision tree, we can easily find the probability of tossing exactly one H. We are then interested in the combination of outcomes consisting of either HT or TH, written as $\{HT, TH\}$. These are the middle two outcomes shown on the right-hand side in Figure 12.1. The probability of this *event* is written $\Pr[\{HT, TH\}]$ or more simply as $\Pr[HT, TH]$. Since there are four equally likely outcomes and two outcomes (HT and TH) are of interest to us, we easily calculate $\Pr[HT, TH] = \frac{2}{4} = \frac{1}{2}$.

DEFINITION **Outcome**

A possible result of a probability experiment (that is, a situation involving uncertainty).

DEFINITION **Event**

A collection of outcomes occurring in a probability experiment.

Example 12.2 Calculate the Probability of at Least One H in Two Coin Tosses

To find this probability, consider the decision tree of Figure 12.1 and focus on the outcomes that have at least one H.

Solution

The event of interest consists of the top three outcomes $\{HH, HT, TH\}$ shown in Figure 12.1. Since all outcomes are equally likely, and there are four outcomes altogether, the probability of at least one head appearing is $\Pr[HH, HT, TH] = \frac{3}{4}$. ◁

(Related Exercises: 1–3)

Counting the outcomes was easy here, but it is not always so. Counting is a fundamental tool of a branch of mathematics called *combinatorics*. In some cases, counting can be challenging and can involve fairly large numbers. For example, there are 1,098,240 different ways of obtaining just "one pair" when dealing five cards from a deck of 52 cards.

DEFINITION **Combinatorics**

A branch of mathematics that studies methods of counting collections of objects; for example, determining how many members are in a collection as well as how many different ways we can order and/or rearrange a collection.

12.2 Some Rules of Probability

So far, we have seen how to assign probabilities to individual outcomes and to some simple combinations of outcomes (i.e., events). This section will show how the decision tree can aid us in finding the probabilities of outcomes and events in more complex situations.

Multiplying Probabilities

We can add further information to our decision tree. In the decision tree for tossing two coins shown in Figure 12.2, we have placed along each edge the probability associated with moving between the corresponding two states. For example, the probability of H on the first toss is $\frac{1}{2}$, the probability of T on the first toss is $\frac{1}{2}$, the probability of obtaining H on the second toss (after H on the first toss) is also $\frac{1}{2}$, and so on. In this simple example, all edges receive the same probability $\frac{1}{2}$. But this will not always be the case.

Each outcome (which corresponds to a path from s to a particular endpoint) has a probability that can be calculated by *multiplying* the probabilities of the edges along the path. For example, the path $s \to H \to HT$ has probability $\frac{1}{2} \times \frac{1}{2} = \frac{1}{4}$; in other words $\Pr[HT] = \frac{1}{4}$.

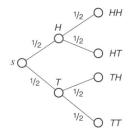

FIGURE 12.2: A weighted decision tree

In fact, each path from s to a final outcome has the same probability $\frac{1}{4}$, which is reassuring since we presumed that all these outcomes are equally likely.

This way of obtaining the probabilities of events by multiplying edge probabilities holds more generally, even when the edge probabilities are not all the same. We state this general rule below.

DEFINITION **Multiplication Rule**
In a decision tree, the probability of an outcome can be found by multiplying the edge probabilities along the path joining s and that outcome.

Example 12.3 Calculate the Probability of T and then H in Two Coin Tosses

Use the decision tree in Figure 12.2 to determine the probability that T appears on the first toss and then H appears on the second toss of a coin.

Solution

The decision tree in Figure 12.3 highlights the associated path $s \to T \to TH$. This path has probability $\Pr[TH] = \Pr[T \text{ and } H] = \frac{1}{2} \times \frac{1}{2} = \frac{1}{4}$. ◁

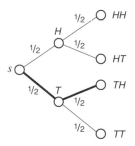

FIGURE 12.3: Path in the decision tree

Example 12.4 Create a Fair Game

Sherlock and Mycroft have just completed dinner at an exclusive restaurant. In order to decide who will pay for this fabulous meal, Mycroft produces a coin from his pocket and suggests that they toss the coin to decide who pays. Sherlock, however, suspects that the coin is biased, with probability $p \neq \frac{1}{2}$ of turning up heads, and believes that Mycroft will use this to his own advantage. In an instant, Sherlock proposes a modification of the coin-tossing procedures that will convert this into a fair game, one not biased toward either person. How did he do this?

Solution

Sherlock proposes tossing the coin twice, which can be viewed using the decision tree in Figure 12.4. Here p represents the probability of H on a toss, so $1 - p$ is the probability of T. The four outcomes are labeled with their probabilities, obtained from the Multiplication Rule. Specifically, the path $s \to H \to HT$ has probability $\Pr[HT] = p(1 - p)$ and the path $s \to T \to TH$ has probability $\Pr[TH] = (1 - p)p$. Since these two outcome probabilities are equal, Sherlock proposes that they each select one of these outcomes (HT or TH) and then toss the coin twice to decide. If either HH or TT occurs, then they start over again. ◁

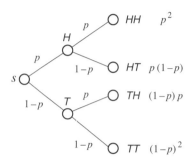

FIGURE 12.4: Tossing a biased coin twice

Adding Probabilities

Often we would like to find the probability of a certain combination of outcomes, such as $\Pr[HT, TH]$, which represents the probability that either HT or TH occurs. The probability of this event E (a *disjunction*, see Chapter 8) is determined by *adding* up the probabilities of the individual outcomes:

$$\Pr[E] = \Pr[HT, TH] = \Pr[HT \text{ or } TH] = \Pr[HT] + \Pr[TH] = \tfrac{1}{4} + \tfrac{1}{4} = \tfrac{1}{2}.$$

Similarly, the probability of obtaining at least one head is found as

$$\Pr[E] = \Pr[HH, HT, TH] = \Pr[HH] + \Pr[HT] + \Pr[TH] = \tfrac{1}{4} + \tfrac{1}{4} + \tfrac{1}{4} = \tfrac{3}{4}.$$

This rule can be stated more generally as follows:

DEFINITION **Addition Rule**
The probability of an event E, where E is a combination of outcomes, can be found by adding the probabilities of all the outcomes represented by E.

We will generalize this rule a bit further when we discuss disjoint events in Chapter 14.

Example 12.5 Calculate the Probability of Different Results in Two Coin Tosses

Determine the probability of obtaining two different results: either H followed by T, or T followed by H.

Solution

In the decision tree shown in Figure 12.5, we highlight the two distinct outcomes HT and TH. Previously we established that each of these outcomes had probability $\frac{1}{4}$. We can apply the Addition Rule to obtain $\Pr[HT, TH] = \Pr[HT \text{ or } TH] = \frac{1}{4} + \frac{1}{4} = \frac{1}{2}.$ ◁

(*Related Exercises: 4–12*)

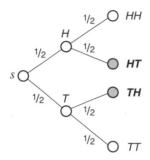

FIGURE 12.5: Two outcomes in a decision tree

Sample Spaces and the Complement of an Event

All the possible outcomes in a particular situation constitute what is called the *sample space*. The sample space for tossing one die is $\{1, 2, 3, 4, 5, 6\}$, while for tossing a coin twice the sample space is $\{HH, HT, TH, TT\}$.

DEFINITION **Sample Space**
The set of all possible outcomes resulting from a probability experiment.

Since every situation results in exactly one outcome occurring, the probabilities of all outcomes in the sample space must sum to 1. For example, the outcome probabilities for tossing one die are each $\frac{1}{6}$ and they sum to

$$\tfrac{1}{6} + \tfrac{1}{6} + \tfrac{1}{6} + \tfrac{1}{6} + \tfrac{1}{6} + \tfrac{1}{6} = 1.$$

In the case of tossing a coin twice, each outcome in the sample space has probability $\frac{1}{4}$ and by the Addition Rule we know that the event $E = \{HH, HT, TH\}$ has probability $\frac{3}{4}$. The *complement* of the event E consists of all the outcomes in the sample space that are *not* included in the event E. The complementary event is written $\overline{E}$ and in this example $\overline{E} = \{TT\}$.

How are the probabilities of these two events related? Since exactly one outcome from either E or from $\overline{E}$ must occur, we have the following rule:

DEFINITION **Complement Rule**
For any event E, $\Pr[E] + \Pr[\overline{E}] = 1$.

Sometimes it is easier to calculate the probability of the complement of an event, instead of the probability of the event itself. This is illustrated in the next example.

Example 12.6 Calculate the Probability of at Least One H in Four Coin Tosses

A single coin is tossed four times. The coin can come up either H or T, and each result is equally likely (given a fair coin) with $\Pr[H] = \Pr[T] = \frac{1}{2}$. How likely is it that a head shows up on one or more of the four tosses?

Solution

It is easier to find the probability of the complementary event $\overline{E}$, which denotes obtaining tails on all four tosses: $\overline{E} = \{TTTT\}$ and $\Pr[\overline{E}] = \frac{1}{2} \times \frac{1}{2} \times \frac{1}{2} \times \frac{1}{2} = \frac{1}{16}$. Since $\Pr[E] + \Pr[\overline{E}] = 1$, we can now conclude that $\Pr[E] = 1 - \Pr[\overline{E}] = 1 - \frac{1}{16} = \frac{15}{16}$. ◁

(Related Exercises: 13–18)

12.3   Counting in Stages

The calculation of probabilities requires the ability to count the number of ways n a certain event can occur, as well as the total number m of possible outcomes. There are different strategies we can use to count events and outcomes, and these strategies will be illustrated through a variety of examples.

Our tools will be the decision tree, as well as the rules for multiplying probabilities, adding probabilities, and using complementary events. To get started, we consider the following example.

Example 12.7 Calculate the Probability of a Sum of 8 When Rolling Two Dice

Even though the dice are rolled simultaneously, it is convenient to think of them as being rolled in succession. The first (say red) die can show any of $1, 2, \ldots, 6$ and each value is equally likely (given a fair die): $\Pr[1] = \Pr[2] = \cdots = \Pr[6] = \frac{1}{6}$. The second (green) die behaves similarly. We are interested in determining how often a sum of 8 will show up when rolling the red and green dice.

Solution

This situation can be illustrated using the two-stage decision tree in Figure 12.6. There are six edges leaving the starting state s and from each of these six states there are six more edges. Altogether there are then $6 \times 6 = 36$ outcomes in the sample space, each corresponding to a path from s to one of the 36 outcomes on the right-hand side of Figure 12.6. These 36 distinct outcomes are also displayed in Figure 12.7, which shows the different face-up combinations for the red and green dice.

We are interested in the event where the sum of the two face values equals 8. There are five of these sum-8 face-up combinations: $E = \{(2, 6), (3, 5), (4, 4), (5, 3), (6, 2)\}$. Thus, $\Pr[E] = \frac{n}{m} = \frac{5}{36}$. ◁

(Related Exercises: 9–12)

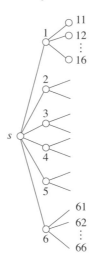

FIGURE 12.6: A two-stage decision tree

green

1 1	1 2	1 3	1 4	1 5	1 6
2 1	2 2	2 3	2 4	2 5	2 6
3 1	3 2	3 3	3 4	3 5	3 6
4 1	4 2	4 3	4 4	4 5	4 6
5 1	5 2	5 3	5 4	5 5	5 6
6 1	6 2	6 3	6 4	6 5	6 6

red

FIGURE 12.7: Outcomes from rolling two dice

In general, when a two-stage process has a_1 outcomes in the first stage and a_2 outcomes in the second stage (regardless of the result of the first stage), then the sample space has $a_1 \times a_2$ outcomes in total.

This same reasoning can be applied when there are k stages and $a_1, a_2, \ldots, a_k$ outcomes to the successive stages. This is called the *Multistage Rule*.

DEFINITION **Multistage Rule**

In a process having k stages, suppose that each stage i can result in a_i outcomes, regardless of the preceding stages. Then the sample space consists of $a_1 \times a_2 \times a_3 \times \cdots \times a_k$ outcomes.

We now use the Multistage Rule to count the outcomes in various sample spaces.

Example 12.8 Use the Multistage Rule to Count Telephone Area Codes

Each area code consists of three digits. The current convention is that the first digit can range from 2–9, the second digit can range from 0–8, and the third digit can range from 0–9. How many telephone area codes are possible?

Solution

The formation of an area code can be viewed as a three-stage decision tree that consists of picking the first digit, then the second digit, and finally the third digit. By the Multistage Rule, there are $8 \times 9 \times 10 = 720$ possible area codes. Included in this count are certain emergency codes, such as 911, which need to be excluded as possible area codes. ◁

(Related Exercise: 19)

Example 12.9 Use the Multistage Rule to Count the Number of Security PINs

A bank requires each new customer to select a four-digit PIN (personal identification number) and insists that it contains no repeated digits. How many different PINs are possible?

Solution

This can be considered as a four-stage decision process. The first stage corresponds to selecting the first digit of the PIN and there are 10 possible choices here. Once the first digit has been selected, the second stage (selecting the second digit) involves only nine choices to ensure that a duplicate digit has not been selected. Once the first and second digits have been selected, there are eight choices for the third digit. Finally, the last digit can only be one of seven possibilities to avoid repeating a previous digit. By the Multistage Rule, there are $10 \times 9 \times 8 \times 7 = 5040$ possible PINs. ◁

(Related Exercises: 20–22)

Example 12.10 Use the Multistage Rule to Count the Number of License Plates

A certain state predicts that in the next several years the number of registered vehicles will approach one million. Currently, the state license plate displays two letters followed by three digits. Will the current license plate design suffice to accommodate one million vehicles?

Solution

Counting the number of possible license plates is a five-stage decision process where the first two stages select one of the 26 letters (A through Z) and the last three stages select one of the 10 digits (0 through 9). Using the Multistage Rule, we obtain $26 \times 26 \times 10 \times 10 \times 10 = 676,000$ different license plates, clearly not enough to accommodate one million vehicles. ◁

(Related Exercise: 23)

12.4   Checking Our Intuition

We now refer back to Section 12.1 in which we presented several examples where our intuition might mislead us, but where basic probability concepts now come to the rescue.

Example 12.11 Calculate a Probability in a License Plate Problem

What is the probability that all four digits appearing in a four-number license plate are different, given a choice of digits 0–9 in each position?

Solution

The sample space here consists of $10 \times 10 \times 10 \times 10 = 10{,}000$ total outcomes, namely $\{0000, 0001, \ldots, 9998, 9999\}$. The number of ways in which all four digits appearing can be different (our desired event E) is $10 \times 9 \times 8 \times 7 = 5040$. So $\Pr[E] = \frac{n}{m} = 5040/10{,}000 = 0.504$. In other words, there are slightly better than even chances that the four digits appearing on the license plate are all different. ◁

(Related Exercises: 24–25)

Example 12.12 Calculate the Coincidence Probability of an Individual Birthday

What is the probability in a class of 30 students that today is someone's birthday?

Solution

We might be tempted to give a quick, but incorrect, answer of $\frac{30}{365} = 0.082$ because there are 30 possible birthday dates out of 365 choices. Instead, since there are 30 students, the sample space is a 30-stage decision process, with the outcome of stage i being the birthday of student i, one of the 365 possible days. The sample space then contains $365 \times 365 \times \cdots \times 365 = 365^{30}$ outcomes.

Now focus on the complementary event $\overline{E} =$ "today is no one's birthday." Event $\overline{E}$ involves selecting in turn 364 birthdays (avoiding today) for each of the 30 students, so there are $364 \times 364 \times \cdots \times 364 = 364^{30}$ outcomes corresponding to the event $\overline{E}$. We can then determine

$$\Pr[\overline{E}] = \frac{364 \times 364 \times \cdots \times 364}{365 \times 365 \times \cdots \times 365} = \left(1 - \tfrac{1}{365}\right)\left(1 - \tfrac{1}{365}\right)\cdots\left(1 - \tfrac{1}{365}\right) = \left(1 - \tfrac{1}{365}\right)^{30},$$

which gives (approximately) $\Pr[\overline{E}] = 0.921$. Using the Complement Rule, $\Pr[E] = 1 - \Pr[\overline{E}] = 1 - 0.921 = 0.079$, or about an 8% chance that today is someone's birthday. ◁

Example 12.13 Calculate the Coincidence Probability of Some Matching Birthdays

In a class of 30 students, what is the probability that at least two students have the same birthday?

Solution

Let E be the event that (at least) two students have the same birthday, so the event $\overline{E}$ involves selecting in turn *distinct* birthdays for all 30 students. By the Multistage Rule, there are $365 \times 364 \times \cdots \times 336$ outcomes in event $\overline{E}$ and so

$$\Pr[\overline{E}] = \frac{365 \times 364 \times 363 \times \cdots \times 336}{365 \times 365 \times 365 \times \cdots \times 365} = 1\left(1 - \tfrac{1}{365}\right)\left(1 - \tfrac{2}{365}\right)\cdots\left(1 - \tfrac{29}{365}\right) = 0.294.$$

By the Complement Rule, we conclude that $\Pr[E] = 1 - \Pr[\overline{E}] = 1 - 0.294 = 0.706$, meaning that there is approximately a 71% chance that two people (or more) in a class of 30 students share the same birthday. ◁

(Related Exercises: 26–27, 31)

Example 12.14 Calculate the Probability of Matching Baseball Cards

Each cereal box contains a card picturing one of nine different baseball All-Stars. You can win a cash prize if you purchase several boxes and obtain a "match" (getting the same picture). What are the chances of winning a prize if you purchase five cereal boxes? Sorry, only one prize per customer regardless of the number of matches.

Solution

One might be tempted to answer $\frac{5}{9} = 0.555$, since there are five boxes and nine possible baseball cards. However, this reasoning is incorrect.

Instead, we can view this problem as a five-stage decision process in which each stage represents the result (one of the nine baseball cards) from opening each of the five boxes. By the Multistage Rule, there are $9 \times 9 \times 9 \times 9 \times 9 = 59{,}049$ outcomes.

The event E of interest here represents matching (at least) one pair of the five cards. However, it is easier to compute the probability of the complementary event $\overline{E}$, representing no matches among the five baseball cards. Since duplicates are not allowed, the number of outcomes in $\overline{E}$ is given by $9 \times 8 \times 7 \times 6 \times 5 = 15{,}120$. As a result

$$\Pr[\overline{E}] = \tfrac{15{,}120}{59{,}049} = 0.256.$$

By the Complement Rule, $\Pr[E] = 1 - \Pr[\overline{F}] = 1 - 0.256 = 0.744$, which is quite different from the (tempting but erroneous) 0.555 answer. ◁

(Related Exercise: 28)

Chapter Summary

In inductive logic, we use probabilities to quantify the likelihood of a conclusion (event) following from the premises (evidence). To properly obtain such probabilities, we develop a number of basic rules for combining the probabilities of individual outcomes. Determining these individual probabilities can be usefully modeled as a multistage decision process, and depicted using a decision tree. We can ultimately calculate the probability of complex events by first decomposing them into simpler pieces, which can then be reassembled using a few arithmetic rules for combining event probabilities. While this chapter and the next deal with situations involving equally likely outcomes, Chapter 14 shows how these concepts can be applied to situations in which the outcomes can occur with differing probabilities.

Applications, Representations, Strategies, and Algorithms

Applications			
Gambling (1)		Business (8–10)	

Representations			
Graphs			Tables
Unweighted	Weighted	Tree	Data
2, 7	3–5	2–5, 7	7

Strategies
Rule-Based
2–14

Algorithms		
Exact	Sequential	Numerical
2–14	6, 12–14	2–14

Exercises

Decision Trees

1. A fair coin is tossed three times.

   (a) Display the associated decision tree and label the outcomes.

   (b) Calculate the probability of obtaining exactly two heads.

   (c) Calculate the probability of obtaining at least two tails.

2. A family has three children. Assume that each child is equally likely to be a boy or a girl.

   (a) Display the associated decision tree and label the outcomes.

   (b) Calculate the probability that this family has exactly one girl.

   (c) Calculate the probability that this family has at least one girl.

3. A board game is played in which each player draws a card from a special deck and advances the specified number of squares. After each draw the card is replaced. It is known that the deck contains five cards that allow an advance of three squares, five cards that allow an advance of two squares, and five cards that allow an advance of one square. Sarah draws two cards (with replacement) from the deck.

   (a) Display the associated decision tree and label the outcomes.

   (b) Find the probability that Sarah ends up advancing four squares after two draws.

   (c) Find the probability that Sarah ends up advancing five squares after two draws.

4. A biased coin is tossed three times; the probability of a head turning up is 0.6.

    (a) Display the associated decision tree and show the edge probabilities.

    (b) Calculate the probability of obtaining exactly two tails.

    (c) Calculate the probability of obtaining at least two heads.

5. A biased coin is tossed three times; the probability of a head turning up is 0.7.

    (a) Display the associated decision tree and show the edge probabilities.

    (b) Calculate the probability of obtaining exactly two heads.

    (c) Calculate the probability of obtaining at least two tails.

6. There are seven marbles in a fish bowl. Two of the marbles are purple, one is white, and the other four are orange. You draw one marble and put it back; then you draw a second marble.

    (a) Display the associated decision tree and show the edge probabilities.

    (b) Calculate the probability that you end up drawing an orange marble and a white marble (in any order).

7. You are at a birthday party with three friends and each throws their name in a hat. The hat is shaken up and you draw out a name. If you draw your own name, you stop; otherwise, you draw again. You are allowed up to three attempts to draw your name from the hat. A name once drawn is replaced in the hat.

    (a) Display the associated decision tree and show the edge probabilities.

    (b) Calculate the probability that you are successful in drawing your name within three attempts.

Multiplication and Addition Rules

8. Three bills are in a cash register: two are $100 bills and one is a $50 bill. Without looking, you draw a bill out of the register and then replace it. You then draw another bill. Calculate the probability that you end up with $150 after two draws.

9. A pair of (six-sided) dice are rolled. We are interested in the probability that the sum of the numbers showing on the two dice equals 5.

    (a) How many outcomes are in the sample space?

    (b) List the pairs that sum to 5.

    (c) Calculate the probability that the sum of the numbers rolled equals 5.

10. A pair of (six-sided) dice are rolled. We are interested in the probability that the sum of the numbers showing on the two dice equals 7.

    (a) How many outcomes are in the sample space?

    (b) List the pairs that sum to 7.

    (c) Calculate the probability that the sum of the numbers rolled equals 7.

11. A pair of (six-sided) dice are rolled. We are interested in the probability that the two numbers showing on the dice differ by exactly 2.

    (a) How many outcomes are in the sample space?

    (b) List the pairs that produce a difference of exactly 2.

    (c) Calculate the probability that the numbers rolled differ by exactly 2.

12. A pair of (six-sided) dice are rolled. We are interested in the probability that the two numbers showing on the dice differ by exactly 1.

    (a) How many outcomes are in the sample space?

    (b) List the pairs that produce a difference of exactly 1.

    (c) Calculate the probability that the numbers rolled differ by exactly 1.

Complement Rule

13. A license plate consists of five digits. Determine the probability that such a license plate contains some repeated digits.

14. In the 17th century, it was common for gamblers to bet on obtaining at least one 1 in four rolls of a die. Calculate the probability of this event.

15. Some friends are playing the game *Settlers of Catan*^{TM}, in which each player rolls two dice to determine the resources he or she receives. After 20 turns, the sum 8 has never been rolled. Thinking this is impossible, you want to know how likely is this occurrence. Calculate the probability that the sum 8 is not rolled in the first 20 turns.

16. While playing *Settlers of Catan*^{TM}, you have selected a position in which you receive a resource if you achieve a sum of 5, 6, or 8 (when rolling two dice).

    (a) Calculate the probability of rolling a sum of 5, 6, or 8 on your next turn, and thus receiving a resource on your next turn.

    (b) Calculate the probability of receiving no resource during the next 10 rolls.

17. You are planning a hike over the upcoming long holiday weekend. You look at the forecast and determine that the chance of rain today is 50%, the chance of rain tomorrow is 60%, and the chance of rain on the third day is 70%. Calculate the probability that there is rain on at least one of the next three days.

18. You are texting a friend on your cell phone. Your friend's phone receives the message as a string of 0s and 1s (e.g., 00011010101110101...). Suppose the probability of a transmission error in a single bit position is 0.01, and suppose the text message is 50 bits long. How likely is it that there will be at least one error in this string of 50 bits?

Multistage Rule

19. You are planning a flight from Tampa to Tulsa. There are no direct flights between these cities, but there are five airlines flying from Tampa to Atlanta, eight from Atlanta to Dallas, and three from Dallas to Tulsa. How many different flight combinations are possible from Tampa to Tulsa?

20. A bank requires each new customer to select a six-character PIN (personal identification number) consisting of two letters followed by four digits, and insists that it contains no "X" and no repeated digits.

(a) How many different PINs are possible? [*Hint*: Think in terms of a six-stage decision tree.]

(b) If the bank has four million customers, will there be enough different PINs so that no two customers have the same PIN?

21. A university requires each student to select a six-character password consisting of three letters followed by three digits, and insists that it contains no repeated letters or repeated digits. How many different passwords are possible in this system?

22. A company requires each employee to select a password consisting of seven digits, and insists that successive digits be different. For example, 6762735 is acceptable but 6772148 is not. How many different passwords are possible in this system?

23. A license plate consists of five digits. How many such license plates contain all different digits?

Testing Our Intuition

24. You and some friends are playing *Yahtzee*™, in which each player rolls five dice.

   (a) What is the probability that all five dice show the same number?

   (b) What is the probability that all five dice show different numbers?

25. In the game of *Farkle*™, each player rolls six dice. If all six dice show the same number, you score 3000 points; if all six dice show different numbers, you score 1500 points.

   (a) What is the probability that all six dice show the same number?

   (b) What is the probability that all six dice show different numbers?

26. An honors literary criticism class has 12 students.

   (a) Calculate the chance that today is the birthday of someone in the class.

   (b) Calculate the chance that two or more of the students in the class have the same birthday.

27. An honors public policy class has 22 students.

   (a) Calculate the chance that today is the birthday of someone in the class.

   (b) Calculate the chance that two or more of the students in the class have the same birthday.

28. In Example 12.14, we calculated the probability of finding at least one matching All-Star card when you purchase five cereal boxes.

   (a) What are the chances of a match if you purchase only four cereal boxes?

   (b) What are the chances of a match if you purchase six cereal boxes?

Projects

29. Simulate the rolling of a single die by using the Dice Experiment available at the web site [OR1]. Use $n = 1$ die and vary the number of times the die is rolled. In particular, observe the number of times a 5 appears and calculate the empirical probability of obtaining a 5. Investigate how these empirical probabilities compare with the theoretical probability as you vary the number of die rolls over a large range.

30. Simulate the rolling of two dice by using the Dice Experiment available at the web site [OR1]. Use $n = 2$ dice and vary the number of times the dice are rolled. In particular, observe the number of times a sum of 6 appears and calculate the empirical probability of obtaining the sum 6. Investigate how these empirical probabilities compare with the theoretical probability as you vary the number of dice rolls over a large range.

31. Develop a spreadsheet program to calculate the birthday coincidence probabilities considered in Example 12.13 for different numbers n of students in the class. Vary the value of n over the range 25 to 45 and summarize the results obtained.

Bibliography

[Brualdi 2010] Brualdi, R. 2010. *Introductory combinatorics*, Chapter 3. Upper Saddle River, NJ: Pearson Prentice Hall.

[Olofsson 2007] Olofsson, P. 2007. *Probabilities: The little numbers that rule our lives*, Chapter 1. Hoboken, NJ: John Wiley & Sons.

[Orkin 1991] Orkin, M. 1991. *Can you win?*, Chapters 1–3. New York: W. H. Freeman.

[Packel 1981] Packel, E. 1981. *The mathematics of games and gambling*, Chapter 2. Washington, D.C.: Mathematical Association of America.

[OR1] http://www.socr.ucla.edu/htmls/SOCR_Experiments.html (accessed November 11, 2013).

Chapter 13

Counting and Unordered Outcomes

Misunderstanding of probability may be the greatest of all general impediments to scientific literacy.

—Stephen Jay Gould, author and evolutionary biologist, 1941–2002

Chapter Preview

This chapter considers situations in which selections are made, yet the order of selecting items is not relevant. Examples of such situations include selecting the members of a committee, choosing numbers for a weekly lottery, and dealing cards from a deck of playing cards. We develop a useful principle for counting outcomes in which the order of selection doesn't matter. The resulting Combinations Formula can then be applied to calculate the probability of complex events. Having such a generally applicable formula is especially useful since our intuition may not be a reliable guide in assessing probabilities in more complicated situations.

13.1 Unordered Outcomes and Lotteries

In several instances we would like to calculate the probability of events in which objects are selected without regard to order. For example, Ravi might want to assess the chances of being chosen to attend a conference in the case where his company randomly chooses only two out of 30 possible employees. In such a situation, it doesn't matter whether Ravi is selected first or second; that is, the order of selection doesn't matter. The following example illustrates another instance in which order doesn't matter.

Example 13.1 Calculate the Number of Clinking Glasses

There are six guests present at a dinner party. A toast is proposed and every pair of guests "clink" their glasses together. How many clinks can be heard?

Solution

One might reason as follows. Each person will clink with five other people and so (since there are six guests) there should be $6 \times 5 = 30$ clinks in total.

To appreciate the flaw in this logic, let us label the guests A, B, C, D, E, F. We can now pair up the clinking guests: A clinks with B, C, D, E, and F; B also clinks with A, C, D, E ,and F; and so on. However, A clinking with B is the same as B clinking with A, so we have double counted the total number of clinks. Similarly, C and D clinking should only be counted once: not both as C clinking with D and again as D clinking with C.

So the total number of clinks is in fact $(6 \times 5)/2 = 15$. There are 15 such *unordered* pairs:

$$AB, \ AC, \ AD, \ AE, \ AF$$
$$BC, \ BD, \ BE, \ BF$$
$$CD, \ CE, \ CF$$
$$DE, \ DF$$
$$EF$$

In this list, there is no order implied in the way each pair is shown. For example, AB here simply indicates that A and B have clinked glasses.

We can extend this reasoning to any number of guests. Suppose that there are n guests at the party and each guest clinks glasses with all the other $n - 1$ guests. The actual number of clinks heard is then $n(n - 1)/2$. In other words, there are $n(n - 1)/2$ different ways to list n pairs of numbers without regard to order. ◁

Example 13.2 Calculate the Probability of Winning the *Match 5* Lottery

To play the Maryland state lottery game *Match 5*, five different numbers are selected from $1, 2, \ldots, 39$. Prizes are awarded if a player matches at least three of the five winning numbers. What are the chances of matching all five numbers (i.e., being a first-place $50,000 winner)?

Solution

To solve this problem, we need to count the number of outcomes in the sample space, which involves all the different ways of choosing five numbers from 1–39. For this lottery game, the order of drawing the numbers does not matter. If the winning lottery numbers are 23, 11, 15, 9, 38 and you wrote your selection as 9, 11, 38, 15, 23 or as 38, 11, 15, 23, 9 then you still win! But if you choose 23, 11, 15, 9, 37 (changing only the last number), you lose; you have made a different five-number selection. To determine your chances of being a first-place winner, we first need to count how many different five-number selections are possible when there are 39 numbers.

It turns out that there are 575,757 different groups of five distinct numbers; this is the total number of possible entries for the lottery. If you choose 23, 11, 15, 9, 38 and it matches all five numbers chosen by the lottery commission, then you win first place. So the probability of having this single winning ticket (our event E) is

$$\Pr[E] = \Pr[\text{winning first place in the 5-number lottery}] = \tfrac{1}{575,757}.$$

How did we calculate the 575,757 unordered distinct arrangements of five numbers?

(a) If we consider all possible *ordered* arrangements of five numbers out of 39 choices, we would have 39 choices for what is listed first, 38 choices for the second number, and so on. Using the Multistage Rule, there are $39 \times 38 \times 37 \times 36 \times 35 = 69{,}090{,}840$ ordered arrangements. (One can think of this value as the number of distinct license plates where there are 39 symbols available in each of the five positions and no symbol is repeated.)

(b) After you choose five specific numbers (such as 23, 11, 15, 9, 38), how many different ways could you rearrange them? You have five choices for what you listed first (9, 11, 38, 15, or 23), four choices for what is listed second, etc. As a result, there are $5 \times 4 \times 3 \times 2 \times 1 = 120$ different orderings for these five numbers.

(c) Because each selection of five specific numbers will appear 120 times in our overall listing of 69,090,840 orderings, we divide 69,090,840 by the 120 different orderings that appear for our specific set of five numbers: $69,090,840/120 = 575,757$. This final value eliminates the double counting in the same way as occurred in Example 13.1. So, the number of different unordered selections can be written as

$$\frac{39 \times 38 \times 37 \times 36 \times 35}{5 \times 4 \times 3 \times 2 \times 1} = \frac{69,090,840}{120} = 575,757.$$

These 575,757 different unordered five-number selections then comprise the relevant sample space. Thus, winning the top $50,000 prize in this lottery should occur just once in every 575,757 tries. Put another way, if we bought 100 $1 tickets each and every week, it would take (on average) over 110 years for us to win the prize: $100 \times 52 \times 110 = \$572,000$. In the process, we would have expended $572,000 in purchases of $1 tickets to win the $50,000 prize. This is not such a good investment! ◁

In Example 13.2, it was necessary to compute the expression $5 \times 4 \times 3 \times 2 \times 1$. There is a useful shorthand for such a product of falling factors. Namely, we write $5! = 5 \times 4 \times 3 \times 2 \times 1$. This factorial notation arose previously in Chapter 5 in estimating the number of possible tours for the Traveling Salesman Problem.

DEFINITION **Factorial**
If n is a positive integer, we define $n! = n \times (n-1) \times (n-2) \times \cdots \times 2 \times 1$, the product of the first n positive integers. For technical reasons, we also define $0! = 1$.

Combinations Formula

Using the factorial function, we state a general formula to aid in probability calculations.

DEFINITION **Combinations Formula**
The number of *unordered* selections of r distinct objects taken from a total of n objects is commonly denoted by $C(n, r)$, indicating the number of *combinations* of n objects taken r at a time. It is given by the formula

$$C(n, r) = \frac{n!}{(n-r)!r!} = \frac{n \times (n-1) \times (n-2) \times \cdots \times (n-r+1)}{r \times (r-1) \times (r-2) \times \cdots \times 1}.$$

The important thing to notice in the Combinations Formula is that there are r successively decreasing factors in the numerator and r successively decreasing factors in the denominator. Of course, the numerator starts with n while the denominator starts with r. As a simple illustration of the formula

$$C(5, 2) = \frac{5!}{(5-2)!2!} = \frac{5 \times 4 \times 3 \times 2 \times 1}{(3 \times 2 \times 1)(2 \times 1)} = \frac{5 \times 4}{2 \times 1} = \frac{20}{2} = 10.$$

Returning to Example 13.2 we can determine the number of ways of selecting five numbers from a total of 39 numbers, without regard to order, as $C(39, 5) = 575,757$ using the

Combinations Formula:

$$C(39,5) = \frac{39!}{34!5!} = \frac{39 \times 38 \times 37 \times 36 \times 35}{5 \times 4 \times 3 \times 2 \times 1} = \frac{69,090,840}{120} = 575,757.$$

We can then write the probability of a first-place win in the *Match 5* lottery as

$$\frac{C(5,5)}{C(39,5)} = \frac{1}{575,757}.$$

Here the numerator expresses the fact that we must choose all five numbers from the five winning numbers.

(Related Exercises: 1–5)

Example 13.3 Calculate the Probability of a Second-Place Win in the *Match 5* Lottery

In the Maryland state lottery game *Match 5* discussed in Example 13.2, five different numbers are selected from $1, 2, \ldots, 39$. What are the chances of matching exactly four out of the five numbers (second place)? Remember that the sample space is the same as in Example 13.2, consisting of 575,757 equally likely outcomes.

Solution

Here we focus on the event E of matching exactly four of the winning numbers. To be concrete, if the winning numbers happen to be 12, 14, 25, 34, 35, then matching exactly four numbers would occur when we obtain an outcome such as $\{X, 14, 25, 34, 35\}$ where X is any number other than 12, 14, 25, 34, 35. There are 34 such outcomes since there are $39 - 5 = 34$ allowable values for X. Similarly, there are 34 outcomes for each of the additional situations that correspond to matching exactly four winning numbers:

$$\{12, X, 25, 34, 35\}, \{12, 14, X, 34, 35\}, \{12, 14, 25, X, 35\}, \{12, 14, 25, 34, X\}.$$

Altogether there are $5 \times 34 = 170$ outcomes in E so $\Pr[E] = 170/575,757$; this corresponds to approximately one chance in 3387. Said in another way, the odds are 3386:1 against your matching exactly four numbers.

Using the Combinations Formula, we can write the solution to this problem as follows:

$$\frac{C(5,4) \times C(34,1)}{C(39,5)} = \frac{5 \times 34}{575,757} = \frac{170}{575,757}.$$

The numerator corresponds to selecting four matching numbers from the five winning numbers and then selecting one non-matching number from the 34 losing numbers. ◁

Example 13.4 Calculate the Probability of a Third-Place Win in the *Match 5* Lottery

What are the chances of matching exactly three of the five numbers in this Maryland lottery?

Solution

Now we focus on the event E of matching exactly three winning numbers. If the winning numbers again happen to be 12, 14, 25, 34, 35, then matching exactly three of them would mean you select two numbers X, Y that are *not* 12, 14, 25, 34, 35: for example, $\{X, Y, 25, 34, 35\}$.

The number of possible distinct pairs of *losing* numbers (without regard to order) can be found by calculating $C(34, 2)$, the number of ways of selecting two numbers out of the 34 losing numbers:

$$C(34, 2) = \frac{34 \times 33}{2 \times 1} = 561.$$

So there are 561 outcomes in E corresponding to $\{X, Y, 25, 34, 35\}$. But we are not quite finished since there are other ways of matching exactly three numbers: these include $\{X, 14, Y, 34, 35\}, \{12, X, 25, Y, 35\}$, and $\{12, 14, 25, X, Y\}$. In fact, the number of such "patterns" is the number of ways of selecting (without regard to order) the three *matching* numbers from among the five winning numbers, or $C(5, 3) = (5 \times 4 \times 3)/(3 \times 2 \times 1) = 10$. Altogether there are $10 \times 561 = 5610$ outcomes in E so that $\Pr[E] = 5610/575{,}757$; this translates into the approximate odds of 102:1 against matching exactly three numbers.

The final answer here can be written as follows:

$$\frac{C(5, 3) \times C(34, 2)}{C(39, 5)} = \frac{10 \times 561}{575{,}757} = \frac{5610}{575{,}757}.$$

The numerator corresponds to selecting three matching numbers from the five winning numbers and then selecting two non-matching numbers from the 34 losing numbers. ◁

(*Related Exercises: 13–20, 33*)

Example 13.5 Calculate Probabilities in the Composition of a Committee

A legislative committee consists of seven Republicans and five Democrats. Four individuals are to be selected from this committee for a fact-finding overseas mission. To be fair, the names of all twelve committee members are placed in a hat and the members selected for this fact-finding mission are determined by pulling out four names. (a) How likely is it that all four selected members are Republicans? (b) How likely is it that all four selected members are Democrats? (c) How likely is it that equal numbers of Republicans and Democrats are selected?

Solution

First, we determine the number of (unordered) selections of four individuals from the twelve committee members as $C(12, 4)$, which is calculated using the Combinations Formula:

$$C(12, 4) = \frac{12 \times 11 \times 10 \times 9}{4 \times 3 \times 2 \times 1} = 495.$$

This represents the size of the sample space for all parts of this problem.

(a) The number of ways to select four individuals from the group of seven Republicans is

$$C(7, 4) = \frac{7 \times 6 \times 5 \times 4}{4 \times 3 \times 2 \times 1} = 35$$

so that the required probability is

$$\frac{C(7, 4)}{C(12, 4)} = \frac{35}{495} = 0.071.$$

(b) The number of ways to select four individuals from the group of five Democrats is

$$C(5, 4) = \frac{5 \times 4 \times 3 \times 2}{4 \times 3 \times 2 \times 1} = 5$$

so that the required probability is

$$\frac{C(5,4)}{C(12,4)} = \frac{5}{495} = 0.010.$$

(c) This can be viewed as a two-stage process: first selecting the Republican members and then the Democratic members for the fact-finding mission. The number of ways of selecting two individuals from the group of seven Republicans is

$$C(7,2) = \frac{7 \times 6}{2 \times 1} = 21,$$

while the number of ways of selecting two individuals from the group of five Democrats is

$$C(5,2) = \frac{5 \times 4}{2 \times 1} = 10.$$

By the Multistage Rule, there are $C(7,2) \times C(5,2) = 21 \times 10 = 210$ ways of choosing equal numbers of Republicans and Democrats, so the likelihood of obtaining such equal numbers is given by

$$\frac{C(7,2) \times C(5,2)}{C(12,4)} = \frac{210}{495} = 0.424.$$

The numerator corresponds to selecting two members from the seven Republicans and then selecting two members from the five Democrats. The denominator reflects the fact that we are selecting four individuals in total from a committee with twelve members. ◁

(*Related Exercises: 6–12*)

13.2 Poker Hands

The next examples involve the card game of five-card poker and illustrate how we can repeatedly apply the Multistage Rule, when order is not important, to calculate the probability of fairly complex events. Throughout we will again rely upon the Combinations Formula.

Recall some facts about a standard deck of 52 cards. Each card has thirteen possible *denominations*—2, 3, 4, 5, 6, 7, 8, 9, 10, J, Q, K, A—where J, Q, K, A stand for Jack, Queen, King, and Ace. There are four different *suits*: Hearts, Diamonds, Clubs, and Spades. Thus, AH and 2S denote the Ace of Hearts and the two of Spades, respectively. The thirteen denominations times the four suits totals 52 cards.

Example 13.6 Determine the Probability of Being Dealt Four of a Kind

Determine the likelihood that you will be dealt *four of a kind*, a hand that contains four cards with the same denomination (one in each suit). The fifth card can have any denomination (except the one already used) in any suit. Figure 13.1 illustrates such a dealt hand.

Solution

To begin, we define the sample space: it consists of all (unordered) selections of five cards

FIGURE 13.1: Example of four of a kind

from the 52-card deck. The number of outcomes in the sample space is then $C(52, 5)$, which we calculate using the Combinations Formula:

$$C(52, 5) = \frac{52 \times 51 \times 50 \times 49 \times 48}{5 \times 4 \times 3 \times 2 \times 1} = 2{,}598{,}960.$$

Next, we count the number of outcomes that comprise event E, corresponding to drawing four of a kind. An example of such an outcome would be {9H, 9C, 9D, 9S, 4D}, where all four suits of the nine appear and a different denomination (a four) also appears. More generally, instead of the nine, we could choose any denomination from the thirteen different values $2, 3, \ldots, 10$, J, Q, K, A. Think of this as the first stage of our selection process. We then choose all four of the different suits, which is our second stage.

In the third stage we choose any card (the four of Diamonds in our example) different from the four cards defined by the first stage. There are $52 - 4 = 48$ different choices possible for the selection at the third stage, so the Multistage Rule yields $13 \times 1 \times 48 = 624$ outcomes (different four-of-a-kind hands) comprising event E. The probability of obtaining four of a kind is then $\Pr[E] = 624/2{,}598{,}960 = 0.00024$. Stated in another way, we would expect to be dealt such a hand just one time in every 4165 deals from a well-shuffled deck.

This answer can be written as follows, which parallels the three-stage construction process:

$$\frac{C(13, 1) \times C(4, 4) \times C(48, 1)}{C(52, 5)} = \frac{13 \times 1 \times 48}{2{,}598{,}960} = \frac{624}{2{,}598{,}960}. \quad \triangleleft$$

Example 13.7 Determine the Probability of Being Dealt a Full House

Determine the likelihood that you are dealt a *full house*, a hand with one denomination making three of a kind, and another denomination constituting a pair. An example of such a hand appears in Figure 13.2, which displays a three of a kind (eights) and a pair (Kings).

FIGURE 13.2: Example of a full house

Solution

Let event E correspond to obtaining a full house. The construction of such a hand can be viewed as a four-stage selection process. We select any denomination for the three of a kind (13 choices) in the first stage, and in the second stage we select the corresponding three suits from {H, D, C, S} in $C(4, 3) = 4$ ways. In the third stage, we select a different denomination for the pair (twelve choices remain), and in the fourth stage we select their two suits in $C(4, 2) = 6$ ways. By the Multistage Rule, there are $13 \times 4 \times 12 \times 6 = 3744$ outcomes in event E so that $\Pr[E] = 3744/2{,}598{,}960 = 0.0014$; this means the likelihood of being dealt a full house is about once in every 694 deals. Stated another way, the odds of getting a full house are 693:1 against you.

We can write this answer as follows:

$$\frac{C(13,1) \times C(4,3) \times C(12,1) \times C(4,2)}{C(52,5)} = \frac{13 \times 4 \times 12 \times 6}{2{,}598{,}960} = \frac{3744}{2{,}598{,}960}. \quad \triangleleft$$

Example 13.8 Determine the Probability of Being Dealt a Flush

A *flush* consists of five cards, all in the same suit, such as the five clubs shown in Figure 13.3. How likely is it that you will be dealt such a hand?

FIGURE 13.3: Example of a flush

Solution

The suit of a flush can be any one of four suits and the values of the selected cards can be any five chosen from the thirteen denominations. Using the Combinations Formula, there are $C(13,5) = 1287$ unordered selections of five values from the thirteen denominations.

An interesting wrinkle occurs here, however. Among these 1287 are selections such as $\{5,6,7,8,9\}$—which would give us an even better hand (called a *straight flush*). Since we only want to determine the chances of a (regular) flush, we need to remove all such "consecutive" occurrences from our count. In fact there are 10 of these special cases: $\{A,2,3,4,5\}, \{2,3,4,5,6\}, \{3,4,5,6,7\}, \ldots, \{10,J,Q,K,A\}$. As a result, there are $1287 - 10 = 1277$ valid selections for the denominations.

Now apply the Multistage Rule to our two-stage selection process: there are 1277 choices for the denominations (first stage) and four choices for the suit (second stage), giving $1277 \times 4 = 5108$ outcomes comprising event E. The probability of obtaining a flush is then $\Pr[E] = 5108/2{,}598{,}960 = 0.0020$, signifying chances of about 1 in 500.

We can write this answer as follows:

$$\frac{(C(13,5) - C(10,1)) \times C(4,1)}{C(52,5)} = \frac{(1287 - 10) \times 4}{2{,}598{,}960} = \frac{5108}{2{,}598{,}960}. \quad \triangleleft$$

(*Related Exercises: 21–31*)

Chapter Summary

We first develop an approach for counting outcomes in which the order of selection is not relevant. The resulting Combinations Formula enables us to find the number of ways to select a set of r items from a larger set of n items. More complicated counting problems can be analyzed by decomposing the given situation into its various components and applying the Combinations Formula to each component. The Multistage Rule is then applied to combine these counts into an overall number. Using the ratios of desired counts to the total number of possibilities in the sample space then produces the probability of occurrence of complex events, such as those arising in lotteries and poker hands.

Applications, Representations, Strategies, and Algorithms

Applications		
Gambling (2–4)		

Representations
Sets
Lists
1

Strategies		
Brute-Force	Rule-Based	Composition/Decomposition
1	2–8	3–4

Algorithms			
Exact	Enumeration	Sequential	Numerical
1–8	1	2–8	2–8

Exercises

Combinations Formula

1. Fifteen citizens of a town have been summoned as possible jurors for a criminal trial. The defense lawyer can act to remove three of the potential jurors from consideration. In how many ways can these three jurors be identified?

2. A cross-country track team consists of 10 runners. The coach needs to select six of the runners (without regard to their past performance on the team) for an out-of-state track meet. In how many ways can this be done?

3. Stanley needs to choose three flavors for his sundae from the 28 flavors available at the local ice cream store. How many possible selections are possible for him?

4. Serena is driving from Charlottesville to her home in Atlanta and wants to take along four music CDs to play on her trip. In how many ways can she select the CDs from her collection of eleven favorite music CDs?

5. Sheldon is part of a committee that provides refreshments at a computer club that holds monthly meetings. He needs to select three months out of the upcoming year to provide donuts. In how many ways can this be done?

6. A polling company wants to interview six likely voters at a coffee shop, with equal numbers of men and women selected. In how many ways can the voters be selected if there are eleven women and nine men at the coffee shop?

7. Leonard is taking a final exam in Advanced Quantum Physics. The test is structured into two parts: the theory part contains six questions and the applications part contains five questions. Three questions are to be answered from the theory portion and two questions are to be answered from the applications portion. In how many different ways can Leonard answer the final exam questions?

8. A department in a university is selecting a new chair, so a search committee needs to be constituted. The committee will consist of four faculty and two non-faculty members. There are 20 faculty and 14 non-faculty members in the department. In how many ways can the committee be formed?

9. Suppose that the faculty in Exercise 8 consists of 9 tenured faculty and 11 tenure-track faculty. The search committee will now consist of two tenured faculty, two tenure-track faculty, and two non-faculty members. In how many ways can the committee be formed?

10. In 2013, the U.S. Senate Select Committee on Intelligence consisted of eight Democrats and seven Republicans. A subcommittee is being formed to investigate a new intelligence leak; it is to consist of three Democrats and three Republicans. In how many ways can this subcommittee be formed?

11. A University Grievance Committee consists of four students and eight faculty members. Two students and three faculty members need to be assigned from this committee to a new grievance case. In how many ways can this be done?

12. A sociology class contains 9 out-of-state students and 12 in-state students. The instructor randomly selects six students to work on a project.

    (a) How many different selections of six students are possible?

    (b) How many different selections of six out-of-state students are possible?

    (c) How many different selections of six students are possible that contain equal numbers of in-state and out-of-state students?

Lottery Probabilities

13. To play a certain lottery game, you choose six numbers from 1–40. Find the probability of matching exactly four of the six winning numbers for a third-place prize.

14. In a particular Education Lottery, a player chooses five numbers from 1–38.

    (a) Calculate the probability of matching all five winning numbers.

    (b) Calculate the probability of matching exactly four of the five winning numbers.

    (c) Calculate the probability of matching exactly three of the five winning numbers.

15. In the *Powerball* lottery, 5 white balls are chosen from a group of 59 white balls and 1 red ball is chosen from another group of 35 red balls. Find the probability of winning the jackpot, which involves matching all five of the selected white balls and also matching the red ball.

16. In the *Powerball* lottery, 5 white balls are chosen from a group of 59 white balls and 1 red ball is chosen from another group of 35 red balls. Find the probability of winning the second prize, which involves matching all five of the selected white balls but not the red ball.

17. In the *Powerball* lottery, 5 white balls are chosen from a group of 59 white balls and 1 red ball is chosen from another group of 35 red balls. Find the probability of matching the red ball only, and none of the white balls.

18. In the *Powerball* lottery, 5 white balls are chosen from a group of 59 white balls and 1 red ball is chosen from another group of 35 red balls. Find the probability of matching four of the selected white balls but not the red ball.

19. A state is considering a lottery in which numbers are selected from 1–35. It expects to sell around 300,000 tickets each week. In order to make the lottery interesting, it is desirable for several weeks to pass before a winner is determined (but not too many weeks). Based on these considerations, the lottery commission would like the odds for winning in each week to be about 1 in 1.6 million. How many numbers from 1–35 should a player be instructed to choose?

20. Another state is considering a lottery in which seven numbers are selected from some range $1, 2, \ldots, n$. It expects to sell around 300,000 tickets each week. In order to make the lottery interesting, it is desirable for several weeks to pass before a winner is determined (but not too many weeks). Based on these considerations, the lottery commission would like the odds for winning in each week to be about 1 in 1.5 million. What value of n should be used to achieve the commission's objectives?

Poker Probabilities

Except as noted, the following problems refer to five-card poker. The *face cards* are J, Q, K, A while the *number cards* are $2, 3, \ldots, 10$.

21. Determine the probability of being dealt a *royal flush*: $\{10, \text{J}, \text{Q}, \text{K}, \text{A}\}$ all in the same suit. A royal flush would then be expected to occur once in about how many hands?

22. Determine the probability of being dealt a *straight flush*: five cards in the same suit that form a consecutive sequence of denominations. This answer should include drawing a royal flush.

23. Determine the probability of being dealt a *straight*: five cards forming a consecutive sequence of denominations. This answer should exclude drawing a straight flush.

24. Determine the probability of being dealt a five-card hand that contains no face cards.

25. Determine the probability of being dealt a five-card hand that contains all face cards.

26. Determine the probability of being dealt a five-card hand that contains at least one number card.

27. Determine the probability of being dealt a *small flush*: exactly four cards in the same suit and one card in a different suit. [*Hint*: Think of this as a three-stage decision problem.]

28. Determine the probability of being dealt a five-card hand with two pairs: two cards have one denomination, two cards have another denomination, and the remaining card's denomination is different from those of the other four cards. [*Hint*: Think of this as a four-stage decision problem.]

29. Determine the probability of being dealt three of a kind, in which three cards have the same denomination, while the remaining two cards have different denominations from each other and the other three cards. [*Hint*: Think of this as a five-stage decision problem.]

30. Determine the probability of being dealt a five-card hand with only one pair: two of the cards have the same denomination, but the remaining three cards do not have this denomination and have different denominations from one another. [*Hint*: Think of this as a six-stage decision problem.]

31. Determine the probability of a *super-full house* (not official terminology) in a dealt hand of six cards. In such a hand, one denomination forms three of a kind while the other denomination forms a different three of a kind.

Projects

32. The book *Bringing Down the House* [Mezrich 2003], and the movie *21* adapted from this book, follow the adventures of six MIT students who used card counting methods to gain advantages over the standard probabilities calculated for the game of blackjack. Investigate card counting schemes and the improved winning probabilities these schemes can provide.

33. The game *Keno* is a popular game played in casinos. To play, you select 10 numbers from $1, 2, \ldots, 80$ and submit your numbers. Then 20 winning numbers are chosen from the 80 numbers by a random selection process, which can be viewed live. You receive a monetary payoff for your $2 bet based on matching at least five of the selected winning numbers. Use the Combinations Formula to calculate the probability of matching exactly $5, 6, \ldots, 10$ of the winning numbers.

Bibliography

[Hartley 2004] Hartley, J. 2004. *The little black book of poker*. White Plains, NY: Peter Pauper Press.

[Mezrich 2003] Mezrich, B. 2003. *Bringing down the house*. New York: Free Press.

[Orkin 1991] Orkin, M. 1991. *Can you win?*, Chapter 7. New York: W. H. Freeman.

[Packel 1981] Packel, E. 1981. *The mathematics of games and gambling*, Chapter 4. Washington, D.C.: Mathematical Association of America.

[Thorpe 1966] Thorpe, E. O. 1966. *Beat the dealer*. New York: Random House.

[OR1] http://www.cwu.edu/~glasbys/poker.htm (accessed November 18, 2013).

Chapter 14

Independence and Conditional Probabilities

The 50-50-90 rule: Anytime you have a 50-50 chance of getting something right, there's a 90% probability you'll get it wrong.

—Andy Rooney, American radio and television writer, 1919–2011

Chapter Preview

This chapter studies independent events, where the occurrence of one event is not influenced by the occurrence of the other. It also presents the notion of disjoint events, those that cannot possibly happen together. It turns out that a number of uncertain situations can be analyzed by decomposing them into independent and disjoint events. Appropriate generalizations of the Multiplication Rule and the Addition Rule are involved. However, in many situations we encounter dependent events (e.g., forecasting today's weather having observed yesterday's weather). In such circumstances, we require the more general concept of conditional probability. Understanding conditional probabilities can be aided by the use of decision trees, introduced in Chapter 12.

14.1 Independent Events and Disjoint Events

In Example 12.3, we found that in tossing a fair coin twice, the probability of obtaining tails on the first toss and heads on the second toss could be calculated as $\Pr[TH] = \frac{1}{4} = \frac{1}{2} \times \frac{1}{2}$. That is, the probability of this *outcome* equals the probability of obtaining a tail on the first toss multiplied by the probability of obtaining a head on the second toss. This relationship holds because the result on the second toss is *independent* of the result on the first toss: the result on the second toss is not affected by the result on the first toss. More generally, we will be interested in whether two *events* E and F are independent of one another.

DEFINITION **Independent Events**

Events E and F are independent when the probability of event E followed by event F equals the probability of event E multiplied by the probability of event F:

$$\Pr[E \text{ and } F] = \Pr[E] \times \Pr[F].$$

The above formula is a natural generalization of the Multiplication Rule stated in Chapter 12. The probability of events E and F both occurring is called a *joint event probability* and is sometimes designated as $\Pr[EF]$ instead of $\Pr[E \text{ and } F]$. For our purposes, independence of two events means the knowledge that one event has occurred does not affect the probability of the other event occurring. The event "It will be sunny today" would not be independent of the event "It was sunny yesterday," since the previous day's weather does have an effect on the likelihood of the current day's weather.

In tossing a coin twice, we have also seen that the probability of obtaining exactly one head is given by $\Pr[HT, TH] = \Pr[HT \text{ or } TH] = \frac{1}{2} = \frac{1}{4} + \frac{1}{4}$. That is, the probability of exactly one head is obtained by adding the probabilities of two outcomes: a head occurring only on the first toss (HT), and a head occurring only on the second toss (TH). This simplification is based on the Addition Rule of Chapter 12, which states that the probability of an event is obtained by adding together the probabilities of its constituent outcomes. What makes this Addition Rule valid is that the constituent outcomes are *disjoint*: they cannot happen together. More generally, if *events E and F have no outcomes in common*, then the probability $\Pr[E \text{ or } F]$ can be easily calculated by summing their respective probabilities.

DEFINITION **Disjoint Events**

Events E and F are disjoint if they both cannot happen simultaneously. In this case, the probability that either E or F occurs is given by

$$\Pr[E \text{ or } F] = \Pr[E] + \Pr[F].$$

In terms of the decision tree, being disjoint means that the events E and F contain no common outcomes. This difference between disjoint and overlapping events is shown schematically by the Venn diagrams in Figure 14.1. Events E and F are disjoint in (a) while they overlap in (b).

FIGURE 14.1: Disjoint and overlapping events

The following examples illustrate the concepts of independence and disjointness.

Example 14.1   Calculate the Probability of Rolling a Sum of 7 or 11 with a Pair of Dice

A pair of (six-sided) dice is rolled. We would like to determine the probability that the sum of the numbers showing on the two dice equals 7 or 11. Such a calculation is important in certain casino games. (a) List the outcomes constituting the event E in which the face values sum to 7, and list the outcomes constituting the event F in which the face values sum to 11. (b) List the outcomes for the combined event G in which the face values sum to either 7 or 11. (c) Calculate the probability $\Pr[G] = \Pr[E \text{ or } F]$ and compare it with $\Pr[E] + \Pr[F]$.

Solution

(a) Event $E = \{(1,6), (2,5), (3,4), (4,3), (5,2), (6,1)\}$ consists of outcomes in which the sum of the face values is 7, and event $F = \{(5,6), (6,5)\}$ consists of outcomes in which the sum of the face values is 11. These events are represented by the Venn diagram in Figure 14.2.

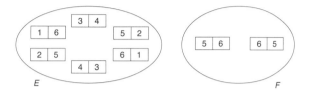

FIGURE 14.2: Face values that sum to 7 or 11

(b) The event of interest $G = \{(1,6), (2,5), (3,4), (4,3), (5,2), (5,6), (6,1), (6,5)\}$ contains all of the outcomes shown in Figure 14.2.

(c) Since G contains eight equally likely outcomes, $\Pr[G] = \frac{8}{36}$. Similarly, $\Pr[E] = \frac{6}{36}$ and $\Pr[F] = \frac{2}{36}$. Notice that $\Pr[E] + \Pr[F] = \frac{6}{36} + \frac{2}{36} = \frac{8}{36} = \Pr[G]$. Here the events E and F cannot happen simultaneously, so the rule for disjoint events applies: $\Pr[G] = \Pr[E \text{ or } F] = \Pr[E] + \Pr[F]$. ◁

Example 14.2 Calculate the Probability of Rolling at Least One 5 with a Pair of Dice

In the context of successively rolling a pair of dice, we would like to determine the probability that the number 5 shows up on at least one of the faces. (a) List the outcomes constituting the event E in which the number 5 appears on the first die, and list the outcomes constituting the event F in which the number 5 appears on the second die. (b) List the outcomes for the combined event G in which the number 5 appears during the roll of the two dice. (c) Calculate the probability $\Pr[G] = \Pr[E \text{ or } F]$ and compare it with $\Pr[E] + \Pr[F]$.

Solution

(a) The respective events here are $E = \{(1,5), (2,5), (3,5), (4,5), (5,5), (6,5)\}$ and $F = \{(5,1), (5,2), (5,3), (5,4), (5,5), (5,6)\}$.

(b) The event $G = \{(1,5), (2,5), (3,5), (4,5), (5,1), (5,2), (5,3), (5,4), (5,5), (5,6), (6,5)\}$.

(c) Since events E and F each contain six outcomes, $\Pr[E] = \frac{6}{36}$, $\Pr[F] = \frac{6}{36}$, and $\Pr[E] + \Pr[F] = \frac{12}{36}$. Since event G consists of eleven outcomes, $\Pr[G] = \frac{11}{36}$. Notice that $\Pr[G] \neq \Pr[E] + \Pr[F]$. The reason for this is that events E and F are not disjoint. They share the common outcome $(5,5)$ which is counted twice in the sum $\Pr[E] + \Pr[F]$ but only once in $\Pr[G]$. This overlap between events E and F is seen in the Venn diagram of Figure 14.3. ◁

(Related Exercises: 1–11)

In Example 14.2, consider the probability of event E followed by event F. This is simply the joint probability $\Pr[EF] = \Pr[E \text{ and } F]$ and is thus the probability of obtaining $(5,5)$: a roll of 5 on the first die and a roll of 5 on the second die. Therefore, $\Pr[EF] = \frac{1}{36} =$

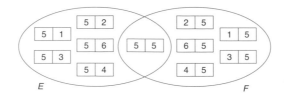

FIGURE 14.3: Face values in which a 5 appears

$\frac{1}{6} \times \frac{1}{6} = \Pr[E] \times \Pr[F]$. This is to be expected, since the knowledge of E happening does not affect the likelihood of F happening: namely, E and F are independent events.

Now we are ready to tackle two problems that require judicious application of both disjoint and independent events.

Real-time financial transactions (such as those carried out at a bank's ATM) need to be extremely accurate. Unfortunately, the processor carrying out these transactions can be subject to occasional errors in its numerical calculations. One way to increase the accuracy of such transactions is to have three independent processors (call them 1, 2, 3) carry out the calculations separately. The numerical results from processors 1, 2, and 3 are then compared and a "majority vote" is taken to decide the correct answer.

Specifically, if processor 1 and processor 3 both return the same value, but processor 2 gives a different answer, then the answer from 1 and 3 is accepted. If all three agree, then that common answer is accepted. The only problem occurs if 1, 2, and 3 give different results; in this case, the system does not work and manual intervention will be needed.

Example 14.3 Calculate the Probability of Obtaining a Correct Consensus Answer

Suppose each processor yields a correct answer C with probability p and yields a wrong answer W with probability $1 - p$. The sample space consists of all possible answers from processors 1, 2, and 3: that is, $\{CCC, CCW, CWC, CWW, WCC, WCW, WWC, WWW\}$. We would like to determine the probability that the majority vote protocol will produce an accepted answer that is correct (C) as opposed to the wrong answer (W). (a) Calculate the probability that all three processors return the correct answer. (b) Calculate the probability that exactly two of the processors return the correct answer. (c) Calculate the probability that a majority vote of all processors will return the correct answer.

Solution

(a) The event of interest is $E = \{CCC\}$. Since the answers from processors 1, 2, and 3 are considered as independent, $\Pr[CCC] = \Pr[C] \times \Pr[C] \times \Pr[C] = p^3$.

(b) The event of interest is $F = \{CCW, CWC, WCC\}$. Again by independence, $\Pr[CCW] = \Pr[C] \times \Pr[C] \times \Pr[W] = p^2(1 - p)$, Similarly, $\Pr[CWC] = \Pr[WCC] = p^2(1 - p)$. By the Addition Rule, $\Pr[F] = 3p^2(1 - p)$.

(c) We are interested in the event G in which the correct answer is produced by at least two of the processors. So we can express $\Pr[G] = \Pr[E \text{ or } F]$. Since E and F are disjoint (both cannot occur together), $\Pr[G] = \Pr[E] + \Pr[F] = p^3 + 3p^2(1 - p)$.

As a numerical illustration, if $p = 0.95$ then using a single processor results in a correct answer 95% of the time. If instead we use a majority vote among three independent processors, the reliability of the answer increases to $(0.95)^3 + 3(0.95)^2(0.05) = 0.99275$, thus producing a correct answer in 99.3% of the cases. ◁

(Related Exercises: 12–15)

Example 14.4 Calculate the Probability of Winning a Tournament with Four Teams

Teams A, B, C, D are playing in a single-elimination tournament, which is represented by the bracket in Figure 14.4. Here, A and B are paired in the first round, as well as C and D. The winners then advance to the next round, where the tournament champion is decided.

From past meetings of teams, it is known that there is an 80% chance that A will beat B, a 70% chance that A will beat C, and a 65% chance that A will beat D. Also, there is a 60% chance that B will beat C, a 70% chance that B will beat D, and a 55% chance that C will beat D. Given this information, how likely is it that team A will win the tournament?

FIGURE 14.4: A tournament with four teams

Solution

One way that A can emerge the winner is if A beats B and C beats D, followed by A beating C. Assuming that these contests are independent of one another, the probability of this compound event E is given by $\Pr[E] = 0.80 \times 0.55 \times 0.70 = 0.308$. A second way that A can emerge the winner is if A beats B and D beats C, followed by A beating D. Assuming independence, the probability of this compound event F is given by $\Pr[F] = 0.80 \times (1 - 0.55) \times 0.65 = 0.234$. Events E and F are disjoint, since C beats D in the first event, while the reverse holds in the second event. Using the rule for disjoint events then gives $\Pr[A \text{ wins}] = \Pr[E] + \Pr[F] = 0.308 + 0.234 = 0.542$. ◁

(Related Exercise: 16)

14.2 Conditional Probabilities

When two events are independent, the knowledge of one event does not convey any information about the likelihood of the other event. However, not all events are independent. Indeed, in many situations, the occurrence (or nonoccurrence) of some event provides information that can be used to refine our estimates of other event probabilities. For instance, the life expectancy of an individual depends on his or her current age (how many years the individual has already lived). To illustrate, recent data from the National Center for Health Statistics show that a white female aged 20 can expect to live an additional 61.9 years (to age 81.9), while a 60-year-old white female can expect to live an additional 24.5 years (to age 84.5).

These considerations lead us to the concept of *conditional probabilities*—probabilities that incorporate prior information. Many of the paradoxes and misunderstandings concerning probabilities occur through the misapplication of conditional probabilities. The following example introduces the concept of conditional probabilities.

Example 14.5 Calculate the Probability that Two Democrats are Chosen

A local school board committee consists of five Democrats and seven Republicans. Two committee members are to be chosen at random to be national delegates. What is the probability that both are Democrats?

Solution

For ease of identification, let D represent the event that a Democrat is chosen and let R represent the event that a Republican is chosen. As done in Chapter 12, it is convenient to think of this situation as a two-step decision process: selecting the first delegate and then selecting the second delegate. The probability that the first delegate is a Democrat is $\Pr[D] = \frac{5}{12}$. However, the probability that the second delegate is a Democrat, given the additional information that the first delegate is already a Democrat, is no longer $\frac{5}{12}$. Rather, there are now four Democrats and seven Republicans left from which to choose, so that the probability of selecting a second Democrat is now $\frac{4}{11}$. We write this as the *conditional probability* $\Pr[D|D] = \frac{4}{11}$, the probability of event D *given* that event D has already occurred.

We can use the decision tree in Figure 14.5 to show this two-stage process, explicitly specifying the conditional probabilities of each second choice, given the occurrence of the first choice. In particular, the highlighted path in Figure 14.5 from the starting state s to the outcome DD represents the result of choosing a first event (namely D) and then a second event (D). By the Multiplication Rule, the probability of outcome DD (selecting two Democrats) is the product of the edge probabilities along this path:

$$\Pr[DD] = \Pr[D] \times \Pr[D|D] = \frac{5}{12} \times \frac{4}{11} = \frac{20}{132}. \quad \triangleleft$$

(Related Exercises: 17–18)

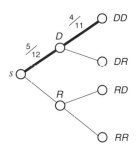

FIGURE 14.5: Decision tree for delegate selection

Example 14.6 Calculate Other Probabilities in Delegate Selection

As in Example 14.5, two delegates are to be randomly selected from a group of five Democrats and seven Republicans. (a) What is the probability that both are Republicans? (b) What is the probability that one Republican and one Democrat are chosen?

Solution

Figure 14.6 displays all edge probabilities for the decision tree. For example, $\Pr[D|R] = \frac{5}{11}$ since after the selection of one Republican, there are five Democrats and six Republicans remaining. Moreover, use of the Complement Rule gives $\Pr[R|D] = 1 - \Pr[D|D] = 1 - \frac{4}{11} = \frac{7}{11}$ and $\Pr[R|R] = 1 - \Pr[D|R] = 1 - \frac{5}{11} = \frac{6}{11}$.

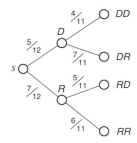

FIGURE 14.6: Enhanced decision tree for delegate selection

(a) Applying the Multiplication Rule gives

$$\Pr[RR] = \Pr[R] \times \Pr[R|R] = \tfrac{7}{12} \times \tfrac{6}{11} = \tfrac{42}{132}.$$

(b) To find the probability of selecting one Democrat and one Republican, we first calculate

$$\Pr[DR] = \Pr[D] \times \Pr[R|D] = \tfrac{5}{12} \times \tfrac{7}{11} = \tfrac{35}{132},$$
$$\Pr[RD] = \Pr[R] \times \Pr[D|R] = \tfrac{7}{12} \times \tfrac{5}{11} = \tfrac{35}{132}.$$

The event E of interest involves the two outcomes DR and RD. By the Addition Rule,

$$\Pr[E] = \Pr[DR] + \Pr[RD] = \tfrac{35}{132} + \tfrac{35}{132} = \tfrac{70}{132}. \quad \triangleleft$$

(Related Exercises: 19–21)

In the decision tree of Figure 14.6, the edge probabilities are not all the same, and the conditional probabilities (at the second stage) crucially depend on what happened at the first stage. The Multiplication Rule allows us to find *outcome probabilities* by multiplying the edge probabilities along the path to that outcome. This observation generalizes, and we can find *joint event* probabilities by multiplying the appropriate (conditional) probabilities.

DEFINITION **Joint Probability Rule**
The probability of the joint event EF (that is, event E followed by event F) can be calculated as $\Pr[EF] = \Pr[E] \times \Pr[F|E]$.

We illustrate this general rule for calculating joint probabilities in the following examples.

Example 14.7 Calculate the Probability of Drawing Two Aces

Two cards are drawn at random from a deck of 52 cards. What is the probability that both selected cards are Aces?

Solution

Again we can use the idea of conditional probabilities. Let E represent the event that the first card drawn is an Ace and let F represent the event that the second card drawn is an Ace. Then $\Pr[E] = \tfrac{4}{52}$ since the deck contains four Aces. Also, $\Pr[F|E] = \tfrac{3}{51}$ since drawing a first Ace leaves only three Aces among the 51 remaining cards. Applying the Joint Probability Rule

$$\Pr[EF] = \Pr[E] \times \Pr[F|E] = \tfrac{4}{52} \times \tfrac{3}{51} = 0.00452,$$

which represents less than a 0.5% chance. Here the events E and F are not independent of one another: the knowledge that an Ace occurred as the first selection affects the probability that a second Ace is drawn next. Specifically, $\Pr[F|E] = \frac{3}{51} \neq \Pr[F] = \frac{4}{52}$. ◁

If $\Pr[F|E] = \Pr[F]$, then the knowledge that E occurred has no effect on the probability that F occurs, so the events should be independent. The Joint Probability Rule does indeed cover this case. Namely, if $\Pr[F|E] = \Pr[F]$, then

$$\Pr[EF] = \Pr[E] \times \Pr[F|E] = \Pr[E] \times \Pr[F],$$

which is just our previous definition for independent events. We now use this alternative representation for independence to explain an apparent paradox in the following situation.

Example 14.8 Calculate Probabilities in Drawing a Single Card

(a) A single card is drawn from a standard deck of 52 cards. Let E be the event that the card drawn is an Ace, and let F be the event that the card drawn is a Diamond. Using the definition of independence, determine if E and F are independent. (b) Suppose that the Queen of Hearts is removed from a standard deck of 52 cards, and then a single card is drawn. Let E now be the event that the card drawn is an Ace, and let F be the event that the card drawn is a Diamond. Using the definition of independence, determine if E and F are independent.

Solution

(a) Since the deck contains 4 Aces and 13 Diamonds, $\Pr[E] = \frac{4}{52} = \frac{1}{13}$ and $\Pr[F] = \frac{13}{52} = \frac{1}{4}$. Also, $\Pr[EF] = \Pr[E \text{ and } F] = \Pr[\text{Ace of Diamonds}] = \frac{1}{52} = \frac{1}{13} \times \frac{1}{4} = \Pr[E] \times \Pr[F]$. This shows that E and F are independent events.

(b) Since the deck of 51 cards contains 4 Aces and 13 Diamonds, $\Pr[E] = \frac{4}{51}$ and $\Pr[F] = \frac{13}{51}$. Also, $\Pr[EF] = \Pr[E \text{ and } F] = \Pr[\text{Ace of Diamonds}] = \frac{1}{51} \neq \frac{4}{51} \times \frac{13}{51} = \Pr[E] \times \Pr[F]$. This shows that E and F are *not* independent events. That is, removing a card that is neither an Ace nor a Diamond has affected their mutual independence! To explain this, notice that $\Pr[\text{Ace}] = \Pr[E] = \frac{4}{51}$ whereas $\Pr[\text{Ace}|\text{Diamond}] = \Pr[E|F] = \frac{1}{13}$. So knowledge that a Diamond was drawn does indeed affect the probability that the card drawn was an Ace. ◁

We now revisit Example 13.5 and show it can be solved using conditional probabilities.

Example 14.9 Calculate Probabilities in the Composition of a Committee

A legislative committee consists of seven Republicans and five Democrats. Four individuals are to be selected from this committee for a fact-finding overseas mission. To be fair, the names of all twelve committee members are placed in a hat and the members selected for this fact-finding mission are determined by pulling out four names. (a) How likely is it that all four selected members are Republicans? (b) How likely is it that all four selected members are Democrats? (c) How likely is it that equal numbers of Republicans and Democrats are selected?

Solution

(a) The probability that the first selected member is a Republican is $\frac{7}{12}$. The conditional probability that the second member is a Republican, given that the first is a Republican, is $\frac{6}{11}$. In a similar way, the conditional probability that the third member is a Republican, given that the first two are Republicans, is $\frac{5}{10}$; the conditional probability that the fourth member is a Republican, given that the first three are Republicans, is $\frac{4}{9}$. By repeated application of the Joint Probability Rule, the probability that all four selections happen to be Republican is given by the product $\frac{7}{12} \times \frac{6}{11} \times \frac{5}{10} \times \frac{4}{9} = \frac{35}{495}$, just as found in Example 13.5.

(b) To determine the probability that all four selected members are Democrats, we compute the product $\frac{5}{12} \times \frac{4}{11} \times \frac{3}{10} \times \frac{2}{9} = \frac{5}{495}$, again agreeing with the answer found in Example 13.5.

(c) We need to determine the probability of obtaining two Republicans and two Democrats. To begin, let's compute $\Pr[RDDR]$, the probability of selecting a Republican followed by two Democrats and then one more Republican, using the Joint Probability Rule:

$$\Pr[RDDR] = \Pr[R] \times \Pr[D|R] \times \Pr[D|RD] \times \Pr[R|RDD] = \tfrac{7}{12} \times \tfrac{5}{11} \times \tfrac{4}{10} \times \tfrac{6}{9} = \tfrac{35}{495}.$$

However, there are other sequences of two Rs and two Ds that need to be considered, such as $DDRR$, occurring with probability

$$\Pr[DDRR] = \Pr[D] \times \Pr[D|D] \times \Pr[R|DD] \times \Pr[R|DDR] = \tfrac{5}{12} \times \tfrac{4}{11} \times \tfrac{7}{10} \times \tfrac{6}{9} = \tfrac{35}{495}.$$

In fact, there are a total of $C(4,2) = 6$ such sequences involving two Rs and two Ds, as this is the number of ways of selecting (without order) the two positions for the Rs in the sequence of four letters. Moreover, each such sequence has the same probability $\frac{35}{495}$, meaning that the overall probability of obtaining two Rs and two Ds in some order is $6 \times \frac{35}{495} = \frac{210}{495}$. ◁

(*Related Exercises: 22–25*)

Example 14.10 Calculate the Probability of Passing a Certification Test

Tonya needs to pass a certification test in order to practice medicine in her state. The pass rate is 65% for students who take the test for the first time. If a student fails, then the test can be retaken and the pass rate increases to 75%. Assuming that a student can take the test at most twice, what is the probability that Tonya passes the certification test?

Solution

In the decision tree of Figure 14.7, the first stage represents taking the test for the first time, and the second stage represents the retest. The abbreviations S and F are used to indicate the results at each stage (success, failure). The edges of the tree are labeled with the associated probabilities.

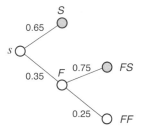

FIGURE 14.7: Decision tree for a certification test

Notice that we use the Complement Rule to deduce at the first stage that $\Pr[F] = 1 - \Pr[S] = 1 - 0.65 = 0.35$ and at the second stage that $\Pr[F|F] = 1 - \Pr[S|F] = 1 - 0.75 = 0.25$. We do not need to extend the decision tree any further, since we are only interested in the event E of passing the test (on either try).

Using the Joint Probability Rule, $\Pr[FS] = \Pr[F] \times \Pr[S|F] = 0.35 \times 0.75 = 0.2625$. Since the event of interest $E = \{S, FS\}$ contains disjoint outcomes, $\Pr[E] = \Pr[S] + \Pr[FS] = 0.65 + 0.2625 = 0.9125$, representing an overall pass rate of 91.25%. ◁

Example 14.11 Solve a *Car Talk* Puzzler

Zeke and Zack were two high-school buddies who lived close to each other, and one day they went off in search of after-school employment. Neither of them had a particularly reliable car so they decided to get a job at the same place so they could help each other get to work. Finding employment together proved harder than they had thought, until they happened upon a pig farmer who was impressed with their manure-shoveling prowess. "Can you start tomorrow?" he asked. "Yes, sir!" they said. "Do you have a car?" "Yes, in fact each of us has a car," said Zeke. "But mine has a touchy fuel pump and a probability of starting of 80%. And Zack's starter has been acting up and the probability of his car starting is only 70%." The farmer thought for a minute and said, "Sorry, boys, but my pigs and I need both of you here at least 90% of the time. Things pile up, you know. I'm afraid I can't hire you." Well, with long faces they walked away bemoaning their bad luck. Who wouldn't want to spend his free time shoveling pig manure, right? But in an instant, they figured something out. They would have those dream jobs after all. What did they say to the farmer that got them hired?

Solution

To solve this puzzle [OR1], we construct the decision tree of Figure 14.8. The first stage represents the condition of Zeke's car and the second stage represents the condition of Zack's car. The abbreviations S and F show the results at each stage: either the car starts (S) or fails to start (F). The edges of the tree are labeled with the associated probabilities, and outcome probabilities are also shown. Zeke and Zack will have a working car in either of the two situations represented by S or FS, highlighted in Figure 14.8. Since these outcomes are disjoint, the probability that either occurs is given by $\Pr[S \text{ or } FS] = \Pr[S] + \Pr[FS] = 0.8 + 0.14 = 0.94$. So Zeke and Zack can confidently assert that they will have a working car 94% of the time. ◁

(Related Exercises: 26–28)

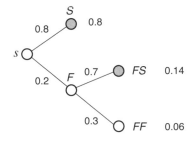

FIGURE 14.8: Decision tree for the *Car Talk* puzzler

Conditional probabilities can also be useful in explaining an apparent paradox in searching through a deck of cards.

Example 14.12 Determine a Most Likely Occurrence in a Card Drawing Dilemma

Deshawn shuffles a standard deck of 52 cards and places it face down. He then continues to draw one card at a time from the top of the pile, stopping when the first Red Queen appears. (a) How likely is it that the first card drawn is a Red Queen? (b) How likely is it that the second card drawn is the first Red Queen? (c) How likely is it that the first Red Queen occurs on the third draw? (d) Generalize the previous results and thereby determine on which draw a Red Queen is most likely to appear.

Solution

We can answer the first three parts by constructing the decision tree of Figure 14.9, in which S indicates success (the appearance of a Red Queen) and F indicates failure (a Red Queen does not appear on that draw). Since there are two Red Queens, the probability of drawing one from the original deck of 52 cards is $\frac{2}{52}$. This represents the probability of success at the first stage, so the probability of failure is $1 - \frac{2}{52} = \frac{50}{52}$. After a failure at the first stage, there are now 51 cards, so the (conditional) probability of success is $\frac{2}{51}$ while the probability of failure is $1 - \frac{2}{51} = \frac{49}{51}$. Figure 14.9 displays the probabilities for the first three stages.

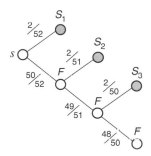

FIGURE 14.9: Searching for the first Red Queen

(a) The probability of success at the first draw (first stage) is then $\frac{2}{52} = 0.0385$.

(b) The probability of success at the second draw (second stage) is given by the product of the edge probabilities on the path to outcome S_2: $\frac{50}{52} \times \frac{2}{51} = 0.0377$.

(c) The probability of success at the third draw (third stage) is given by the product of the edge probabilities on the path to outcome S_3: $\frac{50}{52} \times \frac{49}{51} \times \frac{2}{50} = 0.0370$.

(d) The probabilities in parts (a)–(c) are decreasing, so it seems that this trend will continue. To verify this, we can rewrite these probabilities as

$$\Pr[\text{success on first draw}] = \frac{2}{52},$$
$$\Pr[\text{success on second draw}] = \frac{2}{52} \times \frac{50}{51},$$
$$\Pr[\text{success on third draw}] = \frac{2}{52} \times \frac{49}{51}.$$

In general, $\Pr[\text{success on draw } n] = \frac{2}{52} \times \frac{52-n}{51}$, $n \geq 1$. Since the maximum value of $\frac{52-n}{51}$ occurs for $n = 1$, a Red Queen is most likely to appear on the very first draw! ◁

(Related Exercises: 29–30)

It is sometimes useful to restate the formula for joint event probabilities to enable the direct calculation of conditional probabilities.

DEFINITION **Conditional Probability Rule**
The conditional probability $\Pr[F|E]$ can be calculated using $\Pr[F|E] = \frac{\Pr[EF]}{\Pr[E]}$.

Example 14.13 Calculate the Probability that a Head Shows on the Fifth Coin Toss

A fair coin is tossed four times and each time it comes up tails. A gambler is sure that the next toss is quite likely to be a head, since in the long run the number of heads and the number of tails should balance out. How should we advise the gambler before he bets his life savings on the next toss of the coin? What is the probability that a head will come up on the fifth toss?

Solution

Here the appropriate sample space consists of all outcomes involving five tosses of a fair coin: $HHHHH, HHHHT, \ldots, TTTTH, TTTTT$. There are $2^5 = 32$ outcomes in the sample space; since the coin is fair we assign the same probability $\frac{1}{32}$ to each outcome. Let E denote the event that a tail turns up on the first four tosses and let F denote the event that a head follows four tails: $E = \{TTTTT, TTTTH\}$ and $F = \{TTTTH\}$.

What is relevant here is $\Pr[F|E]$, the probability that a head follows a run of four tails. This can be calculated using the Conditional Probability Rule:

$$\Pr[F|E] = \frac{\Pr[EF]}{\Pr[E]} = \frac{\Pr[F]}{\Pr[E]} = \frac{1/32}{2/32} = \frac{1}{2}.$$

Therefore, there is still just a 50% chance of the next toss being a head, so the gambler should curb his enthusiasm. ◁

There seems to be a paradox here. Our intuition says that the total number of heads and tails should balance out in the long run. So, don't we expect a head to occur after this run of four tails? This thinking is referred to as the *Gambler's Fallacy* [OR2]. One explanation is as follows. Suppose we looked at two, rather than five, tosses of the coin. If a tail occurred on the first toss, then couldn't we argue that a head needs to occur next by the same balancing argument? Clearly this is an invalid argument. It is based on trying to apply a characteristic of an infinitely long sequence of coin tosses to a very short sequence of tosses.

DEFINITION **Gambler's Fallacy**
This is the mistaken belief that the likelihood of a random event increases after it has been observed to happen less frequently in the past.

Example 14.14 Calculate the Probability of Guessing a Card's Hidden Side

A gambler approaches you with a deck of three cards: one card (A) is Green on both sides; one (B) is Red on both sides; and one (C) is Green on one side and Red on the other. You draw a card and place it on the table and a red side shows face up. Neither you nor the gambler knows the color of its other side. The gambler now proposes the following bet: if the hidden side of this card is red, he wins; if it is not, you win. Is this a fair bet? Should you play this game?

Solution

Figure 14.10 displays the decision tree for this problem. The first stage of the decision tree shows the probability of drawing either card A, card B, or card C; all these choices are equally likely. The second stage shows the probability of either of the two sides turning up; again these are equally likely. As seen in the tree, each of the six outcomes is equally likely.

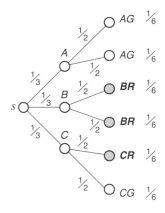

FIGURE 14.10: A game with green and red cards

To assess the gambler's offer, we would like to calculate the conditional probability $\Pr[B|R]$, the probability that the card with two red sides is selected, given that one red side faces up. The condition that one red side faces up (event R) is shown as the highlighted set of outcomes in Figure 14.10; given this condition, two out of these three outcomes correspond to the event B. Since all outcomes are equally likely, we obtain $\Pr[B|R] = \frac{2}{3}$, meaning that the gambler (who bets on the other side being red) has the clear upper hand. So this is not a fair offer for us.

We can also obtain this result by applying the Conditional Probability Rule. First, note that the probability a red side will show is given by $\Pr[R] = \frac{3}{6} = \frac{1}{2}$. Also, $\Pr[BR] = \frac{1}{6} + \frac{1}{6} = \frac{1}{3}$. Applying the Conditional Probability Rule results in

$$\Pr[B|R] = \frac{\Pr[BR]}{\Pr[R]} = \frac{1/3}{1/2} = \frac{2}{3}.$$

Again this calculation shows that the bet is not an even one—the gambler can be expected to win two times out of three. ◁

(Related Exercises: 31–32)

14.3   Intuition and Randomness

Consider the following situation [Dawes 1988] confronted by a clinical psychologist:

> A patient with an anxiety disorder is asked to keep a log of days on which an attack occurs for the next 21 days. From past experience, this patient has on average an attack once every two days. Which of the following sequences indicate

some external pattern to the attacks over time, and which might the clinician consider to be simply random? An X indicates an attack, whereas an O indicates that no attack occurred on that day.

(a) X X O X O X X O X O O X O O X O X O X O X

(b) O X X X O O X O O O X X O O X X O O X O O

Most people consider the second sequence to be indicative of some external pattern. Let's analyze these sequences using some simple probability ideas.

Suppose a sequence is truly random. Then, as in the Gambler's Fallacy (Example 14.13), an X or an O following any particular symbol will be equally likely, with probability $\frac{1}{2}$; recall that the patient has an attack about once every two days. This means that the probability of an alternation (an X following an O, or an O following an X) should also be $\frac{1}{2}$.

In 21 days, there are 20 pairs of consecutive symbols or 20 possible alternations, so a truly random sequence ought to contain about 10 alternations. Sequence (a) contains 15 alternations, whereas sequence (b) contains only 10. That is, sequence (a) contains just too many alternations to believe that it is random. The second sequence was in fact created by flipping a coin 21 times [OR3] and represents a truly random sequence.

Our intuition is sometimes fooled by randomness. Sequence (b) contains two long "runs" of three identical symbols in a row; we tend to view this as nonrandom, when in fact it is a natural occurrence of random sequences. The next example is taken from the realm of sports.

Example 14.15 Assess a hot hand in basketball shooting

In a contest for making three-point baskets, one player's results for 23 attempts are shown below. Hits and misses are indicated by the symbols H and M, respectively. We observe runs of six and eight baskets in a row. Is this indicative of a "hot hand", an unusually long streak of high performance, or could it be explained as a chance occurrence?

M H H H H H H M M H H H H H H H H M H H M M H

Solution

We can use the online calculator [OR4], which tests whether the data are representative of an unusual series of successive events (long "runs") or is a consequence of randomness. Using this calculator, the likelihood of the observed runs in this basketball shooting example is around 0.15; this is not so unusual and so we can conclude that there is no real evidence against having observed a random occurrence. ◁

(*Related Exercise: 33*)

The next chapter will delve deeper into situations in which our intuition is not a reliable guide in situations that contain uncertainty. The key to making solid decisions will be the proper use of conditional probabilities.

Chapter Summary

This chapter develops an approach for calculating the joint probability $\Pr[E \text{ and } F]$ when the events E and F are independent of each other (e.g., the probability of rolling a 2 on a die and then rolling a 5). We also learn how to calculate the probability $\Pr[E \text{ or } F]$ of disjoint events when these events cannot occur simultaneously (e.g., the probability of obtaining a 2 or a 5 on a single roll of the die). These two rules can be applied to more complicated situations, by decomposing them into disjoint events as well as independent events. More generally, probability calculations may be needed when the events involved are dependent; in this case, conditional probabilities are required. For example, the occurrence of rain tomorrow will likely depend on whether it has rained today. We illustrate how conditional probabilities can be usefully viewed in terms of a decision tree. Understanding conditional probabilities enables us to make rational decisions in the face of uncertain information.

Applications, Representations, Strategies, and Algorithms

Applications				
Business (3)	Sports (4, 15)		Society/Politics (5–6, 9)	
	Medicine (10)		Gambling (13)	

Representations				
Graphs		Sets		Symbolization
Weighted	Tree	Lists	Venn Diagrams	Algebraic
5–6, 10–12, 14	5–6, 10–12, 14	3, 13, 15	1–2, 4	3

Strategies
Rule-Based
1–15

Algorithms				
Exact	Enumeration	Sequential	Numerical	Algebraic
1–14	1–2	3-4, 7, 9–12	3–15	12

Exercises

Independent and Disjoint Events

1. Suppose that two fair dice are rolled. Let E be the event in which the sum of the face values is even, and let F be the event in which the first die rolled shows a 5.

   (a) What is the appropriate sample space for this experiment?

(b) List the outcomes comprising event E and determine $\Pr[E]$.

(c) List the outcomes comprising event F and determine $\Pr[F]$.

(d) List the outcomes comprising the event $\{E \text{ and } F\}$ and determine $\Pr[E \text{ and } F]$. Are events E and F independent?

2. Suppose that a fair coin is tossed three times. Let E be the event in which at least one tail appears in the first two tosses, let F be the event in which more heads than tails appear in the three tosses, and let G be the event in which the results of the first and third tosses are identical.

   (a) What is the appropriate sample space for this experiment?

   (b) List the outcomes comprising events E, F, and G.

   (c) Calculate $\Pr[E], \Pr[F], \Pr[G]$.

   (d) Calculate $\Pr[E \text{ and } F], \Pr[E \text{ and } G], \Pr[F \text{ and } G]$.

   (e) Using the definition of independence, determine whether E and F are independent, whether E and G are independent, and whether F and G are independent.

3. A card is chosen at random from a deck of 52 cards. It is then replaced and a second card is chosen.

   (a) What is the probability of choosing a Heart on the first card and then a seven on the second card?

   (b) Using the definition of independence, determine whether these two events are independent.

4. Suppose that two fair dice are rolled. Let E be the event in which the sum of the face values is at least 9, and let F be the event in which both dice show the same number.

   (a) List the outcomes comprising event E and determine $\Pr[E]$.

   (b) List the outcomes comprising event F and determine $\Pr[F]$.

   (c) List the outcomes comprising the event $\{E \text{ and } F\}$ and determine $\Pr[E \text{ and } F]$. Are events E and F independent?

5. Suppose that two fair dice are rolled. Let E be the event in which the sum of the face values is 4, 5, or 6. Let F be the event in which both dice show the same number.

   (a) List the outcomes comprising event E and determine $\Pr[E]$.

   (b) List the outcomes comprising event F and determine $\Pr[F]$.

   (c) List the outcomes comprising the event $\{E \text{ and } F\}$ and determine $\Pr[E \text{ and } F]$. Are events E and F independent?

6. A pair of dice are rolled. We are interested in when the face values produce a sum of 4 (event E) or a sum of 5 (event F).

   (a) List the outcomes comprising event E and the outcomes comprising event F.

   (b) List the outcomes comprising the event $\{E \text{ or } F\}$, in which the face values sum to either 4 or 5.

   (c) Calculate $\Pr[E \text{ or } F]$ and compare it with $\Pr[E] + \Pr[F]$. Are events E and F disjoint?

7. Suppose that two fair dice are rolled. Let E be the event in which the sum of the face values is 7. Let F be the event in which the two numbers showing on the dice differ by exactly 2.

   (a) List the outcomes comprising event E and the outcomes comprising event F.

   (b) List the outcomes comprising the event $\{E \text{ or } F\}$.

   (c) Calculate $\Pr[E \text{ or } F]$ and compare it with $\Pr[E] + \Pr[F]$. Are events E and F disjoint?

8. Suppose that two fair dice are rolled. Let E be the event in which the sum of the face values is 6. Let F be the event in which the two numbers showing on the dice differ by exactly 2.

   (a) List the outcomes comprising event E and the outcomes comprising event F.

   (b) List the outcomes comprising the event $\{E \text{ or } F\}$.

   (c) Calculate $\Pr[E \text{ or } F]$ and compare it with $\Pr[E] + \Pr[F]$. Are events E and F disjoint?

9. Suppose that a fair coin is tossed three times. Let E be the event in which exactly one head appears in the first two tosses, and let F be the event in which the results of the first and third tosses are identical.

   (a) List the outcomes comprising event E and the outcomes comprising event F.

   (b) List the outcomes comprising the combined event $\{E \text{ or } F\}$.

   (c) Calculate $\Pr[E \text{ or } F]$ and compare it with $\Pr[E] + \Pr[F]$. Are events E and F disjoint?

10. A card is chosen at random from a deck of 52 cards. Let E be the event in which the card drawn is a Red Queen, and let F be the event in which the card drawn is a Red face card. Recall that the face cards are J, Q, K, A.

    (a) List the outcomes comprising event E and the outcomes comprising event F.

    (b) List the outcomes comprising the combined event $\{E \text{ or } F\}$.

    (c) Calculate $\Pr[E \text{ or } F]$ and compare it with $\Pr[E] + \Pr[F]$. Are events E and F disjoint?

11. A card is chosen at random from a deck of 52 cards.

    (a) What is the probability of choosing either a Heart or a seven?

    (b) Determine whether these two events are disjoint.

12. Sergio is playing in a tennis match and in a game of football this week. The probability that his team will win the tennis match is 0.7, and the probability that his team will win the football game is 0.9. Assume that the results of the sporting events are independent—they do not affect each other. What is the probability that his team will win both events?

13. The probability that you go on a walk on any given day is 0.3 and the probability that it rains on any given day is 0.4. Consider three different scenarios.

    (a) The probability that you walk on a rainy day is 0.05. Do you walk independently of whether or not it rains? Give a mathematical explanation.

(b) You never walk on rainy days, so the probability that you walk on a rainy day is 0. Do you walk independently of whether or not it rains? Give a mathematical explanation.

(c) The probability that you walk on a rainy day is 0.12. Do you walk independently of whether or not it rains? Give a mathematical explanation.

14. A five-member jury is deliberating on whether to render a "guilty" or "not-guilty" verdict. In order to reach a "guilty" judgment, unanimous agreement of the jurors is needed. After deliberating, each of the jurors has probability 0.4 of voting "not-guilty" via a secret ballot. Each member votes independently of the others.

(a) What is the probability that a "guilty" verdict is reached?

(b) What is the probability that a "not-guilty" verdict is reached?

15. A five-member panel of experts must decide whether to fund a proposed software project. In order to receive funding, the project must be approved by a majority of members of the panel. Each panel member has probability 0.8 of voting to fund the project, and votes independently of the others.

(a) What is the appropriate sample space for this problem?

(b) Determine the probability that the project will be funded.

16. Use the data supplied in Example 14.4 to answer the following questions.

(a) Determine the probability that team B wins the tournament.

(b) Determine the probability that team C wins the tournament.

(c) Determine the probability that team D wins the tournament.

Conditional Probabilities

17. A debate team consists of six women and two men. Two members of this team are to be randomly selected to participate in the next debate event. We are interested in the gender composition of the two selected debaters.

(a) Draw a two-stage decision tree that represents this situation.

(b) Determine the probability that both selected debaters are women.

(c) Determine the probability that both selected debaters are men.

18. You have invested in a particular pharmaceutical company because it is rumored that the company's new drug will shortly be approved by the FDA. There is an 80% chance of FDA approval, and a 95% chance that the value of the stock you hold will double if FDA approval is given. If FDA approval is not given, then there is only a 10% chance of your stock doubling.

(a) Draw an appropriate decision tree and determine the probabilities of each of the four outcomes.

(b) Calculate the probability that the FDA approves the drug but the stock price does not double.

(c) Calculate the probability that the FDA does not approve the drug yet the stock price doubles.

19. In a jar of jelly beans, seven are orange, five are purple, and four are green. You scoop out two of the jelly beans at random.

   (a) Draw an appropriate decision tree for this problem. How many outcomes are there?

   (b) Calculate the probability that both jelly beans are the same color.

   (c) Calculate the probability that the jelly beans have two different colors.

20. There is a 10% chance that the prime interest rate will be increased at the next meeting of the Federal Reserve, a 20% chance that it will be decreased, and a 70% chance that it will remain the same. If the rate increases, there is an 80% chance that your investment portfolio will decrease (otherwise, it will increase in value). If the rate decreases, there is a 90% chance that your investment portfolio will increase in value. If the rate stays the same, there is a 60% chance that your investment portfolio will increase in value.

   (a) Draw an appropriate decision tree for this problem. How many outcomes are there?

   (b) Calculate the probability that your investment portfolio will increase in value.

21. Nine travelers are journeying through Middle Earth; four are hobbits and the rest are dwarfs. You select three of the company at random.

   (a) Using a decision tree, determine the probability that three hobbits are selected.

   (b) Using a decision tree, determine the probability that three dwarfs are selected.

   (c) Using a decision tree, determine the probability that one hobbit and two dwarfs are selected.

22. Three cards are drawn at random from a deck of 52 cards. Using the Joint Probability Rule, calculate the probability that all three are number cards. Recall that the number cards are $2, 3, \ldots, 10$.

23. You are dealt two cards from a shuffled deck of 52 cards. Using the Joint Probability Rule, calculate the probability that both cards dealt are face cards. Recall that the face cards are J, Q, K, A.

24. You are dealt three cards from a shuffled deck of 52 cards. Using the Joint Probability Rule, calculate the probability that all three cards are Hearts.

25. A manufacturing company is conducting a quality control survey on its product. Each box contains 50 items. From each box, five items are randomly chosen and tested. If one of the items is found to be defective, the entire box is rejected.

   (a) What is the probability that quality control will accept the box if the box contains three defective parts?

   (b) What is the probability that quality control will reject the box if the box contains 10 defective parts?

26. The pass rate is 40% for students taking the LSAT for the first time. It is also known that the pass rate increases to 80% for students taking it a second time. Assuming that a student can take the test at most twice, what is the overall probability of passing the LSAT?

27. A statistics professor gives students up to three chances to pass a brutal midterm exam. The success rate on the first attempt is 60%, it increases to 65% for the second attempt, and it increases to 70% for the third attempt. Determine the overall probability of passing the statistics midterm exam.

28. Returning from a long vacation, Ralph has forgotten the password for his bank account. He is pretty sure though that it is one of seven possibilities. Unfortunately, the banking web site only allows three attempts at entering a password; after that it locks out the user. Assume that Ralph initially tries a random password from the seven possibilities, and then continues to try a random password from the remaining possibilities. Determine the probability that Ralph will be able to successfully access his bank account.

29. Tom and Ray are auto mechanics in a small repair shop. Tom is much faster at repairing cars (namely, he repairs twice as many cars as Ray). However, Tom does a good repair job three out of four times, while the more methodical Ray does a good repair job four out of five times. Carlos takes in his car to the repair shop and leaves it to be done by whichever mechanic is free.

    (a) Draw an appropriate decision tree [*Hint*: There are four outcomes: Tom does a good repair (TG), Tom does a poor repair (TP), Ray does a good repair (RG), Ray does a poor repair (RP).]

    (b) What is the probability that a good repair job is done on Carlos' car?

30. A concession stand at an amusement park sells hot chocolate, water, and lemonade. On a rainy day, the sales average 50% hot chocolate, 30% water, and 20% lemonade; on a sunny day, the sales average 15% hot chocolate, 25% water, and 60% lemonade. Over the entire year, it is sunny 70% of the time and rainy 30% of the time .

    (a) Draw a decision tree to represent the outcomes in the sample space and determine their probabilities.

    (b) What proportion of the total beverages sold over the entire year is hot chocolate and what proportion is water?

31. The weather forecast for the weekend indicates that there is a 30% chance of rain, a 20% chance of high winds, and a 50% chance of clear skies. From past observations, we know that Norman will sail his boat 80% of the time it is clear, 70% of the time it is windy, and 60% of the time it is rainy.

    (a) Draw a decision tree to represent the outcomes in the sample space and compute their probabilities.

    (b) What is the probability that it is windy and Norman will sail his boat?

    (c) What is the probability that Norman will sail his boat on the weekend?

    (d) On Monday, Norman tells Verna that he sailed his boat. Use the Conditional Probability Rule together with parts (b) and (c) to find the probability that the weather was windy over the weekend.

Projects

32. Read the chapter "The Case of the Unmarked Graves" in *Conned Again, Watson! Cautionary Tales of Logic, Math, and Probability* [Bruce 2001] and express the various puzzles found there in terms of conditional probabilities. Verify the solutions proposed by Reverend Dodgson and Sherlock Holmes by drawing decision trees and calculating the appropriate probabilities.

33. Streaks in sports on occasion reach amazing lengths. Are these in fact extraordinary occurrences or can they be explained as the result of naturally occurring random phenomenon? Investigate this topic using the concept of conditional probabilities, as presented in *Hot Hand: The Statistics Behind Sports' Greatest Streaks* [Reifman 2012].

Bibliography

[Bruce 2001] Bruce, C. 2001. *Conned again, Watson! Cautionary tales of logic, math, and probability*, Chapter 5. Cambridge, MA: Perseus.

[Dawes 1988] Dawes, R. M. 1988. *Rational choice in an uncertain world*. Orlando, FL: Harcourt Brace & Company.

[Reifman 2012] Reifman, A. 2012. *Hot hand: The statistics behind sports' greatest streaks*. Dulles, VA: Potomac Books.

[OR1] http://www.cartalk.com/content/pig-poke (accessed December 28, 2013).

[OR2] http://www.youtube.com/watch?v=K8SkCh-n4rw (accessed December 28, 2013).

[OR3] http://www.socr.ucla.edu/htmls/SOCR_Experiments.html (accessed December 28, 2013).

[OR4] http://home.ubalt.edu/ntsbarsh/Business-stat/otherapplets/Randomness.htm (accessed January 9, 2014).

Chapter 15

Bayes' Law and Applications of Conditional Probabilities

Medicine is a science of uncertainty and an art of probability.

—William Osler, Canadian physician, 1849–1919

Chapter Preview

We may have some estimate about whether a certain event will occur (e.g., that a particular stock will go up today). However, additional information can often be used to refine our estimates. If the Federal Reserve Board has just announced an increase in the prime lending rate, then we may be less optimistic that our stock will go up today. In this chapter, we will see how it is possible to revise our prior estimates in view of new information to obtain updated estimates. Calculation of conditional probabilities allows us to incorporate such new information. We also illustrate the difference between conditional probabilities and their inverses, which is frequently the cause of serious misinterpretations. When inferences go wrong in the courtroom, forensics lab, or doctor's office, they can have devastating effects on people's lives. The proper application of conditional probabilities can assist us in making logically defensible decisions in the face of seemingly paradoxical situations.

15.1 Bayes' Law

The probability estimate of an event can be revised to incorporate newly acquired information. This procedure for revising estimates involves using what is called *Bayes' Law*, named after Thomas Bayes.

HISTORICAL CONTEXT **Thomas Bayes (1702–1761)**
An English clergyman who set out a theory of probability in "An Essay Towards Solving a Problem in the Doctrine of Chances," published (after his death) in 1764. Today his name lives on in the concept of *Bayesian spam filtering*, a technique that calculates the probability of a message being spam based on the occurrence of certain key words and phrases contained in the message.

Bayes' Law is based on the application of the Conditional Probability Rule encountered in Section 14.2:

$$\Pr[E|F] = \frac{\Pr[EF]}{\Pr[F]}.$$

The following example shows how this formula can be used to update our estimates of uncertain events.

Example 15.1 Calculate the Probability of a Fraudulent Tax Return

It is known from historical data that 6% of all income tax returns for self-employed individuals are fraudulent. Fraudulent returns are very likely to contain a deduction for a home office. Specifically, 70% of fraudulent returns contain this deduction, whereas only 40% of honest (non-fraudulent) returns contain this deduction. Suppose that a self-employed individual's tax return is randomly selected and is found to contain a home office deduction. How likely is it that this return is in fact fraudulent?

Solution

We begin by constructing a two-stage decision tree, where the first stage corresponds to whether the return is fraudulent (F) or honest (H). The second stage corresponds to whether the return has a home office deduction (D) or not (N). This tree is shown in Figure 15.1.

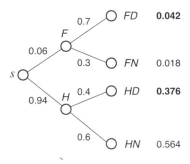

FIGURE 15.1: Decision tree for the tax return problem

Stage One: Since 6% of all returns are fraudulent, we have labeled with probability 0.06 the edge leading to F; by the Complement Rule, the edge leading to H has probability $1 - 0.06 = 0.94$.

Stage Two: Since 70% of all fraudulent returns claim a deduction for a home office, the edge from F to FD is assigned the conditional probability $\Pr[D|F] = 0.7$; by the Complement Rule, the edge leading to FN has probability $1 - 0.7 = 0.3$. Since 40% of all honest returns claim an office deduction, the edge from H to HD is assigned the conditional probability $\Pr[D|H] = 0.4$; by the Complement Rule, the edge to HN has probability $1 - 0.4 = 0.6$.

Given these values, we can compute the probability of each of the four outcomes as the product of the edge probabilities; these outcome probabilities are shown in Figure 15.1. To solve the problem—how likely is it that a random return with an office deduction is fraudulent—we must determine the conditional probability $\Pr[F|D]$. There are two approaches:

– **Intuitive approach.** The condition specified (taking an office deduction) involves only the outcomes FD and HD; the corresponding outcome probabilities are shown in bold in Figure 15.1. Together these two outcomes occur almost 42% of the time: $0.042 + 0.376 = 0.418$. Among these outcomes, we are interested only in when an office deduction is fraudulent (FD), which is about 4% of the time. So the proportion of fraudulent returns among those taking an office deduction should be the ratio of fraudulent office deductions to all office deductions. That is given by the probability 0.042 (FD) divided by the probability 0.418 ($FD + HD$):

$$\Pr[F|D] = \frac{0.042}{0.418} = 0.100.$$

– **Formal approach.** Here we directly apply the Conditional Probability Rule

$$\Pr[F|D] = \frac{\Pr[FD]}{\Pr[D]}.$$

The numerator $\Pr[FD] = 0.06 \times 0.7 = 0.042$. Since the event $D = \{FD, HD\}$ consists of disjoint outcomes, we have $\Pr[D] = \Pr[FD] + \Pr[HD] = 0.042 + 0.376 = 0.418$. Therefore,

$$\Pr[F|D] = \frac{\Pr[FD]}{\Pr[D]} = \frac{0.042}{0.418} = 0.100.$$

Again we determine that there is a 10% chance that a randomly selected return with an office deduction is fraudulent. This provides an improved estimate of a fraudulent return, compared with the overall 6% rate of fraudulent returns. ◁

In a similar way we can calculate the probability that a randomly selected return with *no office deduction* (N) is fraudulent by focusing in the decision tree on the outcomes FN and HN. These outcomes constitute in effect our new sample space and we calculate $\Pr[F|N]$ by dividing the probability that FN occurs by $\Pr[N]$:

$$\Pr[F|N] = \frac{\Pr[FN]}{\Pr[N]} = \frac{0.018}{0.018 + 0.564} = 0.031.$$

There is then a 3.1% chance that a randomly selected return containing no office deduction is fraudulent.

(Related Exercises: 1–5)

Example 15.2 Calculate the Probability of a Correct Medical Diagnosis

A laboratory test for Q fever is known to be 95% accurate in diagnosing this disease in an infected person; in other words, it has a *true positive rate* of 95%. On the other hand, if a person is healthy, the test has a 1% *false positive rate*; it incorrectly classifies a healthy person as having Q fever 1% of the time. Suppose 0.5% of the population has Q fever. If Sally walks into a doctor's office and tests positive for Q fever, how likely is it that she is actually infected with Q fever?

DEFINITION **True Positive Rate (Sensitivity)**
The proportion of time that a test correctly diagnoses the status of a patient having a certain condition.

DEFINITION **False Positive Rate**
The proportion of time that a test incorrectly diagnoses the status of a patient not having a certain condition.

Solution

We proceed by first constructing the two-stage decision tree shown in Figure 15.2. Since 0.5% of the population has Q fever, we have labeled with probability 0.005 the edge leading to Q; by the Complement Rule, the edge leading to H (healthy, not having Q fever) has probability $1 - 0.005 = 0.995$. The second stage indicates whether the test produces a positive (P) or a negative (N) result. The edge probabilities are assigned using the given information, enabling us to calculate the four outcome probabilities shown in Figure 15.2.

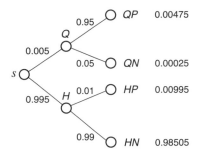

FIGURE 15.2: Decision tree for the medical diagnosis problem

The solution to the problem requires finding $\Pr[Q|P]$, the probability of having the disease, given a positive test result. We calculate this as the ratio of $\Pr[QP]$, which is 0.00475, to $\Pr[P] = \Pr[QP] + \Pr[HP] = 0.00475 + 0.00995 = 0.0147$, giving

$$\Pr[Q|P] = \frac{\Pr[QP]}{\Pr[P]} = \frac{0.00475}{0.0147} = 0.323.$$

So, there is only a 32.3% chance that Sally has Q fever. ◁

(*Related Exercises: 6–7, 30*)

There seems to be a paradox here. After all, the test is fairly accurate with a true positive rate of 95% and a false positive rate of only 1%. Specifically, $\Pr[P|Q] = 0.95$ signifies a fairly high probability. The apparent paradox is that the *inverse* conditional probability $\Pr[Q|P] = 0.323$ is not especially high.

Conditional probabilities are often confused with their inverse probabilities, assuming that if one is high (or low) then the other is likewise high (or low). We will investigate other examples of this confusion in Sections 15.2 and 15.3. Next, however, we investigate another common usage of conditional probabilities and Bayes' Law.

Example 15.3 Calculate the Probability That an Email is Spam

Spam email, also referred to as junk email or unsolicited bulk email, often contains links taking users to phishing or malware web sites that attempt to acquire personal information or harm your computer. *Bayesian spam filtering* is a statistical method that assigns probabilities to certain words and phrases, based on the past history of these words occurring either in spam or legitimate email. It uses conditional probabilities to classify an email as spam or not, and moves it to a junk email folder if required.

Specifically, suppose that 4% of spam emails have the phrase "award money" in the subject line, only 0.3% of non-spam emails have this phrase, and that 45% of all emails are spam. Given these facts, calculate the probability that this spam filter correctly classifies a random email as spam (S) when its subject line contains the phrase "award money" (A): in other words, find $\Pr[S|A]$.

Solution

We proceed by first constructing the two-stage decision tree shown in Figure 15.3. Since 45% of the emails are spam, we have labeled with probability 0.45 the edge leading to S; by the Complement Rule, the edge leading to legitimate mail (L) has probability $1 - 0.45 = 0.55$. The second stage indicates whether the email contains the phrase "award money" (A) or does not (N). The edge probabilities are assigned using the given information, allowing us to calculate the four outcome probabilities shown in Figure 15.3.

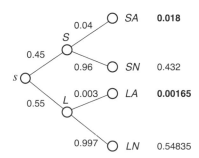

FIGURE 15.3: Decision tree for the spam filtering problem

Using the outcome probabilities shown in bold, we can compute the conditional probability

$$\Pr[S|A] = \frac{\Pr[SA]}{\Pr[A]} = \frac{0.018}{0.018 + 0.00165} - \frac{0.018}{0.01965} = 0.916.$$

So, there is a 92% chance that a random email containing the phrase "award money" will be spam. ◁

Example 15.4 Evaluate the Reliability of an Eyewitness Report

A cab was seen speeding away after a hit-and-run accident. The city has cabs that are either Green (15%) or Blue (85%). An eyewitness identifies the cab as being Green. Tests under similar conditions (lighting, etc.) show that the eyewitness can reliably distinguish Green from Blue cabs 80% of the time. How likely is it that the speeding cab was in fact Green?

Most people believe that there is an 80% chance that the cab in question was Green. Analyze this situation using conditional probabilities. Define the events $G =$ "the speeding cab was Green" and $B =$ "the speeding cab was Blue". Also, let G' and B' denote the events that the observed cab was *reported* to be Green and *reported* to be Blue, respectively. Calculate the conditional probability $\Pr[G|G']$, the likelihood that the cab is in fact Green given that the eyewitness reported it as such.

Solution

It is helpful to draw a decision tree in which the first stage corresponds to our initial estimates of Green and Blue cabs, without eyewitness information: namely, $\Pr[G] = 0.15$ and $\Pr[B] = 0.85$. The second stage refers to the reported color of the cab; we can label edges in this stage using the "true positive" identification probabilities $\Pr[G'|G] = \Pr[B'|B] = 0.8$, and the "false positive" identification probabilities $\Pr[B'|G] = \Pr[G'|B] = 0.2$. Using these values, it is straightforward to compute the outcome probabilities displayed in Figure 15.4.

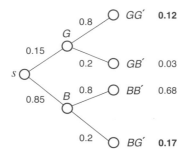

FIGURE 15.4: Decision tree for the taxicab problem

Since we are given the information G' that the cab was reported to be Green, our sample space is restricted to the outcomes whose probabilities are shown in bold, so $\Pr[G'] = \Pr[GG'] + \Pr[BG'] = 0.12 + 0.17 = 0.29$; this gives

$$\Pr[G|G'] = \frac{\Pr[GG']}{\Pr[G']} = \frac{0.12}{0.29} = 0.41.$$

As a result, there is only a 41% chance that the eyewitness evidence is reliable—just half the value (80%) that many people mistakenly believe. ◁

(Related Exercise: 8)

Example 15.4 was devised and studied by two Nobel laureates, Daniel Kahneman and Amos Tversky.

HISTORICAL CONTEXT **Daniel Kahneman (1934–) and Amos Tversky (1937–1996)**
Winners of the 2002 Nobel Prize in Economics for their groundbreaking study of decision making under uncertainty and how human decisions often depart from those predicted by standard economic theory.

Example 15.5 Calculate the Advantage of Switching Doors in "Let's Make a Deal"

A fabulous prize is hidden behind one of three doors A, B, or C. The contestant selects door A and the game show host then opens door B, revealing no prize. The contestant is now offered the opportunity to change her guess from door A to door C. Should she switch her guess?

Three assumptions are made: (1) the prize is equally likely to be hidden behind each of the doors; (2) the host will not open a door if the actual prize is there; (3) if the prize is in fact behind the selected door, the host will be equally likely to open either of the other two doors.

Solution

We construct a decision tree and then use the Conditional Probability Rule. In the first stage, a door is randomly *selected* for hiding the prize, indicated by the events A, B, C. By our assumption, the prize is equally likely to be behind each door, so every edge in the first stage has probability $\frac{1}{3}$. In the second stage, the host opens a door, according to the assumed behavior. Since the contestant has chosen door A, the host will only choose to open the other two doors; we use B' and C' to denote the *opening* of doors B and C in the second stage, respectively. This gives the decision tree shown in Figure 15.5.

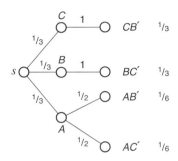

FIGURE 15.5: Decision tree for the door-switching problem

Notice that if the prize is really hidden behind door C, then the host must certainly (with probability 1) open door B; similarly if the prize is behind door B, then he must (with probability 1) open C. However, if door A really hides the prize, the host is equally likely to open either B or C. These considerations produce the edge probabilities and the outcome probabilities shown in Figure 15.5.

We now want to calculate $\Pr[A|B']$ and $\Pr[C|B']$, easily done using the probabilities displayed in the decision tree:

$$\Pr[A|B'] = \frac{\Pr[AB']}{\Pr[B']} = \frac{\Pr[AB']}{\Pr[AB'] + \Pr[CB']} = \frac{\frac{1}{6}}{\frac{1}{6} + \frac{1}{3}} = \frac{1}{3}$$

$$\Pr[C|B'] = \frac{\Pr[CB']}{\Pr[B']} = \frac{\Pr[CB']}{\Pr[AB'] + \Pr[CB']} = \frac{\frac{1}{3}}{\frac{1}{6} + \frac{1}{3}} = \frac{2}{3}.$$

Thus, there is only a 1 in 3 chance of the prize being behind door A, whereas there is a 2 in 3 chance of it being behind door C. Consequently, it is advisable for the contestant to change her guess from A to C. ◁

(Related Exercises: 9–10)

This particular puzzle, known as the *Monty Hall problem*, has been well publicized and there are several web sites available in which one can play repeated games against the computer [OR1, OR2]. Empirically you will find that it is indeed advantageous to switch your guess.

Example 15.6 Calculate an Updated Probability in Biology

Over the past 60 years, a group of biologists in Russia have been working to domesticate the silver fox. Over time they began to notice that the domesticated foxes tended to exhibit mottled fur patterns in contrast to the wild variety that favored solid red or gray colors. In collecting data for 2000 foxes, they found that 55% were domesticated. Among the domesticated foxes, 64% were mottled, whereas among the wild foxes 82% were solid colored.

Now suppose that you encounter on your walk in Russia a silver fox with mottled fur. How likely is it that this particular fox is domesticated?

Solution

Rather than using a decision tree to solve this problem, we construct a 2×2 table with rows corresponding to type of fur and columns corresponding to type of fox. Table 15.1 displays these rows and columns, and also the total number of foxes (2000). We can use the given information to fill in the individual cells of this table as well as the row and column totals.

TABLE 15.1: Fur types and fox breeds

	Domesticated	Wild	
Mottled Fur			
Solid Color			
			2000

Since 55% of the foxes are domesticated, the sample consists of $0.55 \times 2000 = 1100$ domesticated foxes and $0.45 \times 2000 = 900$ wild foxes. These column totals are shown in Table 15.2. Next, among the 1100 domesticated foxes, 64% have mottled fur, or $0.64 \times 1100 = 704$ mottled and $0.36 \times 1100 = 396$ have solid color. Similarly, among the 900 wild foxes, 82% have solid color, or $0.82 \times 900 = 738$, and so there are $0.18 \times 900 = 162$ with mottled fur. Finally, we fill in the row totals by simple addition.

TABLE 15.2: Completed table for silver foxes

	Domesticated	Wild	
Mottled Fur	**704**	162	**866**
Solid Color	396	738	1134
	1100	900	**2000**

Using Table 15.2, we can now find the conditional probability of a domesticated fox (D), given that it has mottled fur (M). We simply consult the bold entries in this table to get

$$\Pr[D|M] = \frac{\Pr[DM]}{\Pr[M]} = \frac{704/2000}{866/2000} = \frac{704}{866} = 0.813.$$

There is then an 81.3% chance that the mottled fox we encounter will be domesticated. ◁

(Related Exercises: 11–19)

15.2 Conditional Probabilities and Their Inverses

Conditional probabilities arise in making informed decisions—whether to be concerned about a positive medical test, or whether to switch our choice of doors in the Monty Hall problem. As noted earlier, there is a tendency to confuse the conditional probability $\Pr[E|F]$ with its inverse probability $\Pr[F|E]$, assuming either that they are the same or of similar magnitude. We now investigate this relationship.

Below are the formulas for a conditional probability and its inverse probability:

$$\Pr[E|F] = \frac{\Pr[EF]}{\Pr[F]}, \qquad \Pr[F|E] = \frac{\Pr[EF]}{\Pr[E]}.$$

From these we can derive the *Ratio Rule* for conditional probabilities.

RESULT **Ratio Rule**
Conditional probabilities and their inverses satisfy $\frac{\Pr[E|F]}{\Pr[F|E]} = \frac{\Pr[E]}{\Pr[F]}$.

The Ratio Rule shows us that the relative frequencies of the events E and F (i.e., $\Pr[E]$ and $\Pr[F]$) have a direct bearing on how the probability $\Pr[E|F]$ relates to $\Pr[F|E]$.

In an age of increased security against terrorism, we must come to grips with the advantages and disadvantages of profiling (not to mention the ethics of such discrimination). The next example shows one of the unfortunate side effects of profiling.

Example 15.7 Investigate an Argument Regarding Muslim Terrorists

The following statements have been broadcast on a talk-radio show:

> Most terrorists have been found to be Muslim. So it is imperative to thoroughly interrogate all Muslims before they are allowed to board an airplane.

Let M denote "being a Muslim" and let T denote "being a Terrorist." The first statement asserts that $\Pr[M|T]$—the proportion of Muslims among known terrorists—is a high value. But the stated implication is that $\Pr[T|M]$ is also high. Investigate this conclusion.

Solution

The probability $\Pr[M]$ of being Muslim is relatively high, while the probability $\Pr[T]$ of being a Terrorist is quite small, meaning that the ratio $\Pr[T]/\Pr[M]$ is indeed small:

$$\frac{\Pr[T|M]}{\Pr[M|T]} = \frac{\Pr[T]}{\Pr[M]} = \text{small.}$$

Since $\Pr[T|M] = \Pr[M|T] \times \text{small}$, it is quite conceivable that $\Pr[T|M]$ is small even though $\Pr[M|T]$ is large. In any case, the stated conclusion is not warranted. ◁

Example 15.8 Investigate an Argument Regarding Pet Ownership

An article in a magazine for executives made the following statement:

> Results of a survey of 74 Chief Executive Officers (CEOs) indicate that there may be a link between childhood pet ownership and future career success. Fully 94% of the CEOs surveyed, all of them employed within Fortune 500 companies, had possessed a dog, cat, or both as youngsters.

The article goes on to suggest that pet ownership may be important in developing good character traits (generosity, empathy, etc.) that are important for a CEO. If we let C denote being a CEO and let O indicate childhood pet ownership, the survey has found that $\Pr[O|C] = 0.94$, a very high value. However, the article goes on to suggest that the inverse probability $\Pr[C|O]$—the probability of a childhood pet owner eventually becoming a CEO—is also high. Investigate using the Ratio Rule.

Solution

The frequency $\Pr[O]$ of childhood pet ownership is likely high while the frequency $\Pr[C]$ of being a CEO is quite small, meaning that $\Pr[C]/\Pr[O]$ is small. By the Ratio Rule

$$\frac{\Pr[C|O]}{\Pr[O|C]} = \frac{\Pr[C]}{\Pr[O]} = \text{small},$$

so it is quite likely that $\Pr[C|O]$ is small even though $\Pr[O|C]$ is large. Therefore, the stated conclusion is not supported. ◁

(Related Exercises: 20–23, 27)

Example 15.9 Investigate an Argument Regarding Drug Usage

The headline in a California newspaper stated "Most on Marijuana Using Other Drugs." In the article itself, a study is cited that concludes

> When (hard) drugs are used by teenagers they are also very likely to be marijuana users.

Letting M stand for marijuana user and D for (hard) drug user, the study found that $\Pr[M|D]$ is high. However, the newspaper headline asserts the inverse: that $\Pr[D|M]$ is high. Investigate this claim.

Solution

Since the ratio $\Pr[D]/\Pr[M]$ of hard drug users to marijuana users is fairly small, applying the Ratio Rule yields

$$\frac{\Pr[D|M]}{\Pr[M|D]} = \frac{\Pr[D]}{\Pr[M]} = \text{small}.$$

Consequently, it is quite likely that $\Pr[D|M]$ is small as well, meaning that the claim cannot be justified without further evidence. ◁

This situation can also be represented using a Venn diagram, introduced in Section 7.4. In Figure 15.6, the set of marijuana users is shown as the oval labeled M while the set of hard drug users is shown as the oval labeled D. The relative sizes of the respective ovals are suggestive of the respective sizes of the two populations.

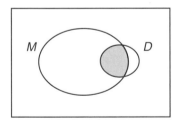

FIGURE 15.6: Venn diagram for the drug usage example

The shaded area represents those individuals using both marijuana and hard drugs. The probability $\Pr[M|D]$ is high, as shown by the large size of the shaded area relative to the area of D. However, the probability $\Pr[D|M]$ is not high, as shown by the small size of the shaded area compared to the area of M. Again, the reason for this is the relatively large prevalence of marijuana users compared to hard drug users.

(Related Exercises: 24–25)

Example 15.10 Investigate an Argument Linking Age and Accidents

Statistics on traffic incidents reveal that two-thirds of the accidents in a certain community are caused by individuals aged 65 years or older. Therefore, the insurance company believes it should raise the automobile insurance rates for older individuals in this community. Investigate this claim.

Solution

What we are given is that $\Pr[O|A] = \frac{2}{3}$ where O indicates an older person and A indicates having an accident in the community. However, the decision of the insurance company should be based on $\Pr[A|O]$, the accident rate of older drivers. Here again is a fallacious argument based on confusing conditional probabilities and their inverses.

For example, what if the community happens to be a retirement community in Arizona with a preponderance of older individuals? Let us again apply the Ratio Rule:

$$\frac{\Pr[A|O]}{\Pr[O|A]} = \frac{\Pr[A]}{\Pr[O]} = \text{small},$$

since in such a retirement community, the number of traffic accidents is fairly small while the number of older residents is large. So it is likely that $\Pr[A|O]$ is small, even though $\Pr[O|A]$ is large. As a result, the insurance company would need more information to draw a supportable conclusion. ◁

To quantitatively illustrate the difference between $\Pr[O|A]$ and $\Pr[A|O]$, consider a retirement community with 3000 individuals, where 85% of the population are older residents and 10% of the population was involved in traffic accidents during a six-month period. From these specifications and $\Pr[O|A] = \frac{2}{3}$, we can fill in Table 15.3.

TABLE 15.3: Accidents and age groups

	Accidents	No Accidents	
Older Residents	200	2350	2550
Younger Residents	100	350	450
	300	2700	3000

The first row of this table shows that

$$\Pr[A|O] = \frac{200}{2550} = 0.078,$$

so the accident rate for older individuals is fairly low. By contrast, the accident rate for younger individuals (Y) is determined to be

$$\Pr[A|Y] = \frac{100}{450} = 0.222,$$

which is almost three times the accident rate for older individuals. Clearly it would not be justifiable to raise the insurance rates for older individuals in this community.

(Related Exercise: 26)

15.3   Probability and the Law

A famous example of fallacious probabilistic reasoning in the courtroom is the Sally Clark case.

HISTORICAL CONTEXT **Sally Clark (1964–2007)**
In 1999, Sally Clark was convicted of murdering her two small infants. The defense claimed that the deaths were actually instances of SIDS (sudden infant death syndrome), yet she was wrongly convicted of murder based on a faulty probabilistic analysis. The decision was overturned in 2003, after she had spent more than three years in prison.

Sir Roy Meadow, an expert witness called by the prosecution, asserted that the probability of two such deaths in the same family from natural causes is highly unlikely, approximately 1 in 73 million. Based in large part on this argument, Sally Clark was convicted by the British court of causing her children's deaths.

Assuming for the moment that the reported figure "1 in 73 million" is accurate, this small value represents the probability of two deaths in the same family from natural causes or Pr[two deaths|innocent]. However, what the prosecution conveyed to the jury was that the probability she was innocent, Pr[innocent|two deaths], was also "1 in 73 million". This is yet another case of confusing conditional probabilities and their inverses. Namely, the truth of the first claim is insufficient by itself to support the truth of the second.

The figure "1 in 73 million" was arrived at by citing empirical data indicating that a single SIDS death to a family such as the Clarks (professional and nonsmoking) was only 1 in 8500, and then arguing that the probability of two SIDS deaths is therefore $(1/8500) \times (1/8500)$ or approximately 1 in 73 million. The assumption of independence (Section 14.1) is being implicitly used here. Yet, in view of common environmental and genetic conditions, it is much more reasonable to suspect that a second SIDS death in a family, given that one has already occurred, is not independent of the first death and is far more likely than $1/8500$. This (implicit) assumption of independence also contributes to making a serious error in judgment about the innocence of Sally Clark.

The Prosecutor's Fallacy

The *Prosecutor's Fallacy* occurs in the courtroom when the prosecuting attorney argues (erroneously) that the probability is quite small that the defendant is innocent (I) given the evidence (E). This is done by ignoring (mistakenly or intentionally) the difference between a conditional probability and its inverse. Specifically, it involves the substitution of $\Pr[E|I]$ for $\Pr[I|E]$.

We have seen that $\Pr[E|I]$ can be small while $\Pr[I|E]$ is much larger. Judges, prosecutors, and defense attorneys need to consider not only whether evidence is admissible and true, but also whether it is presented in a way that facilitates insight. Too often statistical evidence is presented in a way that clouds thinking or hides the truth.

> DEFINITION **Prosecutor's Fallacy**
> This is a fallacy used by prosecuting attorneys to argue for the guilt of a defendant based on the small probability of evidence occurring in a random member of the population. It is based on a misinterpretation of conditional probabilities.

Example 15.11 Use a Table to Assess the Reliability of a Polygraph Test

Consider a polygraph test that is given to 5000 people, where 8% are lying ($0.08 \times 5000 = 400$ people) and 92% are telling the truth ($0.92 \times 5000 = 4600$ people). The test is 93% accurate. Namely, it will correctly identify 93% of those individual lying as being liars ($0.93 \times 400 = 372$ people), producing a *positive* polygraph result; it will correctly identify 93% of those individuals telling the truth as being truthful ($0.93 \times 4600 = 4278$ people), producing a *negative* polygraph result. This information is displayed in Table 15.4.

TABLE 15.4: Reliability of a polygraph test

	Person is Lying	Person is Truthful	
Positive Result	372		
Negative Result		4278	
	400	4600	5000

Suppose now that a suspect takes this polygraph test and it indicates that he is lying (producing a positive test result). Given the 93% accuracy of the test, the prosecutor confidently argues that the suspect is lying (L) because the suspect tested positive (P). Explain how this example illustrates the Prosecutor's Fallacy.

Solution

First, we fill in the missing entries in Table 15.4 by using the fact that the entries in each column sum to the column total. We then compute the row totals, giving Table 15.5. The prosecutor argues that the suspect is lying because he tested positive on a polygraph test known to be 93% accurate. That is, the prosecutor emphasizes the conditional probability $\Pr[P|L] = 0.93$, which can be verified directly from the table entries. Namely, given that someone is a liar (L), the first column defines the relevant subpopulation; of these 400 people, 372 test positive (P) so that

$$\Pr[P|L] = \frac{372}{400} = 0.93.$$

TABLE 15.5: Completed table for the polygraph test

	Person is Lying	Person is Truthful	
Positive Result	372	322	694
Negative Result	28	4278	4306
	400	4600	5000

The prosecutor is ignoring the inverse probability $\Pr[L|P]$, which specifies the fraction of the time a suspect really is lying given a positive polygraph result. Here the table again assists us in this calculation. The relevant subpopulation consists of those with a positive polygraph result (694 individuals in the first row); of these, 372 are in fact lying so that

$$\Pr[L|P] = \frac{372}{694} = 0.536.$$

This is an illustration of the Prosecutor's Fallacy, since the inverse conditional probability is ignored because it would weaken the case against the suspect. That is, given the evidence (a positive polygraph result) there is approximately a 54% chance that the suspect is lying and thus a 46% chance that the suspect is really telling the truth. There is no compelling case for conviction in these circumstances. ◁

Another opportunity for erroneous decision making in the courtroom arises when the discussion involves comparing DNA samples collected at a crime scene with DNA housed in a particular database. For example, the FBI's DNA Laboratory analyzes crime-scene blood samples and generates DNA profiles. Their Combined DNA Identification System (CODIS) examines 13 regions in the human genome. These "loci" are such that it is virtually impossible for two randomly chosen DNA samples (from unrelated people) to match completely at all of these 13 loci. However, state and other DNA databases of offender profiles may be based on fewer than 13 loci or the crime-scene blood sample may be degraded, resulting in fewer valid loci to compare. This complicates the issue of what a match should signify in the courtroom since several individuals can produce a match in the database.

Example 15.12 Investigate the Probability of Guilt Based on a DNA Database Search

Consider the following situation. A DNA sample is obtained at a crime scene and a DNA profile on four loci is generated. It is compared against 200,000 offender profiles in the databank of the state where the crime was committed and several matches are found. Suppose that the probability of a match at any single loci is $\frac{1}{10}$. Then, assuming independence, the probability that someone's DNA matches the DNA profile from the crime scene at all four different loci is 1 in 10,000:

$$\frac{1}{10} \times \frac{1}{10} \times \frac{1}{10} \times \frac{1}{10} = \frac{1}{10,000}.$$

This *random match probability* is a reliable indicator of the rarity of the obtained DNA profile in the population at large (barring contamination during sample collection or laboratory errors).

A suspect is brought to trial and the prosecutor argues that, since the DNA matches at all four loci, the suspect has only a 1/10,000 chance of being innocent. Explain the fallacy of this argument.

Solution

It is misleading to ignore the number of records searched and the number of matches found when presenting the evidence. Since 1 in every 10,000 people share the crime-scene DNA profile, in a population of 200,000 profiles, we would expect to find approximately 200,000 × (1/10,000) = 20 matches at the four loci. If the actual perpetrator of the crime is in the database, then he, as well as 19 others, would display a match on all four loci. Using the given data, we can construct Table 15.6.

The prosecutor might emphasize to the jury the probability of a match (M) given that someone is innocent (I), which can be found from the first column in Table 15.6:

$$\Pr[M|I] = \frac{19}{199,999} = 0.000095.$$

TABLE 15.6: DNA database search

	Innocent	Guilty	
Match	19	1	20
No Match	199,980	0	199,980
	199,999	1	200,000

However, the pertinent information is the probability of an individual's innocence given that his DNA produces a match, found from the first row of the table:

$$\Pr[I|M] = \frac{19}{20} = 0.95.$$

Again, we see that the proper use and interpretation of conditional probabilities can completely change the assessment of innocence or guilt in the courtroom. ◁

(Related Exercises: 28, 31)

Example 15.13 Use a Table to Calculate Conditional Probabilities for a Blood Test

Blood collected at the scene of a crime shows that the perpetrator of the crime has a particular genetic disorder. In a population of 20,000 individuals, only 11% (2200 people) are known to have the disorder. A laboratory test for this disorder has 97.5% *sensitivity*: with probability 0.975 someone will test positive P for the disorder, given that they really do have the disorder D. The test has 90% *specificity*: with probability 0.90 someone will test negative for the disorder, given that they really don't have the disorder. Using these facts, you can fill in the remaining entries in Table 15.7.

TABLE 15.7: Testing for a gene disorder

	Gene Disorder	No Gene Disorder	
Positive Test			
Negative Test			
	2200		20,000

Now suppose that a suspect's blood tests positive for this disorder and the prosecutor argues that he has the right person because the accuracy of the test is quite high. Explain how this example illustrates the Prosecutor's Fallacy by comparing $\Pr[P|D]$ and $\Pr[D|P]$.

Solution

As 2200 people among the 20,000 have the disorder, there are 17,800 who do not. Using the specificity of 90%, there are then $0.90 \times 17,800 = 16,020$ persons without the disorder who test negative. Using the sensitivity of 97.5%, there are $0.975 \times 2200 = 2145$ persons with the disorder who test positive. We can then fill in the remaining entries, producing Table 15.8.

The prosecutor argues for the guilt of the suspect based on the high 97.5% test sensitivity:

$$\Pr[P|D] = \frac{2145}{2200} = 0.975.$$

TABLE 15.8: Completed table for the gene disorder problem

	Gene Disorder	No Gene Disorder	
Positive Test	2145	1780	3925
Negative Test	55	16,020	16,075
	2200	17,800	20,000

However, we need instead to consider the probability that the suspect has the disorder (D) given a positive test result (P). This can be found by considering the subpopulation defined by the first row of Table 15.8:

$$\Pr[D|P] = \frac{2145}{3925} = 0.546.$$

This inverse conditional probability shows that there is only about a 55% chance that someone who tests positive really has the disorder, which means that there is a 45% chance that the suspect doesn't have the disorder even with a positive test result. This considerably weakens the prosecutor's case. ◁

(*Related Exercise: 29*)

Chapter Summary

Conditional probabilities allow us to incorporate new information when we are assessing uncertain events. In this chapter, we apply Bayes' Law to update given probabilities in light of additional information. In particular, decision trees are used to display known probabilities and known conditional probabilities. Bayes' Law then amounts to "renormalizing" outcome probabilities in the decision tree, by focusing on just those outcomes consistent with the given information. Several paradoxical situations can be easily explained by proper calculation of conditional probabilities; here we see that the relative frequency of events affects the magnitude of such conditional probabilities. Certain fallacious arguments arise when one intentionally or mistakenly confuses a conditional probability with its inverse probability. We apply a simple rule, the Ratio Rule, as well as Venn diagrams and tables, to reveal that there can be significant differences between the probabilities $\Pr[E|F]$ and $\Pr[F|E]$.

Applications, Representations, Strategies, and Algorithms

Applications				
Business (1, 8, 10) Medicine (2) Technology (3)				
Psychology (4) Biology (6) Law Enforcement (7, 9, 11–13)				
Representations				
Graphs		Tables	Sets	Symbolization
Weighted	Tree	Data	Venn Diagrams	Algebraic
1–5	1–5	6, 10–13	9	7–10
Strategies				
Rule-Based				
1–13				
Algorithms				
Exact	Approximate	Numerical	Algebraic	Geometric
1–6, 10–13	7–10	1–6, 10–13	7–10	9

Exercises

Bayes' Law and Decision Trees

1. You are presented with three coins, one of which is biased so that it always turns up heads. These three coins all look alike, so you have no way of telling which is biased (B). You select one of the coins; there is a one-third chance that it is the biased coin. To gather more evidence, you toss this coin three times. It produces the result (R) of turning up heads all three times. You would now like to revise your estimate of whether this coin is the biased one.

   (a) Construct a two-stage decision tree, in which the first stage indicates whether the selected coin is biased (B) or normal (N). The second stage indicates whether the result (R) of three heads was observed or was not $(O = \text{Otherwise})$.

   (b) Calculate the outcome probabilities for the decision tree and use these values to calculate $\Pr[B|R]$. Interpret what this value means.

2. The weather forecast for the weekend indicates that there is a 60% chance of sun (S), a 30% chance of high winds (W), and a 10% chance of rain (R). From past observations, we know that Mohan will play frisbee 85% of the time it is sunny, 60% of the time it is windy, and 30% of the time it is rainy. Otherwise, he plays cards indoors.

   (a) Construct a two-stage decision tree, in which the first stage represents the type of weather (S, W, R) and the second stage indicates whether Mohan will play frisbee (F) or will play cards (C).

   (b) Mohan tells Janet that he played cards over the weekend. What is the probability that the weather was windy?

3. It is known that 35% of students take advantage of a study session before an important midterm exam. From past data, 70% of the students utilizing the study session score an A on the midterm, 20% score a B, and the rest score a C. For those students not utilizing the study session, 30% score an A, 50% score a B, and the rest score a C. Use a decision tree to answer the following questions.

   (a) If Devi scored an A on the midterm, how likely is it that she attended the study session?

   (b) If Pietro scored a C on the midterm, how likely is it that he did not attend the study session?

4. Anneke is shopping online for a new flat screen TV. From past experience, she buys from the Amazing Electronics site 30% of the time, the Better Buy site 50% of the time, and the Circuit Circus site 20% of the time. At Amazing Electronics, 30% of the TV sales are LCDs and the rest are LEDs; at Better Buy, 40% of the TV sales are LCDs and the rest are LEDs; and at Circuit Circus, 80% of the TV sales are LCDs and the rest are LEDs. Use a decision tree to answer the following questions.

   (a) If Anneke is seen to have purchased an LCD TV, how likely is it that it came from Better Buy?

   (b) If Anneke is seen to have purchased an LED TV, how likely is it that it came from Circuit Circus?

5. To fulfill their college math requirement, liberal arts students take Statistics, Calculus, or Discrete Math; 35% of these students take Statistics, 15% take Calculus, and the remainder take Discrete Math. The passing rate for students in the Statistics course is 40%, while in the Calculus course it is 30% and in the Discrete Math course it is 55%. Use a decision tree to answer the following questions.

   (a) If Raj passed his required math course, how likely is it that he was taking Discrete Math?

   (b) If Devon did not pass his required math course, how likely is it that he was taking Statistics?

6. A laboratory test for tularemia has a true positive rate of 98% and a false positive rate of 1%. Approximately 0.3% of the population has tularemia. If Ralph walks into a doctor's office and tests positive for tularemia, how likely is it that he is actually infected with tularemia? Draw a decision tree and use it to compute this probability.

7. A mammogram test has 80% sensitivity in detecting cancerous tumors and 90% specificity in correctly diagnosing the absence of cancerous tumors. The incidence rate for these types of tumors in the general public is 0.4%. An otherwise healthy woman has a mammogram as part of her yearly checkup and this test shows a positive result. How likely is it that she does in fact have a cancerous tumor? Draw a decision tree and use it to compute this probability.

8. During the Vietnam War, an enemy fighter plane made a raid on a U.S. camp. Both Cambodian and Vietnamese jets were known to operate in the area. One witness to the pre-dawn raid identified the fighter as being Cambodian. Subsequent tests of this witness established that he could correctly identify the aircraft (as Cambodian or as Vietnamese) 85% of the time. It is known that in that area, 80% of the aircraft were Vietnamese. Use a decision tree to represent this information and to determine the reliability of the reported evidence—namely, the probability that the aircraft observed to be Cambodian was in fact Cambodian.

9. In a jar of jelly beans, five are orange and five are green. You scoop out with your right hand one jelly bean and then another with the left hand.

    (a) Draw an appropriate decision tree for this problem.

    (b) You open your right hand and reveal an orange jelly bean. What is the probability that the other one is orange also?

    (c) Suppose you place both jelly beans in your right hand, without looking at their colors. Now you take out one jelly bean and find that it is orange. What is the probability that the other one is orange also?

10. Four identical sealed envelopes are on a table. One of them contains a $100 bill; each of the rest contains a coupon for 50 cents off your favorite beverage. You select one of the envelopes and hold onto it, without opening it.

    The host of this contest (who knows where the $100 prize is located) takes two of the envelopes and holds onto them. You are assured they do not contain the $100 prize.

    The host now offers you the opportunity to switch your selection to the one envelope remaining on the table. You need to decide whether to switch.

    (a) Draw a decision tree to model this situation.

    (b) Using the decision tree, compute the probability that you win the $100 prize *if you switch*.

    (c) Using the decision tree, compute the probability that you win the $100 prize *if you don't switch* (keep your original choice).

    Should you switch? Explain.

Bayes' Law and Data Tables

11. A test for Lyme disease is 96% accurate: it will correctly detect 96% of patients who have been infected and will also correctly identify 96% of patients who have not been infected. There are 3000 people who took the test and 100 actually have Lyme disease. Use this information to complete the table below, and then calculate the probability that a random person has Lyme disease, given a positive test result.

	Infected	Not Infected	
Positive Test			
Negative Test			
	100		3000

12. A screening test for a rare blood disorder (which affects only 2% of individuals) is given to 5000 participants. The test will correctly detect 96% of people who have the disorder and will also correctly detect 98% of people who do not have this disorder. Construct an appropriate two-way table and use it to calculate the probability that a person who tests positive for the blood disorder actually has it.

13. In a study of 2000 older adults, 12% had severe osteoporosis. A diagnostic test for early detection of this disease is 95% accurate in correctly detecting those who have the disease and is 90% accurate in correctly detecting those who do not have the disease. Construct an appropriate two-way table and use it to calculate the probability that a person who tests positive for severe osteoporosis actually has it.

14. A polygraph test has been administered to 3000 individuals and has been found to have 95% sensitivity and 94% specificity. It is known that 90% of the population are truthful. If a person tests positive for lying using this device, how likely is it that he is in fact lying?

15. In a study of 1500 men, it was determined that 21% smoked and 7% had lung cancer. National statistics indicate that 20% of all male smokers contract lung cancer.

    (a) Use this information to complete the following table.

    (b) A man is randomly selected from the 1500 subjects and it is found that he does not have lung cancer. Determine the probability that he is not a smoker.

	Cancer	No Cancer
Smoker		
Non-Smoker		
		1500

16. A company accountant is attempting to collect on 1000 invoices. From past experience, he knows that 5% of all invoices will not be paid on time. Also, 20% of the invoices will be big-ticket items (above $1000) and that only 10% of these big-ticket items will not be paid on time.

    (a) Use this information to complete the table shown below.

    (b) How likely is it that an invoice for less than $1000 will be paid on time?

	< $1000	> $1000
On Time		
Late		
		1000

17. A manufacturer of high quality headphones produces 2500 units each week. Past data show that 5% of all headphones produced are defective. There is a 20% chance that a defective unit will pass inspection and be sent out to stores; there is a 24% chance that a nondefective unit will fail the inspection. If an item passes inspection, how likely is it that it will be defective?

18. Drivers are rated by an insurance company as either poor, good, or excellent based on their driving records. In a group of 10,000 drivers, 4500 were involved in an accident during the year. Of the 5500 drivers in the good category, 3300 of them had no accidents during the past year. Of the 1000 drivers in the excellent category, 800 of them had no accidents during the past year. Use this information to complete an appropriate two-way table. Given that an accident is reported to the insurance company, how likely is it that a driver in the poor category was involved?

19. The probability of a randomly selected licensed driver being involved in an accident during 2011 is categorized by age in the following U.S. Census Bureau table:

    (a) What is the probability that a randomly selected driver is involved in an accident?

    (b) In looking at a random accident in 2011, what is the probability that the driver is less than 25 years old?

Age	Chance of Accident	% of Licensed Drivers
< 25	0.16	27.2
25–34	0.09	19.8
35–44	0.08	17.6
45–54	0.07	16.7
55–64	0.05	10.4
> 64	0.04	8.3

Conditional Probabilities and Their Inverses

20. Analyze the following argument using the Ratio Rule: "A large proportion of fatal accidents are known to involve older vehicles. Therefore, it is in the best interest of public safety to remove older vehicles from the road, perhaps by instituting very strict standards for registering older vehicles."

21. After meeting a former athlete who was over 75 years old, a sports commentator noted that former athletes often live to be over 75. He then observed that it would be quite common to find former athletes among a group of people over 75. Analyze this argument using the Ratio Rule.

22. A headline reads "Most Speeding Tickets Issued to Minors." The article goes on to suggest that police should target minors for special attention in terms of monitoring their driving speed. Analyze this argument using the Ratio Rule.

23. An article in a medical journal reported that in a study of 1200 cases of Hepatitis B, 10% were the result of receiving blood transfusions. Does it follow that you have a risk of 10% of contracting Hepatitis B if you receive a blood transfusion? Explain using the Ratio Rule.

24. A concerned viewer wrote to a network TV executive that almost all of the Italians on their crime shows were portrayed as mobsters. The executive wrote back that the proportion of Italians among the mobsters was not any higher than the proportions of other ethnic groups. Analyze the validity of this response using a Venn diagram.

25. Suppose it is known that if your food has been tainted with salmonella, you have an 80% chance of dying. That is, $\Pr[D|S] = 0.80$. Does this mean that if someone dies after eating a meal, there is then an 80% chance that the individual was poisoned? In other words, does $\Pr[S|D] = 0.80$? Demonstrate that $\Pr[D|S]$ can be quite different from $\Pr[S|D]$ using a Venn diagram.

26. A national association of athletes decides to test 302 random individuals for use of performance-enhancing drugs. The following data were obtained:

	Used Drugs	No Drugs Used
Positive Test	118	50
Negative Test	4	130

302

Use this table to determine the probability $\Pr[P|D]$ that a randomly selected athlete tested positive, given that he actually used drugs. Also, determine the probability $\Pr[D|P]$ that an athlete actually used drugs, given a positive test result.

27. We revisit Example 15.8 and explore its conclusions further. Suppose that the probability $\Pr[O]$ of childhood pet ownership is 0.5 and that the probability $\Pr[C]$ of becoming a CEO of a Fortune 500 company is 0.000002. Recall that $\Pr[O|C] = 0.94$.

   (a) Determine $\Pr[C|O]$. Are C and O independent?

   (b) Do your results indicate that children are more likely to become CEOs when they own a pet, compared to a random child in the population?

   (c) Let O' denote the event that a child does not own a pet. Determine $\Pr[C|O']$. Are C and O' independent?

   (d) Do your results indicate that children are more likely to become CEOs when they own a pet, compared to children who do not own a pet?

Probability and the Law

28. Fingerprints found at a crime scene are compared with a state's databank of 900,000 records. A suspect in the crime is brought in for questioning and his fingerprints match those found at the scene of the crime at five loci. Assume the probability of a match at any single loci is $\frac{1}{10}$. Complete the table below and use this to compare $\Pr[M|I]$ with $\Pr[I|M]$. Which is the more pertinent probability for assessing Innocence (I) based on a fingerprint match (M)?

	Innocent	Guilty
Match		
No Match		
		900,000

29. Crime-scene evidence shows that the perpetrator of a crime has a particular genetic disorder (D). A laboratory test for this disorder has 95% sensitivity and 85% specificity. A suspect's blood tests positive (P) for this disorder and the prosecutor argues that he has the right person because the accuracy of the test is quite high. Pertinent facts: In a population of 50,000 individuals, only 5% (2500 people) are known to have this disorder. Using this information, complete the table below. Explain how this situation illustrates the Prosecutor's Fallacy by comparing $\Pr[P|D]$ and $\Pr[D|P]$.

	Gene Disorder	No Gene Disorder
Positive Test		
Negative Test		
	2500	50,000

Projects

30. Consider the scenario described in Example 15.2, where the laboratory test has a true positive rate of 95% and a false positive rate of 1%. Now vary the proportion of the population having Q fever over the range 1% to 30%. Discuss the results you obtain for the conditional probability that Sally is infected with Q fever.

31. Consider the scenario described in Example 15.12, in which 1 in 10,000 people share the crime-scene DNA profile. Vary the size of the state databank from 20,000 to 200,000 and determine how the probability of innocence varies over this range. Discuss your results.

32. The Federalist papers were written during 1787–1788 to persuade voters of New York to ratify the Constitution. The authorship of twelve of these essays was considered in doubt; they could have been written by either Alexander Hamilton or James Madison. Frederck Mosteller and David Wallace applied Bayes' Law to this problem, using the frequencies of words used by both authors in other writings. Further investigate this large-scale application of Bayesian analysis to a historical dispute.

Bibliography

[Donnelly 2005] Donnelly, P. 2005. "Appealing statistics." *Significance* 2: 46–48.

[Gigerenzer 2002] Gigerenzer, G. 2002. *Calculated risks: How to know when numbers deceive you*, Chapter 9. New York: Simon & Schuster.

[Hastie 2010] Hastie, R., and Dawes, R. 2010. *Rational choice in an uncertain world*, Chapter 8. Los Angeles: Sage.

[Kahneman 2011] Kahneman, D. 2011. *Thinking, fast and slow*. New York: Farrar, Straus and Giroux.

[McGrayne 2011] McGrayne, S. 2011. *The theory that would not die*. New Haven, CT: Yale University Press.

[Olofsson 2007] Olofsson, P. 2007. *Probabilities: The little numbers that rule our lives*, Chapter 4. Hoboken, NJ: John Wiley & Sons.

[Rosenhouse 2011] Rosenhouse, J. *The Monty Hall problem*. Oxford: Oxford University Press.

[OR1] `http://www.shodor.org/interactivate/activities/SimpleMontyHall/` (accessed September 13, 2013).

[OR2] `http://math.ucsd.edu/~crypto/Monty/monty.html` (accessed September 13, 2013).

Chapter 16

Expected Values and Decision Making

Life is largely a matter of expectation.

—Horace, Roman poet, 65–27 BC

Chapter Preview

To analyze and understand a course of action that involves uncertainties, it is helpful to construct a probability model that lists the probabilities of all possible outcomes in a particular situation or experiment. Often we are interested in knowing the gains or losses associated with each outcome, and the expected value that can be anticipated. The expected value indicates the long-run gain or loss that occurs after many independent repetitions. This numerical value can provide valuable information in making rational decisions in the face of uncertainties. In particular, its value can guide the determination of optimal mixed strategies in two-person zero sum games. Moreover, the appropriate use of expected values can explain several paradoxical situations.

16.1 Probability Models and Expected Value

In order to quantify the uncertainties associated with some situation or experiment, we begin by identifying the possible outcomes of the experiment and then assigning probabilities to each outcome. This process more formally consists of constructing a *probability model* for the situation at hand. Recall from Chapter 12 that the set of all outcomes is called the sample space and that the sum of the probabilities of all outcomes in the sample space is 1.

DEFINITION **Probability Model**
A probability model prescribes a sample space S and assigns probabilities to every outcome in S; all such probabilities must sum to 1.

Example 16.1 Construct a Probability Model for Tossing a Fair Coin Twice

Describe a sample space S and probabilities for each outcome in the sample space for the situation in which a fair coin is tossed two times in succession.

Solution

A legitimate probability model is based on the sample space $S = \{HH, HT, TH, TT\}$, which lists the results of the two coin tosses. Since we are tossing a fair coin, each of these four outcomes is equally likely (with probability of occurrence $\frac{1}{4}$). Table 16.1 provides a compact way to list the sample space and the associated probabilities. ◁

TABLE 16.1: Probability model for tossing a coin twice

Outcome	HH	HT	TH	TT
Probability	$\frac{1}{4}$	$\frac{1}{4}$	$\frac{1}{4}$	$\frac{1}{4}$

Table 16.1 is based on partitioning the sample space into outcomes, the simplest form of events. We can also obtain a legitimate probability model if we divide up the sample space into other disjoint events. For instance, we can divide up the sample space $S = \{HH, HT, TH, TT\}$ in Example 16.1 into the disjoint events $\{0 \text{ Heads}, 1 \text{ Head}, 2 \text{ Heads}\}$, with the corresponding probabilities shown in Table 16.2. Notice that the event $E = \{1 \text{ Head}\} = \{HT, TH\}$ has probability $\Pr[E] = \Pr[HT, TH] = \Pr[HT] + \Pr[TH] = \frac{1}{4} + \frac{1}{4} = \frac{1}{2}$, since the outcomes HT and TH are disjoint.

TABLE 16.2: Another probability model for tossing a coin twice

Event	0 Heads	1 Head	2 Heads
Probability	$\frac{1}{4}$	$\frac{1}{2}$	$\frac{1}{4}$

(Related Exercises: 1–9)

Expected Value

A probability model provides the basis for describing the numerical results of uncertain events. For example, consider a simple game in which we toss a coin twice, winning \$1 for each head appearing face up and losing \$1 for each tail appearing. The monetary values associated with each event are tabulated in Table 16.3, which adds these numerical values to the entries in Table 16.1.

TABLE 16.3: Values obtained in a coin-tossing game

Event	HH	HT	TH	TT
Probability	$\frac{1}{4}$	$\frac{1}{4}$	$\frac{1}{4}$	$\frac{1}{4}$
Value	+\$2	\$0	\$0	−\$2

If we play this game many times, how much can we expect to win? Since each event is equally likely (occurring with probability $\frac{1}{4}$), we would anticipate gains and losses to balance out and so we should end up with \$0. Mathematically, we can calculate the *expected value* as the following weighted average:

$$\tfrac{1}{4}(+2) + \tfrac{1}{4}(0) + \tfrac{1}{4}(0) + \tfrac{1}{4}(-2) = 0.$$

More generally, the expected value is defined as the (theoretical) average gain/loss experienced in the long run. We calculate it by adding together the product of the probability p_i of each event and its value v_i.

DEFINITION **Expected Value**
Expected Value $= p_1 v_1 + p_2 v_2 + \cdots + p_n v_n$

If we have a probability model with different probabilities for each possible event, more likely events will be weighted more heavily in the average than less likely events. This occurs in the following example.

Example 16.2 Calculate the Expected Value of the Final Grade in Calculus

Table 16.4 shows a probability model based on the final grades that 1100 students received in all sections of Calculus I during a recent semester (using a 4-point scale with A = 4, B = 3, etc.). Notice that the sample space consists of five outcomes, whose probabilities sum to 1. Determine the average grade-point for this class and interpret what it means.

TABLE 16.4: Grades in the Calculus I course

Event	A	B	C	D	F
Probability	.258	.352	.198	.041	.151
Value	4	3	2	1	0

Solution

The average grade-point for this course is

$$(.258)(4) + (.352)(3) + (.198)(2) + (.041)(1) + (.151)(0) = 2.525, \text{ or a C}^+.$$

If we take a random sample of students in the class, the average grade these students receive in this class should be approximately 2.525; this sample average will get closer to 2.525 as more and more students are included in the sample. ◁

(Related Exercises: 10–13)

Similarly, as we play the coin-tossing game described earlier more and more times, our overall winnings will get closer to $0. This is an important fact of probability, called the *Law of Large Numbers*.

DEFINITION **Law of Large Numbers**
If an experiment is repeated more and more times, the average of the numerical results obtained will approach the expected value computed from the probability model.

In other words, the *empirical* average that we observe will approach the *theoretical* expected value. Gambling casinos make profits because of the Law of Large Numbers. As thousands of people play each game of chance over and over, the casino can depend on the increasing accuracy of the predictions of probability (that is, the expected value of each game).

(Related Exercise: 32)

Example 16.3 Calculate an Expected Value in the Game of Roulette

Consider again the game of roulette, described in Section 12.1. The ball could land in any of 38 slots in the spinning roulette wheel: 18 red, 18 black, and 2 green. As the dealer spins the wheel, gamblers bet on where the ball will land. Suppose you place a $1 bet on red: you will win $1 on red and will lose $1 on black or green. How much would you expect to gain (or lose) in the long run?

Solution

Table 16.5 shows the associated probability model, augmented with the values for each event. The expected value is then $\left(\frac{18}{38}\right)(1) + \left(\frac{20}{38}\right)(-1) = -0.053$. The negative value means that if you place a single $1 bet on red, then over the long run, you will lose an average of 5.3 cents per bet. The casino can count on winning an average of 5.3 cents per bet as more and more such bets are made. ◁

TABLE 16.5: Probability model for the game of roulette

Event	Red	Non-Red
Probability	$\frac{18}{38}$	$\frac{20}{38}$
Value	+$1	−$1

(Related Exercises: 14–16)

A Paradox in Roulette?

Suppose now that a player bets $10 on a single number, for example, on #13. In this game, the player wins $360 if #13 comes up, and receives $0 otherwise. Since there are 38 possible (equally likely) outcomes, the expected payoff is

$$\tfrac{1}{38}(360) + \tfrac{37}{38}(0) = 9.474.$$

By placing a $10 bet, the net earnings of the player are $9.474 - 10 = -0.526$, a loss of approximately 53 cents per play. If a player places a $10 bet 35 times on #13, then the player should expect a loss of $35 \times 0.526 = \$18.41$ over the course of the 35 plays.

Since the payoff for a win is $360, a player actually comes out ahead after 35 plays precisely when that the person wins *at least once*. We can then calculate the probability of coming out ahead after 35 plays using the Complement Rule:

$$\Pr[\text{win at least once}] = 1 - \Pr[\text{lose all 35 times}] = 1 - \left(\tfrac{37}{38}\right)^{35} = 0.607.$$

If people lose on average $18.41 over the course of 35 plays, how then can 61% of the bettors actually come out ahead? This apparent paradox can be easily explained. While 61% of the players win, the amount that they win is typically small (compared to what they pay out to play). On the other hand, the other 39% of the players will lose big: namely, the entire $350 wagered. The expected value is thus heavily influenced by the (minority) of losing players. We will see again (in Section 16.3) that one must be careful in computing and interpreting average values.

Example 16.4 Determine the Expected Value of the *Palmetto Cash 5* Lottery

In the South Carolina lottery game *Palmetto Cash 5*, five different numbers are selected from $1, 2, \ldots, 38$. Prizes are awarded if a player matches two, three, four, or five of the five winning numbers. Each ticket costs $1. The payoffs (and associated odds) are given in Table 16.6. What is your expected gain (or loss) in playing this game?

TABLE 16.6: *Palmetto Cash 5* prizes

Match 5	$100,000
Odds	1 in 501,942
Match 4	$300
Odds	1 in 3042
Match 3	$5
Odds	1 in 95
Match 2	$1
Odds	1 in 9.2

Solution

Table 16.6 indicates the monetary payoffs returned from making a $1 bet. The odds listed in this table can be obtained by using the Combinations Formula (see Section 13.1). Now, however, our objective is to use these payoffs and odds to make an informed decision. First, we calculate the expected gain (expected payoff) using this probability model:

$$\left(\tfrac{1}{501,942}\right)(\$100,000) + \left(\tfrac{1}{3042}\right)(\$300) + \left(\tfrac{1}{95}\right)(\$5) + \left(\tfrac{1}{9.2}\right)(\$1) = \$0.459.$$

Since it costs $1 to play the game, we can expect to lose approximately 54 cents per play. So, this is not a wise investment! ◁

(Related Exercises: 17–18)

In some cases, expected values can be useful guides in allocating resources. The following example illustrates this idea in the context of an interrupted coin-tossing game.

Example 16.5 Determine Payouts in an Interrupted Coin-Tossing Game

Henry and Thelma plan to toss a fair coin seven times in a best-of-seven series. Henry wins a round if the coin turns up H, while Thelma wins if it turns up T. The first player to win four rounds collects a prize of $80. However, the game is interrupted after Henry is ahead two wins to one, and both players want to split the prize in an equitable way. How should the $80 be divided between the two players in this interrupted game?

Solution

We construct a directed graph to display the possible outcomes of the game, starting at the initial state $(2, 1)$ that indicates Henry is ahead two wins to one. Each directed edge from $(2, 1)$ represents an equally likely transition to a new state: either $(3, 1)$ or $(2, 2)$. The full graph of possibilities is shown in Figure 16.1; there are no more transitions needed once a player has won four rounds. We also show at each vertex the probability of reaching that vertex by some path from state $(2, 1)$. For example, state $(3, 2)$ is reachable by two paths:

$$(2, 1) \to (3, 1) \to (3, 2) \quad \text{and} \quad (2, 1) \to (2, 2) \to (3, 2).$$

Each of these paths has probability $\frac{1}{4}$, the product of the probabilities $\frac{1}{2}$ along each (equally likely) edge. Adding these two path values gives $\frac{1}{4} + \frac{1}{4} = \frac{1}{2}$ as the probability of reaching state $(3, 2)$ from the initial state $(2, 1)$.

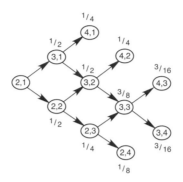

FIGURE 16.1: Graph for the interrupted game

Thus, the probability that Henry wins the game is the probability of reaching any of the outcomes $(4, 1)$, $(4, 2)$, or $(4, 3)$; the probability that Thelma wins the game is the probability of reaching either of the outcomes $(2, 4)$ or $(3, 4)$. These probabilities are given as

$$\Pr[\text{Henry wins}] = \Pr[(4,1)] + \Pr[(4,2)] + \Pr[(4,3)] = \tfrac{1}{4} + \tfrac{1}{4} + \tfrac{3}{16} = \tfrac{11}{16},$$

$$\Pr[\text{Thelma wins}] = \Pr[(2,4)] + \Pr[(3,4)] = \tfrac{1}{8} + \tfrac{3}{16} = \tfrac{5}{16}.$$

Now we can calculate the expected value for Henry as $\frac{11}{16}(\$80) + \frac{5}{16}(\$0) = \$55$ and the expected value for Thelma as $\frac{5}{16}(\$80) + \frac{11}{16}(\$0) = \$25$. So the fair way to split the $80 prize in this interrupted game is by giving $55 to Henry and $25 to Thelma. ◁

Example 16.6 Achieve the Largest Expected Net Gain in Matching Baseball Cards

Each cereal box contains a card picturing one of nine different baseball All-Stars. You win a cash prize of $15 if you purchase several boxes and obtain a match (getting the same picture). At most one $15 prize can be awarded to each person; that is, a person cannot enter more than once with several matching cards. Each cereal box costs $2. How many cereal boxes, if any, should you buy to maximize your net prize earnings?

Solution

Let's begin by seeing what our expected gain/loss will be if two cereal boxes are purchased. In this case, there are $9 \times 9 = 81$ possible outcomes, of which nine are favorable (a pair of matching cards). This then gives the probability of a match as $\frac{9}{81} = \frac{1}{9} = 0.1111$. So the expected prize value for two boxes is $(0.1111)(\$15) + (0.8889)(\$0) = \$1.67$. However, since purchasing two boxes cost $4, this strategy results in a net loss of $2.33.

With three boxes, there are $9 \times 9 \times 9 = 729$ possible outcomes. The probability of no matches is

$$\frac{9 \times 8 \times 7}{9 \times 9 \times 9} = \frac{56}{81} = 0.6914,$$

which gives a matching probability of $1 - 0.6914 = 0.3086$. The expected prize value is then $(0.3086)(\$15) + (0.6914)(\$0) = \$4.63$; since the purchase cost is $6.00, there is a net loss of $1.37.

Similar calculations for the purchase of up to nine cereal boxes are displayed in Table 16.7. The largest net gain is $1.29 and the optimal strategy is then to purchase six boxes. If this strategy is used over and over again, one would expect to gain $1.29. ◁

(*Related Exercises: 19–20*)

TABLE 16.7: Expected gains and losses in a card-matching game

# Boxes	Probability of a Match	Expected Payout	Purchase Cost	Net Gain/Loss
1	0	0	$2.00	−$2.00
2	0.1111	$1.67	$4.00	−$2.33
3	0.3086	$4.63	$6.00	−$1.37
4	0.5391	$8.09	$8.00	+$0.09
5	0.7439	$11.16	$10.00	+$1.16
6	0.8862	$13.29	$12.00	+$1.29
7	0.9621	$14.43	$14.00	+$0.43
8	0.9916	$14.87	$16.00	−$1.13
9	0.9991	$14.99	$18.00	−$3.01

Example 16.7 Explain a Coin-Tossing Paradox Using Expected Values

The following interesting situation has been described in a TED talk by Peter Donnelly [OR1]. Toss a fair coin and count the number of tosses until the pattern HTH first occurs. Likewise, count the number of tosses until the pattern HTT first occurs. Intuition suggests that on average both should take about the same number of tosses. However, the average number of tosses to achieve the first HTH is in fact *larger* than the average number to achieve the first HTT. Use expected value calculations to verify this surprising result.

Solution

We draw a diagram of the eight states after three coin tosses, with transitions between states represented by directed edges. For example, if we are in state THH (the previous three tosses of the coin), then the next toss is equally likely to take us to state HHH or to state HHT; these transitions are shown as the two edges leaving state THH. For convenience we have labeled the states as $1, 2, \ldots, 8$ in the directed graph of Figure 16.2.

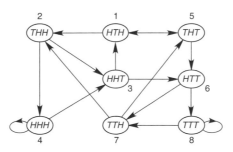

FIGURE 16.2: Graph for the coin-tossing paradox

We want to calculate the expected number of steps until we reach state 1 (HTH), where we can assume that the initial state is equally likely to be any of the vertices $1, 2, \ldots, 8$ in the graph. Let m_i be the expected number of steps to state 1, starting at vertex i. Then we have the following relations:

$$m_1 = 0$$
$$m_2 = 1 + \tfrac{1}{2}m_3 + \tfrac{1}{2}m_4$$
$$m_3 = 1 + \tfrac{1}{2}m_1 + \tfrac{1}{2}m_6$$
$$m_4 = 1 + \tfrac{1}{2}m_3 + \tfrac{1}{2}m_4$$
$$m_5 = 1 + \tfrac{1}{2}m_1 + \tfrac{1}{2}m_6$$
$$m_6 = 1 + \tfrac{1}{2}m_7 + \tfrac{1}{2}m_8$$
$$m_7 = 1 + \tfrac{1}{2}m_2 + \tfrac{1}{2}m_5$$
$$m_8 = 1 + \tfrac{1}{2}m_7 + \tfrac{1}{2}m_8$$

For example, $m_1 = 0$ since we need no additional steps to reach state 1, starting at state 1 (we're already there!). On the other hand, starting at state 2, it takes one step to reach either state 3 or state 4; we are equally likely to be at either and then it takes an additional m_3 or m_4 steps (respectively) to reach state 1. This gives us the second equation. The others follow by similar logic. Solving these equations gives the following values:

$$m_1 = 0, m_2 = m_4 = 8, m_3 = m_5 = 6, m_6 = m_8 = 10, m_7 = 8.$$

The expected number of steps to reach state 1 (HTH) consists of the first three steps plus the average number of steps from the first state reached to arrive at HTH:

$$3 + (0 + 8 + 8 + 6 + 6 + 10 + 10 + 8)/8 = 3 + 7 = 10.$$

In a similar way we can set up equations to determine the expected number of steps to reach state 6 (HTT), solve for the m_i, and then compute the average

$$3 + (6 + 6 + 6 + 6 + 4 + 4 + 0 + 8)/8 = 3 + 5 = 8.$$

To conclude, the average number of steps to reach HTH is indeed larger than the average number to reach HTT. ◁

(*Related Exercises: 21–22*)

16.2   Expected Value and Game Theory

Let's revisit the coin-matching game from Chapter 11, where two players decide to show each other either a head or a tail.

Example 16.8 Calculate the Expected Value in a Coin-Matching Game

The payoffs in dollars to Player A from Player B are listed in Table 16.8. Recall that a negative value indicates a gain to Player B. As the payoffs indicate, Player A wins if the coins match while Player B wins if they don't. Since Player A gains $5 + 1 = 6$ if they match, while Player B wins $3 + 3 = 6$ if they do not, one might think this is a "fair game"—not in favor of either player in the long run. (a) Demonstrate that there is no saddlepoint in this game. (b) Find the optimal strategies for each player. (c) Using these optimal strategies, calculate the value of this game and interpret what it means.

TABLE 16.8: Payoffs in a coin-matching game

		Player B	
		H	T
Player A	H	5	−3
	T	−3	1

Solution

(a) The row minimum and column maximum values are shown in Table 16.9. The maximin value for Player A is −3, while the minimax value for Player B is +1; so the game does not have a saddlepoint. We suspect then that the true value of the game should lie somewhere between −3 and +1.

TABLE 16.9: Checking for a saddlepoint in a coin-matching game

		Player B		Row Minimum
		H	T	
Player A	H	5	−3	−3
	T	−3	1	−3
Column Maximum		5	1	

(b) Since there is no saddlepoint, we seek optimal mixed strategies for this game: specifically, the probabilities $(q, 1 - q)$ to be used by Player A, as shown in Table 16.10.

TABLE 16.10: Determining probabilities in a coin-matching game

		Player B		Probability
		H	T	
Player A	H	5	−3	q
	T	−3	1	$1 - q$

Player A's expected value if Player B shows H is then $EV_H = q(5) + (1 - q)(-3) = 8q - 3$, while Player A's expected value if Player B shows T is $EV_T = q(-3) + (1 - q)(1) = 1 - 4q$. Since Player A can't predict which strategy Player B will choose, it seems reasonable to equate these two algebraic expressions and then solve for q. This will produce probabilities that Player A can use in order to do as well as possible against Player B, regardless of what the latter does:

$$8q - 3 = 1 - 4q, \text{ so } 12q = 4 \implies q = \tfrac{1}{3}.$$

Player A should then play the two strategies with respective probabilities $\tfrac{1}{3}$ and $\tfrac{2}{3}$. By the symmetry of the payoff table, Player B should also adopt the same mixed strategy.

(c) By showing H one-third of the time and T two-thirds of the time, Player A will guarantee the following expected value:

$$8q - 3 = 8(\tfrac{1}{3}) - 3 = -\tfrac{1}{3}.$$

Using this $(\tfrac{1}{3} H, \tfrac{2}{3} T)$ mixed strategy, Player A is guaranteed to lose at most $33\tfrac{1}{3}$ cents on average per play. Moreover, this is an *optimal* mixed strategy—meaning that Player A cannot guarantee any better outcome on average. Interestingly, it is optimal for Player A to randomly select the second strategy (T) twice as often as the first strategy (H), even though the payoff for matching HH is much larger than the payoff for matching TT. This game is not fair; it is weighted in favor of Player B by $33\frac{1}{3}$ cents on average per play. ◁

Geometric Approach

Why did we equate the expected values EV_H and EV_T in Example 16.8? It is useful to look at this geometrically, by plotting the graphs of $8q - 3$ and $1 - 4q$, which are the respective expected values if Player B plays each of the two strategies. The colored broken line in Figure 16.3 represents the *minimum* of the two lines at each value of q: it shows the *guaranteed* expected value for Player A at each value of q. Since Player A is adopting a maximin strategy, Player A wishes to have this guaranteed minimum value as large as possible, and so should select the value q corresponding to the highest peak of the broken line curve—namely, where the two expected value lines intersect. The x-coordinate at this intersection gives $q = \frac{1}{3}$ while the y-coordinate provides the value of the game $-\frac{1}{3}$.

(Related Exercises: 23–24)

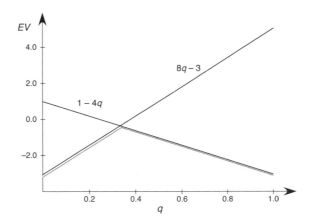

FIGURE 16.3: Plot of two expected values

This geometric approach can be used whenever one player has two strategies, even though the other player may have many more strategies. However, we can't just arbitrarily equate expected values. The next example shows how a geometric point of view can guide us.

Example 16.9 Calculate the Expected Value Given Three Strategies for Player B

Table 16.11 shows the payoffs to Player A. Again a negative value indicates a gain to Player B. (a) Verify that there is no saddlepoint for this game. (b) Using probabilities $(q, 1 - q)$ for Player A, determine the expected payoffs to Player A when Player B uses each of the three available strategies. (c) Determine the optimal mixed strategy for Player A and the value of this game.

TABLE 16.11: Payoffs in a two-person game

	Player B		
Player A	5	−1	2
	−2	4	−1

Solution

(a) The row minimum and column maximum values are calculated in Table 16.12. Since the maximin value −1 and the minimax value 2 differ, there is no saddlepoint.

TABLE 16.12: Checking for a saddlepoint

	Player B			Row Minimum
Player A	5	−1	2	−1
	−2	4	−1	−2
Column Maximum	5	4	2	

(b) Based on the mixed strategy $(q, 1-q)$ for Player A indicated in Table 16.13, the expected payoffs to Player A, when Player B uses each of the three strategies $(1, 2, 3)$, are

$$EV_1 = q(5) + (1-q)(-2) = 7q - 2$$
$$EV_2 = q(-1) + (1-q)(4) = 4 - 5q$$
$$EV_3 = q(2) + (1-q)(-1) = 3q - 1$$

TABLE 16.13: Finding the optimal mixed strategy

	Player B			Probability
Player A	5	−1	2	q
	−2	4	−1	$1 - q$

(c) These expected values have been plotted in Figure 16.4. The colored broken line, showing the minimum expected payoff at each q, is also indicated; it shows the amount Player A can guarantee for each q. The peak (maximum) of this broken line occurs when $EV_2 = EV_3$:

$$4 - 5q = 3q - 1, \text{ so } 8q = 5 \implies q = \tfrac{5}{8}.$$

The optimal strategy for Player A is then to use Strategy 1 with probability $\frac{5}{8}$ and Strategy 2 with probability $\frac{3}{8}$. Using this q, the value of the game is given by $4 - 5q = 4 - 5(\frac{5}{8}) = \frac{7}{8}$. Notice that $3q - 1 = 3(\frac{5}{8}) - 1 = \frac{7}{8}$ also gives the value of the game. Consequently, this game is biased in favor of Player A by $\frac{7}{8}$. ◁

(*Related Exercises: 25–27*)

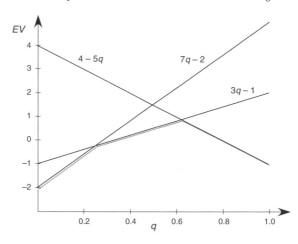

FIGURE 16.4: Plot of three expected values

16.3   Simpson's Paradox

Simpson's Paradox illustrates how statistical conclusions can be confounded by a simple misunderstanding of the data. It is an unexpected result that occurs when combining or partitioning groups—especially when the groups are of different sizes. Ultimately, it centers around the improper averaging of averages.

DEFINITION **Simpson's Paradox**
This paradox occurs when a numerical association between variables (e.g., $x > y$) inverts (e.g., $y > x$) when the studied population is viewed both as partitioned and as a whole.

HISTORICAL CONTEXT **Edward H. Simpson (1922–)**
A British statistician and cryptanalyst who described the paradox in a 1951 technical paper. Another British statistician, Udny Yule, is also credited with identifying the paradox.

The following example shows how this paradox can be exploited by advertisers who might artificially partition a population in order to exhibit a correlation that places their product in a more favorable light.

Example 16.10 Identify Simpson's Paradox in Evaluating a Clinical Trial

A drug company has conducted trials of its new cholesterol-lowering drug (*NEW*) and has compared it with an existing drug on the market (*OLD*), both for individuals with active lifestyles and for those who are sedentary. Table 16.14 gives the number of individuals in each lifestyle category who show a reduction in cholesterol level after treatment with one of the two drugs. In the Active group, 5 out of 20 subjects (25%) showed a significant reduction in cholesterol using the *OLD* drug, whereas 54 out of 180 subjects (30%) showed a reduction with the *NEW* drug. So for this group the *NEW* drug is more effective. In the Sedentary group, 90 out of 150 (60%) improved using the *OLD* drug, whereas 26 out of 40 (65%) improved using the *NEW* drug. Again the *NEW* drug is more effective for this group.

TABLE 16.14: Effects of two drugs on individual groups

	Active Lifestyle			Sedentary Lifestyle		
	Reduction	None	Percent	Reduction	None	Percent
OLD	5	15	25.0%	90	60	60.0%
NEW	54	126	30.0%	26	14	65.0%

These data show that the *NEW* drug is better for both groups. But can it truthfully be marketed as superior? Demonstrate that Simpson's Paradox arises in this example.

Solution

Simpson's Paradox is revealed when you combine the Active and Sedentary groups together. Table 16.15 shows that 95 out of 170 (55.9%) improved using the *OLD* drug, while 80 out of 220 (36.4%) improved using the *NEW* drug. Viewed in this way, the *NEW* drug now appears to be clearly inferior to the *OLD* drug!

TABLE 16.15: Effects of two drugs on combined groups

	Reduction	None	Percent
OLD	95	75	55.9%
NEW	80	140	36.4%

Notice that the *OLD* and *NEW* drugs are given to 20 and 180 subjects, respectively, in the Active group, and are given to 150 and 40 subjects in the Sedentary group. Perhaps the division into Active and Sedentary may be artificial and is done just to make the drug maker's *NEW* formulation look superior. ◁

Explaining the Paradox

Simpson's Paradox occurs when we try to *average* a set of averages or percents. To explain this, look at the previous table of results and focus on the percents shown in bold in Table 16.16.

TABLE 16.16: Explaining Simpson's Paradox

	Active Lifestyle			Sedentary Lifestyle		
	Reduction	None	Percent	Reduction	None	Percent
OLD	5	15	**25.0%**	90	60	**60.0%**
NEW	54	126	*30.0%*	26	14	*65.0%*

For the *OLD* drug, the overall improvement rate is an average of 25% and 60%, weighted more towards 60% since there are more subjects in the Sedentary group (150) compared to the Active Group (20):

$$[20(.25) + 150(.60)]/170 = 0.559.$$

This averaging process is shown visually in Figure 16.5(a). Similarly, with the *NEW* drug, the overall improvement rate is an average of the italicized entries 30% and 65% in Table 16.16, weighted more towards 30% since there are more subjects in the Active group (180) compared to the Sedentary Group (40):

$$[180(.30) + 40(.65)]/220 = 0.364.$$

Figure 16.5(b) visually shows this averaging process for the *NEW* drug. In other words, even though the *NEW* drug beats the *OLD* drug in each group (30 > 25) and (65 > 60), the correct weighted average clearly favors the *OLD* drug over the *NEW* drug (55.9 > 36.4).

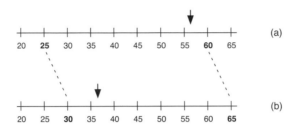

FIGURE 16.5: Averaging of averages

Example 16.11 Explain Simpson's Paradox in Comparing Tax Rates

Table 16.17 compares the total amount of income (in thousands of dollars) and total U.S. personal income tax collected (also in thousands of dollars) by income level for both 1974 and 1978. Also computed are the tax rates by income level. We see that for every income level, the tax rate decreased from 1974 to 1978. Identify and explain how these data illustrate Simpson's Paradox.

TABLE 16.17: Tax rates by income level

Adjusted Gross Income	1974			1978		
	Income	Tax	Tax Rate	Income	Tax	Tax Rate
Under $5000	41,651,643	2,244,467	.054	19,879,622	689,318	.035
$5000 to $9999	146,400,740	13,646,348	.093	122,853,315	8,819,461	.072
$10,000 to $14,999	192,688,922	21,449,597	.111	171,858,024	17,155,758	.100
$15,000 to $99,999	470,010,790	75,038,230	.160	865,037,814	137,860,951	.159
$100,000 or more	29,427,152	11,311,672	.384	62,806,159	24,051,698	.383

Solution

Table 16.18 extends Table 16.17 by adding the total income and total tax over all income levels for both years. Now we observe an increase in the overall tax rate, from 14.1% to 15.2%, between these years. To explain this, notice that there was a substantial shift toward higher income levels from 1974 to 1978. The percent of income attributed to the top two categories increased from 56.7% to 74.7% between these years—and these higher income levels are taxed at higher rates. This weighting toward higher tax rates causes the overall average tax rate to increase from 14.1% to 15.2% between these years, even though tax rates decreased for each income level. ◁

(Related Exercises: 28–31)

TABLE 16.18: Tax rates with combined income levels

Adjusted Gross Income	1974			1978		
	Income	Tax	Tax Rate	Income	Tax	Tax Rate
Under $5000	41,651,643	2,244,467	.054	19,879,622	689,318	.035
$5000 to $9999	146,400,740	13,646,348	.093	122,853,315	8,819,461	.072
$10,000 to $14,999	192,688,922	21,449,597	.111	171,858,024	17,155,758	.100
$15,000 to $99,999	470,010,790	75,038,230	.160	865,037,814	137,860,951	.159
$100,000 or more	29,427,152	11,311,672	.384	62,806,159	24,051,698	.383
Total	880,179,247	123,690,314	.141	1,242,434,934	188,577,186	.152

Chapter Summary

In situations with uncertain outcomes, expected value calculations provide us with the average gain or loss that will occur in the long run. It is convenient to formulate such situations using a probability model that captures the gains and losses associated with each outcome. In addition, a directed graph representation can prove useful in determining both probabilities and expected values. We see that expected values can be a valuable aid to making rational decisions. This chapter also revisits game theory scenarios, and illustrates a geometric approach that guides the determination of optimal mixed strategies. Simpson's Paradox shows that one must be careful in trying to average expected values.

Applications, Representations, Strategies, and Algorithms

Applications			
Education (2)	Gambling (3–4)	Medicine (10)	Business (11)

Representations				
Graphs	Tables		Sets	Symbolization
Directed	Data	Decision	Lists	Algebraic
5, 7	1–4, 10–11	6, 8–9	1	8–9

Strategies	
Rule-Based	Composition/Decomposition
1–4, 6, 8–11	5, 7

Algorithms				
Exact	Enumeration	Numerical	Algebraic	Geometric
1–11	1	1–6, 10–11	7–9	8–9, 11

Exercises

Probability Models

1. A fair coin is tossed three times. Specify a probability model that assigns probabilities to the number of heads obtained in the three tosses.

2. A biased coin, with a head appearing 60% of the time, is tossed twice. Specify a probability model that assigns probabilities to the number of heads obtained in the two tosses.

3. A biased coin, with a head appearing 60% of the time, is tossed three times. Specify a probability model that assigns probabilities to the number of heads obtained in the three tosses.

4. A four-sided die (with values 1, 2, 3, 4 on the faces) is rolled twice. Specify a probability model that assigns probabilities to the sum of the face values obtained in the two rolls.

5. A four-sided die (with values 1, 2, 3, 4 on the faces) is rolled twice. We are interested in the absolute difference between the face values observed on the two rolls: the nonnegative difference between the two face values. Specify a probability model that assigns probabilities to the (nonnegative) differences obtained in the two rolls.

6. In industrialized countries, it is known that the probability of a male birth is 0.512. Specify a probability model that assigns probabilities to the number of male births in a family of three children, assuming independence of the three births.

7. Two delegates are randomly selected from a congressional committee that is composed of four Republicans and six Democrats. We are interested in the number of Republicans selected as delegates. Specify an appropriate probability model. [*Hint*: Draw a two-stage decision tree to represent the selection of both delegates.]

8. A jar contains five orange jelly beans and three purple jelly beans. Two jelly beans are randomly drawn from the jar. We are interested in the number of orange jelly beans obtained. Specify an appropriate probability model. [*Hint*: Draw a two-stage decision tree to represent the two drawings.]

9. A deck of 16 cards contains only face cards (J, Q, K, A) representing all four suits. Two cards are randomly selected from this deck. We are interested in the number of Aces appearing in the two cards drawn. Specify an appropriate probability model. [*Hint*: Draw a two-stage decision tree in which each stage indicates whether an Ace is selected.]

Expected Value

10. Warren has $1000 to invest in a new tech stock he has heard about on a blog. Based on the performance of similar tech stocks, Warren estimates that there is a 10% chance his $1000 investment will be worth only $500 at the end of one year, a 17% chance

it will be worth $600, and a 30% chance it will still be worth $1000. However, there is a 25% chance it will be worth $1500 and an 18% chance it will be worth $2000. Determine the expected value of a $1000 initial investment at the end of one year.

11. A final exam will consist of 25 true-false questions, 10 multiple-choice questions, and 5 short-answer questions. Each true-false question is worth 2 points, each multiple-choice question is worth 3 points, and each short-answer question is worth 4 points. Fiona estimates that she has a 70% chance of correctly answering a true-false question, a 65% chance of correctly answering a multiple-choice question, and an 80% chance of correctly answering a short-answer question. What is her expected score on the final?

12. An automobile insurance policy (for a particular make and model of car) costs $800 per year. The insurance company estimates that there is a 6% chance that it will need to pay out $5000 on claims during the year, a 2.5% chance that it will need to pay out $7500 on claims during the year, and a 1.5% chance that it will need to pay out $10,000 on claims during the year. Otherwise, it does not anticipate any claims for this type of policy. Determine the average amount the insurance company will pay out for this policy each year, and then calculate its annual net profit on this policy.

13. An insurance company determines that the probability that a randomly selected customer in a certain class of drivers will have an accident in the next year is 0.1. If the customer from this class has an accident, the insurance company must pay $5000. The insurance company pays nothing if the customer does not have an accident. To cover their operating expenses, make a profit, and protect themselves against an unexpectedly large number of claims, the insurance company decides to add 30% to the expected payout and set that value as their premium. What should be the premium on this policy?

14. You place a $1 bet on number 7 in the game of roulette. Recall that there are 38 different numbers on the roulette wheel. If the number 7 comes up, you win $35; otherwise, you receive nothing. Compute your expected gain (or loss) in making this bet.

15. A fair coin is tossed three times. You make a $3 wager on the number of heads that turn up. Specifically, if exactly one head turns up, you win $1; if exactly two heads turn up, you earn $4; if three heads turn up, you earn $9; otherwise, you receive nothing. Determine the expected gain (or loss) if you make this wager.

16. In a single round of a casino dice game, a player wins $8 if a sum of either 2, 3, or 12 is rolled with a pair of dice. It costs $1 to play a round. Determine the expected gain (or loss) to the player in one round of this game.

17. Louie runs a shooting gallery at a fair where various prizes are offered. The probability of a customer winning a large prize (that costs Louie $25) is 0.005, the probability of a customer winning a medium prize (that costs $10) is 0.01, and the probability of a customer winning a small prize (that costs $5) is 0.05. Each round of the shooting gallery costs the customer $1.

    (a) Determine Louie's expected profit on each customer.

    (b) A customer plays the game four times. What is the probability that the customer wins prizes worth more than he paid?

18. The payoffs and probabilities for playing the game of *Keno* are listed in the following table. For $2 you can pick 10 numbers and will receive a payoff if you have at least five matches. Determine the expected gain (or loss) in playing this game.

Match	Payoff	Probability
5	$4	0.0514277
6	$40	0.0114794
7	$280	0.0016111
8	$1800	0.0001354
9	$8000	0.0000061
10	$50,000	0.0000001

19. Verify the match probabilities listed in Table 16.7 for the purchase of 4, 5, 6, or 7 boxes.

20. A soft drink company has launched a special promotion. It has issued five codes under the bottle caps of all of its bottles. Each bottle cap reveals exactly one code, and each code has equal likelihood of appearing. If two of the codes match, you get $5. You can receive at most one $5 prize and each drink costs $1. How many drinks (if any) should you purchase to maximize your expected gain?

21. In Example 16.7 we computed the expected number of steps until state 1 (HTH) is reached. Set up appropriate equations relating the expected number of steps m_i from vertex i until state 6 (HTT) is reached. Solve these equations for the m_i and then compute the expected number of steps to reach HTT.

22. Suppose that a fair coin is tossed repeatedly and we are interested in the expected number of tosses until we first see the sequence HT.

    (a) Set up a graph with vertices representing the possible states after two coin tosses. Transitions between states correspond to directed edges.

    (b) Write down and then solve a system of equations in the variables m_i representing the average number of tosses to get to state HT, given that we start at vertex i.

    (c) Compute the expected number of tosses until we first see the sequence HT.

Zero-Sum Games

23. Consider the following two-person zero-sum game with payoffs shown to Player A.

	Player B	
Player A	8	−3
	−7	2

    (a) Verify that there is no saddlepoint for this game.

    (b) Using a geometric approach, determine the optimal mixed strategy $(q, 1 - q)$ for Player A.

    (c) Is this game fair, or is it biased in favor of one of the players? If so, by how much?

24. Consider the following two-person zero-sum game with payoffs shown to Player A.

	Player B	
Player A	-12	10
	7	-5

(a) Verify that there is no saddlepoint for this game.

(b) Using a geometric approach, determine the optimal mixed strategy $(q, 1 - q)$ for Player A.

(c) Is this game fair, or is it biased in favor of one of the players? If so, by how much?

25. Consider the following two-person zero-sum game with payoffs shown to Player A.

	Player B		
Player A	-3	-1	4
	4	2	-1

(a) Verify that there is no saddlepoint for this game.

(b) Using a geometric approach, determine the optimal mixed strategy $(q, 1 - q)$ for Player A.

(c) Is this game fair, or is it biased in favor of one of the players? If so, by how much?

26. Consider the following two-person zero-sum game with payoffs shown to Player A.

	Player B		
Player A	9	5	-6
	-15	-6	10

(a) Verify that there is no saddlepoint for this game.

(b) Using a geometric approach, determine the optimal mixed strategy $(q, 1 - q)$ for Player A.

(c) Is this game fair, or is it biased in favor of one of the players? If so, by how much?

27. Consider the following two-person zero-sum game with payoffs shown to Player A.

(a) Verify that there is no saddlepoint for this game.

(b) Using a geometric approach, determine the optimal mixed strategy $(p, 1 - p)$ for Player B.

(c) Is this game fair, or is it biased in favor of one of the players? If so, by how much?

	Player B	
	−5	8
Player A	−4	3
	1	−5

Simpson's Paradox

28. It is common to compute a baseball player's batting average by dividing the number of hits by the number of times at bat (walks being excluded). The following actual data were obtained for the 1989 and 1990 seasons for two players, whom we will call Dave and Andy.

	1989		1990		Combined	
	Hits	Times at Bat	Hits	Times at Bat	Hits	Times at Bat
Dave	12	51	124	439	136	490
Andy	113	476	140	493	253	969

(a) Compute Dave's batting average separately for 1989 and for 1990, and likewise Andy's batting average for 1989 and for 1990. Who has the higher average each year?

(b) Compute Dave's combined batting average for both seasons, and likewise Andy's combined batting average for both seasons. Who has the higher average for the two seasons?

(c) Explain why this is an example of Simpson's Paradox.

29. In the first of two football games between rival teams, Player A gained 16 yards in 5 carries, while Player B gained 123 yards in 30 carries. In the second game, Player A gained 83 yards in 17 carries, while Player B gained 104 yards in 21 carries.

(a) Which player had the highest average yards per carry in the first game? In the second game?

(b) Compute the average yards per carry for each player when the statistics for both games are combined. Who then has the higher rushing average?

(c) Explain why this is an example of Simpson's Paradox.

30. The following data were collected showing the number of flights with on-time departures and the number with delayed departures for Alpha Airlines and Beta Airlines.

Airport	Alpha		Beta	
	On Time	Delayed	On Time	Delayed
JFK	501	58	695	120
Charlotte	223	11	3830	420
National	210	20	391	61
Atlanta	490	104	325	126
Chicago	1810	308	203	59

(a) Compute the on-time percentage at each airport, separately for each airline. Which airline displays the higher on-time percentage for each airport?

(b) Compute the on-time percentage for each airline by combining all airports. Which airline now displays the higher on-time percentage overall?

(c) Explain how this paradoxical situation arises.

31. A small liberal arts college receives applications from both in-state and out-of-state students. The number of applications and admissions are shown in the table below, both for female and male applicants.

	In-State		Out-Of-State	
	Accepted	Applied	Accepted	Applied
Female	42	200	76	100
Male	30	200	208	320

(a) Compute the acceptance rates for in-state females and for out-of-state females.

(b) Compute the acceptance rates for in-state males and for out-of-state males.

(c) Compute the overall acceptance rate for females (regardless of resident status) and do likewise for males. Why does this illustrate Simpson's Paradox? Discuss how the paradox can be explained.

Projects

32. Simulate the tossing of a single biased coin by using the Binomial Coin Experiment available at the web site [OR2]. We are interested in the number of heads that appear in 10 tosses, when the probability of a head on any single toss is $p = 0.6$. Use $n = 10$ tosses and the setting *Stop 1000* to determine the average number of heads over 1000 repetitions of the experiment. Repeat this experiment of 1000 repetitions several times and record the average number found for each repetition. How do your observed averages compare with the theoretical expected value of 6?

33. Game theory has been applied to many real-world situations, ranging from modeling the arms race to understanding the behavior of biological species. Explore some of these applications (to anthropology, international relations, and business) by reading relevant chapters in [Schelling 1981, Straffin 2002].

Bibliography

[Olofsson 2007] Olofsson, P. 2007. *Probabilities: The little numbers that rule our lives*, Chapter 5. Hoboken, NJ: John Wiley & Sons.

[Schelling 1981] Schelling, T. C. 1981. *The strategy of conflict*. Cambridge, MA: Harvard University Press.

[Simpson 2010] Simpson, E. 2010. "Bayes at Bletchley Park." *Significance* 7: 76–80.

[Straffin 2002] Straffin, P. 2002. *Game theory and strategy*. Washington, D.C.: Mathematical Association of America.

[OR1] `http://www.ted.com/talks/peter_donnelly_shows_how_stats_fool_juries.html` (accessed December 18, 2013).

[OR2] `http://www.socr.ucla.edu/htmls/SOCR_Experiments.html` (accessed June 7, 2014).

Unit IV

Counting: Voting Methods and Apportionment

Chapter 17

Voting Methods

It's not the voting that's democracy, it's the counting.

—*Jumpers*, Tom Stoppard, British playwright, 1937–present

Chapter Preview

In this chapter we present and analyze the most prominent methods for counting votes when every entity (or voter) has just one vote. An extension is also considered to cases in which support for multiple candidates is incorporated into the voting scheme. It becomes clear that how we decide to count the votes—our voting method—plays an influential role in determining the outcome of elections. Moreover, certain situations can arise that render these voting methods less than ideal; such paradoxical situations are identified and discussed for each of the methods presented.

17.1   Methods for Counting Votes: One Person–One Vote

Suppose we need to select a political candidate for office, or the Best Picture of the Year at the Academy Awards, or the winner of the Heisman trophy in football, or the winner of a TV dance competition. To determine the winner (or, more generally, to rank the candidates), we distribute a ballot, collect votes, and then use a *voting system* to aggregate the choices expressed by the votes.

DEFINITION **Voting System**
A system that uses individual preferences to select the winning candidate for the group as a whole. Each system has specific rules that describe how the votes are to be counted in order to yield a final result.

When there are just two candidates, it is common to have an election decided by *majority rule*. Namely, the winner is the candidate receiving strictly more than 50% of the votes. When there are more than two candidates, two strategies come to mind. One is to apply the majority rule and if no candidate receives more than 50%, we eliminate losing candidates and conduct some type of run-off election. Alternatively, we might consider the winner to be the one who receives a *plurality* of the votes cast: that is, more votes than any other candidate.

Voters are assumed to have *preferences* for each candidate: first, second, third, etc. What we want to do is to use these individual preferences to select the winning candidate for the entire group of voters. With information about each voter's preferences (first choice, second choice, third choice, etc.), we can make a more informed decision in choosing an overall winner, rather than simply relying on the first choices of each voter.

Suppose we have 15 voters and three candidates (A, B, C), and the preferences of the 15 voters can be grouped into four *profiles*, each representing a ranking of the three candidates. These profiles can be listed in the columns of a *preference table* (see Table 17.1). For example, the second profile column of this table shows that five voters prefer B over A, and A over C. It follows that these five voters would necessarily choose B over C in any contest (this is called "transitivity"). Altogether there are $6 + 5 + 3 + 1 = 15$ voters represented by this preference table.

TABLE 17.1: Profiles in a preference table

	Voter Profiles			
	6	5	3	1
First	A	B	C	C
Second	B	A	B	A
Third	C	C	A	B

Using the concept of a preference table, we can now study common voting methods that have been proposed for converting individual preferences for the candidates into an outcome for the election as a whole. The six methods we will first focus on are: Majority, Plurality, Plurality Run-Off, Instant Run-Off, Sequential Run-Off, and Pairwise Comparison.

Majority Method

This voting method only concentrates on the first-place votes, ignoring the remaining voter preferences.

DEFINITION **Majority Method**
The winner is the candidate having *more than* 50% of the first-place votes.

A problem with this method is that no single candidate may receive more than 50% of the votes. Or it might happen, in the case of two candidates, that both receive exactly 50% of the votes. In either situation, a decision must be made on how to proceed. For example, we might be forced into a run-off election; several run-off methods are discussed later in this section.

Plurality Method

When there are more than two candidates in an election, no candidate might receive more than 50% of the votes. We could then declare the winner to be the one who receives a plurality of the votes cast: that is, more votes than any other candidate.

DEFINITION **Plurality Method**
The candidate having the most first-place votes is declared the winner.

Example 17.1 Determine the Plurality Winner

Table 17.2 shows a preference table for 15 voters having four profiles, each indicating how many voters have a particular ranking of A, B, and C. Determine the Plurality winner.

TABLE 17.2: Finding the Plurality winner for 15 voters

	Voter Profiles			
	6	5	3	1
First	A	B	C	C
Second	B	A	B	A
Third	C	C	A	B

Solution

Candidate A receives 6 first-place votes, B receives 5 votes, and C receives $3 + 1 = 4$ votes. So Candidate A is deemed the overall winner using the plurality method. Notice however that Candidate A does not receive a majority of the 15 votes. ◁

(Related Exercises: 1–3)

Example 17.1 reveals an unwanted consequence of the Plurality method.

PARADOX (Plurality): Here 60% of the people don't want Candidate A, but A wins anyway.

A striking example of how a plurality candidate can win without majority support occurred in the Minnesota gubernatorial race of 1998. In that election, Jesse Ventura won with a plurality of votes, but only 37% of the voters supported him. He was not the top choice of 63% of the voters (see Table 17.3).

TABLE 17.3: 1998 Minnesota governor's race

Candidate	Party	Votes	%
Jesse Ventura	Reform	773,713	37.0
Norm Coleman	Republican	717,350	34.3
Hubert Humphrey III	Democratic	587,528	28.1
Ken Pentel	Green	7034	0.3

There is another serious drawback to using the plurality method. To illustrate this, let's return to Example 17.1 with 15 voters. Recall that Candidate A is determined to be the winner using the Plurality method. Now suppose that at the last moment, Candidate C decides to drop out of the race. Under the reasonable assumption that voters' preferences remain unchanged (except that C no longer is considered), we have the updated Table 17.4.

Now A receives 7 first-place votes while B receives 8 first-place votes, making Candidate B the winner. In other words, the removal of an "irrelevant candidate" (the losing Candidate C) changes the outcome of the election!

TABLE 17.4: Preference table after Candidate C withdraws

	Voter Profiles			
	6	5	3	1
First	A	B	B	A
Second	B	A	A	B

PARADOX (Plurality): A losing candidate C drops out of the race and thereby changes the winner of the election.

One of the defects of the plurality method is that it allows a "spoiler candidate" C to enter the race and take votes away from the favored candidate A so that a less favored candidate B now wins. This occurred in the 2000 Presidential election in Florida in which Ralph Nader (Green Party candidate) is claimed to have diverted votes away from Al Gore; as a result George W. Bush won Florida and, consequently, the very close 2000 Presidential election.

In view of this phenomenon, some states conduct a run-off between the top vote-getters. We next study such run-off methods.

Plurality Run-Off Method

This method is often used in primary voting in order to prevent "spoiler candidates" from concealing how the voters would rank the main candidates.

DEFINITION **Plurality Run-Off Method**
When no candidate gets a majority of first-place votes, this method holds a new election between the top two candidates (those receiving the largest number of first-place votes). That is, the top two vote-getters in the election participate in a run-off election; now the candidate who receives a majority of the first-place votes is declared the winner.

Even when using the Plurality Run-Off method, interesting situations can occur. In the 1966 Democratic gubernatorial primary in Georgia, there were five major candidates running, with the voting results listed in Table 17.5.

TABLE 17.5: 1966 Georgia gubernatorial primary election

Candidate	Votes	%
Ellis Arnall	231,480	29.4
Lester Maddox	185,672	23.6
Jimmy Carter	164,562	20.9
James Gray	152,973	19.4
Garland Byrd	39,994	5.1

The top two candidates, Arnall and Maddox, were paired in the run-off election, which Maddox won with 54% of the votes—despite the fact that more voters preferred Arnall to Maddox in the primary. One possible explanation is that voters for Carter, Gray, and Byrd played a pivotal role in reallocating their votes in the run-off.

Example 17.2 Determine the Plurality Run-Off Winner

Using the preferences expressed in Table 17.6, determine the Plurality Run-Off winner.

TABLE 17.6: Finding the Plurality Run-Off winner for 36 voters

	Voter Profiles			
	12	11	8	5
First	A	B	C	C
Second	C	A	B	A
Third	B	C	A	B

Solution

Looking at the first-place votes, A receives 12 votes, B receives 11, and C receives 13. Since no candidate receives a majority of the 36 votes, there is a run-off between A and C. Table 17.7 shows the updated preference table after Candidate B is eliminated.

TABLE 17.7: Preference table after Candidate B is eliminated

	Voter Profiles			
	12	11	8	5
First	A	A	C	C
Second	C	C	A	A

Candidate A receives $12 + 11 = 23$ votes while C receives $8 + 5 = 13$ votes. Candidate A is then declared the clear winner. ◁

(Related Exercises: 4–7)

Now suppose that the losing Candidate C drops out of the initial race. Table 17.8 shows the modified preference table for Candidates A and B.

TABLE 17.8: Preference table after Candidate C withdraws

	Voter Profiles			
	12	11	8	5
First	A	B	B	A
Second	B	A	A	B

Candidate B receives more first-place votes (19) than Candidate A (17), so Candidate B is declared the winner. Again, we have the paradox involving an "irrelevant candidate" seen earlier.

PARADOX (Plurality Run-Off): A losing candidate C drops out of the race and thereby changes the winner of the election.

This example also reveals another dilemma. Going back to the original preferences for the three candidates (Table 17.6), suppose that the five voters with preference C > A > B decide to elevate the winning Candidate A in their estimation, giving their new preference A > C > B. This changes the voting profiles to those displayed in Table 17.9.

In this new situation, Candidates A and B receive the most first-place votes (17 and 11

TABLE 17.9: Preferences after a change in one voter profile

	Voter Profiles			
	12	11	8	5
First	A	B	C	A
Second	C	A	B	C
Third	B	C	A	B

votes, respectively) and are paired in a run-off election: A then receives $12 + 5 = 17$ votes while B receives $11 + 8 = 19$ votes. So, the winner is Candidate B. Here is another paradox.

PARADOX (Plurality Run-Off): The winner of the initial contest (A) receives additional support, yet loses in the new contest!

Another run-off method that has recently gained in popularity is Instant Run-Off Voting (IRV).

Instant Run-Off (IRV) Method

Rather than eliminating all but the top two candidates as in Plurality Run-Off, this method repeatedly eliminates the "weakest link" at each stage until a majority winner emerges.

DEFINITION **Instant Run-Off Method**
Eliminate the candidate with the fewest first-place votes repeatedly, until one candidate emerges with the *majority* of first-place votes. When a candidate is eliminated, the votes for that candidate are redistributed to the candidate preferred next among those remaining.

HISTORICAL CONTEXT **Thomas Hare (1806–1891)**
An English political and legal scholar who was an active proponent of electoral reform. The Instant Run-Off method is also referred to as the Hare Method.

A desirable feature of IRV is that someone can vote for a candidate who is his first choice, but has little chance to win, since that vote will be used again in succeeding rounds. In short, our vote is not "wasted" by voting our true preferences.

This method is widely used for elections held in Ireland. In addition, the 2010 Academy Award for Best Picture was decided using the IRV method. Approximately 6000 Film Academy members participated in ranking all 10 Academy Award nominees for Best Picture. If no film received a majority of votes, the film with the fewest first-place votes was eliminated and ballots were redistributed according to the second-place choice, and so on until one nominee received a majority.

Example 17.3 Determine the Instant Run-Off Winner

Using the voting profiles listed in Table 17.10, determine the Instant Run-Off winner.

Solution

Here A receives 5 first-place votes, B receives 4, C receives 3, and D receives only 2. As no

TABLE 17.10: Finding the Instant Run-Off winner for 14 voters

	Voter Profiles			
	5	4	3	2
First	A	B	C	D
Second	B	A	D	C
Third	C	C	A	A
Fourth	D	D	B	B

candidate achieves the needed majority (at least 8 votes), D is eliminated. This produces the modified preferences shown in Table 17.11, where we see that the two first-place votes for D have been allocated to C.

TABLE 17.11: Preference table after Candidate D is eliminated

	Voter Profiles			
	5	4	3	2
First	A	B	C	C
Second	B	A	A	A
Third	C	C	B	B

Now A receives 5 first-place votes, B receives 4, and C receives 5. Since no candidate achieves a majority, B is eliminated, resulting in the updated Table 17.12.

TABLE 17.12: Preference table after Candidate B is eliminated

	Voter Profiles			
	5	4	3	2
First	A	A	C	C
Second	C	C	A	A

Now Candidate A obtains $5 + 4 = 9$ first-place votes (a majority) and so is declared the IRV winner. ◁

(Related Exercises: 8–12)

Unfortunately, the Instant Run-Off method suffers from similar difficulties as the Plurality Run-Off method. In particular, the same scenario given in Example 17.2 shows that a losing candidate (namely, C) can drop out of the race and thereby affect the outcome.

PARADOX (Instant Run-Off): A losing candidate drops out of the race and thereby changes the winner of the election.

An alternative voting method prescribes a sequence of one-on-one contests.

Sequential Run-Off Method

This method appears in various forms in athletic competitions in which winning teams progress through a tournament system, where specific teams face each other in an order determined by agreed-upon rules.

DEFINITION **Sequential Run-Off Method**

A sequence of one-on-one contests is held, using a pre-defined order (called an *agenda*). The winner, determined using first-place votes, then advances to the next round and has a contest with the next candidate on the agenda. The winner of the last contest is declared the overall winner of the election.

Example 17.4 Determine the Sequential Run-Off Winner

Consider the preferences shown in Table 17.13 for three voters and four candidates A, B, C, D. Here each profile represents just one voter. The given agenda specifies considering candidates in the order [A, B, C, D]. Identify the Sequential Run-Off winner.

TABLE 17.13: Finding the Sequential Run-Off winner

	Voter Profiles		
First	A	B	C
Second	B	D	A
Third	D	C	B
Fourth	C	A	D

Solution

In the contest between A and B, A wins with 2 first-place votes and B is eliminated. Then in the contest between A and C, C wins with 2 first-place votes and A is eliminated. In the final contest between C and D, D wins with 2 first-place votes so Candidate D is the overall winner. ◁

(Related Exercises: 13–15)

PARADOX (Sequential Run-Off): Candidate D wins, yet every voter prefers B to D!

Notice that if we change the agenda order to [D, C, B, A] then in turn C, then D, then B are eliminated, leaving Candidate A as the winner. Consequently, the agenda itself can have a profound effect on the outcome. For example, it becomes harder for the first candidate in the agenda order to become the eventual winner (since that candidate will have to win all contests), while the last candidate in the agenda order only needs to win one contest.

For example, suppose that we have four equally qualified candidates, each with likelihood 0.5 of beating any other candidate in a head-to-head matchup. Assuming the specified agenda is [A, B, C, D], then $\Pr[\text{A wins}] = \Pr[\text{A beats B, C, D}] = 0.5 \times 0.5 \times 0.5 = 0.125$, assuming independence of the matches. On the other hand, $\Pr[\text{B wins}] = \Pr[\text{B beats C, D}] = 0.5 \times 0.5 = 0.25$, while $\Pr[\text{C wins}] = \Pr[\text{C beats D}] = 0.5$ and $\Pr[\text{D wins}] = \Pr[\text{D beats C}] = 0.5$. So, positioning in the agenda order is quite important.

The following method avoids the problem of specifying a particular agenda order.

Pairwise Comparison Method

The method of pairwise comparisons is encountered in situations where it is desirable to have every candidate compete with every other candidate one-on-one. We see this in round-robin tournaments in sports, and in evaluation of competing proposals for funding.

DEFINITION **Pairwise Comparison (Condorcet) Method**
One-on-one contests are held between every pair of candidates and a point is awarded to the winner of each contest. The candidate with the most points (who wins the most contests) is declared the winner.

HISTORICAL CONTEXT **Marquis de Condorcet (1743–1794)**
A French mathematician, political scientist, and philosopher who also became a leader of the French Revolution. The method of pairwise comparisons is also known as the Condorcet Method.

Example 17.5 Determine the Pairwise Comparison Winner

Consider the preference table from Example 17.1, with 15 voters and three candidates, shown again as Table 17.14. There are three possible pairings A-B, A-C, B-C of the three candidates, so there are three one-on-one contests to analyze. Who is the Pairwise Comparison winner?

TABLE 17.14: Finding the Pairwise Comparison winner

	\multicolumn{4}{c}{Voter Profiles}			
	6	5	3	1
First	A	B	C	C
Second	B	A	B	A
Third	C	C	A	B

Solution

In the contest between A and B, A is preferred to B by $6+1 = 7$ voters, while B is preferred to A by $5+3 = 8$ voters. (Note that the sum $7+8 = 15$, the total number of voters.) Thus, B is the winner of this one-on-one contest and gets one point. In the A versus C contest, A gets $6+5 = 11$ votes while C only gets 4 votes, so A gets one point. In the B versus C contest, B gets $6+5 = 11$ votes while C only gets 4 votes, so B gets another point. Since B accumulates 2 points, A gets one point, and C gets none, Candidate B is declared the winner using the Pairwise Comparison method. ◁

(Related Exercises: 16–19, 30)

In this example, we needed to analyze three contests between the three candidates. With four candidates, it turns out there will be six contests to analyze. These numbers come in fact from the Combinations Formula in Section 13.1: namely, there are $C(n, 2)$ ways of selecting pairs of candidates for a matchup from n candidates. You can check that $C(3, 2) = 3$ and $C(4, 2) = 6$.

Example 17.6 Show There May Be No Pairwise Comparison Winner

Let's consider another example, with just three individual voters and with preferences given by Table 17.15. Show that there is no Pairwise Comparison winner.

TABLE 17.15: Determining if a Pairwise Comparison winner exists

	Voter Profiles		
First	A	B	C
Second	B	C	A
Third	C	A	B

Solution

In the contest between A and B, A is preferred to B by 2 voters, while B is preferred to A by 1 voter, so A gets one point. In the A versus C contest, A gets 1 vote while C gets 2 votes, so C gets one point. In the B versus C contest, B gets 2 votes while C gets 1 vote, so B gets one point. All candidates get one point, so there is no Pairwise Comparison winner.
◁

PARADOX (Pairwise Comparison): In some elections, there is no Pairwise Comparison winner.

Another way to express this finding is to say that whereas the preferences of each individual are transitive, the group as a whole need not have transitive preferences. As seen in Example 17.6, A is preferred to B and B is preferred to C, yet C is preferred to A.

We now discuss an interesting situation [Bruce 2001] with an outcome that is analogous to having no Pairwise Comparison winner. A hapless gambler is presented with three curious dice designated Red, Black, and White (see Figure 17.1); each of these dice produces the same roll on average (namely 4). The gambler can select one of these to roll and then his opponent selects another and rolls, with the higher number rolled determining the winner. As it turns out, the hapless gambler regularly seems to be on the losing end of this wager.

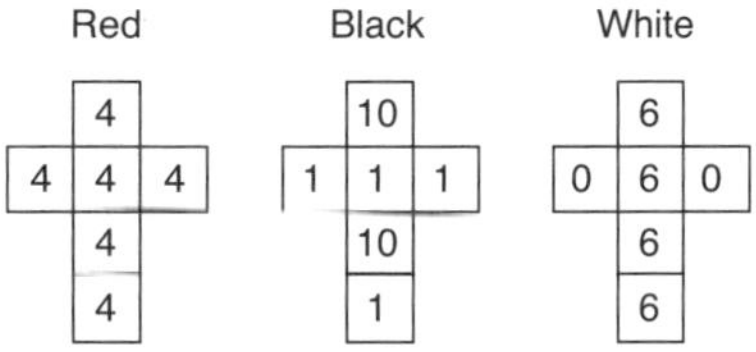

FIGURE 17.1: Three curious dice

We analyze this situation using probability concepts, and determine that on average Red (R) beats Black (B), Black (B) beats White (W), yet White (W) beats Red (R).

1. R vs. B: Since R always shows a 4, R wins whenever B shows a 1. So R wins $\frac{2}{3}$ of the time against B.

2. R vs. W: Since R always shows a 4, W wins whenever W shows a 6. So W wins $\frac{2}{3}$ of the time against R.

3. B vs. W: Using conditional probabilities (see Section 14.2),

$$Pr[B \text{ wins}] = Pr[B \text{ shows a } 10] + Pr[B \text{ shows a 1 and wins}]$$
$$= \tfrac{1}{3} + Pr[B \text{ shows a } 1] \times Pr[B \text{ wins}|B \text{ shows a } 1]$$
$$= \tfrac{1}{3} + \left(\tfrac{2}{3} \times \tfrac{1}{3}\right) = \tfrac{5}{9}$$

So, B wins $\frac{5}{9}$ of the time against W.

A way of visually representing this situation is given by the directed circuit in Figure 17.2. Since the hapless gambler goes first, a clever opponent can always choose a "better" die. For example, if the gambler selects W then you should select B (as B immediately precedes W on the circuit in Figure 17.2). Similarly, if the gambler selects B then you should select R. In the long run, the clever opponent can ensure coming out ahead.

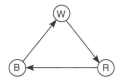

FIGURE 17.2: Circuit formed by playing the curious dice

17.2 Voting Systems Permitting Support for Multiple Candidates

In the previous section, we looked at some common, and seemingly reasonable, methods of counting votes. In each case, we found some inherent weakness in the voting system. Here we study two other systems where individual voters can express preference for more than one candidate at a time. These systems are the Borda Count and the Approval Voting methods.

Borda Count Method

This method is analogous to point systems used to select winners of sporting events (e.g., the Heisman Trophy in football and the MVP in baseball) or to rank universities. It is more familiarly used in computing a student's grade-point average (where A is given four points, B three points, C two points, D one point, and F zero points).

DEFINITION **Borda Count Method**
First-place votes, second-place votes, third-place votes, etc. are all counted. First-place votes weigh more heavily than second-place votes, and second-place votes weigh more heavily than third-place votes, and so on. Specifically, if there are n candidates, the candidate first on a voter's list receives $n - 1$ points, the second receives $n - 2$ points, and so on, with the last receiving 0 points.

HISTORICAL CONTEXT **Jean-Charles de Borda (1733–1799)**
A French mathematician, physicist, and political scientist who is credited with devising a voting system that was used to elect members to the French Academy of Sciences for two decades. Recently discovered historical documents show that this method of counting votes was proposed much earlier by the Majorcan writer and philosopher, Ramon Llull (1232–1315).

Example 17.7 Determine the Borda Count Winner

Table 17.16 shows the profiles for 107 voters and three candidates, with 2 points awarded for first-place votes, 1 point for second-place votes, and 0 points for third-place votes. Identify the Borda Count winner.

TABLE 17.16: Finding the Borda Count winner

	Voter Profiles		
	55	50	2
First (2)	A	B	C
Second (1)	B	C	B
Third (0)	C	A	A

Solution

Candidate A receives 55 first-place votes and 52 third-place votes for a total of $(55 \times 2) + (52 \times 0) = 110$ points; B receives 50 first-place votes and 57 second-place votes for a total of $(50 \times 2) + (57 \times 1) = 157$ points; and C receives 2 first-place votes, 50 second-place votes, and 55 third-place votes for a total of $(2 \times 2) + (50 \times 1) + (55 \times 0) = 54$ points. So Candidate B, with 157 total points, is deemed the clear winner using the Borda Count method. ◁

(Related Exercises: 20–22, 31)

Notice that Candidate A has more first-place votes than Candidate B (in fact, a majority of the first-place votes), yet Candidate B wins using the Borda Count method. This reveals another paradox.

PARADOX (Borda Count): Candidate B wins, yet the losing Candidate A receives a majority of the first-place votes.

Another system allows voting for multiple candidates in an election.

Approval Voting Method

Unlike plurality voting, which forces people to choose between the candidate whom they think is viable (such a Main Party candidate) and the candidate they truly prefer (perhaps an Independent Party candidate), Approval Voting allows them to vote for both. Approval Voting was formally proposed by several people in the 1970s. Since then, it has been adopted by several governments and organizations around the world, most notably by the United Nations in its election of the Secretary-General. Motions to adopt it have also been considered by several U.S. states and is currently used by a number of universities and professional societies.

DEFINITION **Approval Voting Method**
Each voter can cast one vote for as many candidates as deemed acceptable. That is, if the voter approves of three out of five candidates, then three votes are cast, one for each acceptable candidate. The candidate most widely approved of (having the most votes) is considered to be the winner.

In Approval Voting, each voter is given the opportunity to select those candidates considered to be acceptable. In effect, each voter separates the candidates into acceptable and unacceptable, with each acceptable candidate earning one vote. If any voter were to approve of all candidates, this would essentially be equivalent to not voting at all. Or if every voter refuses to vote for anyone except their favorite candidate, then Approval Voting would be no different from plurality voting.

However, certain paradoxical situations can occur, which we explore next.

Example 17.8 Determine the Winner Using Approval Voting

Consider the case of 100 voters and three candidates A, B, C. Table 17.17 lists those candidates who are approved by each voter. To be consistent with earlier preference tables, this table ranks the approved choices of each voter. Determine the winner selected by Approval Voting.

TABLE 17.17: Finding the winner by Approval Voting

Voter Profiles				
26	18	25	15	16
A	C	A	B	C
B	B			

Solution

Table 17.17 indicates that 26 voters like A, then B (the rest are unacceptable); 18 like C, then B; 25 like A alone; 15 like B alone; and 16 like C alone. If we count the number of acceptable votes that each candidate receives, then we find that A gets $26 + 25 = 51$ votes; B gets $26 + 18 + 15 = 59$ votes; and C gets $18 + 16 = 34$ votes. Thus, Candidate B is declared the winner by Approval Voting. ◁

(Related Exercises: 23–24)

Note however that $26 + 25 = 51$ of the 100 first-place votes go to Candidate A—indeed, a majority—yet Candidate A does not win using Approval Voting. So we encounter a problem similar to that found with the Borda Count method.

PARADOX (Approval): Candidate B wins, yet the losing Candidate A has the majority of the first-place votes.

This paradoxical situation arises because we have assumed that each voter profile presents ranked choices. If we relax this requirement and only ask voters to specify their acceptable candidates, then Table 17.17 becomes Table 17.18 and the paradox disappears. By treating Approval Voting as a non-preferential system, the paradox observed in conjunction with Sequential Run-Off cannot arise either. Namely, this paradox—every voter prefers B to D and yet D wins—is impossible under Approval Voting since the voters do not rank the candidates; they either approve of a candidate or some set of candidates or they do not.

TABLE 17.18: Acceptable candidates for each voter profile

Voter Profiles				
26	18	25	15	16
{A, B}	{B, C}	{A}	{B}	{C}

17.3 Application: Hospital Location

To illustrate several of the prior voting methods, suppose that a hospital is to be located in one of four towns: North Point (N), Easton (E), Westfield (W), or Southern Village (S). Residents of each town would prefer that the hospital be located in their town, or if not, to be as close as possible to their town. Figure 17.3 displays a map showing the location of the four towns as well as the town populations (measured in thousands).

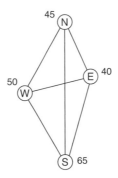

FIGURE 17.3: Locations and populations of four towns

Example 17.9 Design the Voter Profiles

Based on the map distances indicated in Figure 17.3, create voter profiles for towns N, S, W, E. For example, Town N ranks its preferred locations from most desirable to least desirable as N, E, W, S. We associate the population 45 with this set of voters.

Solution

Using the geometric distances shown in the map, Town S has the ranking S, W, E, N; Town W has the ranking W, E, N, S; and Town E has the ranking E, N, W, S. These yield the voting profiles shown in Table 17.19. ◁

TABLE 17.19: Preference table for hospital location

	N	S	W	E
	45	65	50	40
First	N	S	W	E
Second	E	W	E	N
Third	W	E	N	W
Fourth	S	N	S	S

Example 17.10 Determine Winners According to Different Voting Systems

For the hospital location problem in Example 17.9, determine (a) the Plurality winner, (b) the Plurality Run-Off winner, (c) the Borda Count winner, and (d) the IRV winner.

Solution

(a) Using the Plurality method, Town S would be selected since it has the largest number of first-place votes (65).

(b) Using Plurality Run-Off, we identify S and W as having the most first-place votes, so these enter the run-off contest. In this second election, S gets 65 votes while W gets the remaining $45 + 50 + 40 = 135$ votes. The Plurality Run-Off winner is then Town W.

(c) Using the Borda Count, with values 3, 2, 1, 0 for the various rankings, gives $(45 \times 3) + (40 \times 2) + (50 \times 1) + (65 \times 0) = 265$ points for N. Similar calculations yield 195 points for S, 365 for W, and 375 for E, so that Town E is declared the Borda Count winner.

(d) Using IRV, no town receives a majority of the 200 votes, so E (with the fewest first-place votes) is eliminated, giving the updated Table 17.20. Notice that in effect the 40 first-place votes for E have been reallocated to N.

TABLE 17.20: Preferences after Town E is eliminated

	N	S	W	E
	45	65	50	40
First	N	S	W	N
Second	W	W	N	W
Third	S	N	S	S

No town has attained a majority, so W is next eliminated, yielding the new preferences shown in Table 17.21. Town N now has 135 first-place votes (a majority) and is declared the IRV winner. ◁

(Related Exercises: 25–29)

TABLE 17.21: Preferences after Town W is eliminated

	N	S	W	E
	45	65	50	40
First	N	S	N	N
Second	S	N	S	S

Notice that this location example produces four *different* winners (S, W, E, N), depending on which voting procedure is used! It is not clear then where the hospital should be located.

In the next chapter, we investigate paradoxes that can occur in these and other popular voting methods. We are then led to inquire whether there might be a more desirable voting scheme that can be used in situations such as the hospital location problem of this section.

Chapter Summary

Situations arising in politics, boardrooms, athletic competitions, and the world of entertainment all require the determination of an overall winner. More generally, it is useful to produce a ranking of the various candidates, based on the preferences expressed by the different voters. Here we study several common methods used to derive a final ranking of candidates using the known voter profiles. Such profiles are commonly represented using tables, and the voting methods all numerically manipulate the information in such tables to produce a ranking of candidates. Two important conclusions are established by exploring these common voting schemes. First, the final ranking can be strongly influenced by the voting method chosen. Second, each of the voting methods studied possesses certain undesirable features. Both of these considerations lead us to a search, in the next chapter, for an ideal voting method.

Applications, Representations, Strategies, and Algorithms

Applications		
Society/Politics (1–8)	City Planning (9–10)	
Representations		
Tables	Symbolization	
Data	Numerical	Geometric
1–10	1–10	9

Strategies		
Rule-Based		
1–10		

Algorithms		
Exact	Sequential	Numerical
1–10	2, 3–6, 10	1–10

Exercises

Preference Tables

1. Fourteen voters are asked to rank the four candidates A, B, C, D. These voters turn in the following (ranked) ballots:

ACDB	BACD	BDCA	CDAB	DCAB	ACDB	BDCA
DCAB	ACDB	BACD	BDCA	ACDB	DCAB	CDAB

   (a) Construct a preference table for these ballots.

   (b) How many votes constitutes a majority? Is there a Majority winner?

   (c) Is there a Plurality winner?

2. Thirteen voters are asked to rank the four candidates A, B, C, D. These voters turn in the following (ranked) ballots:

CDAB	CBAD	BACD	CDAB	ACDB	CBAD	DABC
ACDB	BACD	CDAB	ACDB	CBAD	CDAB	

(a) Construct a preference table for these ballots.

(b) How many votes constitutes a majority? Is there a Majority winner?

(c) Is there a Plurality winner?

3. Suppose that Candidate D drops out of the race described in Exercise 2. Construct the modified preference table for the 13 voters.

Plurality and Plurality Run-Off

4. Three candidates are being considered for chair of the English Department. A poll of the 56 faculty members resulted in the following preference table.

	Voter Profiles			
	25	20	10	1
First	A	B	C	C
Second	B	C	B	A
Third	C	A	A	B

(a) Determine the top candidate using the Plurality method.

(b) Determine the top candidate using the Plurality Run-Off method.

5. Supporters of an art institute are asked to rank four possible exhibits available on loan from various museums for the next calendar year. The exhibits come from the following museums: High (H), Shanghai (S), Tate (T), and Uffizi (U). Preference profiles for the 41 supporters are listed in the following table.

	Voter Profiles			
	13	12	9	7
First	S	U	T	H
Second	H	H	H	T
Third	U	T	U	S
Fourth	T	S	S	U

(a) Which exhibit is the favorite using the Plurality method?

(b) Which exhibit is the favorite using the Plurality Run-Off method?

6. In a marketing study, 98 shoppers were asked to rank three types of detergent: Hope (H), Joy (J), and Refresh (R). Results are tabulated in the following preference table.

	Voter Profiles					
	28	23	21	15	7	4
First	H	R	R	H	J	J
Second	J	J	H	R	H	R
Third	R	H	J	J	R	H

(a) Which detergent is the favorite using the Plurality method?

(b) Which detergent is the favorite using the Plurality Run-Off method?

7. Seven coaches are asked to rank four top quarterbacks A, B, C, D in order to select the league's most valuable offensive player. The coaches have the following preferences.

	Voter Profiles				
	2	2	1	1	1
First	A	B	B	C	D
Second	D	D	A	A	A
Third	B	C	D	D	B
Fourth	C	A	C	B	C

(a) Which player is selected using the Plurality method?

(b) Which player is selected using the Plurality Run-Off method?

IRV (Instant Run-Off)

8. Use the following voter profiles to determine the IRV winner.

	Voter Profiles			
	4	3	3	2
First	B	A	C	D
Second	A	B	D	C
Third	C	C	A	A
Fourth	D	D	B	B

9. Use the following voter profiles to determine the IRV winner.

	Voter Profiles				
	8	7	5	5	4
First	D	A	B	C	C
Second	B	B	D	A	D
Third	C	C	A	D	B
Fourth	A	D	C	B	A

10. Consider the case of 33 voters having the following preferences.

	Voter Profiles			
	11	9	8	5
First	C	B	A	A
Second	A	C	B	C
Third	B	A	C	B

(a) Determine the IRV winner.

(b) Suppose that the five voters with the last profile in the preference table change their mind and decide to rank C above A. Now determine the IRV winner.

(c) What do you conclude from the results in parts (a) and (b)?

A variant of the IRV method is the *Coombs method*. It successively eliminates the candidate who has the most last-place votes until one candidate emerges with a majority of the first-place votes.

11. Determine the Coombs winner using the preference table in Exercise 8.

12. Determine the Coombs winner using the preference table in Exercise 9.

Sequential Run-Off

13. Four candidates are running for chair of the Board of Trustees, who will be elected by a vote of the Trustees. The following table lists the preferences of the 29 Trustees.

	Voter Profiles			
	11	8	6	4
First	B	A	D	D
Second	A	C	B	A
Third	C	D	C	C
Fourth	D	B	A	B

(a) Determine the Sequential Run-Off winner using the agenda [A, D, C, B].

(b) Determine the Sequential Run-Off winner using the agenda [B, C, D, A].

14. Four individuals A, B, C, D are being considered for a one-year traineeship with a prestigious Silicon Valley firm. These individuals have been ranked by 21 evaluators, giving the following preference table.

	Voter Profiles			
	7	6	5	3
First	C	B	A	D
Second	D	A	C	C
Third	B	D	B	A
Fourth	A	C	D	B

(a) Determine the Sequential Run-Off winner using the agenda [A, B, C, D].

(b) Determine the Sequential Run-Off winner using the agenda [A, C, D, B].

(c) Is there an agenda so that A will emerge as the winner? Explain.

(d) Is there an agenda so that D will emerge as the winner? Explain.

15. Consider the following preference table for 17 voters.

(a) Determine the Sequential Run-Off winner using the agenda [A, C, B, D].

(b) Determine the Sequential Run-Off winner using the agenda [D, C, B, A].

	Voter Profiles						
	5	4	3	2	1	1	1
First	C	B	A	B	A	C	C
Second	D	C	D	C	B	B	D
Third	B	A	B	D	C	D	A
Fourth	A	D	C	A	D	A	B

Pairwise Comparison

16. Use the following preference table for 50 voters to determine the Pairwise Comparison winner (if any) in a contest that involves three essays, written on the topics of Economics (E), Politics (P), and Religion (R).

	Voter Profiles					
	13	12	10	6	5	4
First	E	E	R	P	P	R
Second	P	R	P	R	E	E
Third	R	P	E	E	R	P

17. Use the following preference table for 43 voters to determine the Pairwise Comparison winner (if any).

	Voter Profiles			
	13	12	10	8
First	A	D	B	C
Second	B	C	C	B
Third	D	A	A	D
Fourth	C	B	D	A

18. Use the following preference table for 36 voters to determine the Pairwise Comparison winner (if any).

	Voter Profiles				
	13	9	8	4	2
First	A	B	D	B	D
Second	B	A	A	C	C
Third	C	C	B	D	B
Fourth	D	D	C	A	A

19. Using the preference table given in Example 17.9, determine the Pairwise Comparison winner (if any).

Borda Count

20. A department uses the Borda Count method to elect their representative to the Graduate Student Council. The 32 graduate students in the department have the following preferences for Miguel, Nash, and Omar, who are running for this position. Determine the Borda Count winner.

	Voter Profiles			
	10	9	8	5
First (2)	O	N	M	N
Second (1)	M	O	N	M
Third (0)	N	M	O	O

21. One city out of four competitors (Atlanta, Boston, Cleveland, Denver) is to be selected for the location of a company's new distribution center. Use the Borda Count method to determine the winning city given the following preference table for 15 voters.

	Voter Profiles			
	5	4	4	2
First (3)	C	A	D	B
Second (2)	A	B	C	D
Third (1)	B	C	B	A
Fourth (0)	D	D	A	C

22. Use the following preference table for 34 voters to determine the Borda Count winner.

	Voter Profiles				
	11	9	7	4	3
First (3)	A	C	B	D	C
Second (2)	B	A	C	A	B
Third (1)	D	B	D	B	D
Fourth (0)	C	D	A	C	A

Approval Voting

23. Consider the case of 83 voters and four candidates A, B, C, D. We list below the candidates approved by each voter. Determine the winner using Approval Voting.

Voter Profiles				
22	19	18	15	9
{A, C}	{A, D}	{B}	{B, C, D}	{B, D}

24. A corporate board consisting of 18 members needs to elect their next chair from among five nominees A, B, C, D, E. Below are the candidates approved by each board member. Determine the winner using Approval Voting.

Voter Profiles				
6	5	4	2	1
{B, C}	{A, D, E}	{A, C, D}	{B, E}	{C, D, E}

Combined Methods

25. At the end of the semester, students in a discrete math class completed a survey in which they ranked topics presented in the course: Graphs (G), Logic (L), Probability (P), and Voting (V). Their preference profiles are listed in the following table.

	Voter Profiles				
	14	12	10	8	7
First	G	P	V	L	L
Second	L	V	L	V	V
Third	P	G	G	G	P
Fourth	V	L	P	P	G

(a) Determine the favorite topic using the Plurality method.

(b) Determine the favorite topic using the Plurality Run-Off method.

(c) Determine the favorite topic using the Sequential Run-Off method with agendas (i) [G, L, P, V]; (ii) [L, P, V, G]; (iii) [L, G, V, P]; (iv) [G, V, L, P].

(d) Determine the favorite topic using the Borda Count method.

(e) Is there a Pairwise Comparison winner? If so, identify.

26. The following table lists the preferences of 15 voters for five candidates.

	Voter Profiles				
	5	4	3	2	1
First	A	B	C	E	D
Second	B	D	E	D	C
Third	D	C	D	C	E
Fourth	C	A	B	A	B
Fifth	E	E	A	B	A

(a) Determine the winner using the Plurality method.

(b) Determine the winner using the Plurality Run-Off method.

(c) Determine the winner using the IRV method.

(d) Determine the winner using the Borda Count method.

(e) Is there a Pairwise Comparison winner? If so, identify.

27. The following table lists the preferences of 21 voters for five candidates.

	Voter Profiles			
	8	7	4	2
First	A	D	E	C
Second	B	C	B	D
Third	C	E	D	A
Fourth	D	A	A	B
Fifth	E	B	C	E

(a) Determine the winner using the Plurality method.

(b) Determine the winner using the Plurality Run-Off method.

(c) Determine the Sequential Run-Off winner using the agenda [C, A, D, B, E].

(d) Determine the winner using the Borda Count method.

(e) Is there a Pairwise Comparison winner? If so, identify.

28. The following table lists the preferences of 21 voters for four candidates.

	Voter Profiles				
	6	5	4	3	3
First	C	A	B	A	C
Second	D	C	D	B	A
Third	B	B	C	D	B
Fourth	A	D	A	C	D

(a) Determine the Plurality Run-Off winner.

(b) Determine the Pairwise Comparison winner.

(c) Determine the Borda Count winner.

29. Four states (Florida, Georgia, Hawaii, South Carolina) are being considered for the location of a pharmaceutical company's annual winter convention. Preferences for the four states are shown in the following table.

	Voter Profiles				
	12	10	10	8	5
First	S	H	H	F	G
Second	G	F	F	G	S
Third	F	G	S	S	F
Fourth	H	S	G	H	H

(a) Determine the winning state using the Plurality method.

(b) Determine the winning state using the Plurality Run-Off method.

(c) Determine the winning state using the Instant Run-Off method.

(d) Determine the winning state using the Borda Count method.

(e) Is there a Pairwise Comparison winner? If so, identify.

Projects

30. Exercise 27(e) requires that you determine the winner of each head-to-head contest. Create a directed graph G in which an edge (i, j) is placed if candidate i beats candidate j in a one-on-one contest. Show that there is a (directed) Hamiltonian circuit in G. Explain how the existence of this circuit means that you can devise agendas so that each of the candidates will be a Sequential Run-Off winner using one of these agendas. Generalize this observation and construct other examples in which it can be applied.

31. Create a spreadsheet that will calculate the Borda points for each candidate and then output the Borda Count winner.

Bibliography

[Brams 2008] Brams, S. 2008. *Mathematics and democracy: Designing better voting and fair-division procedures*, Chapter 1. Princeton, NJ: Princeton University Press.

[Brams 2007] Brams, S., and Fishburn, P. 2007. *Approval voting*, second edition, Chapter 1. New York: Springer.

[Bruce 2001] Bruce, C. 2001. *Conned again, Watson! Cautionary tales of logic, math, and probability*, Chapter 7. Cambridge, MA: Perseus.

[Szpiro 2010] Szpiro, G. 2010. *Numbers rule: The vexing mathematics of democracy, from Plato to the present.* Princeton, NJ: Princeton University Press.

Chapter 18

Fairness Criteria and Arrow's Impossibility Theorem

Arthur: I am your King.

Peasant woman: Well I didn't vote for you.

Arthur: You don't vote for kings.

Woman: Well how'd you become king, then?

Arthur: The Lady of the Lake, her arm clad in the purest shimmering samite, held aloft Excalibur from the bosom of the water, signifying by divine providence that I, Arthur, was to carry Excalibur. That is why I'm your king.

Dennis: Listen. Strange women lying in ponds distributing swords is no basis for a system of government. Supreme executive power derives from a mandate from the masses, not from some farcical aquatic ceremony.

—*Monty Python and the Holy Grail*, 1975 film

Chapter Preview

Building on the previous chapter, the current chapter describes six fairness criteria and argues that they should be satisfied by any reasonable voting method. Instances are provided of voting methods that always satisfy a particular criterion as well as examples where voting methods fail to satisfy a particular criterion. Each of the preferential voting schemes presented in Chapter 17 is seen to violate at least one of the six fairness criteria. Naturally this leads to the search for an "ideal" voting method—one that satisfies all the fairness criteria. More broadly, we hope to identify a social ranking method (for collective decision making) that satisfies all of these fairness criteria. In a truly ingenious demonstration, Kenneth Arrow showed mathematically that it is *impossible* to have a social ranking method that meets all the fairness criteria. Consequently, we must consider the strengths and weaknesses of each social ranking method in order to identify one that best meets our needs. This chapter also introduces the idea of insincere voting, in which voters can purposely misrepresent their voting preferences in order to gain at the expense of other voters in the system.

18.1　Fairness Criteria for Voting Methods

Each of the *preferential voting methods* discussed in Chapter 17 was shown to be less than ideal, suffering from certain voting paradoxes.

DEFINITION **Preferential Voting Method**
Rather than asking each voter to select a single preferred candidate, a preferential voting method is based on asking voters to rank all candidates in order, from most preferred to least preferred.

Voting paradoxes arise when a method that has appealing features can nonetheless allow situations that we find objectionable or unfair. In this section, we discuss some fairness criteria that reasonable voting systems would be expected to satisfy. Examples are presented to illustrate when voting methods satisfy these criteria in all cases or when they fail in some situations.

In discussing voting methods in Chapter 17, we indicated that each method can produce situations that are undesirable. We described paradoxes—surprising and unwanted outcomes—that fly in the face of seemingly reasonable criteria that political scientists agree should be satisfied. Here we review six different fairness criteria, and the reasoning behind them, as well as the success of various voting methods in satisfying (in every case) such criteria.

DEFINITION **Monotonicity Criterion**
If A, the winner of the initial contest, receives additional support from some voters and nothing else changes, then A should win in a second contest.

The reasonableness of this criterion appeals to us intuitively: a winner who receives additional support from some voters should not be hurt by this additional support. Otherwise, we have the paradoxical situation in which "more means less." In Chapter 17, we saw however that the Plurality Run-Off method violates this criterion. That is, there are situations in which the criterion does not hold. Let's look at another method that also violates the Monotonicity criterion.

Example 18.1 Show the Instant Run-Off Method Violates the Monotonicity Criterion

Table 18.1 displays a preference table involving four candidates (A, B, C, D) and 35 voters. (a) Verify that A is the winner using the Instant Run-Off method. (b) Determine the Instant Run-Off winner if the six voters with preferences D > C > A > B change their ordering to D > A > C > B. (c) Explain why the Monotonicity criterion has been violated.

Solution

(a) Since A has 11 first-place votes, B has 10, C has 8, and D has 6, Candidate D is eliminated, giving the updated preferences shown in Table 18.2.

Now A has 11 first-place votes, B has 10 votes, and C has 14. So, B is eliminated, giving the updated preferences shown in Table 18.3.

TABLE 18.1: Preference table for 35 voters

	Voter Profiles			
	11	10	8	6
First	A	B	C	D
Second	D	A	B	C
Third	C	D	A	A
Fourth	B	C	D	B

TABLE 18.2: Preference table after Candidate D is eliminated

	Voter Profiles			
	11	10	8	6
First	A	B	C	C
Second	C	A	B	A
Third	B	C	A	B

TABLE 18.3: Preference table after Candidate B is eliminated

	Voter Profiles			
	11	10	8	6
First	A	A	C	C
Second	C	C	A	A

Since A has 21 votes and C has only 14, Candidate A is the Instant Run-Off winner.

(b) If the six voters represented in the last profile change their ranking to D > A > C > B, then the new preferences are given in Table 18.4.

TABLE 18.4: Preference table after six voters change their rankings

	Voter Profiles			
	11	10	8	6
First	A	B	C	D
Second	D	A	B	A
Third	C	D	A	C
Fourth	B	C	D	B

As before, Candidate D with 6 first-place votes is eliminated, resulting in Table 18.5. Now A has 17 first-place votes, B has 10, and C has 8. So C is eliminated, giving Table 18.6. In the final contest between A and B, Candidate B is the Instant Run-Off winner, with 18 first-place votes (a majority of the 35 votes).

(c) In this example, the Monotonicity criterion is not satisfied because A received additional support from the six voters in part (b) and then loses! This shows that the Instant Run-Off method violates the Monotonicity criterion. ◁

(Related Exercises: 2–3)

On the other hand, some methods are guaranteed to satisfy the Monotonicity criterion.

TABLE 18.5: Preference table after Candidate D is eliminated

	Voter Profiles			
	11	10	8	6
First	A	B	C	A
Second	C	A	B	C
Third	B	C	A	B

TABLE 18.6: Preference table after Candidate C is eliminated

	Voter Profiles			
	11	10	8	6
First	A	B	B	A
Second	B	A	A	B

Example 18.2 Show the Plurality Method Always Satisfies the Monotonicity Criterion

An election is held and Candidate A is declared the winner using the Plurality method. A single column of the preference table is illustrated in Table 18.7(a), representing the profile of v voters. Now suppose these v voters give additional support to Candidate A, as seen in Table 18.7(b), but do not change the relative rankings of the other candidates. (a) Show that Candidate A continues to win by the Plurality method if the only change occurs in the profile of these v voters. (b) More generally, explain why the Monotonicity criterion is satisfied for the Plurality method.

TABLE 18.7: Candidate A receives additional support

	v			v
First	B		First	A
Second	C		Second	B
Third	A	$\longrightarrow$	Third	C
Fourth	D		Fourth	D
Fifth	E		Fifth	E
	(a)			(b)

Solution

(a) Table 18.7(b) shows that Candidate A receives v additional first-place votes, compared to Table 18.7(a). So if Candidate A won by the Plurality method in scenario (a), then A will certainly win in scenario (b) with the additional support.

(b) This same reasoning applies in general to any preference table: if Candidate A moves up in the rankings of a particular column, this will increase (or at worst keep the same) the number of first-place votes. As a result, Candidate A continues to win by the Plurality method and the Monotonicity criterion is satisfied. ◁

(Related Exercises: 22–23)

DEFINITION **Unanimity Criterion**
If every voter prefers A to B, then B should not win.

This too seems an intuitively reasonable criterion. That is, an election would be judged unfair if B wins even though everyone prefers A to B. In Example 17.4, we saw however that the Sequential Run-Off method can violate this condition. Let's explore a different voting method that can never violate the Unanimity criterion.

Example 18.3 Show the Borda Count Method Always Satisfies the Unanimity Criterion

Table 18.8 shows one column of a preference table, in which the five candidates are ranked by v voters. Notice that A is ranked higher than B for these v voters. (a) Compute the contribution toward the Borda count for Candidate A and for Candidate B. Which is larger? (b) More generally, argue that if everyone prefers Candidate A to Candidate B, then B cannot possibly be the Borda Count winner, so the Unanimity criterion is always satisfied for the Borda Count method.

TABLE 18.8: A single column of a preference table

	v
First (4)	C
Second (3)	A
Third (2)	D
Fourth (1)	B
Fifth (0)	E

Solution

(a) To calculate the Borda points for A in the single column of Table 18.8, we multiply the third-place value (3) times the number of voters v, giving $3v$ points. Similarly, the Borda points for B in this column are found by multiplying the fourth-place value (1) by the number of voters v, giving $1v = v$ points, which is certainly less than $3v$ points.

(b) More generally, suppose that everyone prefers Candidate A to Candidate B. Since A will then be ranked higher than B in each column, A will accumulate more Borda points than B in that column. Since A is ranked higher than B in every column, A's total Borda points will always exceed B's total Borda points. As a result, B can never be the Borda Count winner since A will always accumulate more total points. ◁

(Related Exercise: 4)

DEFINITION **Irrelevancy Criterion**
If one or more losing candidates drop out of the race, this should not affect the winner of the election.

The reasonableness of this criterion can be illustrated by the following scenario:

> Victor is ready to order dessert from the menu at his favorite restaurant and sees the choices of apple or blueberry pie. He decides to order apple pie. However,

when the waitperson appears to take his order, Victor is informed that the chef has just prepared a cherry pie. Victor now switches his order to blueberry pie!

We would likely consider this to be a paradoxical situation, since the introduction of the irrelevant alternative C (cherry) has changed the choice from A (apple) to B (blueberry). Phrased in terms of voting, when choices A, B, C are presented, voter V chooses B; but when C drops out, then voter V chooses A. This is just the Irrelevancy criterion applied to the set of candidates as a whole.

A closely related situation occurred in the 1995 Women's World Championship Skating Competition. Near the end of the competition, the three skaters Chen Lu, Nicole Bobek, and Surya Bonaly were ranked first, second, and third, respectively. Then the fourteen-year-old Michelle Kwan skated magnificently and leaped to fourth place. Her score caused Bobek and Bonaly's positions to reverse: after Kwan skated, Bonaly was ranked second and Bobek third. That is, Kwan (the irrelevant candidate) changed the final ranking between Bobek and Bonaly, even though no judge changed their ranking of Bobek over Bonaly.

These paradoxical situations lend support to requiring the Irrelevancy criterion to hold for voting methods. However, it turns out that every voting system (except Approval Voting) can violate this condition.

Example 18.4 Show the Pairwise Comparison Method Violates the Irrelevancy Criterion

(a) Find the Pairwise Comparison winner among Candidates A, B, C, D, E given the voter preferences in Table 18.9. (b) Find the Pairwise Comparison winner if Candidates B and C drop out. (c) Explain how this situation violates the Irrelevancy criterion.

TABLE 18.9: Preference table for 11 voters

	Voter Profiles			
	5	2	2	2
First	A	B	C	E
Second	D	E	E	D
Third	C	A	A	A
Fourth	B	D	B	C
Fifth	E	C	D	B

Solution

(a) We first find the Pairwise Comparison winner. In a one-on-one contest between A and B, A is preferred to B by $5 + 2 + 2 = 9$ voters, while B is preferred to A by 2 voters. (Note that the sum $9 + 2 = 11$, the total number of voters.) Thus, A wins this one-on-one contest.

Recall that there are $C(5,2) = 10$ different ways to choose 2 competitors out of 5 for a matchup:

$$C(5,2) = \frac{5 \times 4}{2 \times 1} = \frac{20}{2} = 10.$$

We record the winner of each of these 10 one-on-one contests in Table 18.10. For example, the (A, B) entry lists A as the winner of the A vs. B contest by 9 votes to 2.

We can now tally the points given to each competitor. A won three contests, so A gets 3 points; similarly B gets 1 point, C gets 2 points, D gets 2 points, and E gets 2 points. Thus, A is the Pairwise Comparison winner.

TABLE 18.10: Winners of the 10 one-on-one contests

	A	B	C	D	E
A		A(9/2)	A(9/2)	A(9/2)	E(6/5)
B			C(9/2)	D(7/4)	B(7/4)
C				D(9/2)	C(7/4)
D					E(6/5)
E					

TABLE 18.11: Preference table after Candidates B and C drop out

	Voter Profiles			
	5	2	2	2
First	A	E	E	E
Second	D	A	A	D
Third	E	D	D	A

(b) If Candidates B and C drop out, the revised preferences are shown in Table 18.11.

Notice that there are $C(3,2) = 3$ different ways to choose 2 competitors out of 3 for a matchup. We record the winners of these three one-on-one contests in Table 18.12. Consequently, E is the Pairwise Comparison winner with 2 points.

TABLE 18.12: Winners of the three one-on-one contests

	A	D	E
A		A(9/2)	E(6/5)
D			E(6/5)
E			

(c) We see that when the losing Candidates B and C drop out, the winner is E instead of A. This situation violates the Irrelevancy criterion. ◁

(Related Exercises: 5–10)

DEFINITION **Majority Criterion**
If a majority of voters cast first-place votes for A, then A should win.

This criterion seems reasonable since a majority of voters will be satisfied with the outcome. In Chapter 17, we saw however that the Borda Count method and Approval Voting can fail to satisfy the Majority criterion. Let's now explore a method that can never violate the Majority criterion.

Example 18.5 Show the Sequential Run-Off Method Must Satisfy the Majority Criterion

Table 18.13 shows an incomplete preference table in which Candidate A receives a majority of the first-place votes cast, receiving 20 of the 39 total votes. Suppose that we use an agenda for pairing the winners of successive contests. (a) Explain why, regardless of the agenda used, Candidate A must be declared the winner by the Sequential Run-Off method. (b) More generally, explain why the Sequential Run-Off method will always satisfy the Majority criterion.

TABLE 18.13: A candidate with a majority vote

	Voter Profiles			
	20	10	5	4
First	A			
Second	D			
Third	B			
Fourth	C			

Solution

(a) It doesn't matter if any of the other candidates B, C, or D is ranked higher than A in the last three columns of Table 18.13 because they can never have enough votes to beat A in a one-on-one contest. This is true since Candidate A will always have enough votes (a majority) to defeat any other candidate in a one-on-one contest, regardless of the agenda.

(b) Suppose that Candidate A receives a majority of first-place votes based on several profiles in the preference table. Then in any one-on-one contest, Candidate A will win against any other candidate and so must emerge as the overall winner using any specified agenda. So the Sequential Run-Off method can never violate the Majority criterion. ◁

(Related Exercises: 11, 16–18)

DEFINITION **Transitivity Criterion**
If voters collectively prefer A to B, and also prefer B to C, then they should prefer A to C.

Notice that this criterion applies to methods that allow an overall ranking of all the candidates, not just the selection of a single winner. One such voting scheme is the Borda Count method, which assigns numerical scores to every candidate and so can be used to rank all the candidates.

Example 18.6 Show the Borda Count Method Always Satisfies the Transitivity Criterion

Demonstrate that the Borda Count method will always satisfy the Transitivity criterion. Namely, verify that if (i) voters collectively prefer A to B, and (ii) they prefer B to C, it is necessary that (iii) they also prefer A to C.

Solution

Recall that the Borda Count is based on calculating total points for each candidate. Suppose A receives x points, B receives y points, and C receives z points. Since A is preferred to B, we have $x > y$; and since B is preferred to C, $y > z$. Since x is larger than y, which in turn is larger than z, it must be the case that x is larger than z. In other words, A is collectively preferred to C; this shows that the Transitivity criterion must hold for any election decided by the Borda Count method. ◁

> DEFINITION **Condorcet Criterion**
> If A wins in one-on-one contests with every other candidate, then A should be declared the winner.

This criterion seems eminently reasonable. For example, in a round-robin tournament in which one team (A) beats every other team, it should be the consensus that A is declared the overall tournament winner. However, not every voting method will satisfy this criterion, as shown in the following example.

Example 18.7 Show the Plurality Method Can Violate the Condorcet Criterion

Table 18.14 gives a preference table for 30 voters and four candidates. (a) Determine the Plurality winner. (b) Is there a candidate who wins all one-on-one contests? Explain how this provides a violation of the Condorcet criterion.

TABLE 18.14: Preference table for 30 voters

	Voter Profiles				
	8	7	6	5	4
First	A	D	C	B	C
Second	B	B	B	C	A
Third	C	A	D	D	B
Fourth	D	C	A	A	D

Solution

(a) The Plurality winner is C with 10 first-place votes.

(b) Candidate B wins in one-on-one matches with the other candidates A, C, and D. Specifically, in the contest between A and B, B is preferred to A by 18 to 12 voters. In the contest between B and C, B wins 20 to 10. In the contest between B and D, B wins 23 to 7. However, B is *not* the winner determined by the Plurality method. That is, the Plurality method violates the Condorcet criterion in this example. ◁

(Related Exercises: 12–15)

18.2 Arrow's Impossibility Theorem

Violating the Fairness Criteria

We have discussed six preferential voting methods and each one violates at least one of the six fairness criteria. (Note that Approval Voting is not a preferential method, since voters do not rank their choices.) The six fairness criteria discussed in Section 18.1 are listed below:

1. **Monotonicity:** If A, the winner of the initial contest, receives additional support from some voters and nothing else changes, then A should win in a second contest.
2. **Unanimity:** If every voter prefers A to B, then B should not win.

3. **Irrelevancy:** If one or more losing candidates drop out of the race, this should not affect the winner of the election.

4. **Majority:** If a majority of voters cast first-place votes for A, then A should win.

5. **Transitivity:** If voters collectively prefer A to B, and also prefer B to C, then they should prefer A to C.

6. **Condorcet:** If A wins in one-on-one contests with every other candidate, then A should be declared the winner.

In previous examples, we have seen instances in which a voting method always satisfies a certain fairness criterion and we have seen instances in which a voting method can violate a certain fairness criterion. Table 18.15 shows how the six preferential voting methods discussed fare relative to the six fairness criteria. Notice that Approval Voting is not listed in the table since it is not a preferential system. Entries in the table are explained as follows:

- A check mark indicates that a method *always* satisfies the indicated fairness criterion.
- *Violates* indicates that there are cases in which the indicated property does not hold.
- The examples listed appear in Chapters 17 and 18; they demonstrate that a particular voting method can violate a specific criterion, or will always satisfy it.

TABLE 18.15: Voting methods and the fairness criteria

	Monotonicity	Unanimity	Irrelevancy	Majority	Transitivity	Condorcet
Plurality	$\checkmark$ Ex.18.2	$\checkmark$	*Violates* Ex.17.1	$\checkmark$	$\checkmark$	*Violates* Ex.18.7
Plurality Run-Off	*Violates* Ex.17.2	$\checkmark$	*Violates* Ex.17.2	$\checkmark$	$\checkmark$	*Violates* Ex.18.10
Instant Run-Off	*Violates* Ex.18.1	$\checkmark$	*Violates* Ex.17.3	$\checkmark$	$\checkmark$	*Violates*
Sequential Run-Off	$\checkmark$	*Violates* Ex.17.4	*Violates* Ex.18.8	$\checkmark$ Ex.18.5	$\checkmark$	$\checkmark$
Pairwise Comparison	$\checkmark$	$\checkmark$ Ex.18.11	*Violates* Ex.18.4	$\checkmark$	$\checkmark$	$\checkmark$
Borda Count	$\checkmark$	$\checkmark$ Ex.18.3	*Violates* Ex.18.9	*Violates* Ex.17.7	$\checkmark$ Ex.18.6	*Violates*

We now investigate several entries in Table 18.15 that have not been previously discussed. Additional entries in this table will be addressed in the Chapter Exercises.

(Related Exercises: 14–23)

Example 18.8 Show the Sequential Run-Off Method Violates the Irrelevancy Criterion

Demonstrate that the Sequential Run-Off method violates the Irrelevancy criterion by using the agenda [B, C, A, D, E] and the preference table for 49 voters shown in Table 18.16.

Solution

Following the agenda [B, C, A, D, E], we first consider the B versus C contest, where C wins with 26 votes; in C versus A, the next contest on the agenda, A wins with 43 votes; in the A versus D contest, A wins with 43 votes; and finally A is preferred to E by all 49 votes. So Candidate A is the Sequential Run-Off winner.

TABLE 18.16: Preference table for 49 voters

	Voter Profiles			
	20	13	10	6
First	A	B	B	C
Second	C	A	A	D
Third	E	D	C	B
Fourth	B	C	D	A
Fifth	D	E	E	E

Now if C (with only 6 first-place votes) drops out, the updated agenda is [B, A, D, E] and the revised preferences are shown in Table 18.17.

TABLE 18.17: Preference table after Candidate C drops out

	Voter Profiles			
	20	13	10	6
First	A	B	B	D
Second	E	A	A	B
Third	B	D	D	A
Fourth	D	E	E	E

Using the agenda [B, A, D, E], B wins over A with 29 votes; B wins over D with 43 votes; and B wins over E with 29 votes. Now B is declared the Sequential Run-Off winner. This situation violates the Irrelevancy criterion because a losing candidate dropped out, the preference order among the remaining candidates did not change, yet the previous winner A now loses. ◁

Example 18.9 Show the Borda Count Method Violates the Irrelevancy Criterion

Table 18.18 shows the preferences for 43 voters. (a) Find the Borda Count winner. (b) Let the losing Candidate C drop out of race, and update the preference table to determine whether the Borda Count winner has now changed.

TABLE 18.18: Preference table for 43 voters

	Voter Profiles			
	14	13	10	6
First (4)	A	B	B	C
Second (3)	C	A	A	D
Third (2)	E	D	C	B
Fourth (1)	B	C	D	A
Fifth (0)	D	E	E	E

Solution

(a) Candidate A emerges as the Borda Count winner with 131 total points:

A: $(14 \times 4) + (23 \times 3) + (6 \times 1) = 131$
B: $(23 \times 4) + (6 \times 2) + (14 \times 1) = 118$

C: $(6 \times 4) + (14 \times 3) + (10 \times 2) + (13 \times 1) = 99$
D: $(6 \times 3) + (13 \times 2) + (10 \times 1) = 54$
E: $(14 \times 2) = 28$

(b) Table 18.19 shows the updated preferences if losing Candidate C (with only 6 first-place votes) drops out. Notice that the point values for each ranking have changed. Candidate B now emerges as the Borda Count winner with a total of 95 votes:

A: $(14 \times 3) + (23 \times 2) + (6 \times 1) = 94$
B: $(23 \times 3) + (6 \times 2) + (14 \times 1) = 95$
D: $(6 \times 3) + (23 \times 1) = 41$
E: $(14 \times 2) = 28$

TABLE 18.19: Preference table after Candidate C drops out

	Voter Profiles			
	14	13	10	6
First (3)	A	B	B	D
Second (2)	E	A	A	B
Third (1)	B	D	D	A
Fourth (0)	D	E	E	E

This example shows that the Borda Count method can violate the Irrelevancy criterion because a losing candidate drops out, all preferences remain unchanged, and yet the initial winner A now loses. ◁

Example 18.10 Show the Plurality Run-Off Method Violates the Condorcet Criterion

Table 18.20 shows the preferences of 100 voters for three candidates. (a) Determine the Plurality Run-Off winner. (b) Is there a candidate who wins all one-on-one contests? Explain how this provides a violation of the Condorcet criterion.

TABLE 18.20: Preference table for 100 voters

	Voter Profiles		
	40	35	25
First	A	C	B
Second	B	B	A
Third	C	A	C

Solution

(a) First, B is eliminated (fewest first-place votes) and then A defeats C by 65 to 35 votes. So the Plurality Run-Off winner is A.

(b) Candidate B wins in a one-on-one contest with A (60 to 40), and wins in a one-on-one contest with C (65 to 35). However, B is not the winner determined by the Plurality Run-Off method. So the Plurality Run-Off method can violate the Condorcet criterion. ◁

Example 18.11 Show the Pairwise Comparison Method Always Satisfies the Unanimity Criterion

Verify that the Unanimity criterion will necessarily be satisfied by an election conducted using the Pairwise Comparison method. In other words, show that if every voter prefers A to B, then B should not be elected by the Pairwise Comparison method.

Solution

Suppose that B defeats X in a one-on-one contest. Since every voter prefers A to B, then A will also defeat X in a one-on-one contest. This means that A defeats every candidate that B defeats. Moreover, since every voter prefers A to B, then A will also defeat B in a one-on-one contest. As a result, A accumulates more wins than does B. Consequently, B cannot be elected by the Pairwise Comparison method, since this voting procedure only elects a candidate having the largest number of wins. ◁

Is There a Perfect Voting Method?

We might wonder if there is any voting method that is ideal. Is there a perfect voting method that we have overlooked, or perhaps we are requiring too many fairness criteria to be satisfied? With two candidates, we can simply use the Majority method, which satisfies all the fairness criteria (see Exercise 18.1). It is with three or more candidates that the situation becomes more interesting—and as a practical matter, we certainly need a voting method that can be used with any number of candidates.

In Section 18.1 (and in Chapter 17) we focused on identifying the winner of an election according to different voting methods. But many social judgments and public decisions depend on taking individual preferences in the form of voter profiles (rankings) and from them determining an overall *social ranking* (as in athletic competitions where recognition is given for finishing first, second, third, etc.). For example, the Borda Count and Pairwise Comparison methods, which yield numerical scores, automatically provide numerical rankings of all candidates. The other methods we have discussed can be suitably modified to produce a ranking of all candidates, rather than just identifying a single winner.

(Related Exercise: 27)

DEFINITION **Social Ranking Method**
A voting procedure that transforms the preferences of all voters into a ranked order of the candidates.

Our question, more generally, becomes whether there exists a social ranking method that meets all of our fairness criteria. The economist Kenneth Arrow, in his PhD dissertation and in his 1951 book *Social Choice and Individual Values*, conclusively proved that such paradoxes are inherent in any social ranking method.

HISTORICAL CONTEXT **Kenneth Arrow (1921–)**
An American economist and joint winner (with John Hicks) of the 1972 Nobel Prize in Economics for his pioneering work in mathematical social choice theory. To date, he is the youngest person to have received this award, at age 51.

Arrow proved in essence that a fair method for making collective (social) choices is impossible. A restatement of his "impossibility theorem" is as follows.

RESULT **Arrow's Impossibility Theorem**
When there are at least three candidates, the only social ranking method satisfying the transitivity, unanimity, and irrelevancy criteria is a dictatorship.

Namely, for a social ranking method (applicable to three or more candidates) that is not *despotic*, at least one of these three fairness criteria will be violated.

DEFINITION **Despotic Voting Method**
A despotic voting method will simply output the preferences of a single voter; it is a dictatorship.

Arrow's theorem reduces our list of six criteria to just three and shows that even these three criteria are sufficient to eliminate any (nondespotic) social ranking method. Since we are now considering voting methods that provide a societal ranking of all candidates, rather than just selecting a winner, the criteria mentioned in Arrow's theorem are appropriately generalized versions of the ones we have studied earlier:

1. **Transitivity:** If all voters prefer A to B, and also prefer B to C, then the societal ranking should prefer A to C.
2. **Unanimity:** If every voter prefers A to B, then the societal ranking should prefer A to B.
3. **Irrelevancy:** The societal ranking of A and B should not be affected if another candidate C is removed from the ballot.

How can we escape this conclusion about the fruitless task of finding any fair or meaningful method for determining elections? One way would be to focus on the fact that in most real-world settings, there will be several methods that generally produce acceptable outcomes. Or we could abandon the idea of a preferential voting method altogether, using for example Approval Voting, which avoids the paradoxes plaguing preferential voting. Or we might believe that certain of the criteria are too strong (especially troublesome is the Irrelevancy criterion, which is not satisfied by any of the preferential voting methods presented).

18.3 Voting Methods in Practice

Though we cannot expect to find a preferential voting method that respects all of Arrow's fairness conditions, we may judge certain voting methods to be better than others based on the specific goals that an organization, city, state, or country may have. It is useful to remember that different voting methods are chosen for different purposes; representative applications are listed in Table 18.21.

TABLE 18.21: Voting methods and applications

Plurality	Elections for Mayors, Governors, State Representatives, Presidents
Plurality Run-Off	Primary Elections
Instant Run-Off	Academy Awards; Australian House of Representatives; Olympic Committee (to select host nations); City Council Elections
Sequential Run-Off	Athletic Tournaments
Pairwise Comparison	Round-Robin Tournaments; Proposal Evaluation
Borda Count	Heisman Trophy; MVP in Baseball; Football Polls; Figure Skating; Gymnastics
Approval	United Nations (Secretary-General); Professional Societies

Another consideration in deciding on an appropriate voting method might be more psychological. Voters might dislike voting systems in which their votes appear to be wasted, as in the Plurality method where only the first-place votes are counted and all other preferences are ignored. Plurality Run-Off and to a greater extent Instant Run-Off avoid this problem by reallocating votes after each round. The Pairwise Comparison and Borda methods explicitly take into account all of the voter preferences. A disadvantage of the Pairwise Comparison method is that it is computationally intensive: $C(n, 2) = n(n - 1)/2$ pairwise comparisons are required when there are n candidates. By contrast, Approval Voting just requires a simple count of votes (unlike the Borda method, which requires weighting the votes before adding them together). However, Approval Voting does not use any information about the relative desirability of the candidates marked as approved, and so it may ignore important additional preferences among the voters.

Strategic Voting

Strategic voting takes place when voters do not vote in accordance with their true preferences, but rather vote insincerely in order to influence the election in their favor. As the next example shows, the Plurality method is vulnerable to strategic voting.

Example 18.12 Show the Plurality Method is Vulnerable to Strategic Voting

Table 18.22 shows the preferences of nine voters for three candidates. (a) Determine the winner by the Plurality method. (b) Suppose the three voters in the middle profile insincerely elevate C ahead of B. Now determine the winner by the Plurality method.

Solution

(a) Candidate A receives the most first-place votes and is declared the winner.

(b) In the revised Table 18.23, Candidate C now receives the most first-place votes and is declared the winner. By voting strategically (not according to their true preferences), voters

TABLE 18.22: Preference table to illustrate strategic voting

	Voter Profiles		
	4	3	2
First	A	B	C
Second	B	C	A
Third	C	A	B

in the middle profile have achieved a better outcome since C (the new winner) is ranked higher than A (the original winner) in their true ranking of candidates B > C > A. ◁

TABLE 18.23: Revised preference table

	Voter Profiles		
	4	3	2
First	A	C	C
Second	B	B	A
Third	C	A	B

As another example of the manipulability of the Plurality method, suppose that three candidates are running for the United States Senate—a Democrat, a Republican, and an Independent. A voter may prefer the Independent candidate over the other two, but realizes (because of the pre-election polls) that it will be a close race between the Democrat and the Republican, and that the Independent doesn't have a chance. So the voter may vote strategically (and insincerely) for his second choice (say, the Democrat) just to ensure that the Republican doesn't win. It is believed that strategic considerations in plurality voting heavily favor a two-party system by excluding Independent candidates; in fact, this effect is called *Duverger's law* [OR1] in political science.

Since the Plurality method only relies on first-place rankings, it is not surprising that it is highly susceptible to strategic voting. Yet even methods that take into account all voter preferences can be manipulated. As we will now see, the Borda Count method can be gamed: voters can insincerely rank someone lower just to help a more preferred candidate win.

Example 18.13 Show the Borda Count Method is Vulnerable to Strategic Voting

Table 18.24 represents the sincere preferences of the voters, while the voters in the second profile of Table 18.25 have insincerely ranked Candidate C higher. Determine the Borda Count winner for the original and for the modified preferences. What do you conclude?

TABLE 18.24: Preference table for sincere voting

	Voter Profiles			
	9	7	2	2
First (2)	A	B	C	C
Second (1)	B	A	A	B
Third (0)	C	C	B	A

TABLE 18.25: Preference table for insincere voting

	Voter Profiles			
	9	7	2	2
First (2)	A	B	C	C
Second (1)	B	C	A	B
Third (0)	C	A	B	A

Solution

The Borda Count winner using sincere preferences is A (with 27 votes), while the Borda Count winner using insincere preferences is now B (with 25 votes). Here the seven voters in the second profile have gotten their top candidate B elected by insincerely elevating candidate C, whom they preferred the least. ◁

(Related Exercises: 24–26)

A Strategy-Free Voting System?

While we might hope to find a voting method that is not susceptible to strategic voting, a result analogous to Arrow's theorem demonstrates that this is not possible. Specifically, the *Gibbard-Satterthwaite theorem* (1973) states that it is impossible to design a preferential voting system for electing a winner that is both strategy-free and nondespotic. This result also assumes that there are at least three candidates and that the system is flexible enough to allow any of the candidates to win, given a suitable preference table.

In Approval Voting, where voters simply classify the candidates into acceptable and unacceptable, it is not advantageous for a voter to move a candidate from one category to another. For example, moving Candidate B from unacceptable to acceptable only gives B more votes and does not advance the wishes of this voter. However, there are strategic opportunities available in choosing where sets of voters "draw the line" between acceptable and unacceptable.

Example 18.14 Show Approval Voting is Vulnerable to Strategic Voting

Table 18.26 indicates the true voter preferences for three candidates. In the first profile, both A and B would be considered acceptable, but there is a slight preference for A. Similarly in the second profile there is a slight preference for B over A, though both are acceptable. (a) What would be the result of using Approval Voting? (b) What would be the result if one voter from the first profile decided to approve only A?

TABLE 18.26: Acceptable candidates for 42 voters

Voter Profiles		
11	11	20
A	B	C
B	A	

Solution

(a) Using Approval Voting, both A and B would be tied with 22 votes, beating C who receives only 20 votes. Some type of run-off would then be required.

(b) Using Approval Voting in the modified situation, A would get 22 votes, B would get 21, and C would get 20. The single voter would advance his (slightly) preferred candidate A by insincerely demoting B on his ballot. ◁

What Example 18.14 shows is that we have an *unstable equilibrium* (see Section 11.3): there is an incentive for a single voter to change his or her voting preference. Something quite interesting will then occur if this (insincere) strategy is adopted independently by several voters in the first two profiles. Suppose for instance that five voters from the first profile decided only to list A and five voters from the second profile decided only to list B. This gives the revised Table 18.27.

TABLE 18.27: Revised acceptable candidates for 42 voters

Voter Profiles				
5	6	5	6	20
A	A	B	B	C
	B		A	

Then the final tally would give 17 approval votes to both A and B, with C getting 20 votes. So now C emerges as the winner! This shows that where voters draw the line (between acceptable and unacceptable) can have a profound effect on the results of an election.

Summary

This chapter has studied the advantages and disadvantages of using six preferential voting schemes commonly in use. The preferences of each voter are represented using tables, and voting methods process this tabular information to identify the winner of the election, or more generally to produce an overall ranking of all candidates. The voting methods are assessed by the extent to which they satisfy (or fail to satisfy) certain fairness criteria. The search for an ideal voting method, one that satisfies all these fairness criteria, is brought to an abrupt halt by Arrow's impossibility theorem, which shows that the only viable method is a dictatorship! Adding to this state of affairs is a companion result that demonstrates that all voting methods (except a dictatorship) can be gamed by voters to their advantage.

Applications, Representations, Strategies, and Algorithms

Applications		
Society/Politics (1–14)		
Representations		
Tables	Symbolization	
Data	Numerical	Algebraic
1–5, 7–10, 12–14	1–5, 7–10, 12–14	3, 6
Strategies		
Rule-Based		
1–14		
Algorithms		
Exact	Sequential	Numerical
1–14	1, 4, 8, 10–11	1–14

Exercises

Elections with Two Candidates

1. Suppose an election is held with just two candidates (A, B). With two candidates, there are only two columns in the following preference table. An appropriate voting procedure is to use the Majority method.

	Voter Profiles	
	9	7
First	A	B
Second	B	A

   (a) Show that the Majority method satisfies the Majority criterion.

   (b) Show that the Majority method satisfies the Monotonicity criterion.

   (c) Show that the Majority method satisfies the Irrelevancy criterion (where the losing candidate drops out).

   (d) Generalize the results in (a)-(c) to the case where A has more votes than B.

Violating the Fairness Criteria

2. Consider the following preference table for 29 voters.

Voter Profiles			
9	8	7	5

	9	8	7	5
First	C	A	B	B
Second	B	C	A	C
Third	A	B	C	A

(a) Determine the winner using the Plurality Run-Off method.

(b) Suppose that the five voters in the last profile change their preference ranking to C > B > A. Who is then the winner using the Plurality Run-Off method?

(c) Explain why this example shows that Plurality Run-Off can violate the Monotonicity criterion.

3. Consider the following preference table for 36 voters.

	Voter Profiles				
	12	10	7	4	3
First	A	B	C	D	B
Second	B	C	A	C	D
Third	D	A	D	B	A
Fourth	C	D	B	A	C

(a) Verify that A is the winner using the Instant Run-Off method.

(b) Suppose that the three voters in the last profile change their preference ranking to A > B > D > C. Who is then the winner using the Instant Run-Off method?

(c) Explain why this example shows that Instant Run-Off can violate the Monotonicity criterion.

4. Consider the following preference table for nine voters.

	Voter Profiles		
	4	3	2
First	B	C	D
Second	D	A	C
Third	C	B	A
Fourth	A	D	B

(a) Determine the Sequential Run-Off winner using the agenda [D, C, B, A].

(b) Explain why this example shows that Sequential Run-Off can violate the Unanimity criterion.

5. Consider the following preference table for 21 voters.

(a) Determine the winner using the Plurality method.

(b) Now suppose that Candidate C drops out of the race. Calculate the winner of the new election.

	Voter Profiles			
	7	6	5	3
First	A	B	C	D
Second	B	A	B	C
Third	C	C	D	A
Fourth	D	D	A	B

(c) Explain why this example shows that the Plurality method can violate the Irrelevancy criterion.

6. Consider the following preference table for 13 voters.

	Voter Profiles		
	6	4	3
First	C	B	A
Second	A	A	B
Third	B	C	C

(a) Determine the winner using the Plurality Run-Off method.

(b) Now suppose that Candidate C drops out of the race. Calculate the winner of the new election.

(c) Explain why this example shows that the Plurality Run-Off method can violate the Irrelevancy criterion.

7. Consider the following preference table for 10 voters.

	Voter Profiles			
	4	3	2	1
First	A	C	B	D
Second	B	B	D	C
Third	D	A	A	B
Fourth	C	D	C	A

(a) Determine the winner using the Instant Run-Off method.

(b) Now suppose that Candidate C drops out of the race. Calculate the winner of the new election.

(c) Explain why this example shows that the Instant Run-Off method can violate the Irrelevancy criterion.

8. Consider the following preference table for three voters.

(a) Determine the Sequential Run-Off winner using the agenda [A, B, C, D].

(b) Now suppose that Candidate C drops out of the race. Calculate the winner of the new election using the revised agenda [A, B, D].

(c) Explain why this example shows that the Sequential Run-Off method can violate the Irrelevancy criterion.

	Voter Profiles		
	1	1	1
First	A	C	D
Second	B	B	C
Third	D	A	A
Fourth	C	D	B

9. Consider the following preference table for 68 voters.

	Voter Profiles			
	30	18	16	4
First	A	C	D	D
Second	B	B	A	A
Third	C	D	C	B
Fourth	D	A	B	C

(a) Determine the winner using the Pairwise Comparison method.

(b) Now suppose that Candidates B and C drop out of the race. Calculate the winner of the new election.

(c) Explain why this example shows that the Pairwise Comparison method can violate the Irrelevancy criterion.

10. Consider the following preference table for 15 voters.

	Voter Profiles		
	6	5	4
First (3)	A	D	B
Second (2)	D	B	A
Third (1)	B	A	C
Fourth (0)	C	C	D

(a) Determine the winner using the Borda Count method.

(b) Now suppose that Candidate D drops out of the race. Calculate the winner of the new election.

(c) Explain why this example shows that the Borda Count method can violate the Irrelevancy criterion.

11. Consider the following preference table for 15 voters.

	Voter Profiles			
	5	4	3	3
First (3)	A	B	A	B
Second (2)	B	D	B	C
Third (1)	C	C	D	D
Fourth (0)	D	A	C	A

(a) Determine the winner using the Borda Count method.

(b) Explain why this example shows that the Borda Count method can violate the Majority criterion.

12. Consider the following preference table for 46 voters.

	Voter Profiles				
	14	10	9	8	5
First	D	A	B	C	C
Second	A	C	C	B	B
Third	C	B	A	D	A
Fourth	B	D	D	A	D

(a) Determine the winner using the Plurality method.

(b) Is there a candidate who defeats every other candidate in all one-on-one contests?

(c) Explain why this example shows that the Plurality method can violate the Condorcet criterion.

13. Consider the following preference table for five voters.

	Voter Profiles		
	2	2	1
First	A	B	C
Second	C	C	A
Third	B	A	B

(a) Determine the winner using the Plurality Run-Off method.

(b) Is there a candidate who defeats every other candidate in all one-on-one contests?

(c) Explain why this example shows that the Plurality Run-Off method can violate the Condorcet criterion.

14. Consider the following preference table for 70 voters.

	Voter Profiles			
	30	23	10	7
First	A	B	D	C
Second	D	D	A	D
Third	B	A	C	A
Fourth	C	C	B	B

(a) Determine the winner using the Instant Run-Off method.

(b) Is there a candidate who defeats every other candidate in all one-on-one contests?

(c) Explain why this example shows that the Instant Run-Off method can violate the Condorcet criterion.

15. Consider again the preference table for 46 voters shown in Exercise 12.

   (a) Determine the winner using the Borda Count method.

   (b) Is there a candidate who defeats every other candidate in all one-on-one contests?

   (c) Explain why this example shows that the Borda Count method can violate the Condorcet criterion.

Satisfying the Fairness Criteria

16. Explain why the Plurality method must satisfy the Majority criterion for any preference table.

17. Explain why the Instant Run-Off method must satisfy the Majority criterion for any preference table.

18. Explain why the Pairwise Comparison method must satisfy the Majority criterion for any preference table.

19. Explain why the Plurality method must satisfy the Unanimity criterion for any preference table.

20. Explain why the Sequential Run-Off method with any agenda must satisfy the Condorcet criterion for any preference table.

21. Explain why the Pairwise Comparison method must satisfy the Condorcet criterion for any preference table.

22. We want to study the Pairwise Comparison method and the Monotonicity criterion by using the following preference table.

	Voter Profiles			
	8	7	4	2
First	D	A	C	A
Second	B	E	B	E
Third	E	C	A	D
Fourth	A	D	D	B
Fifth	C	B	E	C

   (a) Verify that A is the Pairwise Comparison winner.

   (b) Now suppose the eight voters in the first profile elevate A (without changing the relative ranking of the other candidates) by changing their preferences to D > A > B > E > C. How do the points (number of one-on-one wins) for A change? How do the points (number of one-on-one wins) for the other candidates change? Verify that A is still the Pairwise Comparison winner.

   (c) More generally, show that the Pairwise Comparison method will satisfy the Monotonicity criterion for any preference table.

23. We want to study the Borda Count method and the Monotonicity criterion by using the following preference table.

	Voter Profiles			
	8	5	4	2
First (3)	A	C	B	B
Second (2)	B	A	A	D
Third (1)	C	B	D	C
Fourth (0)	D	D	C	A

(a) Compute the Borda points for all candidates and verify that A is the Borda Count winner.

(b) Now suppose the two voters in the last profile elevate A (without changing the relative ranking of the other candidates) by changing their preferences to B > A > D > C. How do the new Borda points change? Verify that A is still the Borda Count winner.

(c) More generally, show that the Borda Count method will satisfy the Monotonicity criterion for any preference table.

Strategic Voting

24. Consider the following preference table for nine voters.

	Voter Profiles			
	3	3	2	1
First	A	B	C	B
Second	B	C	A	A
Third	C	A	B	C

(a) Determine the winner using the Plurality Run-Off method.

(b) Now suppose that all voters in the second profile insincerely modify their ranking to C > B > A. Who is then the Plurality Run-Off winner?

(c) Explain why this example demonstrates that Plurality Run-Off is vulnerable to strategic voting.

25. Consider the following preference table for 28 voters.

	Voter Profiles			
	11	8	7	2
First	A	C	B	D
Second	B	D	C	C
Third	D	A	D	B
Fourth	C	B	A	A

(a) Determine the winner using the Instant Run-Off method.

(b) Now suppose that all voters in the first profile insincerely modify their ranking to B > A > D > C. Who is then the Instant Run-Off winner?

(c) Explain why this example demonstrates that Instant Run-Off is vulnerable to strategic voting.

26. Consider the following preference table for five voters.

	Voter Profiles		
	2	2	1
First	A	B	C
Second	B	C	A
Third	C	A	B

(a) Determine the Sequential Run-Off method winner using the agenda [A, B, C].

(b) Now suppose that all voters in the first profile insincerely modify their ranking to B > A > C. Using the same agenda [A, B, C], who is then the Sequential Run-Off winner?

(c) Explain why this example demonstrates that Sequential Run-Off is vulnerable to strategic voting.

Project

27. Section 18.2 introduced the idea of a social ranking method, one that enables all candidates to be ranked. The Borda Count method and the Pairwise Comparison method are easily seen to produce numerical rankings of all candidates. What about the other voting methods studied here? Investigate the topic of *recursive rankings* [OR2] and apply this technique to several of the examples presented in this chapter.

Bibliography

[Brams 2007] Brams, S., and Fishburn, P. C. 2007. *Approval voting*, second edition, Chapter 1. New York: Springer.

[Poundstone 2008] Poundstone, W. 2008. *Gaming the vote: Why elections aren't fair (and what we can do about it)*. New York: Hill & Wang.

[Robinson 2011] Robinson, E. A., and Ullman, D. H. 2011. *A mathematical look at politics*, Chapters 3 and 4. Boca Raton, FL: CRC Press.

[Saari 2001] Saari, D. G. 2001. *Chaotic elections!: A mathematician looks at elections*. Providence, RI: American Mathematical Society.

[OR1] http://www.princeton.edu/~achaney/tmve/wiki100k/docs/Duverger_s_law. html (accessed February 3, 2014).

[OR2] www.ctl.ua.edu/math103/voting/recursiv.htm (accessed February 3, 2014).

Chapter 19

Weighted Voting Systems and Voting Power

The greater the power, the more dangerous the abuse.

—Edmund Burke, Irish statesman and philosopher, 1729–1797

Chapter Preview

This chapter studies voting systems in which the preferences of participants are not treated equally. Examples of such weighted voting systems are found in local, national, and global decision-making organizations. Typically, these systems are used to decide whether to pass a particular measure or motion, based on reaching or exceeding a particular threshold number of votes. In weighted voting systems, it is appropriate and more meaningful to measure the influence or power of each participant. This is illustrated using the Banzhaf Power Index. Significantly, we see that power need not proportionally reflect the voting weight allocated to each participant.

19.1 Weighted Voting Systems

In the voting systems considered in Chapters 17 and 18, each individual voter casts a single ballot and can influence the election the same as any other voter. Even in approval voting, though one voter may approve of more candidates than another, all voters have the same options. However, there are several situations in which some voters are allocated more votes than others. One such example occurs in the corporate setting, in which shareholders have a number of votes equal to the number of shares held in the company. Other examples are found in the political arena: the Electoral College (where U.S. states cast varying numbers of votes for President depending on their representation in the House and the Senate), legislative decisions in multi-party European countries, and the United Nations Security Council.

As noted in Chapter 18, various paradoxes arising in voting systems disappear when there are only two candidates. As a result, we concentrate here on such a case—where there is a single motion to be decided upon and there are only two choices (support or not). The motion will pass when it receives a certain minimum number of the votes cast (e.g., a majority). This situation is termed a *weighted voting system.*

DEFINITION **Weighted Voting System**
A decision-making process in which each participant has a known *weight* (number of votes). A motion will pass when it receives at least a specified number of votes (called the *quota*).

It is also important to have a way of naming a set of participants all voting in the same way relative to a particular motion.

DEFINITION **Coalition**
A coalition C consists of all participants voting for a motion. The weight $w(C)$ of coalition C is the total number of votes represented by the individuals in C.

DEFINITION **Winning and Losing Coalitions**
The coalition C is a winning coalition if its weight $w(C)$ is at least the quota q. The coalition C is a losing coalition if its weight $w(C)$ is less than the quota q.

Example 19.1 Formulate a Weighted Voting System for Shareholders in a Corporation

Greenback Corporation has four shareholders A, B, C, D holding 5, 3, 2, 1 votes, respectively. A proposal to change the investment strategy of the corporation has been put forward for a vote. In order for this proposal to pass, a majority of affirmative votes must be cast. (a) What is the quota for this weighted voting system? (b) Classify the following coalitions as either winning or losing: {A}; {A, C}; {B, D}; and {B, C, D}.

Solution

(a) The total number of votes is $5 + 3 + 2 + 1 = 11$, so a majority is 6 votes; thus, $q = 6$.

(b) $w(\{A\}) = 5 < q$ so {A} is a losing coalition; $w(\{A, C\}) = 7 \geq q$ so {A, C} is a winning coalition; $w(\{B, D\}) = 4 < q$ so {B, D} is a losing coalition; $w(\{B, C, D\}) = 6 \geq q$ so {B, C, D} is a winning coalition. ◁

Example 19.2 Analyze a Revised Voting System for Shareholders in a Corporation

Suppose that Shareholder A in Example 19.1 now is allocated seven votes, with B, C, D maintaining the same number of votes 3, 2, 1. Again, a majority of affirmative votes will be needed for a new investment policy to pass. (a) What is the quota for this weighted voting system? (b) Discuss the role of Shareholder A in this weighted voting system.

Solution

(a) The total number of votes is $7 + 3 + 2 + 1 = 13$, so a majority is $q = 7$.

(b) Notice that 7, the weight of A, is at least as large as the quota q. This means that A determines whether any proposal passes, regardless of the voting choices of the other shareholders. In other words, we have a *dictatorship*. ◁

Example 19.3 Represent Voting in the UN Security Council as a Weighted Voting System

The United Nations Security Council consists of five permanent members and 10 regular members (serving two-year terms). In order for a measure to be passed by the Security Council, it must be approved by at least nine members, including all permanent members. Express this voting procedure as a weighted voting system.

Solution

Since all the regular members are treated equally, let's assign them a weight of 1. The coalition R consisting of all regular members then has weight $w(R) = 10$. However, the coalition R cannot pass a measure by itself, suggesting that we give each permanent member a weight of 11. The coalition P of all permanent members then has weight $w(P) = 55$. In order for a measure to pass, it must receive the votes of all permanent members and at least four more regular members; this gives a quota of $q = w(P) + 4 = 59$. In summary, we have a weighted voting system with 15 participants, 10 having weight 1 and five having weight 11, and with a quota of 59. ◁

(Related Exercises: 1–5)

As illustrated in the previous examples, a coalition C is winning if $w(C) \geq q$ while it is losing if $w(C) < q$. Sometimes a winning coalition can become a losing coalition if a single member of the coalition defects to the other side and leaves the coalition. Such a situation defines a *critical* voter for that coalition.

DEFINITION **Critical Voter**
A critical voter for a winning coalition C is a member of C, who by leaving coalition C, makes it a losing coalition.

Example 19.4 Identify Critical Voters in a Coalition

Consider again a corporation with four shareholders A, B, C, D having 5, 3, 2, 1 votes and a quota of $q = 6$. (a) Find the critical voters for the coalition {A, B, D}. (b) Find the critical voters for the coalition {B, C, D}.

Solution

(a) $w\{A, B, D\}) = 9 \geq q$ so $K = \{A, B, D\}$ is a winning coalition. If A leaves K, then $w(\{B, D\}) = 4 < q$ so $\{B, D\}$ is a losing coalition and A is critical to coalition K. If B leaves K, then $w(\{A, D\}) = 6 \geq q$ so $\{A, D\}$ is still winning and B is not critical to coalition K. If D leaves K, then $w(\{A, B\}) = 8 \geq q$ so $\{A, B\}$ is still winning and D is not critical to coalition K. In summary, only A is critical to the winning coalition {A, B, D}.

(b) $w(\{B, C, D\}) = 6 \geq q$ so $K = \{B, C, D\}$ is a winning coalition. If B leaves K, then $w(\{C, D\}) = 3 < q$ so $\{C, D\}$ is a losing coalition and B is critical. If C leaves K, then $w(\{B, D\}) = 4 < q$ so $\{B, D\}$ is losing and C is critical. If D leaves K, then $w(\{B, C\}) = 5 < q$ so $\{B, C\}$ is losing and D is critical. In summary, each member of {B, C, D} is critical to the winning coalition {B, C, D}. ◁

(Related Exercises: 6–10)

19.2 Banzhaf Power Index

In the one person–one vote systems studied in Chapters 17 and 18, each individual voter can exert the same influence on the outcome of the election. However, in weighted voting systems this is not the case. In this section, we study the influence of each participant in a weighted voting system. We discuss one popular method of measuring the power of voters in a weighted voting system. It is based on the ideas of coalitions and critical voters to a coalition introduced in Section 19.1.

DEFINITION **Banzhaf Power Index**

This measure calculates a voter's power based on the number of coalitions in which that voter is critical. When divided by the total number of times that all participants are critical, this gives the Banzhaf Power Index (BPI).

HISTORICAL CONTEXT **John F. Banzhaf III (1940–)**

An American law professor who introduced the power index in a 1965 article for the *Rutgers Law Journal*. This measure of power was originally invented by Roger Penrose (English mathematician) in 1946, and later reinvented by James S. Coleman (American sociologist) in 1971.

Recall that a critical voter can change a winning coalition into a losing coalition. The BPI is based on first counting the number of winning coalitions in which voters are critical and then normalizing these counts. We illustrate this computation in the following example.

Example 19.5 Calculate the Banzhaf Power Index for Shareholders in a Company

A small startup company has shareholders A, B, C with 8, 5, 3 shares, respectively, and a majority of shares is needed for a proposal to pass. (a) List all the possible coalitions and indicate which are winning coalitions (W) and which are losing (L). (b) Identify the voters in each winning coalition that are critical. (c) Calculate the BPI for each voter.

Solution

(a) Table 19.1 lists all possible coalitions using a truth-table (Section 8.2), where T indicates that a shareholder belongs to the coalition and F indicates that the shareholder does not. With three shareholders there will be $2 \times 2 \times 2 = 2^3 = 8$ possible coalitions; this is an application of the Multistage Rule (from Chapter 12). We abbreviate coalitions such as {A, C} by using AC instead. With 16 shares in total, the quota is $q = 9$ so the winning coalitions are ABC, AB, AC.

(b) If A leaves the winning coalition ABC, then BC is losing since it has weight $8 < q$. However, if either B or C leaves, the resulting coalition is still winning (having weight 11 or 13, respectively). Using similar logic, Table 19.2 lists each winning coalition and its critical voters.

Since A is critical to three winning coalitions, the Banzhaf count for A is 3; B and C are each critical to just one coalition, so they have a Banzhaf count of 1. The total of all Banzhaf

TABLE 19.1: All coalitions with three voters

A	B	C	Coalition	Weight	Type
T	T	T	ABC	16	W
T	T	F	AB	13	W
T	F	T	AC	11	W
T	F	F	A	8	L
F	T	T	BC	8	L
F	T	F	B	5	L
F	F	T	C	3	L
F	F	F	—	0	L

TABLE 19.2: Winning coalitions and critical voters

Coalition	Weight	Critical Voters
ABC	16	A
AB	13	A, B
AC	11	A, C

counts is $3 + 1 + 1 = 5$, so the BPI for A is $3/5 = 0.6$. Similarly, B and C have a BPI of $1/5 = 0.2$. Notice that the sum of the BPI values for all voters is $0.6 + 0.2 + 0.2 = 1$; this is the reason we normalize the Banzhaf counts. ◁

(Related Exercises: 11–15)

Example 19.5 shows that even though B has almost twice the number of shares as C, they both have the same BPI and thus the same power in influencing outcomes. This example also shows that A has three times the power as either of the other two shareholders, even though A has less than three times the number of shares of either B or C. In general, the power of participants need not be proportional to the number of votes they are allocated.

We have calculated the BPI by considering a voter A to be critical whenever A is part of a winning coalition W, but that coalition becomes a losing coalition L once A leaves. Notice however that should voter A join the losing coalition L, then it becomes a winning one W. This means that we should get the same Banzhaf counts if we found all *losing* coalitions and then determined critical voters as those that would change losing coalitions into winning ones by joining. We illustrate this by recalculating the BPI values for Example 19.5.

Example 19.6 Calculate the BPI for Shareholders Using Losing Coalitions

Recall from Example 19.5 that A, B, C have 8, 5, 3 shares, respectively, and the quota for passage of a measure is $q = 9$. (a) List all the possible losing coalitions. (b) Identify the voters that are critical to each coalition. (c) Calculate the BPI for each voter based on criticality to losing coalitions.

Solution

(a) From Table 19.1, we can extract the losing coalitions: those with weight strictly less than $q = 9$. These five losing coalitions are listed in Table 19.3.

(b) For each losing coalition, Table 19.4 lists the associated critical voters (those who would make the coalition winning if they join). For example, coalition A is losing since it has

TABLE 19.3: All losing coalitions with three voters

Coalition	Weight	Type
A	8	L
BC	8	L
B	5	L
C	3	L
—	0	L

weight $8 < 9$; if either B or C join this coalition its weight will be increased to either 13 or 11, both at least as large as q.

TABLE 19.4: Losing coalitions and critical voters

Coalition	Weight	Critical Voters
A	8	B, C
BC	8	A
B	5	A
C	3	A
—	0	—

(c) Voter A is critical for three losing coalitions, while both B and C are critical for only one. So the BPI for A is 3/5 while B and C both have a BPI of 1/5, just as seen earlier. ◁

(Related Exercises: 16–17)

Example 19.7 Calculate the BPI for a Three-Party Legislature

A legislature is composed of three political parties A, B, and C. Recent elections gave seats in the legislature to 18 members of Party A, 17 of Party B, and 3 of Party C. (a) Assuming that all members of a party vote the same on a proposal for increasing taxes, calculate the BPI values if a majority of votes is needed to pass this proposal. (b) Comment on the results found in (a).

Solution

(a) There are 38 total votes, so a majority is $q = 20$ votes. Table 19.5 lists the winning coalitions, their weights, and associated critical voters. We see that each voting group is critical to exactly two winning coalitions, so A, B, and C all have a BPI equal to $2/6 = 1/3$.

TABLE 19.5: Winning coalitions and critical voters

Coalition	Weight	Critical Voters
ABC	38	—
AB	35	A, B
AC	21	A, C
BC	20	B, C

(b) Even though Party C has one-sixth of the votes as Party A, it still has the same influence in passing the proposal, as measured by the BPI. ◁

Example 19.8 Calculate the BPI for a Multi-District Council

A county council has representatives from the four districts A, B, C, D, which are allocated 7, 7, 5, 2 votes based on population. (a) Calculate the BPI values for the districts if a majority of votes is needed to pass a motion. (b) Calculate the BPI values if two-thirds of the votes are needed to pass a motion.

Solution

(a) There are 21 total votes, so a majority is $q = 11$ votes. Table 19.6 shows the winning coalitions, their weights, and the associated critical voters. We see that A, B, C are each critical to four coalitions, while D is critical to none. As a result, A, B, C each have a BPI of $4/12 = 1/3$, whereas D has a BPI of 0. In other words, D has absolutely no influence on the outcome even though it has two votes.

TABLE 19.6: Winning coalitions and critical voters for $q = 11$

Coalition	Weight	Critical Voters
ABCD	21	—
ABC	19	—
ABD	16	A, B
ACD	14	A, C
AB	14	A, B
BCD	14	B, C
AC	12	A, C
BC	12	B, C

(b) There are 21 total votes, so the two-thirds requirement means a quota of $q = 14$ votes. Table 19.7 lists the winning coalitions, their weights, and the associated critical voters. So A and B are each critical to four coalitions, while C and D are each critical to two coalitions. As a result, A and B have a BPI of $4/12 = 1/3$, whereas C and D have a BPI of $2/12 = 1/6$. By increasing the quota from 11 to 14, D now has power equal to that of C. ◁

(Related Exercises: 18–20)

TABLE 19.7: Winning coalitions and critical voters for $q = 14$

Coalition	Weight	Critical Voters
ABCD	21	—
ABC	19	A, B
ABD	16	A, B
ACD	14	A, C, D
AB	14	A, B
BCD	14	B, C, D

In Example 19.8, A and B have the same number of votes (7), so they should end up with the same amount of power—which they do. For purposes of calculating the BPI we can then treat A and B as identical. This idea, together with our previous counting formulas (Chapter 13), can help to simplify the calculation of power indices in more complex settings.

Example 19.9 Calculate the BPI for a Council with a Chair and Four Members

The chair of an Academic Council is allocated two votes, and each of the other four members has a single vote. The quota for adopting a new academic regulation is $q = 4$. (a) Determine all winning coalitions. (b) Count the number of winning coalitions in which the chair is critical. (c) Count the number of winning coalitions in which each ordinary member is critical. (d) Calculate the BPI for the chair and for each ordinary member.

Solution

(a) One winning coalition consists of all four ordinary members, denoted MMMM. There are three types of winning coalitions that contain the chair: PMM, PMMM, PMMMM. For example, PMM denotes the chair plus two of the ordinary members. We can apply the Combinations Formula (Section 13.1) to count the number of such coalitions, since the chair can be paired with two ordinary members in $C(4, 2) = 6$ ways. The other entries in Table 19.8 are found in the same way, giving $1 + 6 + 4 + 1 = 12$ winning coalitions.

TABLE 19.8: Counting winning coalitions in a committee

	Winning Coalition	Coalition Weight	# of Winning Coalitions
(1)	MMMM	4	1
(2)	PMM	4	$C(4, 2) = 6$
(3)	PMMM	5	$C(4, 3) = 4$
(4)	PMMMM	6	$C(4, 4) = 1$

(b) The chair is critical in situations (2) and (3) from Table 19.8, since removal of P from those winning coalitions leaves at most three votes. So P is critical in $6 + 4 = 10$ coalitions.

(c) Let's now focus on a particular member A, who will be critical in situations (1) and (2) from Table 19.8. There is only one coalition ABCD of type (1), where A–D are the ordinary members, so A is critical to this single coalition. For (2), A must be a member of the coalition, which then has the form PAB; since B can be chosen from the other three ordinary members, there are $C(3, 1) - 3$ coalitions in (2) that contain A. In summary, A is critical in $1 + 3 = 4$ coalitions.

(d) The total count of critical voters is $10 + (4 \times 4) = 26$, since the four ordinary members are all treated the same. As a result, the chair has a BPI of $10/26$ and each ordinary member has a BPI of $4/26$. Notice that the chair has power that is $10/4 = 2.5$ times that of any ordinary member, even though the chair has only twice the number of votes as an ordinary member. ◁

(Related Exercises: 21–23)

19.3   Decision Trees

To calculate the BPI, we needed to identify winning coalitions and then count the number of times each voter will be critical to a winning coalition. Unfortunately, the number of winning coalitions can grow rapidly. By using the information that certain voters are identical, this computational task can be lessened, as seen in Example 19.9: we did not need to explicitly list the $2^5 = 32$ possible coalitions and then determine the 12 winning coalitions. The next example shows how to compute the number of times a voter is critical in a systematic way, by appealing to our old friend, the *decision tree*.

Example 19.10 Use a Decision Tree to Determine Coalitions in Which a Voter is Critical

The Rockland County Board of Supervisors has a total of 18 representatives from the five towns in the county. Table 19.9 shows the number of representatives elected from each town, which is based on the size of the town. Calculate the number of winning coalitions in which Ramapo (R) is critical, using a quota of $q = 10$ votes (a majority of the 18 total votes).

TABLE 19.9: Populations and representatives for five towns

Town	Population	Representatives
Stony Point (S)	12,114	1
Haverstraw (H)	23,676	2
Orangetown (O)	52,080	4
Clarkstown (C)	57,883	5
Ramapo (R)	73,051	6
Total	218,804	18

Solution

For R to be a critical voter, it must first be part of a winning coalition. The other members of a winning coalition with R will be denoted by X. Since R has weight 6 and $q = 10$, the set X must have weight at least 4 to form a winning coalition with R. Moreover, if the weight of X were 10 or more, then the coalition X itself (without R) would be a winning coalition and R would not be critical. We now indicate how a decision tree can be used to effectively eliminate all coalitions X that have weight < 4 or weight ≥ 10.

In the tree displayed in Figure 19.1, the weights 6, 5, 4, 2, 1 correspond to towns R, C, O, H, S. The root of the tree is then labeled 6, to indicate that we are only looking at coalitions containing R. The tree will evolve in a way similar to that used in change-making problems (Section 2.2). First, the vertices below 6 correspond to the other voters that can be added to form a coalition X: namely, 5, 4, 2, and 1. Beneath each of these vertices we list any additional vertices of smaller weight that can be added to form a new coalition, without repeating a previous vertex. For example, the vertices listed below 5 are 4, 2, and 1 while those below 4 are 2 and 1.

Each path in the tree from the root then corresponds to a coalition containing R. For example, the path $6 \to 4 \to 2$ indicates that $X = \{O, H\}$ has been added to R, forming the coalition labeled ROH. Now we trim off some of the branches of this tree. Specifically the label **L** signifies that the coalition X has weight < 4 (too low) while **H** signifies that the coalition X has weight ≥ 10 (too high). Once these parts of the tree have been removed, we have just nine vertices remaining, and these correspond to the nine coalitions in which R is critical. To illustrate, the path $6 \to 4 \to 2$ from the root to vertex 2 at level 3 represents the winning coalition ROH in which R is critical. By tracing paths from the root vertex, we can then systematically list all coalitions in which R is critical. ◁

Similar calculations can be made using decision trees to determine that C and O are critical to seven coalitions, while H and S are critical to one coalition. Notice that we previously

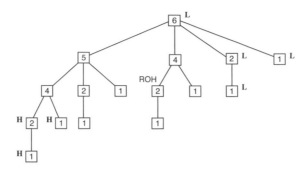

FIGURE 19.1: Decision tree representation for the critical voter R

needed to list each winning coalition and then determine which voters are critical to that coalition. The decision tree approach inverts this logic: we first select a voter and then determine those (winning) coalitions for which that voter is critical.

(*Related Exercises: 24–27*)

19.4   Applications of the Banzhaf Power Index

In this section we discuss the application of the BPI to several real-world situations found in government and politics. Measuring the power of the individual participants may be more revealing than simply counting the number of votes (weight) they possess.

Board of Supervisors

John Banzhaf introduced the use of power indices in a series of court cases involving the Nassau County (New York) Board of Supervisors. Nassau County was composed of six districts with the number of votes proportional to the population; see Table 19.10. A simple majority of the 115 total votes ($q = 58$) was needed for a measure to pass. The last column of Table 19.10 shows the computed BPI values. Using these power measures, Banzhaf argued that the Nassau County voting scheme was unfair since the three smaller districts would have no influence on the outcome of votes. Quite surprisingly, Oyster Bay had no power even though it was allocated 21 votes.

TABLE 19.10: BPI values for six districts in Nassau County

District	Weight	BPI
Hempstead #1	31	1/3
Hempstead #2	31	1/3
North Hempstead	28	1/3
Oyster Bay	21	0
Glen Cove	2	0
Long Beach	2	0

United Nations Security Council

As discussed in Example 19.3, voting in the United Nations Security Council can be viewed as a weighted voting system where the five permanent members have weight 11 and the 10 regular members have weight 1. A quota of $q = 59$ ensures that a measure will pass when it is approved by at least nine members, including all permanent members. Computing the BPI for the two types of participants gives a BPI of $212/1270 = 0.1669$ for each permanent member and $21/1270 = 0.0165$ for each regular member. Each permanent member has power that is $212/21 = 10.095$ times the power of each regular member, slightly less than the corresponding ratio of the number of votes (11).

Scottish Parliament

The Scottish Parliament consists of five political parties: the Scottish National Party, the Labour Party, the Conservative and Unionist Party, the Scottish Liberal Democrats, and the Scottish Green Party. Table 19.11 shows the number of Parliament members in each party (as of 2010).

TABLE 19.11: BPI values for the Scottish Parliament

Party	Members	BPI
Scottish National	46	0.36
Labour	44	0.28
Conservative and Unionist	20	0.28
Liberal Democrats	17	0.04
Green	1	0.04

If we assume that the representatives of each party all vote the same as a bloc, then this can be viewed as a weighted voting system with the weights indicated. In order for a measure to pass the Parliament, it must receive a majority of the 128 total votes, meaning a quota of $q = 65$. The calculated BPI values are indicated in the last column of Table 19.11. Rather surprisingly, the influence of the Scottish National Party (measured by its BPI) is much larger than one would expect from the extra two seats it has relative to the Labour Party. Similarly, the Liberal Democrats only have the same influence as the Green Party, despite the fact that they have 17 times the representation of the Green Party. This example clearly indicates the difference between the number of legislative seats and the power each party can exert.

Electoral College and the U.S. Presidential Election

Election of the U.S. President can be seen as a weighted voting system, with 56 participants: the 50 states, the District of Columbia, three congressional districts in Nebraska, and two congressional districts in Maine. Each state has as many electors as it has Senators and Representatives in the Congress. For example, Delaware has three electoral votes, California has 55, and Washington, D.C. is given three electors, which is equal to the number held by the smallest states. The Electoral College refers to the 538 electors who officially choose the President and the Vice President. In order to be elected, a candidate must have a 270-vote majority of the Electoral College. If no candidate for President wins a majority of the electoral votes, the choice is given to the House; if no candidate for Vice President wins a majority of the electoral votes, the choice is given to the Senate.

The BPI of each U.S. state can be calculated based on its number of electoral votes. Table 19.12 shows a sample of BPI values (converted to percentages) using electoral votes determined by the 2000 Census. Notice that states (such as Kentucky and South Carolina) with the same number of electoral votes have the same power index. Also, an increase in the number of electoral votes generally results in an increase in the power index. For example, after the 1990 Census, Georgia had 13 electoral votes and a power index of 2.376%, which increased to 15 electoral votes and a power index of 2.744% in 2000.

TABLE 19.12: BPI values for the Electoral College

State	Electoral Votes	BPI (%)	Power per Electoral Vote
CA	55	11.407	0.207
TX	34	6.393	0.188
NY	31	5.795	0.187
GA, NC, NJ	15	2.744	0.183
KY, SC	8	1.457	0.182
ME, NE	2	0.364	0.182
Each Cong. Dist.	1	0.182	0.182

The last column of Table 19.12 shows the BPI divided by the number of electoral votes. If power were allocated in a fair and ideal way, we might expect that the "power per vote" should be approximately the same for all states. This does hold for the smaller states, but the largest states get a much larger proportional amount of power for each of their electoral votes. In other words, there seems to be a bias that favors large states.

As a final note, it has been estimated, using the BPI, that the U.S. President has 40 times the power of a Senator and 175 times the power of a Representative. Yet Congress as a whole has 2 1/2 times the power of the President.

European Union

To conclude this section, we apply the idea of power indices to the European Union, which consisted of 15 countries in 2004. (It has since expanded to 28 member states.) The larger weights go to the more populated countries within the Union. For a motion to be passed, the European Union requires a coalition to have at least 62 votes out of the total number of 87 votes, which means at least eight of the member states must support a proposition for it to pass. Table 19.13 shows the number of votes accorded to each of the 15 countries of the 2004 European Union and their respective BPI values using $q = 62$.

TABLE 19.13: BPI values for the 2004 European Union

Country	Votes	BPI % $(q = 62)$	BPI % $(q = 44)$
France, Germany, Italy, UK	10	11.16	11.72
Spain	8	9.24	9.14
Belgium, Greece, Netherlands, Portugal	5	5.87	5.61
Austria, Sweden	4	4.79	4.69
Denmark, Finland, Ireland	3	3.59	3.32
Luxembourg	2	2.26	2.20

It is interesting to see what effect the quota has on these power indices. Suppose the quota had simply been a strict majority of the votes: that is, 44 out of 87. The last column of Table 19.13 shows the resulting BPI values. Notice that all member countries except for the very largest ones (France, Germany, Italy, and the UK) would have their power reduced. This suggests that the choice of quota ($q = 62$) might have been made to ensure that the smaller countries would have increased power, as measured by the BPI.

(Related Exercises: 28–29)

Chapter Summary

This chapter has studied weighted voting systems, in which participants cast varying numbers of votes in favor of a given motion. Using the ideas of a coalition and those voters critical to a coalition, the concept of the Banzhaf Power Index (BPI) is formulated. This index measures the influence of each participant to change the outcome of an election, by seeing how often that participant is critical to a winning coalition. As well as using a tabular representation, it proved useful to employ decision trees to compute the BPI values. Several real-world examples of weighted voting systems and their associated power indices are presented. The surprising result is that this measure of voting power can be highly disproportional relative to the number of votes allocated to each participant.

Applications, Representations, Strategies, and Algorithms

Applications			
Business (1–2, 4–6)	Society/Politics (3, 7–8, 10)		Education (9)

Representations			
Graphs	Tables		Sets
Tree	Data	Decision	Lists
10	10	5–9	1–9

Strategies		
Solution-Space Restrictions	Rule-Based	Composition/Decomposition
10	1–9	9

Algorithms			
Exact	Enumeration	Sequential	Numerical
1–10	10	3	1–10

Exercises

Weighted Voting Systems

1. A committee to select nominees for a national award consists of two college Deans and four faculty members. In order to be nominated by the committee, a candidate needs to be approved by both Deans and at least two of the faculty members. Represent this situation as a weighted voting system by specifying appropriate weights for the committee members and a quota.

2. A student organization consists of a president, vice president, treasurer, and secretary. In order for a motion to pass, the president must be in favor (i.e., the president has absolute veto power). However, in addition to the president's support, there must be a majority of approving votes among the remaining three officers.

   (a) Represent this situation as a weighted voting system by specifying appropriate weights for all four officers as well as a quota.

   (b) Suppose instead that the president acts as a dictator. Represent this situation as a weighted voting system by specifying appropriate weights for all four officers as well as a quota.

3. Recall Example 19.3 (the UN Security Council), with five permanent members and 10 regular members. Passage of a measure requires the approval of at least nine members, including all permanent members.

   (a) Suppose that all regular members receive one vote and all permanent members receive six votes. Show that there is no quota q that satisfies the requirements of the problem.

   (b) Suppose that all regular members receive one vote and all permanent members receive seven votes. Find a quota q that satisfies the requirements of the problem.

4. A corporate Board of Directors consists of a chair and six regular board members. In order for a new policy to be adopted, it must be supported by the chair and at least three of the regular board members, or by at least five of the regular board members. Represent this situation as a weighted voting system by specifying appropriate weights for the chair and the six regular members as well as a quota.

5. The student government at a school consists of nine senators, three of whom are on the finance committee. In order for a budget motion to pass, the finance committee must approve of the motion by a unanimous vote. The motion then goes to the senate, which must pass the motion by a majority vote of all nine senators; we assume that the three members on the finance committee do not change their votes. Represent this situation as a weighted voting system by specifying appropriate weights for the nine senators as well as a quota. [*Hint*: Assign the weight of 1 to all six senators who are not on the finance committee.]

6. Consider the weighted voting system in which participants A, B, C have 21, 15, 10 votes, respectively. Suppose that the quota is 31 votes for a measure to pass.

(a) List all the coalitions and compute their weight.

(b) Determine which are winning and which are losing coalitions.

(c) For each winning coalition, list the critical voters.

7. A corporation's Executive Board consists of the Chief Executive Officer (CEO), Chief Financial Officer (CFO), and Chief Operating Officer (COO). In deciding on strategic actions, these officers have 9, 7, and 3 votes, respectively. Suppose that the quota is 10 votes for a proposed strategic action to be approved.

(a) List all the coalitions and compute their weight.

(b) Determine which are winning and which are losing coalitions.

(c) For each winning coalition, list the critical voters.

8. Consider the weighted voting system in which participants A, B, C, D have 29, 22, 12, 8 votes, respectively. Suppose that the quota is 49 votes for a measure to pass.

(a) List all the coalitions and compute their weight.

(b) Determine which are winning and which are losing coalitions.

(c) For each winning coalition, list the critical voters.

9. A meeting of the White Council consists of Saruman the White, Gandalf the Grey, Radagast the Brown, Galadriel, and Elrond. They are voting on whether to make Saruman head of the White Council (consequently, he is not voting). Consider the weighted voting system in which Gandalf the Grey, Radagast the Brown, Galadriel, and Elrond have 15, 10, 18, and 26 votes, respectively. At least 42 affirmative votes are needed for Saruman to become head of the White Council.

(a) List all the coalitions and compute their weight.

(b) Determine which are winning and which are losing coalitions.

(c) For each winning coalition, list the critical voters.

10. In the context of Exercise 9, suppose that it is now decided that only 40 affirmative votes are needed for Saruman to become head of the White Council.

(a) Determine which are winning coalitions and which are losing coalitions.

(b) For each winning coalition, list the critical voters.

Banzhaf Power Index

11. Compute the BPI for each voter in the system described in Exercise 6, using voters who are critical to winning coalitions.

12. Compute the BPI for each voter in the system described in Exercise 7, using voters who are critical to winning coalitions.

13. Compute the BPI for each voter in the system described in Exercise 8, using voters who are critical to winning coalitions.

14. Compute the BPI for each voter in the system described in Exercise 9, using voters who are critical to winning coalitions.

15. Compute the BPI for each voter in the system described in Exercise 10, using voters who are critical to winning coalitions.

16. For each losing coalition in the system described in Exercise 6, list the critical voters. Use this information to compute the BPI for each voter.

17. For each losing coalition in the system described in Exercise 8 list the critical voters. Use this information to compute the BPI for each voter.

18. A small company has shareholders A, B, C holding 10, 7, 4 shares, respectively. A quota of $q = 11$ shares will be needed for a proposal to pass.

    (a) List all the winning coalitions and their weight.

    (b) For each winning coalition, list the critical shareholders.

    (c) Calculate the BPI for each shareholder.

    (d) Answer the above questions if the quota is raised to $q = 12$.

19. A county council consists of four members A, B, C, D having voting weights 11, 7, 5, 3, respectively. A majority of the total votes is needed for a motion to pass.

    (a) What is the quota q for this system?

    (b) List all the winning coalitions and their weights.

    (c) For each winning coalition, list the critical members.

    (d) Calculate the BPI for each member.

20. Consider a weighted voting system in which participants A, B, C, D, E have weights 6, 4, 1, 1, 1, respectively, and $q = 9$.

    (a) Calculate the BPI values for all participants.

    (b) Suppose that A gives one of his votes to B, resulting in the new set of weights 5, 5, 1, 1, 1. Now calculate the BPI values for all participants, again using $q = 9$.

    (c) Describe the paradox that seems to arise in this situation.

21. Consider a modification of Example 19.9, where now the Academic Council consists of the chair (P) and five ordinary members (M). The chair is allocated two votes, while each ordinary member has a single vote. The quota for adopting a new regulation is $q = 5$.

    (a) Using the symbols P and M, list all the winning coalitions and their weight.

    (b) Count the number of winning coalitions in which P is critical.

    (c) Count the number of winning coalitions in which an ordinary member A is critical.

    (d) Calculate the BPI for each person on the committee.

    (e) How does the power of the chair compare with the power of an ordinary member?

22. A homeowner's association allocates three votes to each of the three regular board members (M), while the chair (P) is given five votes. The quota for a proposed amendment to pass is $q = 8$.

    (a) Using the symbols P and M, list all the winning coalitions and their weight.

(b) Count the number of winning coalitions in which P is critical.

(c) Count the number of winning coalitions in which a regular member A is critical.

(d) Calculate the BPI for each person on the committee.

(e) How does the power of the chair compare with the power of a regular member?

23. In the context of Exercise 22, suppose that the quota is lowered to $q = 6$.

(a) Using the symbols P and M, list all the winning coalitions and their weight.

(b) Count the number of winning coalitions in which P is critical.

(c) Count the number of winning coalitions in which a regular member A is critical.

(d) Calculate the BPI for each person on the committee.

(e) How does the power of the chair compare with the power of a regular member?

Critical Voters and Decision Trees

24. In Example 19.10, we used a decision tree to find all the winning coalitions in which the town of Ramapo (R) is critical. Using a decision tree with $q = 10$, determine all winning coalitions in which Clarkstown (C) is critical.

25. In Example 19.10, we used a decision tree to find all the winning coalitions in which the town of Ramapo (R) is critical. Using a decision tree with $q = 10$, determine all winning coalitions in which Orangetown (O) is critical.

26. Consider a weighted voting system in which participants A, B, C, D, E have weights 5, 7, 8, 9, 11, respectively. A quota of 20 votes is needed for a motion to pass.

(a) Construct a decision tree to determine those coalitions in which E is critical. Label the vertices of this tree with the participant weights 5, 7, 8, 9, 11; the root of the tree will correspond to participant E.

(b) Identify those coalitions X that have weight too low (**L**) and those that have weight too high (**H**).

(c) How many coalitions are there in which E is critical? Identify those coalitions.

27. Consider a weighted voting system in which participants A, B, C, D, E have weights 5, 7, 8, 9, 11, respectively. A quota of 20 votes is needed for a motion to pass.

(a) Construct a decision tree to determine those coalitions in which C is critical. Label the vertices of this tree with the participant weights 5, 7, 8, 9, 11; the root of the tree will correspond to participant C.

(b) Identify those coalitions X that have weight too low (**L**) and those that have weight too high (**H**).

(c) How many coalitions are there in which C is critical? Identify those coalitions.

Projects

28. Using the calculator provided at [OR1], explore the BPI for three voters, who have respective weights of 15, 10, 8. Record the different BPI values that result as you systematically vary the quota q from 1 to 33.

    (a) Why did we specify 33 as the upper limit?

    (b) What patterns do you see as you vary q over this range?

29. Using the calculator provided at [OR1], explore the BPI for three voters, who have respective weights of 15, 10, 8. Record the different BPI values that result as you systematically vary the quota q from 1 to 43.

    (a) Why did we specify 43 as the upper limit?

    (b) What patterns do you see as you vary q over this range?

    (c) What difference in the relations of the three weights might distinguish this case from that in Exercise 28?

Bibliography

[Brams 2008] Brams, S. 2008. *The presidential election game.* Wellesley, MA: A. K. Peters.

[Robinson 2011] Robinson, E. A., and Ullman, D. H. 2011. *A mathematical look at politics,* Chapter 19. Boca Raton, FL: CRC Press.

[Taylor 2008] Taylor, A., and Pacelli, A. 2008. *Mathematics and politics: Strategy, voting, power and proof,* Chapters 2 and 3. New York: Springer.

[OR1] http://cow.math.temple.edu/bpi.html (accessed February 13, 2014).

Chapter 20

Apportionment

In Maine comes and out Maine goes ... God help the state of Maine when mathematics reach for her and undertake to strike her down.

> —Charles E. Littlefield, U.S. Representative (Maine), 1851–1915

Chapter Preview

Previous chapters studied several systems for counting votes in which individual voters are entitled to the same number of votes. We also introduced weighted voting systems and a power index that measures the influence of an individual voter when participants have an unequal number of votes. This chapter examines procedures for assigning, hopefully in a fair way, the number of votes each participant in a weighted voting system should receive. In particular, this problem arises in the House of Representatives in assigning seats (and thus votes) to each state, based on each state's population. The U.S. Constitution does not specify exactly how to do this. If we could allow fractional numbers of seats, then it would be a straightforward task to divide up the 435 House seats in a way proportional to population. The difficulty arises when we require, as is natural, that the number of seats allocated to each state be a whole number. As a result, different methods have competed over the past two centuries to be chosen as the apportionment method for the House of Representatives. These apportionment methods are applicable as well to any situation requiring a fair allocation of limited quantities of items (not necessarily votes) to different-sized groups, for example, distributing a fixed number of computers to the schools within a school district, based on each school's student population. We also introduce certain paradoxes that arise in analyzing apportionment methods.

20.1 Apportionment and the House of Representatives

The Electoral College, which elects the President and Vice President of the United States, consists of 538 electors. Different states might be allocated different numbers of electors and therefore can exert more or less voting power. Each state is allowed as many electors as it has Senators (two per state) and Representatives in the Congress. The number of seats in the House of Representatives is apportioned to each state corresponding to its population as a percentage of the total population of all 50 states:

"Representatives ... shall be apportioned among the several States ... according to their respective Numbers ... The Number of Representatives shall not exceed one for every thirty Thousand, but each State shall have at Least one Representative."

—Article 1, Section 2, *U.S. Constitution*

There are currently 435 seats in the House of Representatives and each seat in the House of Representatives represents one congressional district. Based on the 2010 Census figures, the U.S. population is 308,745,538 and so the *congressional district size* is 308,745,538/435 or approximately 709,760 people (no longer the 30,000 mentioned in the Constitution).

DEFINITION **Congressional District Size**
The average number of people represented by each member in the House of Representatives.

Dividing a state's population by 709,760 will determine how many seats in the House a state receives. For example, North Carolina with a population of 9,535,483 in 2010 has a *state quota* of 9,535,483/709,760 or approximately 13.4348. On the other hand, New York had a 2010 population of 19,378,102 and so would receive a state quota of 27.3023.

DEFINITION **State Quota**
This quantity is each state's proportional share of the 435 representatives, arrived at by dividing the state's population by the congressional district size.

Given North Carolina's state quota of 13.4348, should North Carolina be given 13 seats or 14 seats in the House of Representatives? Should it be given its *lower quota*, the number obtained when rounding down 13.4348 to 13? Or should it be given its *upper quota*, the number obtained when rounding up 13.4348 to 14? As seen earlier, the Constitution does not describe how Congress is to distribute these fractional amounts of state quota, and a variety of *apportionment methods* have been proposed since 1790. An impressive list of prominent historical figures—Thomas Jefferson, John Quincy Adams, Alexander Hamilton, and Daniel Webster—have played important roles in supporting alternative methods for apportioning the seats in the House of Representatives.

DEFINITION **Apportionment Method**
A rule for rounding the fractional portion of each state quota to a whole number so that all 435 House seats are allocated.

We will look closely at the five different apportionment methods listed in Table 20.1, each of which has been used to apportion seats in the House of Representatives [OR1, OR2].

TABLE 20.1: Five apportionment methods

Method	Apportionment Technique
Hamilton (1850)	Largest remainders
Jefferson (1790)	Greatest divisors
Adams (1832)	Smallest divisors
Webster (1832)	Major fractions
Huntington-Hill (1941)	Equal proportions

20.2 Hamilton Apportionment Method

This method, also called the "largest remainders" method, was proposed by Alexander Hamilton, the first Secretary of the Treasury, and was used for 50 years starting in 1850.

Hamilton Apportionment Method

1. Calculate each state quota by dividing the state population by the congressional district size.

2. Give every state its lower quota and then keep assigning the remaining seats (until they run out) to the states, starting with the states having the largest fractional parts.

In 1790 the U.S. Census reported a total population of 3,615,920 and there were only 105 seats to apportion, so the congressional district size equaled $3{,}615{,}920/105 = 34{,}437.33$. After determining state quotas, the sum of the lower quotas was 97, meaning that eight additional seats had to be allocated. When Hamilton's method is applied to this situation, the eight seats are given successively to Kentucky, South Carolina, Rhode Island, Connecticut, Massachusetts, New York, Delaware, and Pennsylvania. As a result, Delaware with a population of 55,540 receives two seats—which exceeds the ratio of one per 30,000 stated in Article 1, Section 2 of the Constitution. This is most likely the reason for the first presidential veto by George Washington.

Example 20.1 Determine the Hamilton Apportionment

Consider a small example, with 23 seats to apportion among three states (A, B, C), having respective populations of 269, 86, and 85. (a) Find the state quota for each state. (b) Find the apportionment to each state, using the Hamilton method.

Solution

We will use Table 20.2 to organize the calculations. First, we list each state population and calculate the total population size. Next, we divide the total population 440 by the 23 seats, giving the congressional district size of $440/23 = 19.13$. We find each state quota by dividing its population by 19.13.

After giving each state its lower quota (obtained by rounding down the state quota), we

TABLE 20.2: Hamilton apportionment for 23 seats

State	A	B	C	Total
Population	269	86	85	440
State Quota	14.062	4.496	4.443	
Lower Quota	14	4	4	22
Apportionment	14	4 + 1 = 5	4	23

have only apportioned $14 + 4 + 4 = 22$ seats. The extra seat is then allocated to B (having the largest fractional part 0.496). We obtain the final apportionment of 14 seats to A, 5 to B, and 4 to C. ◁

We should note that the problem of allocating items to different-sized groups arises in a number of real-world problems, and not just apportioning the House of Representatives. To fit this more general context, the concepts of *congressional district size* and *state quota* can be appropriately redefined.

DEFINITION **Ideal Size**
This represents the average number of people represented by the allocation of one item. This average is obtained by dividing the total population size by the total number of items available for allocation.

DEFINITION **Group Quota**
This represents each group's proportional share of the total items available for allocation. It is obtained by dividing each group's population by the ideal size.

(Related Exercises: 1–7)

Paradoxes of the Hamilton Method

One desirable property of the Hamilton method is that it *preserves quota*: that is, the number of seats allocated to each state is either the quota rounded down or the quota rounded up. However, there are some difficulties with the Hamilton method, notably it permits the so-called *Alabama Paradox*.

DEFINITION **Alabama Paradox**
This situation occurs if a state loses seats in the House of Representatives when the size of the House increases.

This paradox gained its name for the following reason. In 1880, with a 299-seat House, Alabama's quota would be 7.646 and it would be allocated eight seats, given its position in the fractional ranked list of states. However, with a 300-seat House, Alabama's quota would increase to 7.671, yet the number of seats allocated would be seven, because Alabama's position in the ranked list moved down when two other states moved up.

This paradox occurred again in 1900 when it was explained that Maine's allotment of seats would be three for a House size of 350–382, four for a House size of 383–385, three for a House size of 386, four for a House size of 387–388, three for a House size of 389–390, and four for a House size of 391–400. This prompted the comment from Rep. Charles E. Littlefield (R-ME) that is quoted at the beginning of this chapter.

Example 20.2 Show the Hamilton Method Allows the Alabama Paradox

Three courses (Statistics, Calculus, and Discrete Math), with enrollments of 978, 504, and 328 students, respectively, are assigned teaching assistants to help in the class and with grading. (a) Determine the apportionment of 30 teaching assistants to each of the courses using Hamilton's method. (b) Determine the apportionment of 31 teaching assistants to each of the courses using Hamilton's method. (c) Show that when the number of teaching assistants is increased from 30 to 31, the Alabama paradox occurs.

Solution

(a) We begin by constructing Table 20.3 to organize the data. Using the ideal size of $1810/30 = 60.333$, we calculate the group quotas and round down to obtain the lower quotas. For example, Statistics receives a quota of $978/60.333 = 16.210$. This process allocates only 29 teaching assistants, with one more left to apportion. Discrete Math receives an extra seat because it has the highest fractional part 0.436. The final Hamilton apportionment allocates 16 assistants to Statistics, 8 to Calculus, and 6 to Discrete Math.

TABLE 20.3: Hamilton apportionment for 30 assistants

Course	Statistics	Calculus	Discrete Math	Total
Enrollment	978	504	328	1810
Quota	16.210	8.354	5.436	
Lower Quota	16	8	5	29
Apportionment	16	8	6	30

(b) Using the ideal size of $1810/31 = 58.387$, we fill in Table 20.4 and obtain the Hamilton apportionment of 17 assistants to Statistics, 9 to Calculus, and 5 to Discrete Math.

TABLE 20.4: Hamilton apportionment for 31 assistants

Course	Statistics	Calculus	Discrete Math	Total
Enrollment	978	504	328	1810
Quota	16.750	8.632	5.618	
Lower Quota	16	8	5	29
Apportionment	17	9	5	31

(c) The Alabama paradox occurs in this example because the apportionment for Discrete Math has decreased, even though the total number of teaching assistants has increased. ◁

(Related Exercises: 33–34)

Hamilton's method also permits the *Population Paradox*.

DEFINITION **Population Paradox**
This situation occurs if a group's apportionment decreases even though its population increases, or if a group's apportionment increases even though its population decreases.

Example 20.3 Show the Hamilton Method Allows the Population Paradox

A publisher wants to invite 10 faculty members to a conference (all expenses paid) and decides to allocate the invitations to each of three universities based on the number of students using their textbooks. Universities A, B, and C have 1451, 3398, and 5151 students using their textbooks. (a) Determine the Hamilton apportionment to the three universities. (b) Determine the Hamilton apportionment when the overall textbook usage now changes to 1469, 3381, and 4650. (c) Explain how this example illustrates the population paradox.

Solution

(a) Table 20.5 summarizes the calculations. Using the ideal size of $10,000/10 = 1000$, we calculate the group quotas and round down to obtain the lower quotas. This allocates nine invitations, with one more to apportion, which is then distributed to A. The Hamilton apportionment gives A, B, and C the respective allocations of 2, 3, and 5 invitations.

TABLE 20.5: Hamilton apportionment with 10,000 students

School	A	B	C	Total
Textbook Usage	1451	3398	5151	10,000
Quota	1.451	3.398	5.151	
Lower Quota	1	3	5	9
Apportionment	2	3	5	10

(b) Using the new textbook usage values, we obtain the ideal size of $9500/10 = 950$ and complete the entries in Table 20.6. The Hamilton apportionment now gives schools A, B, and C the respective allocations of 1, 4, and 5 invitations.

TABLE 20.6: Hamilton apportionment with 9500 students

School	A	B	C	Total
Textbook Usage	1469	3381	4650	9500
Quota	1.546	3.559	4.895	
Lower Quota	1	3	4	8
Apportionment	1	4	5	10

(c) The population paradox occurs here because A loses one invitation even though its textbook usage increased, whereas B gains one invitation even though its usage actually decreased. ◁

(Related Exercise: 35)

20.3   Divisor Methods

The methods proposed by Thomas Jefferson, John Quincy Adams, Daniel Webster, Edward Huntington, and Joseph Hill share a common approach and are called *divisor methods*.

Rather than using the ideal size as divisor and then ranking the remainders so obtained, a more suitable divisor is used to guide the allocation.

Jefferson Apportionment Method

Thomas Jefferson, third President of the United States, proposed an apportionment method to replace the Hamilton method. This method, called the "greatest divisors" method, was used from 1790 to 1830 instead of the Hamilton method. It begins in the same manner as the Hamilton method, but iteratively refines the divisor used.

Jefferson Apportionment Method

1. Calculate each group's quota by dividing the group population by the ideal size and then *round down* all fractional quotients.

2. If the resulting apportionment is too small, modify the divisor d so that it is smaller; if the apportionment is too large, modify the divisor d so that it is larger. Repeat this until the correct number of items are apportioned.

Example 20.4 Determine the Jefferson Apportionment

Consider again the situation in Example 20.1 with 23 seats to apportion among three states A, B, and C. The total population is 440 so the ideal size is $440/23 = 19.13$. (a) Use $d = 19.13$ for the first trial divisor, round down all state quotas, and determine if the apportionment is too small or too large. (b) Alter the divisor d as necessary to obtain the final apportionment for each state.

Solution

(a) Using 19.13 as the first divisor in Table 20.7 and rounding down gives an apportionment with 22 seats, which is too low. As a result, we want to decrease the divisor d.

TABLE 20.7: Jefferson apportionment for 23 seats

State	A	B	C	Total
Population	269	86	85	440
Quota: $d = 19.13$	14.062	4.496	4.443	
Rounded Down	14	4	4	22
Quota: $d = 17$	15.824	5.059	5.000	
Rounded Down	15	5	5	25
Quota: $d = 17.5$	15.371	4.914	4.857	
Rounded Down	15	4	4	23

(b) If we use the trial divisor $d = 17$, then Table 20.7 shows that a total of 25 seats are allocated, which is too large. So we try $d = 17.5$, a value in between 17 and 19.13; by using this third trial divisor, the sum of the rounded down quotients exactly equals 23 and we can stop. The final allocation gives 15 seats to A, 4 to B, and 4 to C.

Notice that the final divisor $d = 17.5$ is not unique; using any d greater than 17.20 and less than 17.93 will also give the same allocation of exactly 23 seats. ◁

(Related Exercises: 8–13)

When the Jefferson apportionment method is applied to the 1790 Census, larger states tend to get extra seats at the expense of smaller states. This turns out to be a general feature of the Jefferson method. That is, a large state may tend to get more seats than its quota rounded down, while a small state may get fewer seats than its quota rounded up. This property will be explored further in Chapter 21.

Neither the Alabama paradox nor the population paradox can occur with the Jefferson method, but the Jefferson method need not preserve quota: that is, it is possible for a state to receive an apportionment that is different from its group quota rounded up or down.

Example 20.5 Show the Jefferson Method Need Not Preserve Quota

Suppose there are 25 seats to be allocated among the five states A, B, C, D, and E with populations 923, 293, 243, 130, and 73, respectively. (a) Use the ideal size for the first trial divisor d, round down all group quotas, and determine if the apportionment is too small or too large. (b) Alter the divisor d as necessary to obtain the final apportionment for each state. (c) Explain why this final apportionment does not preserve quota.

Solution

(a) Using $d = 1662/25 = 66.48$ as the first divisor results in an allocation of only 22 seats in Table 20.8.

TABLE 20.8: Jefferson apportionment for 25 seats

State	A	B	C	D	E	Total
Population	923	293	243	130	73	1662
Quota: $d = 66.48$	13.884	4.407	3.655	1.955	1.098	
Rounded Down	13	4	3	1	1	22
Quota: $d = 61$	15.131	4.803	3.984	2.131	1.197	
Rounded Down	15	4	3	2	1	25

(b) Since the total number of allocated seats 22 is too small, we decrease d to 61 in the second section of Table 20.8 and now all 25 seats are allocated.

(c) State A has a group quota of 13.884, yet receives 15 seats, violating the quota rule: it does not receive either its lower quota of 13 or its upper quota of 14. ◁

(Related Exercise: 36)

Adams Apportionment Method

This method was suggested in 1832 by John Quincy Adams, sixth President of the United States, to remedy defects in the Hamilton method. This technique, called the "smallest divisors" method, was never actually used for apportioning the House. It proceeds similarly to the Jefferson method, but uses a different method of rounding after each trial divisor d.

Adams Apportionment Method

1. Calculate each group's quota by dividing the group population by the ideal size and then *round up* all fractional quotients.

---

2. If the resulting apportionment is too small, modify the divisor d so that it is smaller; if the apportionment is too large, modify the divisor d so that it is larger. Repeat this until the correct number of items are apportioned.

Example 20.6 Determine the Adams Apportionment

Suppose there are 18 seats to apportion among three states A, B, and C with respective populations of 233, 176, and 131. (a) Use the ideal size for the first trial divisor, round up all group quotas, and determine if the apportionment is too small or too large. (b) Alter the divisor d as necessary to obtain the final apportionment for each state.

Solution

(a) Using $d = 540/18 = 30$ as the trial divisor in Table 20.9 produces an apportionment with 19 seats, which is too high. As a result, we want to increase the divisor d.

TABLE 20.9: Adams apportionment for 18 seats

State	A	B	C	Total
Population	233	176	131	540
Quota: $d = 30$	7.767	5.867	4.367	
Rounded Up	8	6	5	19
Quota: $d = 34$	6.853	5.176	3.853	
Rounded Up	7	6	4	17
Quota: $d = 33$	7.061	5.333	3.970	
Rounded Up	8	6	4	18

(b) Using the larger divisor $d = 34$ in Table 20.9 allocates a total of 17 seats, which is too small. So we try a third divisor $d = 33$, in between 30 and 34. This now allocates the required 18 seats. The final apportionment gives 8 seats to A, 6 to B, and 4 to C. ◁

(Related Exercises: 14–19)

Similar to the Jefferson method, the Adams apportionment method does not permit the occurrence of either the Alabama paradox or the population paradox. In contrast to the Jefferson method (which favors large states), the Adams method tends to favor small states. Also, like the Jefferson method, the Adams method need not maintain quota: it is possible for a state to receive an apportionment that is different from its group quota rounded up or down.

Example 20.7 Show the Adams Method Need Not Preserve Quota

A total of 215 iPads are to be allocated to five courses A, B, C, D, and E having 234, 989, 300, 97, and 144 enrolled students. (a) Use the ideal size for the first trial divisor d, round up all group quotas, and determine if the apportionment is too small or too large. (b) Alter the divisor d as necessary to obtain the final apportionment for each course. (c) Explain why this apportionment does not preserve quota.

Solution

(a) Using $d = 1764/215 = 8.20$ in Table 20.10 produces an apportionment of 217 iPads, which is too large. So we need a larger divisor d.

TABLE 20.10: Adams apportionment for 215 iPads

Course	A	B	C	D	E	Total
Students	234	989	300	97	144	1764
Quota: $d = 8.20$	28.537	120.610	36.585	11.829	17.561	
Rounded Up	29	121	37	12	18	217
Quota: $d = 8.32$	28.125	118.870	36.058	11.659	17.308	
Rounded Up	29	119	37	12	18	215

(b) Experimentation leads to the larger divisor $d = 8.32$, which apportions all 215 iPads.

(c) The Adams apportionment does not preserve quota since B receives 119 iPads, despite the fact that rounding its group quota of 120.61 would produce either 120 or 121 iPads. ◁

(*Related Exercise: 37*)

Webster Apportionment Method

This apportionment method was suggested in 1832 by Daniel Webster, and was used in 1840 following the Jefferson method, yet before the Hamilton method was adopted. It was also used in 1910 and 1930. This method, called the "major fractions" method, allows rounding of the quotient either up or down. Specifically, we round down if the fractional part is less than 0.5 whereas we round up if it is greater than 0.5. In other words, we use the *arithmetic mean* (AM) as our "dividing line" to determine how the quotients are rounded.

Webster Apportionment Method

1. Calculate each group's quota by dividing the group population by the ideal size and then *round* all fractional quotients *to the nearest integer* (either up or down).
2. If the resulting apportionment is too small, modify the divisor d so that it is smaller; if the apportionment is too large, modify the divisor d so that it is larger. Repeat this until the correct number of items are apportioned.

Example 20.8 Determine the Webster Apportionment

A shipment of 22 new printers is to be allocated to three offices A, B, and C with respective employee populations of 235, 78, and 77. (a) Use the ideal size for the first trial divisor d, round all office quotas to the nearest integer (either up or down), and determine if the apportionment is too small or too large. (b) Alter the divisor as necessary to obtain the final apportionment for each office.

Solution

(a) Using $d = 390/22 = 17.73$ in Table 20.11 results in an apportionment of 21 printers, which is too small, so we need a smaller divisor.

TABLE 20.11: Webster apportionment for 22 printers

Office	A	B	C	Total
Employees	235	78	77	390
Quota: $d = 17.73$	13.254	4.399	4.343	
Rounded to AM	13	4	4	21
Quota: $d = 17.4$	13.506	4.483	4.425	
Rounded to AM	14	4	4	22

(b) By decreasing d to 17.4 in Table 20.11, the sum of the rounded quotients is now 22. The final allocation gives 14 printers to A, and 4 each to B and C. ◁

(Related Exercises: 20–24)

Similar to the Jefferson and Adams method, the Webster apportionment method does not permit the occurrence of either the Alabama paradox or the population paradox. Whereas the Jefferson method tends to favor large states and the Adams method tends to favor small states, the Webster method tends to be neutral toward larger/smaller states. However, the Webster method need not maintain quota.

Example 20.9 Show the Webster Method Need Not Preserve Quota

Suppose that 25 votes at a county board meeting are to be allocated among the five townships A, B, C, D, and E according to their respective populations 235, 142, 985, 305, and 100. (a) Use the ideal size for the first trial divisor d, round all township quotas to the nearest integer, and determine if the apportionment is too small or too large. (b) Alter the divisor d to obtain the final apportionment for each township. (c) Explain why this apportionment does not preserve quota.

Solution

(a) Using $d = 1767/25 = 70.68$ in Table 20.12, the resulting apportionment of 24 votes is too small, so we need a smaller divisor.

TABLE 20.12: Webster apportionment for a county board

Township	A	B	C	D	E	Total
Population	235	142	985	305	100	1767
Quota: $d = 70.68$	3.325	2.009	13.936	4.315	1.415	
Rounded to AM	3	2	14	4	1	24
Quota: $d = 67.8$	3.466	2.094	14.528	4.499	1.475	
Rounded to AM	3	2	15	4	1	25

(b) Using $d = 67.8$ as the next trial divisor, all 25 votes are apportioned.

(c) The Webster apportionment violates quota because Township C receives 15 votes, yet its group quota 13.936 rounds up to 14 and down to 13. ◁

(Related Exercise: 38)

Huntington-Hill Apportionment Method

In the Webster method, quotients are rounded (either up or down) based on the arithmetic mean. The Huntington-Hill method (also called "the method of equal proportions"), adopted in 1941 and still in use today [OR3], was proposed by Joseph Hill (chief statistician at the Census Bureau) and refined by his friend Edward Huntington (professor of mathematics at Harvard). It involves the same process as the Webster method, except that it rounds relative to the *geometric* (rather than the arithmetic) mean.

DEFINITION **Geometric Mean (GM)**

The geometric mean of two nonnegative numbers x and y is the square root of their product, namely $\sqrt{xy}$.

Huntington-Hill Apportionment Method

1. Calculate each group's quota by dividing the group population by the ideal size and find the *geometric mean* of the enclosing interval. Round up if the quotient is greater than the geometric mean or round down if the quotient is less than the geometric mean.

2. If the resulting apportionment is too small, modify the divisor d so that it is smaller; if the apportionment is too large, modify the divisor d so that it is larger. Repeat this until the correct number of items are apportioned.

For example, if the quotient is 7.49, the interval enclosing this number is from 7 to 8, so the geometric mean is $\sqrt{7 \times 8} = \sqrt{56} = 7.48$; since $7.49 > 7.48$, the quotient is rounded up to 8.

Example 20.10 Determine the Huntington-Hill Apportionment

Suppose there are 23 laptops to apportion among three offices A, B, and C with respective employee populations of 259, 87, and 84. (a) Use the ideal size for the first trial divisor, and calculate the geometric mean for each quotient. Round down each quotient if it is smaller than the geometric mean or round up if it is greater. Determine if the corresponding apportionment is too small or too large. (b) Alter the divisor as necessary, calculate new geometric means, and obtain the final apportionment for each office.

Solution

(a) We use $d = 430/23 = 18.70$ in Table 20.13 to calculate all quotients. Note that the quotient for C is 4.492 and $4.492 > 4.472$, so we round its allocation up to 5. Similar calculations are shown in the table for the other two offices. The resulting apportionment of 24 laptops is too large, so we need a larger divisor.

(b) For $d = 19$, the sum of the (geometrically) rounded quotients now equals 23. The final apportionment of laptops is 14 to A, 5 to B, and 4 to C. ◁

(Related Exercises: 25–29)

Similar to the other divisor methods, the Huntington-Hill apportionment method does not permit the occurrence of either the Alabama paradox or the population paradox. As with

TABLE 20.13: Huntington-Hill apportionment for three offices

Office	A	B	C	Total
Employees	259	87	84	430
Quota: $d = 18.70$	13.850	4.652	4.492	
GM	13.491	4.472	4.472	
Rounded to GM	14	5	5	24
Quota: $d = 19$	13.632	4.579	4.421	
GM	13.491	4.472	4.472	
Rounded to GM	14	5	4	23

the Webster method, it tends to be neutral toward larger/smaller states. However, like all the other divisor methods, the Huntington-Hill method need not maintain quota.

Example 20.11 Show the Huntington-Hill Method Need Not Preserve Quota

Consider a situation with four states (having populations 86,920, 4320, 5410, and 3350) and 100 seats to apportion. (a) Use the ideal size for the first trial divisor d, calculate the geometric mean for each quotient, and round relative to the GM. Determine if the corresponding apportionment is too small or too large. (b) Alter the divisor as necessary, calculate new geometric means, and obtain the final apportionment for each state. (c) Explain why this apportionment does not preserve quota.

Solution

(a) Using $d = 100,000/100 = 1000$ in Table 20.14 results in an allocation of only 99 seats, which is too small.

TABLE 20.14: Huntington-Hill apportionment for four states

State	A	B	C	D	Total
Population	86,920	4320	5410	3350	100,000
Quota: $d = 1000$	86.920	4.320	5.410	3.350	
GM	86.499	4.472	5.477	3.464	
Rounded to GM	87	4	5	3	99
Quota: $d = 990$	87.798	4.364	5.465	3.384	
GM	87.499	4.472	5.477	3.464	
Rounded to GM	88	4	5	3	100

(b) Using the smaller divisor $d = 990$ in Table 20.14, we can now fully apportion all 100 seats.

(c) The Huntington-Hill method does not preserve quota. State A receives 88 seats, which is neither its lower quota of 86 nor its upper quota of 87. ◁

(Related Exercise: 39)

Dean Apportionment Method

Yet another (theoretical) apportionment method was proposed in 1832 by James Dean, professor of astronomy and mathematics at the University of Vermont. Rather than being based on the arithmetic mean (Webster) or the geometric mean (Huntington-Hill), the Dean method is based on the *harmonic mean*. Though never used for apportionment of the House, this method did play a role in a constitutional challenge to the current apportionment procedure, the Huntington-Hill method. We will see in the next chapter that the arithmetic mean is at least as large as the geometric mean, which in turn is at least as large as the harmonic mean. This fact will be important in understanding the behavior of apportionment methods.

DEFINITION **Harmonic Mean**

The harmonic mean of two positive real numbers x and y is the reciprocal of the average of their reciprocals, which can be expressed as $2/\left(\frac{1}{x}+\frac{1}{y}\right)$.

We now provide an application of the harmonic mean to an apparent real-world paradox.

Example 20.12 Use Harmonic Means to Calculate an Average Speed

A car travels 140 miles from City A to City B at the rate of 50 mph and then returns at the rate of 70 mph. (a) Show that the average speed is not 60 mph, the arithmetic mean of 50 and 70 mph. (b) Compute the harmonic mean of 50 and 70 mph, and show that this is the correct average speed.

Solution

(a) The car takes 140 miles/50 mph = 2.8 hours to go from City A to City B. The return trip takes 140 miles/70 mph = 2 hours. Altogether, the trip covers 280 miles in 2.8 + 2 = 4.8 hours, for an average speed of 280 miles/4.8 hours = 58.33 mph.

(b) Notice that more time is spent on the trip from A to B than from B to A, so the true average speed should be weighted more toward 50 mph than 70 mph. The correct average speed is in fact the harmonic mean of 50 and 70, which is calculated as

$$\frac{2}{\frac{1}{50}+\frac{1}{70}} = \frac{2}{\frac{7}{350}+\frac{5}{350}} = \frac{2}{\frac{12}{350}} = \frac{700}{12} = 58.33 \quad \triangleleft$$

Chapter Summary

Motivated by the need to allocate seats in the House of Representatives, this chapter has studied apportionment methods, in which a fixed number of items are to be distributed to different-sized groups. These methods are based on different approaches for dealing with the fractional nature of each group's respective quota. Tabular representations are used extensively to organize population sizes and the group quotas, to record the results of different rounding methods, to track the results of using different divisors (when using divisor methods), and to list the final apportionments. Several paradoxes that can arise from using these methods are also discussed. A more extensive discussion of the fairness of apportionment methods continues in the next chapter.

Applications, Representations, Strategies, and Algorithms

Applications
Society/Politics (1, 4–6, 9, 11)   Education (2–3, 7)
Business (8, 10)   Transportation (12)

Representations		
Tables	Sets	Symbolization
Decision	Lists	Numerical
1–11	1–11	1–12

Strategies
Rule-Based
1–12

Algorithms		
Exact	Sequential	Numerical
1–12	4–11	1–12

Exercises

Apportionment Concepts

1. A county council is composed of 12 seats, and it makes decisions affecting the towns A, B, and C that comprise the county. These towns have respective populations of 505, 380, and 240 people.

    (a) Determine the total population of all three towns and the ideal size.

    (b) Give an interpretation of what the ideal size means in this context.

    (c) Calculate the group quota for each town.

    (d) Give an interpretation of what the group quota means in this context.

    (e) Compute the associated lower quota and the upper quota.

2. A real estate firm has eight season football tickets to distribute to its four top-selling agents A, B, C, and D. It was decided to allocate the tickets based on the number of homes sold by the agents during the last year. Agents A, B, C, and D have sold 19, 17, 12, and 10 homes, respectively.

    (a) Determine the total number of homes sold and the ideal size (number of sold homes represented by each season ticket).

   (b) Calculate the group quota for each agent.

   (c) Give an interpretation of what the group quota means in this context.

   (d) Compute the associated lower quota and the upper quota.

3. The city of Centerville runs a free bus system composed of five routes. The riderships last year of the Red, Blue, Orange, Purple, and Green routes were, respectively, 980, 705, 685, 650, and 610 passengers. The supervisor of the bus system wants to allocate 13 new low-emission buses to the five routes, based on past ridership data.

   (a) Determine the ideal size and interpret what this number means.

   (b) Calculate the group quota for each route.

   (c) Give an interpretation of what the group quota means in this context.

   (d) Compute the associated lower quota and the upper quota.

Hamilton Apportionment

4. Use the Hamilton apportionment method to allocate the 12 seats in the county council in Exercise 1. Which towns received their lower quota and which received their upper quota?

5. Use the Hamilton apportionment method to allocate the eight season tickets in Exercise 2. Which agents received their lower quota and which received their upper quota?

6. Use the Hamilton apportionment method to allocate the 13 new buses in Exercise 3. Which routes received their lower quota and which received their upper quota?

7. A certain university is composed of five colleges: Agriculture, Business, Education, Humanities, and Sciences. The number of students taught each year in these colleges are, respectively, 540, 690, 325, 272, and 733. The administration wants to distribute 20 new faculty positions to these colleges, based on the number of students taught.

   (a) Calculate the ideal size and the group quotas for this situation.

   (b) Determine the Hamilton apportionment of faculty positions to colleges.

Jefferson Apportionment

8. Use the Jefferson apportionment method to allocate the 12 seats of the county council in Exercise 1.

9. Use the Jefferson apportionment method to allocate the 20 new faculty positions in Exercise 7.

10. Use the Jefferson apportionment method to allocate the 215 iPads in Example 20.7. Compare your answer to the apportionment shown there for the Adams apportionment method.

11. Use the Jefferson apportionment method to allocate the 25 votes in Example 20.9. Compare your answer to the apportionment shown there for the Webster apportionment method.

12. Use the Jefferson apportionment method to allocate the 23 laptops in Example 20.10. Compare your answer to the apportionment shown there for the Huntington-Hill apportionment method.

13. There are 28 seats to apportion among four states A, B, C, and D having respective populations of 445, 290, 260, and 195.

    (a) Determine the ideal size for this situation, calculate the group quota for each state, and find the resulting apportionment.

    (b) Alter the divisor as necessary to obtain the final Jefferson apportionment for each state.

Adams Apportionment

14. Use the Adams apportionment method to allocate the eight season tickets in Exercise 2.

15. Use the Adams apportionment method to allocate the 13 new buses in Exercise 3.

16. Use the Adams apportionment method to allocate the 25 seats in Example 20.5. Compare your answer to the apportionment shown there for the Jefferson apportionment method.

17. Use the Adams apportionment method to allocate the 22 printers in Example 20.8. Compare your answer to the apportionment shown there for the Webster apportionment method.

18. Use the Adams apportionment method to allocate the 23 laptops in Example 20.10. Compare your answer to the apportionment shown there for the Huntington-Hill apportionment method.

19. A bank has 15 temporary staff to assign to its four branch offices A, B, C, and D having 20, 34, 42, and 13 employees, respectively.

    (a) Determine the ideal size for this situation, calculate the group quota for each office, and find the resulting apportionment.

    (b) Alter the divisor as necessary to obtain the final Adams apportionment for each office.

Webster Apportionment

20. Use the Webster apportionment method to allocate the 12 seats of the county council in Exercise 1.

21. Use the Webster apportionment method to allocate the 28 seats in Exercise 13.

22. Use the Webster apportionment method to allocate the 215 iPads in Example 20.7. Compare your answer to the apportionment shown there for the Adams apportionment method.

23. Use the Webster apportionment method to allocate the 23 laptops in Example 20.10. Compare your answer to the apportionment shown there for the Huntington-Hill apportionment method.

24. The mathematics department has 10 graders to assign to four courses (A, B, C, D), which have, respectively, 57, 19, 15, and 11 students enrolled.

    (a) Determine the ideal size for this situation, calculate the group quota for each course, and find the resulting apportionment.

    (b) Alter the divisor as necessary to obtain the final Webster apportionment for each course.

Huntington-Hill Apportionment

25. Use the Huntington-Hill apportionment method to allocate the 15 temporary staff in Exercise 19.

26. Use the Huntington-Hill apportionment method to allocate the 22 printers in Example 20.8. Compare your answer to the apportionment shown there for the Webster apportionment method.

27. Use the Huntington-Hill apportionment method to allocate the 25 votes in Example 20.9. Compare your answer to the apportionment shown there for the Webster apportionment method.

28. Use the Huntington-Hill apportionment method to allocate the 10 graders in Exercise 24. Compare your answer to the apportionment shown there for the Webster apportionment method.

29. The four volunteers Andi, Bev, Charmaine, and Devi have sold 40, 38, 28, and 17 advertisements for an upcoming campus performance. To reward their efforts, the Drama Department has reserved 10 free tickets to distribute to these volunteers, based on the number of advertisements sold.

    (a) Determine the ideal size for this situation, calculate the group quota for each volunteer, and find the resulting apportionment.

    (b) Alter the divisor as necessary to obtain the final Huntington-Hill apportionment for each volunteer.

Comparison of Methods

30. A small country with four regions has 100 seats in their Parliament to allocate. The four regions (A, B, C, D) have respective populations of 21,120, 152,580, 54,150, and 12,150. Apportion the 100 seats using the Hamilton, Jefferson, Adams, and Webster methods.

31. A college senate wishes to fill 25 seats to represent all students in the college. There are 297 Freshmen, 259 Sophomores, 217 Juniors, and 179 Seniors. Apportion the 25 seats to the four different classes using the Hamilton, Jefferson, Adams, and Webster methods.

32. Nine computers are to be allocated to three classes (A, B, C), containing 40, 9, and 8 students, respectively. Apportion the nine computers using the Webster and the Huntington-Hill methods.

Paradoxes and the Hamilton Apportionment

33. There are 100 seats to apportion to states A, B, and C having respective populations of 6314, 2525, and 1131.

    (a) Determine the Hamilton apportionment of the 100 seats to the three states.

    (b) Determine the Hamilton apportionment when the number of seats to be allocated increases to 101.

    (c) Explain why this example illustrates the Alabama paradox.

34. A company with offices in cities A, B, and C wants to distribute new laptop computers on the basis of the number of employees in each office. The three offices have 161, 250, and 489 employees, respectively.

    (a) Determine the Hamilton apportionment when there are 30 laptops to be distributed.

    (b) Determine the Hamilton apportionment when there are 31 laptops to be distributed.

    (c) Explain why this example illustrates the Alabama paradox.

35. A national assembly consists of 16 seats, which are to be allocated to three states A, B, and C, with respective populations of 6820, 2370, and 1050.

    (a) Determine the Hamilton apportionment.

    (b) Determine the Hamilton apportionment after the populations sizes have changed, with State A increasing in size to 6840, State B increasing in size to 2453, and State C decreasing in size to 1047.

    (c) Explain why this example illustrates the population paradox.

Divisor Methods and Preserving Quota

36. Determine the Jefferson apportionment for Example 20.7 and show that the resulting apportionment does not preserve quota.

37. Use the Adams method to apportion 100 laptops to schools with respective populations of 3477, 173, 216, and 134. Show that this apportionment does not preserve quota.

38. Use the Webster method to apportion 25 seats to states with respective populations of 6900, 2100, 1650, 1000, and 700. Show that this apportionment does not preserve quota.

39. Use the Huntington-Hill method to apportion 100 laptops to schools with respective populations of 3477, 173, 216, and 134. Show that this apportionment does not preserve quota.

Projects

40. Create a spreadsheet for the 1790 Census data found in [Balinski 2001, Appendix B], which lists the 15 states and their populations. Apportion the 105 seats in the House to each state using the five methods discussed (Hamilton, Jefferson, Adams, Webster, and Huntington-Hill). Compare the apportionments. Do any of the methods appear to favor smaller states or favor larger states?

41. Using the 2010 Census data found in [OR4], apportion the 435 seats in the House to each state using the five methods discussed (Hamilton, Jefferson, Adams, Webster, and Huntington-Hill). Compare the apportionments. Do any of the methods appear to favor smaller states or favor larger states?

Bibliography

[Balinski 2001] Balinski, M. L., and Young, H. P. 2001. *Fair representation: Meeting the ideal of one man, one vote*, second edition. Washington, D.C.: Brookings Institution Press.

[Robinson 2011] Robinson, E. A., and Ullman, D. H. 2011. *A mathematical look at politics*, Chapters 7–9. Boca Raton, FL: CRC Press.

[Steen 1982] Steen, L. A. 1982. "The arithmetic of apportionment." *Science News* 121: 317–318.

[Taylor 2008] Taylor, A., and Pacelli, A. 2008. *Mathematics and politics: Strategy, voting, power and proof*, Chapter 5. New York: Springer.

[OR1] http://www.census.gov/history/www/reference/apportionment/ (accessed February 26, 2014).

[OR2] http://www.maa.org/publications/periodicals/convergence/apportioning-representatives-in-the-united-states-congress-introduction (accessed February 26, 2014).

[OR3] http://www.youtube.com/watch?v=RUCnb5_HZc0 (accessed February 26, 2014).

[OR4] https://www.census.gov/population/apportionment/data/2010_apportionment_results.html (accessed February 26, 2014).

Chapter 21

Assessing Apportionment Methods

Nothing is particularly hard if you divide it into small jobs.

— Henry Ford, American industrialist, 1863–1947

Chapter Preview

Throughout history, apportionment methods have been rejected because they lead to certain problems, and they have been replaced by other methods that lead to the same or new problems. As seen in the previous chapter, the Hamilton method is susceptible to various paradoxes (the Alabama paradox and the population paradox). On the other hand, certain divisor methods are biased toward larger states (giving them extra seats), while others are biased toward smaller states. As well, divisor methods need not preserve quota (giving some states more or fewer seats than their fractional ranking among state populations would warrant). This chapter explores whether it is possible to design an apportionment method that avoids all unwanted paradoxes and quota violations. We also discuss various criteria that can be used to compare the apportionments produced by different methods. The chapter concludes with a brief discussion of redistricting, the process by which congressional districts are configured to align with the apportioned number of representatives.

21.1 Another Impossibility Theorem

Recall that an apportionment method preserves quota if every state's final apportionment is equal to either its upper quota or its lower quota. We have seen that the Hamilton method preserves quota, but is susceptible to the Alabama and population paradoxes. There is a third paradox to which the Hamilton Method is also susceptible, the *New States Paradox*.

> DEFINITION **New States Paradox**
> Adding a new state along with its "fair share" of seats can affect the number of seats allocated to other states. A fair share of seats is determined by the new state's population divided by the current ideal size.

The new states paradox was discovered in 1907 when Oklahoma became a state. Before Oklahoma became a state, the House of Representatives had 386 seats. The size of the House increased by five seats to 391 because adding five seats was in proportion to Oklahoma's population. That is, when Oklahoma's population was divided by the congressional district size at the time, Oklahoma would be entitled to five seats. The intent was that the number of seats would remain unchanged for the other states, but when the apportionment was recalculated, Maine gained a seat (from 3 to 4) while New York lost a seat (from 38 to 37).

Example 21.1 Show the Hamilton Method Allows the New States Paradox

A county library is going to distribute 100 laptops to its two main libraries A and B, which serve 2145 and 7855 people, respectively. The total population is 10,000 so the ideal size is $10{,}000/100 = 100$. (a) Determine the Hamilton apportionment. (b) Now determine the Hamilton apportionment after a new branch library C (serving 625 people) is added and six more laptops are allocated. Note that since the ideal size is 100, the new branch library is reasonably entitled to six laptops ($625/100 = 6.25$). (c) Compare the two apportionments and explain how the new states paradox has occurred.

Solution

(a) Based on the ideal size of 100, Table 21.1 shows the quota and the lower quota for libraries A and B. The additional laptop is then allocated to B, which has the larger fractional part ($0.55 > 0.45$).

TABLE 21.1: Hamilton apportionment for two libraries

Population	A	B	Total
Enrollment	2145	7855	10,000
Quota	21.45	78.55	
Lower Quota	21	78	99
Apportionment	21	79	100

(b) Table 21.2 now includes library C, with its population of 625, and the six extra laptops. The ideal size is then $10{,}625/106 = 100.24$, and 105 laptops are allocated using the lower quota. Library A receives an additional laptop because it has the largest fractional part.

TABLE 21.2: Hamilton apportionment for three libraries

Library	A	B	C	Total
Population	2145	7855	625	10,625
Quota	21.40	78.36	6.24	
Lower Quota	21	78	6	105
Apportionment	22	78	6	106

(c) While library C receives its fair share of laptops (six), library A gains an extra laptop at the expense of library B. This provides an instance of the new states paradox. ◁

(Related Exercises: 1–4)

An Ideal Apportionment Method?

Whereas the Hamilton method can exhibit the Alabama, population, and new states paradoxes, the divisor methods (Jefferson, Adams, Webster, and Huntington-Hill) are not susceptible to these paradoxes. However, the divisor methods can violate quota. This state of affairs is summarized in Table 21.3.

TABLE 21.3: Quotas, paradoxes, and the five apportionment methods

	Hamilton	Jefferson	Adams	Webster	Huntington-Hill
Violates Quota	No	Yes	Yes	Yes	Yes
Alabama Paradox	Yes	No	No	No	No
Population Paradox	Yes	No	No	No	No
New States Paradox	Yes	No	No	No	No

It is natural then to ask if an apportionment method can be designed that satisfies the quota rule yet does not permit these paradoxes to occur. Similar to the Arrow impossibility theorem (Chapter 18), there is another result referred to as the Balinski and Young impossibility theorem (1982) that proves such an ideal apportionment method is not possible.

RESULT **Balinski and Young's Impossibility Theorem**

An ideal apportionment method does not exist. That is, any method of apportionment that satisfies quota can produce paradoxes, and any method of apportionment that avoids paradoxes can violate quota.

Since apportionments methods will always be susceptible to imperfections (quota violations or paradoxes), we turn to other ways to assess the practical utility of the competing methods of apportionment.

21.2   Measuring Bias

In Chapter 20, we noted that the Jefferson method tends to favor larger states, whereas the Adams method tends to favor smaller states. The next example compares all four divisor methods in terms of their bias toward larger/smaller states.

Example 21.2 Compare the Apportionments Given by Four Divisor Methods

Ten seats are to be allocated to three states (A, B, C), with respective populations of 1385, 2390, and 6225. (a) Compare the apportionments produced by the Adams, Huntington-Hill, Webster, and Jefferson methods. (b) Can you draw any conclusions about which methods favor larger/smaller states in this example?

Solution

(a) Table 21.4 shows the calculations using the ideal size of $10,000/10 = 1000$. Notice that Adams and Huntington-Hill produce the same apportionment: A and B receive two seats while C receives six. Webster and Jefferson produce the same apportionment: A receives one seat, B receives two, and C receives seven.

TABLE 21.4: Four apportionment methods applied to three states

Method	State	A	B	C	Total
	Population	1385	2390	6225	10,000
	Quota: $d = 1000$	1.385	2.390	6.225	
Adams	Rounded Up	2	3	7	12
	$d = 1200$	1.154	1.992	5.188	
	Rounded Up	2	2	6	10
Huntington-Hill	GM	1.414	2.449	6.481	
	Rounded to GM	1	2	6	9
	$d = 977$	1.418	2.446	6.372	
	Rounded to GM	2	2	6	10
Webster	AM	1.5	2.5	6.5	
	Rounded to AM	1	2	6	9
	$d = 957$	1.447	2.497	6.505	
	Rounded to AM	1	2	7	10
Jefferson	Rounded Down	1	2	6	9
	$d = 800$	1.731	2.988	7.781	
	Rounded Down	1	2	7	10

(b) Webster and Jefferson give one seat to the larger state C at the expense of the smaller state A; these two methods exhibit the larger state bias in this example. By contrast, Adams and Huntington-Hill give one seat to the smaller state A at the expense of the larger state C; these two methods exhibit the smaller state bias in this example. ◁

To explore further the larger state/smaller state bias, Table 21.5 looks at over two centuries worth of apportionments and tabulates the number of times that small states have been favored in 22 apportionments of the House of Representatives [Balinski 2001]. Recall that the Dean method was briefly mentioned in Chapter 20 as another divisor method. This table displays a clear ranking in how the different methods favor smaller states versus larger states: namely,

Adams → Dean → Huntington-Hill → Webster → Jefferson

smaller state bias ⟶ larger state bias

It turns out that this observed historical ranking can be explained by understanding the relationship between the *cut-off points* for rounding used by the different apportionment methods.

Comparing Cut-Off Points

Recall that each divisor method uses some way of deciding when to round up fractional remainders and when to round down. To illustrate, consider two states A and B, where the quota for A is 1.45 and the quota for B is 54.45. Using Webster's method we compare the quotas to their associated arithmetic means, which are 1.5 and 54.5, respectively.

TABLE 21.5: Divisor methods and smaller state bias

Method	#Times Small States Favored
Adams	22
Dean	17
Huntington-Hill	15
Webster	10
Jefferson	0

Consequently, both quotas will be rounded down because they are less than their arithmetic mean. So A will be given one representative and B will be given 54. By contrast, the Huntington-Hill method compares the quotas to their associated geometric means, which are 1.414 and 54.498, respectively. Since $1.45 > 1.414$, the quota for A will be rounded up to two representatives; since $54.45 < 54.498$, the quota for B will be rounded down to 54 representatives. Even though both states have the same fractional part 0.45, the smaller state A receives an extra seat. This occurs because the smaller state has a lower threshold (*cut-off point*) for rounding up relative to the geometric mean used by the Huntington-Hill method. This bias toward smaller states holds in general for the Huntington-Hill method, since the geometric mean is always smaller than (or equal to) the arithmetic mean.

As a matter of fact, the arithmetic mean (AM), geometric mean (GM), and harmonic mean (HM) satisfy the following relationship.

RESULT **Comparing Different Means**
For positive numbers x and y, their arithmetic, geometric, and harmonic means satisfy $HM \leq GM \leq AM$.

For example, $x = 1$ and $y = 2$ have the arithmetic mean $AM = 1.5$, the geometric mean $GM = \sqrt{1 \times 2} = 1.414$, and the harmonic mean $HM = 2/\left(\frac{1}{1} + \frac{1}{2}\right) = 1.333$. Here we verify that $HM \leq GM \leq AM$ holds.

Since the relation $HM \leq GM \leq AM$ holds in general, the cut-off points increase as we move from Dean (HM) $\rightarrow$ Huntington-Hill (GM) $\rightarrow$ Webster (AM), and in so moving we start to favor larger states at the expense of smaller states. In addition, the Adams method always rounds up, so we can place its cut-off point as the quota rounded down; the Jefferson method always rounds down, so we can place its cut-off point as the quota rounded up. This produces the same ranking of methods in terms of smaller state/larger state bias observed in our previous examples:

Adams (round up)	$\rightarrow$	Dean (HM)	$\rightarrow$	Huntington-Hill (GM)	$\rightarrow$	Webster (AM)	$\rightarrow$	Jefferson (round down)

Figure 21.1 shows pictorially where the cut-off points fall for the five divisor methods, in the case when the quota is between 3 and 4. Since the Jefferson method always rounds down, its cut-off point is placed at the upper limit of the interval (here, at 4), as we never round a quota up even if it is 3.999. At the other extreme, the Adams method always rounds up, so its cut-off point can be placed at the lower limit of the interval (here, at 3), as we always round a quota up even if it is 3.001.

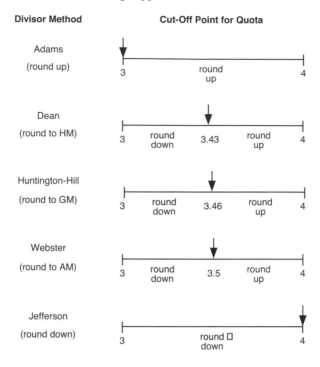

FIGURE 21.1: Five divisor methods and their cut-off points

Huntington-Hill and Webster

The Huntington-Hill and Webster methods both fall in the middle of the larger state/smaller state bias ranking. For other reasons, many people consider them the two best methods among the various divisor methods proposed [OR1]. We now compare them head-to-head.

Webster's method was in use before the Huntington-Hill method became law in 1941. The 1940 Census created a controversy because, as seen in Table 21.6, a Republican state, Michigan, received an extra seat using the Webster apportionment whereas a Democratic state, Arkansas, received an extra seat using the Huntington-Hill apportionment.

TABLE 21.6: 1940 apportionment controversy

State	1940 Population	Webster	Huntington-Hill
Michigan (R)	5,256,106	18	17
Arkansas (D)	1,949,387	6	7

A representative from Arkansas introduced a bill to require that the Huntington-Hill method be used. When it came to a vote, every Democrat except those from Michigan voted for it. Every Republican voted against it. It passed and Huntington-Hill remains the method in use today. This example also illustrates our general ranking of methods in that Webster gives an extra seat to the larger state Michigan, while Huntington-Hill gives an extra seat to the smaller state Arkansas.

Based on analyzing the 1790–2000 apportionments, H. P. Young [OR2] has argued however that Webster is closer to being *unbiased*, whereas Huntington-Hill systematically favors small states by 3 to 4%. He finds it odd that Huntington-Hill was in fact adopted instead of Webster:

> That happened in part because some of the country's leading mathematicians—including John von Neumann, Marston Morse, and Luther Eisenhart—claimed that it was unbiased. In a National Academy of Sciences report to Congress, they claimed that Hill's method was preferable because it "stands in a middle position as compared with the other methods."

HISTORICAL CONTEXT **Constitutional Challenge to Apportionment (1992)**
In 1991 Montana was upset because it lost a seat based on the results of the 1990 Census. It proposed two methods as alternatives to Huntington-Hill: the Dean (harmonic mean) method and the Adams method. In a separate challenge, Massachusetts proposed the Webster method. On March 31, 1992, in a unanimous decision, the U.S. Supreme Court upheld the constitutionality of the Huntington-Hill method.

Example 21.3 Compare the Webster and the Huntington-Hill Apportionments

Nine computers are to be allocated to three offices (A, B, C), with 40, 9, and 8 employees, respectively, for a total of 57 employees. (a) Calculate the apportionment using the Webster method. (b) Calculate the apportionment using the Huntington-Hill method. (c) Draw conclusions about bias toward larger or smaller offices for the two methods.

Solution

(a) For $d = 57/9 = 6.33$ in Table 21.7, the resulting apportionment is too small. With $d = 6.1$, the final Webster apportionment gives seven computers to A, one to B, and one to C.

TABLE 21.7: Webster apportionment for nine computers

Office	A	B	C	Total
Employees	40	9	8	57
Quota: $d = 6.33$	6.319	1.422	1.264	
Rounded to AM	6	1	1	8
Quota: $d = 6.1$	6.557	1.475	1.311	
Rounded to AM	7	1	1	9

(b) For $d = 6.33$ in Table 21.8, all nine computers can be allocated with the Huntington-Hill method: six computers are given to A, two to B, and one to C.

(c) Webster gives a computer to the larger office A at the expense of the smaller office B; Huntington-Hill gives a computer to the smaller office B at the expense of the larger office A. ◁

(Related Exercises: 5–7)

TABLE 21.8: Huntington-Hill apportionment for nine computers

Office	A	B	C	Total
Employees	40	9	8	57
Quota: $d = 6.33$	6.319	1.422	1.264	
GM	6.481	1.414	1.414	
Rounded to GM	6	2	1	9

21.3   Proximity to One Person–One Vote

There are criteria other than larger state/smaller state bias that can be used to assess the fairness of apportionment methods. For example, how close are the Webster and Huntington-Hill apportionments to achieving equal representation, based on population? To answer this question, we first look at the concept of representative share.

DEFINITION **Representative Share**
This measures the number of representatives per person, using the formula
$$\text{representative share} = \frac{\text{state's apportionment}}{\text{state's population}}.$$

Absolute Differences in Representative Share

In order to achieve one person–one vote parity, an apportionment method should try to keep each state's representative share the same. If that is not possible, then the method should keep them as similar as possible. Consequently, it is appropriate to compare alternative apportionment methods by seeing how closely they maintain equality of representative shares for each group. In the following, we concentrate on the largest and smallest values of representative share produced by a particular method and then calculate the *absolute difference* between these representative shares.

For example, Table 21.9 compares the representative shares of Michigan and Arkansas, given their 1940 populations and respective apportionments under both Webster and Huntington-Hill. Specifically, the representative share for Michigan under Webster is obtained by dividing the 18 seats by Michigan's population: $18/5,256,106 = 0.0000034246$ seats per person. Given such a small ratio, it is helpful to move the decimal point six places to the right and write the representative share as 3.4246, which is now interpreted as seats per million people. A similar calculation for Arkansas ($6/1,949,387$) gives 3.0779 seats per million. So, for the Webster apportionment method, the absolute difference between the larger representative share in Michigan and the smaller representative share in Arkansas is 0.3467.

Webster's method gives a smaller absolute difference (0.3467) than Huntington-Hill (0.3566) and so would be considered the more equitable method, using this criterion. In fact, it can be demonstrated that Webster's method will always give a smaller absolute difference in representative shares.

Example 21.4 Calculate and Compare Absolute Differences in Representative Share

TABLE 21.9: Absolute differences in representative share

	Webster	Huntington-Hill
Michigan	$\frac{18}{5,256,106} \rightarrow 3.4246$	$\frac{17}{5,256,106} \rightarrow 3.2343$
Arkansas	$\frac{6}{1,949,387} \rightarrow 3.0779$	$\frac{7}{1,949,387} \rightarrow 3.5909$
Absolute Difference	$3.4246 - 3.0779 = 0.3467$	$3.5909 - 3.2343 = 0.3566$

We revisit Example 21.3, in which nine computers are to be allocated to three offices (with 40, 9, and 8 employees, respectively). Both Webster and Huntington-Hill allocate one computer to office C, but they differ in the allocations to offices A and B. Namely, Webster apportions seven computers to A and one to B, while Huntington-Hill apportions six to A and two to B. (a) Focusing on offices A and B, determine the absolute difference in representative share for each method. (b) Which method is preferable in terms of absolute difference in representative share?

Solution

(a) As seen in Table 21.10, the absolute difference in representative share for Webster is 0.064 and that for Huntington-Hill is 0.072.

TABLE 21.10: Determining representative shares for allocating computers

	Webster	Huntington-Hill
Office A	$\frac{7}{40} = 0.175$	$\frac{6}{40} = 0.15$
Office B	$\frac{1}{9} = 0.111$	$\frac{2}{9} = 0.222$
Absolute Difference	$0.175 - 0.111 = 0.064$	$0.222 - 0.15 = 0.072$

(b) Webster's absolute difference is smaller and so is preferred using this measure. ◁

(*Related Exercises: 8 13*)

Absolute Differences in District Size

Rather than using representative share, we could instead use the concept of *district size* (which is in fact the reciprocal of representative share).

DEFINITION **District Size**
This measures the population represented per seat, using the formula
$$\text{district size} = \frac{\text{state's population}}{\text{state's apportionment}}.$$

An apportionment method should try to equalize the district sizes among the states as much as possible. This leads to comparing methods by computing the absolute difference between the largest and smallest district sizes for each method. To illustrate, Table 21.11

computes the district sizes (in people per seat) for Michigan and Arkansas under the two apportionment methods, as well as the absolute differences in district size.

TABLE 21.11: Absolute differences in district size

	Webster	Huntington-Hill
Michigan	$\frac{5,256,106}{18} = 292,006$	$\frac{5,256,106}{17} = 309,183$
Arkansas	$\frac{1,949,387}{6} = 324,898$	$\frac{1,949,387}{7} = 278,484$
Absolute Difference	$324,898 - 292,006 = 32,892$	$309,183 - 278,484 = 30,699$

The Huntington-Hill method is preferable in this example since it produces a smaller absolute difference in district sizes. (Dean's method is actually the best method using this criterion.)

Example 21.5 Calculate and Compare Absolute Differences in District Size

We return to Example 21.3, in which nine computers are to be allocated to three offices (with 40, 9, and 8 employees, respectively). Webster and Huntington-Hill produce different allocations to offices A and B. Namely, Webster apportions seven computers to A and one to B, while Huntington-Hill apportions six to A and two to B. (a) Focusing on offices A and B, determine the absolute difference in district size for each method. (b) Which method is preferable in terms of absolute difference in district size?

Solution

(a) As seen in Table 21.12, the absolute difference in district size for Webster is 3.286 and that for Huntington-Hill is 2.167. Values are shown as people per computer.

TABLE 21.12: Determining district sizes for allocating computers

	Webster	Huntington-Hill
Office A	$\frac{40}{7} = 5.714$	$\frac{40}{6} = 6.667$
Office B	$\frac{9}{1} = 9$	$\frac{9}{2} = 4.5$
Absolute Difference	$9 - 5.714 = 3.286$	$6.667 - 4.5 = 2.167$

(b) Huntington-Hill's absolute difference is smaller and is preferred using this measure. ◁

(*Related Exercises: 14–19*)

Relative Differences in Representative Share and District Size

When calculating absolute differences, we have encountered a paradoxical situation. Webster's method is better in minimizing absolute differences in representative share, whereas Huntington-Hill is better in minimizing absolute differences in district size. Which of these equally reasonable measures should we use? A way to resolve this conflict is to use *relative differences* rather than absolute differences to compare apportionments.

> DEFINITION **Relative Difference**
>
> $$\text{relative difference} = \frac{\text{larger number} - \text{smaller number}}{\text{smaller number}} \times 100\%.$$

One reason for using a relative rather than an absolute measure can be illustrated as follows. Suppose A has a district size of 100,000 and B has a district size of 50,000. The absolute difference in district sizes is 50,000. Now suppose A has a district size of 75,000 and B has a district size of 25,000. Again the absolute difference is 50,000. However, the inequality in representation is not the same: in the first case, A's district size is 100% larger than that of B, while in the second case A's district size is 200% larger than that of B. Using the relative difference highlights the distinction between these two situations.

In addition, when relative differences are calculated, using either representative share or district size yields the same result. So we avoid the paradoxical situation encountered when using absolute differences. Edward Huntington (a Harvard mathematician) proved that the Huntington-Hill method *always* yields an apportionment that minimizes the relative difference in representative share (or district size). These properties are illustrated in Table 21.13, which shows the calculation of both absolute and relative differences for Michigan and Arkansas. It is seen that the relative differences yield the same result for representative share and for district size. Moreover, Huntington-Hill is preferable to Webster when measured by relative differences in either representative share or district size.

TABLE 21.13: Absolute and relative differences in representative share and district size

	Webster	Huntington-Hill
Absolute Difference (Rep. Share)	$3.42459 - 3.07789 = 0.3467$	$3.59087 - 3.23433 = 0.35654$
Relative Difference (Rep. Share)	$\frac{0.3467}{3.07789} = 0.113$ $\rightarrow 11.3\%$	$\frac{0.35654}{3.23433} = 0.110$ $\rightarrow 11.0\%$
Absolute Difference (District Size)	$324{,}898 - 292{,}006 = 32{,}892$	$309{,}183 - 278{,}484 = 30{,}699$
Relative Difference (District Size)	$\frac{32{,}892}{292{,}006} = 0.113$ $\rightarrow 11.3\%$	$\frac{30{,}699}{278{,}484} = 0.110$ $\rightarrow 11.0\%$

Example 21.6 Compare Relative Differences in Representative Share and District Size

Recall that in Example 21.3, Webster apportions seven computers to A and one to B, while Huntington-Hill apportions six to A and two to B. Example 21.4 calculated the absolute differences in representative share, and Example 21.5 calculated the absolute differences in district size. (a) Focusing on offices A and B, determine the relative difference in representative share for each method. Also, determine the relative difference in district size for each method. (b) Which method is preferable in terms of relative difference in representative share and which is preferable in terms of relative difference in district size?

Solution

(a) The second column of Table 21.14 shows the calculation of relative differences in representative share and district size for Webster. The third column shows similar calculations for Huntington-Hill. Notice that the relative differences for representative share and district size are the same for each method: for Webster it is 57.5%, whereas for Huntington-Hill it is 48.1%.

TABLE 21.14: Absolute and relative differences for allocating computers

	Webster	Huntington-Hill
Absolute Difference (Rep. Share)	$0.175 - 0.1111 = 0.0639$	$0.2222 - 0.15 = 0.0722$
Relative Difference (Rep. Share)	$\frac{0.0639}{0.1111} = 0.575$ $\rightarrow 57.5\%$	$\frac{0.0722}{0.15} = 0.481$ $\rightarrow 48.1\%$
Absolute Difference (District Size)	$9 - 5.7143 = 3.2857$	$6.6667 - 4.5 = 2.1667$
Relative Difference (District Size)	$\frac{3.2857}{5.7143} = 0.575$ $\rightarrow 57.5\%$	$\frac{2.1667}{4.5} = 0.481$ $\rightarrow 48.1\%$

(b) Huntington-Hill's relative difference is smaller than Webster's for both representative share and for district size. So Huntington-Hill is the preferred method. ◁

Now we present a complete example where we first determine the Webster and Huntington-Hill apportionments and then compare the results using all four criteria.

Example 21.7 Compare Absolute and Relative Differences in Representative Share and District Size

Five states have a total population of 610,000. There are 70 seats in the legislature that need to be apportioned to these five states. The Webster and Huntington-Hill apportionments for two of the states (A and B) are calculated in Table 21.15; the other three states receive the same allocation under either apportionment method. Use this information to compare Webster and Huntington-Hill on (a) absolute and relative differences in representative share, and on (b) absolute and relative differences in district size.

Solution

(a) Table 21.16 shows the calculations for representative share. Webster is seen to be preferred in terms of absolute differences, while Huntington-Hill is seen to be preferred in terms of relative differences.

(b) Table 21.17 shows the calculations for district size. Huntington-Hill is seen to be preferred in terms of both absolute and relative differences. ◁

(Related Exercises: 20–25)

Example 21.7 illustrates the general fact that Webster will always do better in terms of absolute difference in representative share, whereas Huntington-Hill will always do better

TABLE 21.15: Webster and Huntington-Hill apportionments for two states

Method	State	A	B
	Population	300,500	21,500
Webster	$d = 8700$	34.540	2.471
	AM	34.5	2.5
	Rounded to AM	35	2
Huntington-Hill	$d = 8715$	34.481	2.467
	GM	34.496	2.449
	Rounded to GM	34	3

TABLE 21.16: Absolute and relative differences in representative share

	Webster	Huntington-Hill
State A	$\frac{35}{300,500} \to 116.473$	$\frac{34}{300,500} \to 113.145$
State B	$\frac{2}{21,500} \to 93.023$	$\frac{3}{21,500} \to 139.535$
Absolute Difference	$116.473 - 93.023 = 23.45$	$139.535 - 113.145 = 26.39$
Relative Difference	$\frac{23.45}{93.023} = 0.252$ $\to 25.2\%$	$\frac{26.39}{113.145} = 0.233$ $\to 23.3\%$

TABLE 21.17: Absolute and relative differences in district size

	Webster	Huntington-Hill
State A	$\frac{300,500}{35} = 8585.7$	$\frac{300,500}{34} = 8838.2$
State B	$\frac{21,500}{2} = 10,750$	$\frac{21,500}{3} = 7166.7$
Absolute Difference	$10,750 - 8585.7 = 2164.3$	$8838.2 - 7166.7 = 1671.5$
Relative Difference	$\frac{2164.3}{8585.7} = 0.252$ $\to 25.2\%$	$\frac{1671.5}{7166.7} = 0.233$ $\to 23.3\%$

for the other three cases. These relationships are summarized in Table 21.18, which shows the preferred method using each of the four assessment measures.

TABLE 21.18: Preferred method using the four measures

	Representative Share	District Size
Absolute Difference	Webster	Huntington-Hill
Relative Difference	Huntington-Hill	Huntington-Hill

(*Related Exercise: 26*)

21.4 Redistricting for Fair Representation

Every 10 years, after each census, each state having two or more Representatives is required to divide itself into the number of Congressional Districts equal to its new House seat apportionment. Such a redistricting typically involves the regrouping of population units (e.g., counties) in order to create new districts that align better with population shifts. This process is carried out either by the state legislature or by another body prescribed by state law, such as the courts when legal issues arise. Legal issues arise if the political party controlling the state legislature tries to manipulate the districting in order to favor its own interests. For example, this manipulation might create a district that is not contiguous (i.e., consists of disjoint pieces) or create a district that is not compact (i.e., the distance between the voters and the geographic center of the district is too great). Such a redistricting plan might violate perhaps the most important criterion emphasized by Supreme Court rulings since 1964: namely, the requirement that Congressional Districts within each state have approximately *equal populations*. Having to meet these conditions of contiguity, compactness, and equal populations is part of judicial efforts to prevent political interest groups from manipulating districts to their own advantage. Such manipulation is called *Gerrymandering* and it has a long history.

HISTORICAL CONTEXT **Gerrymandering**

This term originated in 1812 when a Massachusetts governor (Elbridge Gerry) authorized a redistricting plan that included an unusually spread out, meandering district. This strange-shaped district was said to resemble a salamander, which accounts for the origin of the term *gerrymandering*. In general, gerrymandering may be used to help or hinder particular constituents, such as members of a political, racial, linguistic, religious, or economic group.

In order to measure the degree of adherence to population equality, a relative difference measure is commonly used—one that is a slight variation on the relative difference measures discussed in Section 21.3.

DEFINITION **Population Deviation**

$$\text{population deviation} = \frac{\text{largest population size} - \text{smallest population size}}{\text{average population size}} \times 100\%$$

In the above formula, the numerator computes the difference between the population size of the largest district and the population size of the smallest district. This is then divided by the overall average population size of a district in the state. Typically, a population deviation value of 10% or less is desirable in order to withstand constitutional challenges.

Example 21.8 Compute the Population Deviation for South Carolina in 2000

Table 21.19 shows the population sizes in 2000 of the six Congressional Districts in South Carolina, created for the 106th Congress. Calculate the population deviation for this districting plan. Does this districting plan satisfy the population equality criterion?

TABLE 21.19: South Carolina Congressional District populations, 2000 Census

Congressional District	Population (in 2000)
1	684,765
2	731,022
3	670,139
4	670,335
5	655,525
6	600,226

Solution

First, we compute the average population size of a district:

$$(684{,}765 + 731{,}022 + 670{,}139 + 670{,}335 + 655{,}525 + 600{,}226)/6 = 668{,}669.$$

The largest district has population 731,022 and the smallest has population 600,226 so the population deviation is given by

$$(731{,}022 - 600{,}226)/668{,}669 = 0.196, \text{ or } 19.6\%.$$

Clearly this value exceeds the 10% threshold for acceptance and a new redistricting plan will be needed to account for the population shifts since the last census. ◁

(Related Exercise: 27)

How difficult can it be to partition a state into districts with equal populations? If we were allowed to split up neighborhoods, then it would not be difficult at all. However, it is usually desirable to keep neighborhoods, or census tracts, or even counties from being arbitrarily divided. This "indivisibility" constraint is similar to that encountered in apportionment, where we must maintain an integer number of seats rather than a fractional number. The next example illustrates the complexity of carrying out redistricting in order to achieve approximate population equality.

Example 21.9 Redistrict a State with 16 Counties

Figure 21.2(a) shows the layout of a small state whose 16 counties are arranged in a 4×4 grid. The population sizes are shown for each county. The four current districts are as indicated by the bold lines, being labeled A, B, C, D. (a) Calculate the population deviation for the current districting plan. (b) Find a new districting plan with a reduced population deviation.

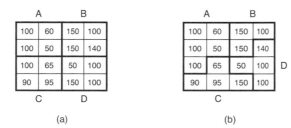

FIGURE 21.2: Two districting plans for a small state

First, we compute the population sizes of the current districts A, B, C, D: 310, 540, 350, 400. The average population size of a district is then $(310 + 540 + 350 + 400)/4 = 400$. In other words, the ideal district contains 400 people. The largest district has population 540 and the smallest has population 310 so the population deviation is given by

$$(540 - 310)/400 = 0.575, \text{ or } 57.5\%.$$

(b) An improved districting plan is obtained by moving one county (with population 140) from district B to district D, one county (with population 150) from district D to district C, and one county (with population 100) from district C to district A. Figure 21.2(b) shows the new districting plan, in which the district sizes are now 410, 400, 400, 390. The population deviation is then given by

$$(410 - 390)/400 = 0.05, \text{ or } 5\%.$$

This is clearly a much more desirable districting plan in terms of satisfying the population equality criterion. ◁

(*Related Exercises: 28–29*)

We can view the situation in Example 21.9 as a *graph partitioning* problem. Namely, we first construct a graph with the counties as vertices and with edges joining adjacent counties (those sharing a common border). This is the same construction used in Section 6.1 for converting from a map to a planar graph. We can also label the vertices by their population sizes. The initial districting plan can be viewed as the partitioning of the entire graph into disjoint districts that is shown in Figure 21.3(a). The improved districting plan is shown in Figure 21.3(b). Since the denominator for the calculation of the population deviation stays the same (it is the ideal district size 400), we seek a partitioning in which the difference between the largest district size and the smallest district size is minimized. This is yet another type of graph optimization problem, although one that is difficult to solve efficiently. In fact, it is an *NP-hard* optimization problem; see Section 5.1.

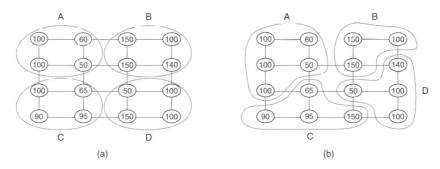

FIGURE 21.3: Two partitionings of a graph

In practice, redistricting is not a strictly mathematical problem, as there are other important political considerations. For example, it may be desirable to create *majority minority* districts—districts in which a majority of the voters come from a particular ethnic or racial minority. This may be done to ensure that minorities, which are spread apart from one another throughout the state, have sufficient numbers to elect a representative who is sensitive to their concerns. Following the 1990 Census, redistricting throughout the country increased the number of African-American and Hispanic majority districts. After the 1992 elections, the number of African-American members in the House of Representatives jumped from 26 to 39. Largely as a result of these new districts, Alabama, Florida, North Carolina, South Carolina, and Virginia elected African-American representatives for the first time since the turn of the century. Supreme Court challenges and districting controversies continue today.

Chapter Summary

This chapter studies historical systems of apportionment, most of which are divisor methods. These methods can be applied to a variety of situations. We found that divisor methods are not susceptible to the Alabama, population, or new states paradoxes, but situations arise where they do not preserve quota. In fact, the Balinski-Young theorem shows that there is no perfect apportionment method, one that preserves quota and yet avoids paradoxes. Consequently, we introduce additional measures that can be used to assess the fairness of apportionment methods. In particular, certain apportionment methods are shown to favor larger states, while others favor smaller states. Webster and Huntington-Hill are two methods that are least susceptible to larger state/smaller state bias, the latter method being the one currently used to apportion seats in the House of Representatives. These two methods are assessed in terms of their ability to achieve fair representation (one person–one vote). Once an apportionment has been selected, the task remains to regroup the population units within a state into districts that satisfy traditional fairness criteria. This is known as the problem of redistricting.

Applications, Representations, Strategies, and Algorithms

Applications		
Education (1)	Society/Politics (2, 7–9)	Business (3–6)

Representations					
Graphs	Tables		Sets	Symbolization	
Weighted	Data	Decision	Lists	Numerical	Geometric
9	8	1–7	1–7	1–9	9

Strategies	
Solution-Space Restrictions	Rule-Based
9	1–8

Algorithms				
Exact	Approximate	Inspection	Sequential	Numerical
1–8	9	9	3	1–8

Exercises

Hamilton Method and the New States Paradox

1. A small country consists of two provinces A and B, with respective populations of 13,400 and 5200.

   (a) Use the Hamilton method to allocate 16 legislative seats to the two provinces.

   (b) Suppose that a new province C is annexed with a population of 3800. Determine the appropriate number of seats to add for an apportionment that includes all three provinces. Then use the Hamilton method to allocate the new quantity of seats to all three provinces.

   (c) Determine if the new states paradox occurs.

2. A concert promoter has 100 tickets to distribute to music stores A and B, which have monthly sales of 1044 and 8956 DVDs.

   (a) Use the Hamilton method to apportion the 100 tickets to the two stores.

   (b) Suppose that a smaller music store C, with sales of 820 DVDs, also wants to receive an allocation of tickets. Determine the "fair share" number of tickets that C should receive. Then use the Hamilton method to apportion the new number of tickets to all three stores.

   (c) Determine if the new states paradox occurs.

3. A library system consists of three branches A, B, and C that serve, respectively, 1080, 6233, and 182 patrons.

   (a) Use the Hamilton method to allocate 100 workstations to the three branches.

   (b) Suppose that a new branch library D is added to the library system, serving an additional 510 patrons. Determine the appropriate number of workstations to add for an apportionment that includes all four branches. Then use the Hamilton method to allocate the new quantity of workstations to all four branches.

   (c) Determine if the new states paradox occurs.

4. A company has 10 temporary staff members to allocate to its main offices A, B, and C, with 990, 680, and 290 employees, respectively.

   (a) Use the Hamilton method to allocate the 10 staff members to the three offices.

   (b) Suppose that a new satellite office, with 525 employees, has now been acquired. Determine the appropriate number of staff members to add for an apportionment that includes all four offices. Then use the Hamilton method to allocate the new number of staff members to all four offices.

   (c) Determine if the new states paradox occurs.

Webster and Huntington-Hill Apportionments

5. A company has 15 items to apportion among four departments. There are 109 employees in the four departments: 20 employees in Dept. A, 34 in Dept. B, 42 in Dept. C, and 13 in Dept. D.

   (a) Use the Webster method to find an apportionment.

   (b) Use the Huntington-Hill method to find an apportionment.

   (c) Compare these two apportionments.

6. Fifteen computers will be distributed to four classes (A, B, C, D) with student populations of 97, 35, 15, and 9, respectively, for a total of 156 students.

   (a) Use the Webster method to find an apportionment.

   (b) Use the Huntington-Hill method to find an apportionment.

   (c) Compare these two apportionments in terms of any bias toward larger/smaller classes.

7. Ten video cameras will be distributed to four crime-scene investigators (A, B, C, D) with case loads of 16, 19, 23, and 50, respectively.

   (a) Use the Webster method to find an apportionment.

   (b) Use the Huntington-Hill method to find an apportionment.

   (c) Compare these two apportionments in terms of any bias toward larger/smaller case loads.

Absolute Difference in Representative Share

8. Using the Webster method, seven new computers are allocated to office A and three are allocated to office B. Offices A and B have 100 and 40 employees, respectively.

   (a) Determine the representative share (computers/person) for office A.

   (b) Determine the representative share (computers/person) for office B.

   (c) Calculate the absolute difference in representative share for the Webster method.

9. Forty new printers are to be distributed to three schools in a district. Webster's method gives 20 printers to school A, 6 to school B, and 14 to school C. Huntington-Hill's method gives 19 printers to school A, 7 to school B, and 14 to school C. Notice that the only differences between these allocations are the numbers given to A (with 1198 students) and B (with 399 students).

   (a) Determine the absolute difference in representative share for Webster's method.

   (b) Determine the absolute difference in representative share for the Huntington-Hill method.

   (c) Which method has the smaller absolute difference in representative share?

10. Teaching assistants are to be allocated to four sections of a large lecture class. When the Webster and Huntington-Hill methods are applied, the allocations only differ for sections A and D. Specifically, Webster assigns two teaching assistants to section A (with 75 enrolled students) and five teaching assistants to section D (with 136 enrolled students). Huntington-Hill assigns three teaching assistants to section A and four teaching assistants to section D.

    (a) Determine the absolute difference in representative share for Webster's method.

    (b) Determine the absolute difference in representative share for the Huntington-Hill method.

    (c) Which method has the smaller absolute difference in representative share?

11. In Exercise 6, 15 computers are distributed to four classes.

    (a) Which classes receive different allocations from the Webster and Huntington-Hill methods?

    (b) Determine the absolute difference in representative share for Webster's method.

    (c) Determine the absolute difference in representative share for the Huntington-Hill method.

    (d) Which method has the smaller absolute difference in representative share?

12. In Exercise 7, 10 video cameras are distributed to four crime-scene investigators.

    (a) Which of the investigators receive different allocations from the Webster and Huntington-Hill methods?

    (b) Determine the absolute difference in representative share for Webster's method.

    (c) Determine the absolute difference in representative share for the Huntington-Hill method.

    (d) Which method has the smaller absolute difference in representative share?

13. A total of 22 votes are to be allocated to three towns on a county council. Town A has a population of 228, Town B has a population of 157, and Town C has a population of 73.

    (a) Determine the Webster and Huntington-Hill apportionments. Which towns receive different allocations from the two methods?

    (b) Determine the absolute difference in representative share for Webster's method.

    (c) Determine the absolute difference in representative share for the Huntington-Hill method.

    (d) Which method has the smaller absolute difference in representative share?

Absolute Difference in District Size

14. In Exercise 8, seven new computers are allocated to office A (with 100 employees) and three are allocated to office B (with 40 employees).

    (a) Determine the district size (people/computer) for office A.

    (b) Determine the district size (people/computer) for office B.

(c) Calculate the absolute difference in district size for this apportionment.

15. In Exercise 9, Webster's method gives 20 printers to school A and 6 to school B, while the Huntington-Hill method gives 19 printers to school A and 7 to school B. School A has 1198 students and school B has 399 students.

   (a) Determine the absolute difference in district size for Webster's method.

   (b) Determine the absolute difference in district size for the Huntington-Hill method.

   (c) Which method has the smaller absolute difference in district size?

16. In Exercise 10, Webster's method assigns two teaching assistants to section A and five to section D, while the Huntington-Hill method assigns three teaching assistants to section A and four to section D. Section A has 75 enrolled students and section D has 136 enrolled students.

   (a) Determine the absolute difference in district size for Webster's method.

   (b) Determine the absolute difference in district size for the Huntington-Hill method.

   (c) Which method has the smaller absolute difference in district size?

17. In Exercise 6, 15 computers are distributed to four classes.

   (a) Which classes receive different allocations from the Webster and Huntington-Hill methods?

   (b) Determine the absolute difference in district size for Webster's method.

   (c) Determine the absolute difference in district size for the Huntington-Hill method.

   (d) Which method has the smaller absolute difference in district size?

18. In Exercise 7, 10 video cameras are distributed to four crime-scene investigators.

   (a) Which of the investigators receive different allocations from the Webster and Huntington-Hill methods?

   (b) Determine the absolute difference in district size for Webster's method.

   (c) Determine the absolute difference in district size for the Huntington-Hill method.

   (d) Which method has the smaller absolute difference in district size?

19. In Exercise 13, a total of 22 votes are allocated to three towns on a county council.

   (a) Which towns receive different allocations from the Webster and Huntington-Hill methods?

   (b) Determine the absolute difference in district size for Webster's method.

   (c) Determine the absolute difference in district size for the Huntington-Hill method.

   (d) Which method has the smaller absolute difference in district size?

Relative Differences in Representative Share and District Size

20. In Exercise 8, seven new computers are allocated to office A (with 100 employees) and three are allocated to office B (with 40 employees).

   (a) Determine the relative difference in representative share for this apportionment.

(b) Determine the relative difference in district size for this apportionment.

(c) Verify that you get the same percentage using relative differences in representative share or district size.

21. Ten IT staff are to be distributed to three district offices A, B, C that serve 900, 234, 500 employees. Webster's method gives six staff to office A and one to office B, while Huntington-Hill's method gives five staff to office A and two to office B. Both methods allocate three staff to office C.

    (a) Determine the relative difference in representative share for Webster's method and also for the Huntington-Hill method.

    (b) Determine the relative difference in district size for Webster's method and also for the Huntington-Hill method.

    (c) Verify that for each method you get the same percentage using representative share or district size.

    (d) Which method achieves the smaller relative difference?

22. Seventeen Super Bowl tickets are to be distributed to three top salespersons A, B, C who made 1000, 817, 1274 sales during the year. Webster's method gives six tickets to A and four to B, while Huntington-Hill's method gives five tickets to A and five to B. Both methods allocate seven tickets to C.

    (a) Determine the relative difference in representative share for Webster's method and also for the Huntington-Hill method.

    (b) Determine the relative difference in district size for Webster's method and also for the Huntington-Hill method.

    (c) Verify that for each method you get the same percentage using representative share or district size.

    (d) Which method achieves the smaller relative difference?

23. In Exercise 6, 15 computers are distributed to four classes. Focus on the two classes that receive different allocations from the Webster and Huntington-Hill methods.

    (a) Determine the relative difference in representative share for Webster's method and also for the Huntington-Hill method.

    (b) Determine the relative difference in district size for Webster's method and also for the Huntington-Hill method.

    (c) Verify that for each method you get the same percentage using representative share or district size.

    (d) Which method achieves the smaller relative difference?

24. In Exercise 7, 10 video cameras are distributed to four crime-scene investigators. Focus on the two investigators that receive different allocations from the Webster and Huntington-Hill methods.

    (a) Determine the relative difference in representative share for Webster's method and also for the Huntington-Hill method.

    (b) Determine the relative difference in district size for Webster's method and also for the Huntington-Hill method.

(c) Verify that for each method you get the same percentage using representative share or district size.

(d) Which method achieves the smaller relative difference?

25. In Exercise 13, a total of 22 votes are allocated to three towns on a county council. Focus on the two towns that receive different allocations from the Webster and Huntington-Hill methods.

    (a) Determine the relative difference in representative share for Webster's method and also for the Huntington-Hill method.

    (b) Determine the relative difference in district size for Webster's method and also for the Huntington-Hill method.

    (c) Verify that for each method you get the same percentage using representative share or district size.

    (d) Which method achieves the smaller relative difference?

26. Suppose that votes are apportioned to states A and B. Show that the relative difference in representative share will be identical to the relative difference in district size. [*Hint*: Let the representative shares for A and B be a and b, respectively, with $a > b$. Then the district shares for A and B will be $\frac{1}{a}$ and $\frac{1}{b}$, respectively, with $\frac{1}{a} < \frac{1}{b}$.]

Redistricting

27. In order to improve the population imbalance in Example 21.8, a new districting plan was devised for the six Congressional Districts in South Carolina. The population sizes for the new plan are shown in the following table. Calculate the population deviation for this new plan.

Congressional District	Population (in 2000)
1	673,610
2	678,048
3	670,421
4	674,305
5	664,705
6	650,923

28. A small state consists of 16 counties arranged in a 4×4 grid, with population sizes shown in the figure below. The four current districts are as indicated by the bold lines, being labeled A, B, C, D.

A		B	
120	115	133	162
78	89	128	94
112	93	107	85
159	122	121	130

C   D

(a) Calculate the population deviation for the current districting plan.

(b) Try to find a new districting plan achieving the smallest population deviation.

29. A small state consists of 16 counties arranged in a 4 × 4 grid, with population sizes shown in the figure below. The four current districts are as indicated by the bold lines, being labeled A, B, C, D.

A   B

43	81	50	68
65	39	67	92
56	58	57	75
67	42	32	48

C   D

(a) Calculate the population deviation for the current districting plan.

(b) Try to find a new districting plan achieving the smallest population deviation.

Project

30. In the 1992 Supreme Court case *U.S. Department of Commerce v. Montana*, the plaintiffs (Montana) argued that the Huntington-Hill method was unconstitutional as it did not respect the one person–one vote principle. Investigate the relative deviation criterion used by the Supreme Court, and compare it to the relative difference criterion used in this chapter. The article [OR3] discusses this issue, as well as the distinction between apportionment and redistricting.

Bibliography

[Balinski 2001] Balinski, M. L., and Young, H. P. 2001. *Fair representation: Meeting the ideal of one man, one vote*, second edition. Washington, D.C.: Brookings Institution Press.

[OR1] http://www.cics.northwestern.edu/documents/nilr/v2n1Bell.pdf (accessed March 5, 2014).

[OR2] http://www.brookings.edu/research/papers/2001/08/politics-young (accessed March 5, 2014).

[OR3] http://faculty.tcu.edu/epark/papers/Apportionment.pdf (accessed March 5, 2014).

[OR4] http://www.democracychronicles.com/proportional-representation-apportionment (accessed March 5, 2014).

[OR5] http://redistrictinggame.org (accessed March 5, 2014).

Unit V

Numbers: Cryptosystems and Security

Chapter 22

Modular Arithmetic and Cryptography

The right of the people to be secure in their persons, houses, papers and effects, against unreasonable searches and seizures, shall not be violated.

—The Fourth Amendment to the U.S. Constitution

Chapter Preview

The topic of cryptography is of interest not only to computer scientists, electrical engineers, and mathematicians, it is of vital interest to everyone. Every time we log into a secure network, make an online credit card payment, have paychecks deposited directly into our checking accounts, or worry about hackers accessing our private information, we should be concerned with cryptography. Our finances, our jobs, and our social lives involve and rely upon information security. Accordingly, this chapter presents the basic concepts of cryptography and illustrates how it can be used to add security to transmitted messages and data. This leads us into the branch of mathematics termed number theory, the study of integers and their properties.

22.1 Goals of Cryptography

What is cryptography? Initial definitions might include references to secret codes, the CIA, and the NSA. The term *cryptography* comes from Latin, meaning "secret writing." It is featured in popular culture through books and movies about spies and code breaking; as well, a sculpture called *Kryptos*, containing four secret codes, stands in front of CIA headquarters in Virginia. A more encompassing definition of cryptography includes references to computer passwords and the security of digital information. Computer-hacking schemes are now familiar that access the credit card information of millions of people for criminal purposes. In addition, governments and businesses are vulnerable to the malicious destruction or unintended alteration of data. Cryptography also involves methods of checking the accuracy or integrity of various forms of digital information.

DEFINITION **Cryptography**

The science of algorithms designed to secure communications and digital information so that this information remains confidential (not accessible to unauthorized individuals) and accurate (not altered by malicious intent or inadvertent system errors). Cryptography is a cross-disciplinary study, involving concepts from mathematics, computer science, and electrical engineering.

Cryptography was once the sole domain of governments and the military, but with the pervasiveness of digital data storage it has become a priority for business and individuals. Today governments must create their own secure systems to hide information from hackers, criminals, spies, and terrorists. As well, they must find ways to break secure systems of others that contain vital information about criminal and terrorist activity. Unfortunately, in breaking these systems, there have been reports of government infringement on citizens' rights to privacy [OR1].

Secrecy (privacy or confidentiality) is historically the primary goal of cryptography. To achieve secrecy, we must create a set of rules or processes that convert information from a readable state into an unreadable state; this is done so that only those who know the specific details of the process (called the *key*) can have access to this information. The process of inputting *plaintext* (a readable message) and outputting *ciphertext* (an unreadable message) is referred to as an *encryption algorithm*.

DEFINITION **Encryption Algorithm**

A list of instructions (agreed upon by sender and receiver) for inputting plaintext and outputting ciphertext. It involves the use of an *encryption key* (specific information) that is unknown to unauthorized individuals.

Only the intended receivers of an encrypted message (the ciphertext) should know how to turn it back into plaintext. They need to know the *decryption algorithm* and the key.

DEFINITION **Decryption Algorithm**

A list of instructions (agreed upon by sender and receiver) for inputting ciphertext and a *decryption key*, and then outputting plaintext. The instructions may be just the reverse of the encryption algorithm or may be completely different; the decryption key may be the same or different from the encryption key.

Can hackers or spies, who don't have the decryption key, break all cryptographic systems? Even using computers and sophisticated mathematics, current systems are unbreakable. In Chapters 23 and 24 we will see that there are some systems that can't be broken theoretically (can't be broken no matter how much time we have) or practically (can't be broken in a meaningful amount of time).

In the pre-computer era, there were many systems that required substantial effort to be broken. A famous encryption system called the *Vigenère Cipher* (to be discussed in Section 22.4) was used for approximately 300 years (including the U.S. Civil War) until it was broken. The German encryption scheme used in WWII involved a cryptographic machine called *Enigma* and was extremely difficult to break. For example, the successive occurrences of the letter E in the message WE NEED FUEL might be replaced by D, G, P, R today and by T, X, O, M tomorrow. This resulted in so many possible combinations for substitutions that it was impossible to break such a system without exploiting mistakes made by those sending the messages.

HISTORICAL CONTEXT **German Enigma Machine (1920–1940)**
The *Enigma* machine was a portable device with keyboard and several rotors (changed daily) that allowed the same typed letter appearing in different positions in a message to be replaced by different letters. It was used during WWII to encrypt strategic messages. Eventually, through Polish and British code-breaking innovations, through espionage, and through errors made by those encrypting the messages, the *Enigma* messages were successfully decrypted and this intelligence effort was credited with shortening the war.

HISTORICAL CONTEXT **Code Breaking in World War II**
– If an operator of the *Enigma* machine sent a message using yesterday's rotor settings, he might correct this mistake by changing the rotors and sending the same message again. This duplication allowed British code breakers to see patterns, crack the first message, and determine the internal wiring of the rotors.

– After intercepting numerous encrypted messages from the Japanese in WWII, U.S. code breakers were aided by the fact that the Japanese repeatedly used certain polite phrases to begin each message, such as "I have the honor to inform your excellency" and other formal titles. This revealed patterns and enabled reliable decryption.

Encryption and decryption became a science after WWII. Mathematicians, engineers, and computer scientists contributed to the development of modern cryptography, which now supports a variety of employment opportunities.

22.2 Modular Arithmetic

An important ingredient of modern cryptosystems is the concept of *modular arithmetic*. In essence, it allows us to work with a limited range of integer values, rather than needing to consider all possible (positive and negative) integer values.

DEFINITION **Modular Arithmetic**
A system of arithmetic in which integer values lie on a circular number line rather than on the infinite number line. Mathematical operations also take place on this circular number line.

One can motivate the ideas of modular arithmetic by considering how we do arithmetic with time. For example, suppose that it is now 11 am and backing up our computer will take four hours. We'd like to know when the backup will be complete. If we compute $11 + 4 = 15$, then this value needs to be converted into standard "clock" time—namely, by subtracting 12—giving us $15 - 12 = 3$, which we can interpret as 3 pm. Technically speaking, we are doing arithmetic on numbers *modulo* 12 and the notation for this operation would be 15 (mod 12) = 3 since 3 is the remainder when 15 is divided by 12.

DEFINITION a (mod n)
This mathematical function produces the remainder when a is divided by n. The remainder is an integer in the range $0, 1, \ldots, n - 1$.

The next example extends this notion further.

Example 22.1 Find the Ending Time of an Event

It is now 10 am and a special Internet bidding will expire in 44 hours. At what time of day will this bidding end?

Solution

Again, we begin with the quantity $10 + 44 = 54$. However, subtracting 12 hours leaves us with $54 - 12 = 42$, still not a valid time. So we continue to subtract 12 hours until we get a value in the appropriate range $0, 1, \ldots, 11$:

$$42 - 12 = 30, \qquad 30 - 12 = 18, \qquad 18 - 12 = 6.$$

So the final remainder is 54 (mod 12) = 6, but is this am or pm? A way to express the value 6 that we have obtained is $54 - (4 \times 12) = 54 - 48 = 6$. So we have subtracted off multiples of 12 until we are in the range of valid times. Moreover, we have subtracted off 12 an *even* number of times; this means that the ending time for the bidding is 6 am. Had we subtracted off 12 an *odd* number of times, it would have changed the original am time to a pm time. ◁

Now suppose that the bidding in Example 22.1 had ended in 50 hours. Following the same logic, we start with the sum $10 + 50 = 60$ and continue to subtract off multiples of 12. However, there are two logical choices:

$$60 - (4 \times 12) = 12, \qquad 60 - (5 \times 12) = 0.$$

If we subtract off 12 four times (an even number), we would get 12 am. So we choose the second approach that subtracts off 12 five times (an odd number) giving us the correct pm designation; the bidding ends at noon. It is our convention then to use 0 instead of 12 to carry out this modular arithmetic, meaning 60 (mod 12) = 0. The wisdom of such a choice will be seen in the following examples.

Example 22.2 Find the Ending Day of a Warranty

Sean bought a new inkjet printer on Friday and the purchase comes with a 60-day warranty. On what day will the warranty expire?

Solution

Because we are interested in days (not hours), we will work modulo 7. For convenience, we encode the days of the week beginning with 0, as seen in Table 22.1.

Since Friday is represented as 5, we compute $5 + 60 = 65$ and then reduce this to an appropriate value by subtracting multiples of 7: $65 - (9 \times 7) = 65 - 63 = 2$. That is, 65 (mod 7) = 2. Now we use Table 22.1 to translate the numerical result 2 back into the correct day, Tuesday. ◁

TABLE 22.1: Encoding the days of the week

Day	Sunday	Monday	Tuesday	Wednesday	Thursday	Friday	Saturday
Encoding	0	1	2	3	4	5	6

Example 22.3 Find a Specific Day of the Year

New Year's Day 2013 was on a Tuesday. On what day will it be in 2015?

Solution

Again we will work modulo 7 since we are interested in days of the week. Since neither 2013 nor 2014 is a leap year, each consists of exactly 365 days. From Table 22.1, Tuesday encodes as the value 2, so we need to compute $2 + 365 + 365 = 732$ in our modulo 7 system. To do this, first divide 732 by 7 to see how many multiples of 7 are present: $732/7 = 104.571$. Next, we subtract off the multiple 104 of 7 from 732 to get $732 - (104 \times 7) = 732 - 728 = 4$. Translating the value 4 using Table 22.1 then gives Thursday as New Year's Day 2015. ◁

(Related Exercises: 1–7)

Table 22.2 summarizes the use of modular remainders in Examples 22.1–22.3:

TABLE 22.2: Using modular remainders

It is now 10 am and bidding will expire in 44 hours. At what time of day will this bidding end?	54 (mod 12) = 6 $54/12 = 4$ (with remainder of 6)	6 am (am because the quotient is even)
It is now 10 am and bidding will expire in 50 hours. At what time of day will this bidding end?	60 (mod 12) = 0 $60/12 = 5$ (with remainder of 0)	12 pm (pm because the quotient is odd)
Sean bought a new inkjet printer on Friday, with a 60-day warranty. On what day will the warranty expire?	65 (mod 7) = 2 $65/7 = 9$ (with remainder of 2)	2 = Tuesday
New Year's Day 2013 occurred on a Tuesday. On what day will it be in 2015?	732 (mod 7) = 4 $732/7 = 104$ (with remainder of 4)	4 = Thursday

To evaluate 23 (mod 3), we divide 23 by 3 to determine the largest multiple of 3 that is less than (or equal to) 23. The largest multiple of 3 that is less than or equal to 23 is 21. Subtracting this multiple gives $23 - 21 = 2$ so that 23 (mod 3) = 2. If negative numbers are involved, we do the same thing. For example, to determine -23 (mod 3), we find that the largest multiple of 3 that is less than or equal to -23 is -24. Subtraction then produces $-23 - (-24) = 1$, so that -23 (mod 3) = 1. Note that in both cases the remainder modulo 3 is an integer in the range $0, 1, 2$.

How do we calculate 6 (mod 7) or -6 (mod 7)? Since 0 is the largest multiple of 7 that is less than or equal to 6 and since $6 - 0 = 6$, we have 6 (mod 7) = 6. This makes sense since the remainder modulo 7 is an integer in the range $0, 1, \ldots, 6$. Similarly, -7 is the largest multiple of 7 that is less than or equal to -6 and since $-6 - (-7) = 1$, we have -6 $(\bmod\ 7) = 1$, again an integer in the required range $0, 1, \ldots, 6$.

Example 22.4 Determine the Number of Rounds in a Dance

A certain dance from the British Isles involves seven couples, with a man and woman constituting each couple. These seven couples are arranged in a circular pattern, as shown in Figure 22.1; the women A, B, C, D, E, F, G form an inner circle and the men J, K, L, M, N, O, P face them from an outer circle. At the end of each round of the dance, the women will have progressed three positions clockwise while the men stay in their places. For instance, woman A will move from facing man J to facing man M after one round. (a) How many rounds will it be before woman A faces man O? (b) How many rounds will it be before woman A faces man N?

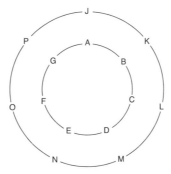

FIGURE 22.1: Circular arrangement of seven couples

Solution

(a) Since the seven men remain in fixed positions, let's label them with the integers $0, 1, \ldots, 6$ as specified in Table 22.3. We see that woman A moves from man J (position 0) to man M (position 3); she next moves to man P (position 6) and then to man L (position 2). In general, her next position is obtained by adding 3 to the current position, modulo 7:

$$0 + 3 \ (\bmod\ 7) = 3, \qquad 3 + 3 \ (\bmod\ 7) = 6, \qquad 6 + 3 \ (\bmod\ 7) = 2, \qquad 2 + 3 \ (\bmod\ 7) = 5.$$

Since 5 is the encoding of the desired final position (man O), it takes exactly four rounds (moving three positions each time) for woman A to reach him. That is, $3 + 3 + 3 + 3$ $(\bmod\ 7) = 12 \ (\bmod\ 7) = 5$.

TABLE 22.3: Encoding the fixed dance positions

Man	J	K	L	M	N	O	P
Encoding	0	1	2	3	4	5	6

(b) Woman A continues her graceful movement from man O (position 5) until she reaches man N (position 4):

$$5 + 3 \ (\bmod\ 7) = 1, \qquad 1 + 3 \ (\bmod\ 7) = 4.$$

As a result, she reaches man N after two more rounds, for a total of six rounds. We can verify this by the calculation $3 + 3 + 3 + 3 + 3 + 3 \pmod 7 = 18 \pmod 7 = 4$. ◁

In answering the questions posed in Example 22.4, we have been solving simple linear equations in our modular system. Specifically, if we let x denote the number of rounds to reach man O (position 5) from man J (position 0), then we are solving the equation

$$3 + 3 + \cdots + 3 \ (x \text{ times}) = 3x \equiv 5 \pmod 7.$$

Rather than writing the above as an ordinary equation $3x = 5$, we write it as the *linear congruence* $3x \equiv 5 \pmod 7$. The congruence symbol $\equiv$ does not indicate equality, but rather that the two numbers have the same remainder, when reduced modulo 7. For example, the linear congruence $60 \equiv 25 \pmod 7$ is valid, since the numbers 60 and 25 both have the same remainder of 4 when divided by 7. That is, $60 \pmod 7 = 4$ and $25 \pmod 7 = 4$.

DEFINITION **Linear Congruence**
The linear congruence $a \equiv b \pmod n$ means that a and b have the same remainder when divided by n. It is termed linear because it involves variables raised only to the first power.

RESULT **Validity of a Congruence**
The linear congruence $a \equiv b \pmod n$ holds whenever the difference $a - b$ is a multiple of n.

For example, the linear congruence $60 \equiv 25 \pmod 7$ holds because the difference $60 - 25 = 35$ is a multiple of 7. Likewise, $10 \equiv -4 \pmod 7$ holds because the difference $10 - (-4) = 14$ is a multiple of 7.

(Related Exercises: 8–12)

Linear congruences respect many of the properties as ordinary equations. The following transformations of a linear congruence do not change its validity:

1. The same integer can be added or subtracted from both sides of a linear congruence.
2. Both sides of a linear congruence can be multiplied by the same integer.
3. A number c appearing in a linear congruence $\pmod n$ can be replaced by another number d that is congruent to it $\pmod n$.

Such manipulations and substitutions can be used to simplify and solve a linear congruence such as $3x \equiv 5 \pmod 7$, which occurs in Example 22.4(a).

The solution turns out to be $x \equiv 4 \pmod 7$, which is not a single solution value but a set of solution values: namely, $4, 11, 18$, etc. Interpreting these solutions in the context of Example 22.4(a), we find that woman A (position 0) will reach man O (position 5) after 4 rounds, or 11 rounds, or 18 rounds, etc. So, she first reaches man O after 4 rounds. How do we find this solution? We cannot isolate the x by dividing both sides of $3x \equiv 5 \pmod 7$ by 3 because division is not a valid operation in modular arithmetic. Instead, we multiply both sides of the equation by a number that produces a multiple of 3 that is congruent to 1, modulo 7. The specific steps are listed next.

1. $3x \equiv 5 \pmod 7$ Given
2. $5(3x) \equiv 5(5) \pmod 7$ 1, Multiplying both sides by 5
3. $15x \equiv 25 \pmod 7$ 2, Simplification
4. $1x \equiv 4 \pmod 7$ 3, Since $15 \equiv 1 \pmod 7$, $25 \equiv 4 \pmod 7$
5. $x \equiv 4 \pmod 7$ 4, Simplification

As a verification we can directly check this solution: $3x \equiv 3(4) \equiv 12 \equiv 5 \pmod 7$.

We can set up a similar linear congruence to solve Example 22.4(b). Remember that if we let x denote the number of rounds for woman A (position 0) to reach man N (position 4), then we are solving the equation $3 + 3 + \cdots + 3$ (x times) $= 3x \equiv 4 \pmod 7$. To isolate the x, we again multiply both sides of the equation by a number that produces a multiple of 3 that is congruent to 1, modulo 7.

1. $3x \equiv 4 \pmod 7$ Given
2. $5(3x) \equiv 5(4) \pmod 7$ 1, Multiplying both sides by 5
3. $15x \equiv 20 \pmod 7$ 2, Simplification
4. $1x \equiv 6 \pmod 7$ 3, Since $15 \equiv 1 \pmod 7$, $20 \equiv 6 \pmod 7$
5. $x \equiv 6 \pmod 7$ 4, Simplification

We now see that woman A first reaches man N after six rounds of the dance. Again, we can directly verify that our solution is correct: $3x \equiv 3(6) \equiv 18 \equiv 4 \pmod 7$.

What was special about the value 5 used to multiply the linear congruence in Step 2? When we multiplied 3 by 5, we created a coefficient 15 of x that is congruent to 1, modulo 7: namely, $15 \equiv 1 \pmod 7$ since $15 - 1 = 14$ is a multiple of 7. The multiplier 5 is called the *inverse* of 3, modulo 7. Note that 5 is the only modular inverse of 3, modulo 7. Table 22.4 shows the results of trying all possible multiples $m = 0, 1, \ldots, 6$ in this modulo 7 system. We see that indeed 5 is the only multiple of 3 that gives a product of 1, modulo 7, and so 5 is the inverse of 3 in this modular system.

(Related Exercises: 13–17)

TABLE 22.4: Finding the inverse of 3, modulo 7

m	$m \times 3 \pmod 7$	Inverse?
0	$0 \times 3 \pmod 7 = 0 \pmod 7 = 0$	No
1	$1 \times 3 \pmod 7 = 3 \pmod 7 = 3$	No
2	$2 \times 3 \pmod 7 = 6 \pmod 7 = 6$	No
3	$3 \times 3 \pmod 7 = 9 \pmod 7 = 2$	No
4	$4 \times 3 \pmod 7 = 12 \pmod 7 = 5$	No
5	$5 \times 3 \pmod 7 = 15 \pmod 7 = 1$	Yes
6	$6 \times 3 \pmod 7 = 18 \pmod 7 = 4$	No

DEFINITION **Modular Inverse**
The inverse of $b \pmod n$ is the (unique) integer a in the range $0, 1, \ldots, n-1$ that gives a product ab that is congruent to 1, modulo n: $ab \equiv 1 \pmod n$.

We now illustrate how the concept of an inverse can aid in solving a puzzle that involves making postage.

Example 22.5 Make Postage Using Two Denominations of Stamps

A postmaster only has 5-cent and 8-cent stamps available. A customer arrives and needs postage for 61 cents. (a) Find all the possible ways of obtaining exactly 61 cents of postage using these two denominations of stamps. (b) What combination involves the fewest number of stamps?

Solution

(a) Let x denote the number of 5-cent stamps used and let y be the number of 8-cent stamps used. We are then led to solving the equation $5x + 8y = 61$, where x and y are nonnegative integers. If this equation holds, then also the related linear congruence $5x + 8y \equiv 61 \pmod 5$ holds. We now try to solve this simpler modular equation.

1. $5x + 8y \equiv 61 \pmod 5$ Given
2. $0x + 3y \equiv 1 \pmod 5$ 1, Since $5 \equiv 0 \pmod 5$, $8 \equiv 3 \pmod 5$, $61 \equiv 1 \pmod 5$
3. $3y \equiv 1 \pmod 5$ 2, Simplification
4. $2(3y) \equiv 2(1) \pmod 5$ 3, Since the inverse of 3 is 2, modulo 5
5. $1y \equiv 2 \pmod 5$ 4, Since $6 \equiv 1 \pmod 5$
6. $y \equiv 2 \pmod 5$ 5, Simplification

Since $y \equiv 2 \pmod 5$ holds, this means that $y - 2$ is a multiple of 5, or $y - 2 = 5k$, for some integer k. The expression $y = 2 + 5k$ can now be substituted into the original equation $5x + 8y = 61$ to give

$$61 = 5x + 8(2 + 5k) = 5x + 16 + 40k \implies 5x = 45 - 40k \implies x = 9 - 8k.$$

We have now obtained a set of solutions for x and y expressed in terms of the integer k: $x = 9 - 8k, y = 2 + 5k$. Table 22.5 shows sample values of x and y obtained when the integer k is varied. Since it is only meaningful to have nonnegative values for x and y, this table identifies two ways of making postage for 61 cents: $(x = 9, y = 2)$ nine 5-cent stamps and two 8-cent stamps giving $9(5) + 2(8) = 61$; $(x = 1, y = 7)$ one 5-cent stamp and seven 8-cent stamps giving $1(5) + 7(8) = 61$.

TABLE 22.5: Possible solutions for making postage

k	-2	-1	0	1	2
x	25	17	9	1	-7
y	-8	-3	2	7	12

(b) The solution involving the fewest number of stamps is $x = 1, y = 7$; so the postmaster should use one 5-cent stamp and seven 8-cent stamps to make postage for 61 cents. ◁

(Related Exercises: 18–20)

22.3   Caesar Cipher

The *Caesar Cipher* was first used over 2000 years ago and it is still used in cryptogram puzzles, but it is no longer taken seriously as a method of encryption. Yet, it will help to introduce the concepts needed in dealing with more complex encryption algorithms.

When using the Caesar Cipher, we shift each letter in a message a fixed number of places to the right in the alphabet. For example, when shifting three places to the right, AUGUST becomes DXJXVW. See Table 22.6. This is called a simple substitution or *monoalphabetic substitution* because it makes just one substitution for each letter over the entire message: U is always replaced with X, S with V, etc. To decrypt DXJXVW, each letter is shifted three places to the left to retrieve AUGUST. In the next section we will describe the Vigenère Cipher; like the multi-rotor *Enigma* machine, it uses multiple substitutions for the same letter when it occurs in different positions in the message (a *polyalphabetic substitution*).

TABLE 22.6: Substitution by shifting three places to the right

A	B	C	D	E	F	G	H	I	J	K	L	M	N	O	P	Q	R	S	T	U	V	W	X	Y	Z
D	E	F	G	H	I	J	K	L	M	N	O	P	Q	R	S	T	U	V	W	X	Y	Z	A	B	C

DEFINITION **Caesar Cipher**
The encryption algorithm is a simple substitution cipher that produces a cipher alphabet by shifting the plaintext alphabet a fixed number of places to the right to encrypt and the same number of places to the left to decrypt. The shift length is the key.

DEFINITION **Monoalphabetic Substitution**
This encryption method uses a fixed substitution over the entire message. For example, if the encryption algorithm replaces G with M, it does so throughout the entire message.

HISTORICAL CONTEXT **Julius Caesar (100–44 BC)**
Julius Caesar was leader of the Roman Empire. He is credited with using monoalphabetic substitution to send encrypted messages to his generals because he did not trust his messengers. According to the Roman historian Suetonius, Caesar used a cipher with a shift length of 3.

The Caesar Cipher is termed a *symmetric* system because the encryption key (the shift length) and the decryption key (the shift length) are the same. The key (the shift length) must be kept hidden or unauthorized individuals will have access to the original plaintext.

DEFINITION **Symmetric Cryptoystem**
In a symmetric system, the same key is used to encrypt and to decrypt messages.

Example 22.6 Encrypt Plaintext Using the Caesar Cipher

We wish to encrypt the plaintext message **ZEPHYR** using a shift length of 5 as the encryption key. (a) Use Table 22.7 to find the numerical position number of each letter in the plaintext. (b) Add the shift length of 5 to each numerical value (moving to the right five places in the alphabet) and then produce the corresponding ciphertext letters. Use modulo 26 arithmetic whenever the shift creates a new position number greater than 25.

TABLE 22.7: Alphabetic position numbers

A	B	C	D	E	F	G	H	I	J	K	L	M	N	O	P	Q	R	S	T	U	V
0	1	2	3	4	5	6	7	8	9	10	11	12	13	14	15	16	17	18	19	20	21
W	X	Y	Z																		
22	23	24	25																		

Solution

(a) The first two rows of Table 22.8 show the plaintext letters and their numerical position values.

TABLE 22.8: Encrypting with the Caesar Cipher

Z	E	P	H	Y	R	Plaintext letter
25	4	15	7	24	17	Alphabetic position
4	9	20	12	3	22	Add a shift of 5
E	J	U	M	D	W	Ciphertext letter

(b) Adding 5 to each entry in the second row of Table 22.8 produces the (shifted) values $30, 9, 20, 12, 29, 22$. Since the first value exceeds 25, we replace it by 30 (mod 26) = 4. Similarly, adding 5 to the numerical position value of 24 for Y produces 29 (mod 26) = 3. Then we again use Table 22.7 to obtain the corresponding encrypted message **EJUMDW**. ◁

Example 22.7 Encrypt Another Plaintext Using the Caesar Cipher

We wish to encrypt the plaintext message **MEET ME IN THE EAST** using a shift length of 10 as the encryption key. (a) Use Table 22.7 to find the numerical position number of each letter in the plaintext. (b) Add the shift length of 10 to each numerical value and then produce the corresponding ciphertext letters. Use modulo 26 arithmetic whenever the shift creates a new position number greater than 25.

Solution

(a) The first two rows of Table 22.9 show the plaintext letters and their numerical position values. Note that spaces between words have been removed.

TABLE 22.9: Encrypting with the Caesar Cipher

M	E	E	T	M	E	I	N	T	H	E	E	A	S	T
12	4	4	19	12	4	8	13	19	7	4	4	0	18	19
22	14	14	3	22	14	18	23	3	17	14	14	10	2	3
W	O	O	D	W	O	S	X	D	R	O	O	K	C	D

(b) Adding 10 to each entry in the second row of Table 22.9 produces the (shifted) values $22, 14, 14, 29, \ldots, 28, 29$. Since the shifted values for S and T exceed 25, we replace them, respectively, by 28 (mod 26) = 2 and 29 (mod 26) = 3. Then we again use Table 22.7 to obtain the corresponding encrypted message WOOD WO SX DRO OKCD. ◁

Example 22.8 Decrypt Caesar Ciphertext Given the Key

We wish to decrypt the message JOLBK encrypted using the Caesar Cipher, assuming we know that a shift length 10 was used as the encryption key. (a) Use Table 22.7 to find the numerical position number of each letter in the ciphertext. (b) Subtract the shift length of 10 from each numerical value (moving 10 places to the left in the alphabet) and then produce the corresponding plaintext letters. Use modulo 26 arithmetic whenever the shift creates a new position number less than 0.

Solution

(a) The first two rows of Table 22.10 show the ciphertext letters and their numerical position values.

TABLE 22.10: Decrypting with the Caesar Cipher

J	O	L	B	K	Ciphertext letter
9	14	11	1	10	Alphabetic position
25	4	1	17	0	Subtract a shift of 10
Z	E	B	R	A	Plaintext letter

(b) Subtracting 10 from each entry in the second row of Table 22.10 produces the (shifted) values $-1, 4, 1, -9, 0$. Since the shifted values for J and B are negative, we replace them, respectively, by -1 (mod 26) = 25 and -9 (mod 26) = 17. Then we again use Table 22.7 to obtain the corresponding decrypted message ZEBRA. ◁

(Related Exercises: 21–25)

We now describe more formally the general procedure used above:

Caesar Cipher Encryption Algorithm

1. Use a table, such as Table 22.7, to assign a position number to each plaintext letter.

2. Add the shift length to the position number of each letter.

3. If any sum is greater than 25, convert it to a position number from 0 to 25 by using modulo 26 arithmetic.

4. Convert these new position numbers to letters using the same table used in Step 1. This produces the ciphertext.

Caesar Cipher Decryption Algorithm

1. Use a table, such as Table 22.7, to assign a position number to each ciphertext letter.

2. Subtract the shift length from the position number of each letter.

---

3. If any difference is less than zero, convert it to a position number from 0 to 25 by using modulo 26 arithmetic.

4. Convert these new position numbers to letters using the same table used in Step 1. This produces the plaintext.

Decryption Using Brute Force

Example 22.8 shows that it is straightforward to decrypt a message encoded using the Caesar Cipher, if we know the shift length key. But how can we decipher such a message if we do not know the key? One way is to carry out an exhaustive key search: that is, we try every possible shift length until we produce some meaningful text. We illustrate this method using a very simple message, encrypted using the Caesar Cipher.

Example 22.9 Use Brute Force to Decrypt Caesar Ciphertext

We wish to decrypt the message BJI encrypted using the Caesar Cipher, assuming we do not know the encryption key (shift length). (a) Use Table 22.7 to find the numerical position number of each letter in the ciphertext. (b) Start trying each possible shift length (starting with 1), stopping once a meaningful translation has been found. We assume that the correct shift length has then been found.

Solution

(a) The first two rows of Table 22.11 show the ciphertext letters and their numerical position values.

TABLE 22.11: Decrypting using brute force

Shift of 1			Shift of 2			Shift of 3			Shift of 4			Shift of 5		
B	J	I	B	J	I	B	J	I	B	J	I	B	J	I
1	9	8	1	9	8	1	9	8	1	9	8	1	9	8
0	8	7	25	7	6	24	6	5	23	5	4	22	4	3
A	I	H	Z	H	G	Y	G	F	X	F	E	W	E	D

(b) The values in the third row are obtained from those in the second row by subtracting the trial shift length. Any negative values obtained are transformed to nonnegative values by using modulo 26 arithmetic. We stop once the meaningful text WED has been obtained using a shift length of 5. It turns out that 5 is the only shift length that will produce an English word. ◁

Remark. It is possible that the brute-force approach of trying all keys (0, 1, 2, ..., 25) will produce more than one meaningful plaintext, but even then, usually only one of the possible plaintexts fits the context.

Example 22.10 Use Brute Force to Decrypt Another Caesar Ciphertext

We wish to decrypt the message KNZX encrypted using the Caesar Cipher, assuming we do not know the encryption key (shift length). (a) Use Table 22.7 to find the numerical position number of each letter in the ciphertext. (b) Start trying each possible shift length, stopping once a meaningful translation has been found. [*Hint*: Start your search with a shift length of 14.]

Solution

(a) The first two rows of Table 22.12 show the ciphertext letters and their numerical position values.

TABLE 22.12: Decrypting using brute force

Shift of 14				Shift of 15				Shift of 16				Shift of 17			
K	N	Z	X	K	N	Z	X	K	N	Z	X	K	N	Z	X
10	13	25	23	10	13	25	23	10	13	25	23	10	13	25	23
22	25	11	9	21	24	10	8	20	23	9	7	19	22	8	6
W	Z	L	J	V	Y	K	I	U	X	J	H	T	W	I	G

(b) The values in the third row are obtained from those in the second row by subtracting the trial shift length. Any negative values obtained are transformed to nonnegative values by using modulo 26 arithmetic. We stop once the meaningful text **TWIG** has been obtained using a shift length of 17. It turns out that 17 is the only shift length that will produce an English word. ◁

(*Related Exercises: 26–27*)

Decryption Using Frequency Analysis

Breaking a Caesar Cipher (deciphering a message without knowing the key) can be done with brute force, but it is quicker if we can eliminate some alternatives through *frequency analysis*. Using this type of analysis, we exploit knowledge that certain letters (and combinations of letters) occur more or less frequently in English. For example, a study of some English language texts might conclude that E, A, and R are the most common letters, while Z, J, and Q are the least common. So, in decrypting a given ciphertext, any letter that appears most frequently is replaced with an E, A, or R. We know that Caesar Ciphers preserve letter frequency; for example, if F has been used to replace B in the plaintext, then the number of Bs in any plaintext will be the same as the number of Fs in the ciphertext.

DEFINITION **Frequency Analysis**

Using a frequency table that lists the relative frequencies of letters in a written language, we determine the most frequently occurring letters in the given ciphertext and use this information to make intelligent guesses as to which plaintext letter some ciphertext letter might represent. Frequency analysis is only useful in cryptosystems that preserve letter frequency.

Table 22.13 lists the relative frequencies (as percentages) for letters appearing in the main entries of the *Concise Oxford Dictionary* [OR2]. The letters are displayed in the order of most common to least common as we proceed down successive columns.

Example 22.11 Use Frequency Analysis to Decrypt Caesar Ciphertext

We wish to decrypt the ciphertext **PHHW PH DW WKH ZDOO** encrypted using the Caesar

TABLE 22.13: Oxford Dictionary frequency table

E	11.2%	S	5.74%	H	3.00%	V	1.01%
A	8.50%	L	5.49%	G	2.47%	X	0.290%
R	7.58%	C	4.54%	B	2.07%	Z	0.272%
I	7.55%	U	3.63%	F	1.81%	J	0.197%
O	7.16%	D	3.38%	Y	1.78%	Q	0.196%
T	6.95%	P	3.17%	W	1.29%		
N	6.65%	M	3.01%	K	1.10%		

Cipher, assuming we do not know the encryption key (shift length). (a) Rank the letters appearing in the ciphertext in order of their observed frequency. (b) Replace the letter with highest observed frequency by the most frequent letter found in Table 22.13. (c) Shift the remaining letters in the ciphertext, consistent with this replacement. If a legitimate English plaintext is produced, stop. Otherwise, try a replacement with the letter having the next highest frequency in Table 22.13, and repeat the previous step.

Solution

(a) Table 22.14 shows the seven distinct letters in the ciphertext, ranked in order of their observed frequency of occurrence among the 15 letters comprising the ciphertext.

TABLE 22.14: Observed frequencies of letter occurrences

H	W	D	P	O	K	Z
4	3	2	2	2	1	1

(b) In the second row of Table 22.15 we have replaced the highest observed frequency letter H by the highest frequency letter E from Table 22.13.

(c) Substituting the ciphertext letter H for the plaintext letter E corresponds to a shift length of 3. Since the message is encoded using a Caesar Cipher, we try decrypting the entire ciphertext by shifting all other letters by three positions to the left. Since this results in a meaningful plaintext MEET ME AT THE WALL, we assume this is the plaintext. ◁

(Related Exercises: 28–29, 39)

TABLE 22.15: Decrypting with a frequency table

Ciphertext	PHHW PH DW WKH ZDOO
Replace top frequency letter: H→ E	PEEW PE DW WKE ZDOO
Shift remaining letters left by three positions	MEET ME AT THE WALL

The Caesar Cipher does not fully highlight the strength of conducting a frequency analysis, because once we make a substitution for one letter based on frequency, we can simply apply the shift length defined by this substitution to all the other letters and then see if the resulting text makes sense. If it does, we are finished; if it does not, we continue to try in succession the next highest frequency letters from Table 22.13.

There are other, more involved, monoalphabetic substitution ciphers whose decryption depends more crucially on frequency analysis.

Cryptograms

Puzzles known as *cryptograms* are not Caesar Ciphers, but represent more general monoalphabetic substitution ciphers. Namely, each letter will be replaced by the same new letter throughout the entire message, but these new letters are not assigned according to any fixed shift length. Specifically, the letters A to Z are mapped to a fixed permutation of these 26 letters. Decrypting a cryptogram is usually based on our intuitive sense of the frequency of letters (rather than a precise frequency table), as well as our knowledge of the frequency of short words and letter patterns in the English language. For example, one way to attack a cryptogram is by recognizing common one- or two-letter words in English and identifying letter patterns such as double letters, vowel combinations, etc.

Puzzle books, magazines, and websites contain a wide variety of cryptograms that can be solved without knowing the key simply using pencil and paper. Since 1929, the American Cryptogram Association has supported recreational interest in these kinds of puzzles [OR3].

Example 22.12 Solve a Cryptogram Using Frequency and Language Patterns

Solve the cryptogram VJGTG KU X BXT TGPVXN using letter frequency, common words, and patterns.

Solution

Table 22.16 shows the observed frequencies of the ciphertext letters. Since G, T, X have the highest observed frequencies, we will try substituting the most common letters from Table 22.13 for them. In particular, the third word is X and there is no English word E, so we try the substitution X→ A, giving the second row of Table 22.17. Which ciphertext letter can correspond to E? Since the fourth word cannot end in AE, we replace G by E, giving the third row of the table. We have now found replacements for G and X, so let's see what substitution can be made for T. Since the most common letters E and A from Table 22.13 have already been assigned, we use the next most common letter R and make the substitution T→ R in the fourth row of Table 22.17. Moving on to V, which has observed frequency 2, we see which letter in the remainder of the first column of Table 22.13 might substitute for it. Given the letters already present in the first word, using I or O doesn't seem likely so we try the substitution V→ T shown in the fifth row. Now it seems reasonable to guess J→ H in the first word, and the second word is likely to be IS, giving the next row. The next most common unused letters in Table 22.13 are O, N, and L. So the last word is likely to be RENTAL, after the substitutions P→ N and N→ L. Finally, only the substitution B→ C seems likely and we end up with the plaintext THERE IS A CAR RENTAL. ◁

(Related Exercises: 30–31, 40)

TABLE 22.16: Observed frequencies of letter occurrences

G	T	X	V	B	J	K	N	P	U
3	3	3	2	1	1	1	1	1	1

We made various assumptions when making guesses in Example 22.12; however, caution is warranted in carrying out this type of frequency analysis. The human brain automatically fills gaps in words and sentences (and may do so in ways that could be wrong). An example from research in the psychology of language [OR4] shows us how quickly we fill in and read text that has been greatly distorted. We can easily reconstruct the following sentence, in

TABLE 22.17: Solving a cryptogram

Original Cryptogram	VJGTG KU X BXT TGPVXN
Substitute X→ A	VJGTG KU A BAT TGPVAN
Substitute G→ E	VJETE KU A BAT TEPVAN
Substitute T→ R	VJERE KU A BAR REPVAN
Substitute V→ T	TJERE KU A BAR REPTAN
Substitute J→ H , K → I, U→ S	THERE IS A BAR REPTAN
Substitute P→ N , N → L	THERE IS A BAR RENTAL
Guess B → C	THERE IS A CAR RENTAL

which the first and last letters of each word are in the right place, although the other letters may not be: "It deosn't mttaer what oredr the ltteers in a wrod are, bcuseae we do not raed ervey lteter by itslef but the wrod as a wlohe."

Security of the Caesar Cipher

The Caesar Cipher is easy to break because it has a *key space* of 26; there are only 26 different shift lengths to try. The smaller the key space, the more likely it is that we or a computer can use brute force and succeed. Key spaces in modern cryptosystems are so large that even the fastest computers cannot check all possible combinations in a reasonable amount of time.

DEFINITION **Key Space**
The set of all possible keys that can be used to initialize an encryption algorithm.

Another problem with Caesar Ciphers is that the secret key must be held by both the sender and the receiver of the message. If the receiver cannot meet in person to learn the key, how can you transmit it to him? An email message could certainly be intercepted. Moreover, how can you be sure that your partner will keep it secure? These concerns rule out the use of Caesar Ciphers in any modern cryptosystem. These ciphers do however provide the basis for more secure systems, such as the one described in the next section.

(Related Exercise: 41)

22.4 Vigenère Cipher

The Vigenère Cipher uses multiple shifts of the alphabet throughout the message (like the German *Enigma* machine discussed in Section 22.1). Instead of always using a single shifted alphabet, it uses several Caesar Ciphers within the same message. Such ciphers use *polyalphabetic substitution.*

DEFINITION **Polyalphabetic Substitution**

In this type of encryption method, substitutions for letters within the plaintext are based on multiple shifted alphabets. The same letter can have different substitutions within the same message.

The various shift lengths are defined by the alphabetic position of each letter in a *keyword*. If the keyword is CAT, then C is in alphabetic position 2 so the shift is 2, A is in position 0 so the shift is 0, and T is in position 19 so the shift is 19. By placing the keyword CAT repeatedly beneath the plaintext message, each letter in the plaintext will automatically be shifted according to its association with C, A, or T. The next two examples illustrate how this encryption procedure can be accomplished in a systematic way.

Example 22.13 Encrypt Plaintext Using the Vigenère Cipher

We want to encrypt the plaintext **EXIT TO THE LEFT** using the Vigenère Cipher with keyword **CAT**. (a) Align every letter of the plaintext with every letter of the keyword by repeating the keyword under the plaintext. (b) Add the alphabetic position value of each letter in the plaintext to the alphabetic position value of each letter in the keyword, working modulo 26. (c) Convert the resulting numerical values into the ciphertext.

Solution

(a) The plaintext **EXIT TO THE LEFT** is written in the second row of Table 22.18 and the numerical position values (from Table 22.7) are written above the plaintext letters. Then the keyword CAT is repeated as needed under the plaintext message. These letters will serve as the three possible shift lengths. (b) The numerical position values of the keyword are listed below the keyword letters and then we form the sum, modulo 26, of rows one and four. (c) The numerical values in row 5 are then converted to ciphertext using Table 22.7, resulting in the encrypted message **GXBV TH VHX NEYV**. ◁

TABLE 22.18: Encrypting a message with the Vigenère Cipher

4	23	8	19	19	14	19	7	4	11	4	5	19	Alphabetic position (plaintext)
E	X	I	T	T	O	T	H	E	L	E	F	T	Plaintext
C	A	T	C	A	T	C	A	T	C	A	T	C	Keyword
2	0	19	2	0	19	2	0	19	2	0	19	2	Alphabetic position (keyword)
6	23	1	21	19	7	21	7	23	13	4	24	21	Add shift lengths to plaintext
G	X	B	V	T	H	V	H	X	N	E	Y	V	Ciphertext

It is important to notice that, unlike the Caesar Cipher, the same letter in the plaintext may be encrypted as different letters in the ciphertext. For example, the three Es in the above plaintext message are successively encoded as G, X, E. This desirable property makes frequency analysis less obvious for the Vigenère Cipher.

DEFINITION **Vigenère Cipher**

A method of encrypting plaintext that uses a series of different Caesar Ciphers; each shift length is determined by the alphabetic position of each keyword letter.

HISTORICAL CONTEXT **Vigenère Cipher**
This method was first described in a book by the Italian cryptographer G. Belaso (1553) and was later misattributed in the 19th century to Blaise de Vigenère, a French cryptographer. The Vigenère Cipher was used by the Confederacy during the Civil War. Charles Babbage, credited with inventing the first mechanical computer, broke the Vigenère Cipher in 1854.

Example 22.14 Encrypt Another Plaintext Using the Vigenère Cipher

Using the Vigenère Cipher with keyword DOG, we want to encrypt the first four words of a Benjamin Frankin quotation: "There never was a good war or a bad peace." (a) Align every letter of the plaintext with every letter of the keyword by repeating the keyword under the plaintext. (b) Add the alphabetic position value of each letter in the plaintext to the alphabetic position value of each letter in the keyword, working modulo 26. (c) Convert the resulting numerical values into the ciphertext.

Solution

(a) The plaintext THERE NEVER WAS A is written in the second row of Table 22.19 and the numerical position values (from Table 22.7) are written above the plaintext letters. Then the keyword DOG is repeated as needed under the plaintext message. These letters will serve as the three possible shift lengths.

(b) The numerical position values of the keyword are listed below the keyword letters and then we form the sum, modulo 26, of rows 1 and 4.

(c) The numerical values in row 5 are then converted to ciphertext using Table 22.7, resulting in the encrypted message WVKUS THJKU KGV O. ◁

TABLE 22.19: Encrypting another message with the Vigenère Cipher

19	7	4	17	4	13	4	21	4	17	22	0	18	0
T	H	E	R	E	N	E	V	E	R	W	A	S	A
D	O	G	D	O	G	D	O	G	D	O	G	D	O
3	14	6	3	14	6	3	14	6	3	14	6	3	14
22	21	10	20	18	19	7	9	10	20	10	6	21	14
W	V	K	U	S	T	H	J	K	U	K	G	V	O

It was fairly easy to encrypt messages using the Vigenère Cipher. But how easy is it to decrypt ciphertext, assuming that we know the keyword? The next example shows how this can be done quite simply by inverting the encryption process.

Example 22.15 Decrypt Vigenère Ciphertext Given the Keyword

We wish to decrypt the message HAECURCE WWICO encrypted using the Vigenère Cipher, assuming we know that the keyword is WAR. (a) Align every letter of the ciphertext with every letter of the keyword by repeating the keyword under the plaintext. (b) Subtract the alphabetic position value of each letter in the keyword from the alphabetic position value of each letter in the ciphertext, working modulo 26. (c) Convert the resulting numerical values into the plaintext.

Solution

(a) The ciphertext HAECURCE WWICO is written in the second row of Table 22.20 and the numerical position values (from Table 22.7) are written above the ciphertext letters. Then the keyword WAR is repeated as needed under the ciphertext message.

(b) The numerical position values of the keyword are listed below the keyword letters. We then compute the difference between rows 1 and 4, modulo 26. For example, in the first column we compute $(7 - 22) \pmod{26} = -15 \pmod{26} = 11$.

(c) The numerical values in row 5 are then converted to plaintext using Table 22.7, resulting in the plaintext message LANGUAGE FAILS. ◁

(Related Exercises: 32–37)

TABLE 22.20: Decrypting a message encoded using the Vigenère Cipher

7	0	4	2	20	17	2	4	22	22	8	2	14	Alphabetic position
H	A	E	C	U	R	C	E	W	W	I	C	O	Ciphertext
W	A	R	W	A	R	W	A	R	W	A	R	W	Keyword
22	0	17	22	0	17	22	0	17	22	0	17	22	Alphabetic position
11	0	13	6	20	0	6	4	5	0	8	11	18	Subtract shifts
L	A	N	G	U	A	G	E	F	A	I	L	S	Plaintext

We can formally state the steps to implement the Vigenère Cipher:

Vigenère Cipher Encryption Algorithm

1. Use a table, such as Table 22.7, to assign a position number to each letter in the plaintext.

2. Position the keyword under the plaintext (repeating as necessary) and find the position number of each letter in the keyword.

3. Add the position numbers of the plaintext and the keyword letters.

4. If any sum is greater than 25, convert it to a position number from 0 to 25 by using modulo 26 arithmetic.

5. Convert these new position numbers to letters using the same table used in Step 1. This produces the ciphertext.

Vigenère Cipher Decryption Algorithm

1. Use a table, such as Table 22.7, to assign a position number to each letter in the ciphertext.

2. Position the keyword under the ciphertext (repeating as necessary) and find the position number of each letter in the keyword.

3. Subtract the position number of each keyword letter from the position number of the corresponding ciphertext letter.

4. If any difference is less that 0, convert it to a position number from 0 to 25 by using modulo 26 arithmetic.

5. Convert these new position numbers to letters using the same table used in Step 1. This produces the plaintext.

Security of the Vigenère Cipher

It is unlikely that a brute-force approach will succeed in deducing the keyword used by a thoughtfully constructed Vigenère Cipher. Consider how the size of the keyword affects the number of possible keywords that would have to be tested. A three-letter keyword results in $26 \times 26 \times 26 = 17{,}576$ possible shifts, using the Multistage Rule from Chapter 12. A five-letter keyword increases the number to $26 \times 26 \times 26 \times 26 \times 26 = 11{,}881{,}376$ possible shifts. So, to prevent an adversary (even with access to a high-speed computer) from deducing our keyword in a reasonable amount of time, we would just use a fairly long keyword.

Rather than using brute force, we can instead attempt to use frequency analysis to reveal the plaintext, encoded using the Vigenère Cipher. Recall the plaintext and ciphertext from Example 22.13 where the letter E in the plaintext message is successively encoded as the letters G, X, E. This then distorts the observed frequency distribution for letters in the encrypted message. Consequently, we cannot use standard frequency analysis on the totality of ciphertext letters in a message produced with the Vigenère Cipher.

On the other hand, suppose that we know the length of the keyword, even though the keyword itself is not known. We would then know which letters have been encrypted with the same Caesar Cipher. For example, if the keyword has length 3 (as in Example 22.13), then the same Caesar Cipher would be used on positions 1, 4, 7, 10, and 13; a second one on positions 2, 5, 8, 11, and 14; and a third one on positions 3, 6, 9, and 12. A frequency analysis could then be done knowing that letters in these positions would be the result of the same shift length. This is a weakness, and unfortunately for the Vigenère Cipher, cryptanalysts have identified simple statistical techniques that can be used to discover the keyword length.

There is one scenario in which the Vigenère Cipher would be unassailable. This occurs when (a) the keyword length is the same as the plaintext length, (b) the keyword is a randomly generated string of letters, and (c) the keyword is used only once. This means that a different Caesar Cipher is needed to encrypt each plaintext letter, and then there is no way to do a frequency analysis since there is only one observed ciphertext letter for each shift length. The impracticality of only using a keyword once, however, is a serious problem.

(Related Exercise: 38)

Chapter Summary

This chapter introduces basic concepts of cryptography as well as the use of modular arithmetic. Attention is then focused on two historical substitution ciphers: the Caesar Cipher and the Vigenère Cipher. Because the Caesar Cipher preserves letter frequency and because its key space is small, it is easy to decrypt ciphertext messages without knowing the key (the shift length). On the other hand, Vigenère Ciphers become progressively harder to break as the size of the keyword increases. However, once we know the length of the keyword, a frequency analysis can be carried out on the component Caesar Ciphers. In subsequent chapters we move on to more sophisticated and more secure methods of encryption.

Applications, Representations, Strategies, and Algorithms

Applications				
Business (1–2, 5)		Time Measurement (3)		Security (6–11, 13–15)

Representations				
Tables		Symbolization		
Data	Decision	Numerical	Algebraic	Geometric
2–4	6–15	1–4, 6–15	5	4

Strategies		
Brute-Force	Solution-Space Restrictions	Rule-Based
9–10	11-12	1–15

Algorithms			
Exact	Sequential	Numerical	Algebraic
1–15	1, 5, 9–12	1–15	5

Exercises

Modular Remainders

1. Evaluate the following mathematical function $a \pmod{n}$ that gives the remainder upon division of a by n. Note that the remainder should be in the range $0, 1, \ldots, n-1$.

   (a) $10 \pmod 6$

   (b) $3 \pmod 6$

   (c) $-13 \pmod 6$

2. Evaluate the following mathematical function $a \pmod{n}$ that gives the remainder upon division of a by n. Note that the remainder should be in the range $0, 1, \ldots, n-1$.

   (a) $35 \pmod 9$

   (b) $6 \pmod 9$

   (c) $-2 \pmod 9$

3. Evaluate the following mathematical function $a \pmod{n}$ that gives the remainder upon division of a by n. Note that the remainder should be in the range $0, 1, \ldots, n-1$.

   (a) $51 \pmod{26}$

   (b) $-51 \pmod{26}$

   (c) $52 \pmod{26}$

4. If there are no leap years, show using modular arithmetic that if New Year's Day is on a Monday in one year, then it is on a Tuesday the following year. (Represent the days of the week using $0 = $ Sun, $1 = $ Mon, $\ldots$, $6 = $ Sat.)

5. Suppose you begin driving to Louisiana on Wednesday. It takes you three days to drive there, including a stop in Alabama for one night. You stay in Baton Rouge for six nights, then in New Orleans for five nights, and again take two days to drive straight home. What day will you return home? (Represent the days of the week using $0 =$ Sun, $1 =$ Mon, ..., $6 =$ Sat.)

6. Suppose a municipal bond matures 250 months after its purchase in July 2012. In what month and year will it mature? (The months of the year can be labeled using $0 =$ Jan, $1 =$ Feb, ..., $11 =$ Dec.)

7. In Example 22.4, woman B starts the dance by facing man K. How many rounds will it take until she first faces man L? What modular equation are we solving?

Linear Congruences

Evaluate whether the following linear congruences are valid. Note that $a \equiv b \pmod{n}$ means that a and b have the same remainder when divided by n. Equivalently, it means that $a - b$ is a multiple of n.

8. $10 \equiv 25 \pmod 5$

9. $26 \equiv 6 \pmod{10}$

10. $14 \equiv 21 \pmod 6$

11. $-10 \equiv 14 \pmod 3$

12. $11 \equiv -5 \pmod 6$

Modular Inverses and Linear Congruences

13. Find the inverse of 7 modulo 9 by trying values $x = 0, 1, \ldots, 8$.

14. Find the inverse of 8 modulo 11 by trying values $x = 0, 1, \ldots, 10$.

15. Find the inverse of 8 modulo 17 by trying values $x = 0, 1, \ldots, 16$.

16. Find all solutions to the linear congruence $3x \equiv 2 \pmod 7$.

17. Find all solutions to the linear congruence $9x \equiv 2 \pmod{10}$.

18. A postmaster only has 4-cent and 7-cent stamps available. A customer arrives and requires postage for 47 cents. (a) Find all the possible ways of obtaining exactly 47 cents of postage with these two denominations of stamps by first solving the congruence $4x + 7y \equiv 47 \pmod 4$ for y. (b) Substitute the expression found for y in order to solve for x. (c) Determine all ways of dispensing exactly 47 cents of postage.

19. A postmaster only has 13-cent and 17-cent stamps available. A customer arrives and requires postage for \$4.53. (a) Find all the possible ways of obtaining exactly \$4.53 with these two denominations of stamps by first solving the congruence $13x + 17y \equiv 453 \pmod{13}$ for y and then using substitution to solve for x. (b) Determine the way to make postage that involves the fewest number of stamps.

20. A postmaster only has 11-cent and 13-cent stamps available. A customer arrives and requires postage for \$5.53. (a) Find all the possible ways of obtaining exactly \$5.53 with these two denominations of stamps by first solving the congruence $11x + 13y \equiv 553 \pmod{11}$ for y and then using substitution to solve for x. (b) Determine the way to make postage that involves the fewest number of stamps.

Caesar Cipher

In the following problems, use Table 22.7 for the alphabetic position numbers.

21. Suppose Michael's password at work must be a series of seven letters that do not form a word. So he uses a Caesar Cipher with a shift length of 14 to encrypt the plaintext SEAGULL anytime he needs to retrieve the password. Encrypt SEAGULL and then determine his password (the ciphertext).

22. Suppose Joan's password at work must be a series of seven letters that do not form a word. So she uses a Caesar Cipher with a shift length of 20 to encrypt the plaintext PELICAN anytime she needs to retrieve the password. Encrypt PELICAN and then determine her password (the ciphertext).

23. Decrypt the ciphertext NASHRK in order to find the missing word in the following phrase: *Be ____, you could be wrong.* The key is a shift length of 6.

24. Decrypt the ciphertext RULJLQDO in order to find the missing word in the following phrase: *If you are not prepared to be wrong, you'll never come up with anything ____.* The key is a shift length of 3.

25. Decrypt the ciphertext JAAREN in order to find the missing word in the following phrase: *I love to travel, but I hate to ____.* The key is a shift length of 9.

26. Decrypt the ciphertext FRZ from a Caesar Cipher by using brute force (since you do not know the shift length). Try each possible shift length $(1, 2, \ldots, 25)$ and stop the search once a meaningful translation is found.

27. Decrypt the ciphertext JMVI from a Caesar Cipher by using brute force (since you do not know the shift length). Try each possible shift length $(1, 2, \ldots, 25)$ and stop the search once a meaningful translation is found.

Frequency Analysis

28. Decrypt the ciphertext F XJ IXQB from a Caesar Cipher by using frequency analysis (since you do not know the shift length). (a) Rank the letters appearing in the ciphertext in order of their observed frequency. (b) Replace the letter with highest observed frequency by the most frequent letter found in Table 22.13. (c) Shift the remaining letters in the ciphertext, consistent with this replacement. If a legitimate English word is produced, stop. Otherwise, try a replacement with the letter having the next highest frequency in Table 22.13, and repeat the previous step.

29. Decrypt the ciphertext U ZQQP M OMD from a Caesar Cipher by using frequency analysis (since you do not know the shift length). (a) Rank the letters appearing in the ciphertext in order of their observed frequency. (b) Replace the letter with highest

observed frequency by the most frequent letter found in Table 22.13. (c) Shift the remaining letters in the ciphertext, consistent with this replacement. If a legitimate English word is produced, stop. Otherwise, try a replacement with the letter having the next highest frequency in Table 22.13, and repeat the previous step.

30. Use a frequency analysis, based on Table 22.13, common English words, and letter patterns, to decode the cryptogram JG YKNN DG CV VJG PGY CTGC.

31. Use a frequency analysis, based on Table 22.13, common English words, and letter patterns, to decode the cryptogram DQ DQW LV VPDOO.

Vigenère Cipher

32. Encrypt the plaintext QUANTUM using the keyword WAVE.

33. Encrypt the plaintext PHOTON using the keyword ZERO.

34. Encrypt the plaintext PARTICLE using the keyword LASER.

35. Decrypt the ciphertext HQPWWMKR using the keyword DICE.

36. Decrypt the ciphertext ZNBUTG using the keyword STATIC.

37. Decrypt the ciphertext ALWYCG using the keyword LAW.

38. The Vigenère ciphertext IR VR VKG KQXUH was produced using a keyword of length 2, but we don't know the key. Since the keyword has length 2, we do know that there are two different Caesar Ciphers involved. To illustrate the use of brute force, try using the following keywords of length 2: AB, AC, CD, DE. See whether one produces a meaningful plaintext message.

Projects

39. Use a spreadsheet (or computer program) to analyze several paragraphs of ordinary English plaintext. Compare the (normalized) frequency of occurrence of letters you find in this portion of text to the values found in Table 22.13.

40. Investigate different kinds of cryptograms and proposed methods to solve them, for example by consulting [OR3, OR5]. Apply these techniques to solve some sample cryptograms.

41. The *multiplication-shift cipher* is a variation of the Caesar Cipher. This method encrypts a letter of plaintext using a key that includes both a multiplier and a shift length. Explore the details of how this method can be used to encode and decode messages; see, for example, the book [Bauer 2013, Ch. 2] as well as [OR6].

Bibliography

[Bauer 2013] Bauer, C. 2013. *Secret history: The story of cryptology*, Chapters 1–3. Boca Raton, FL: CRC Press.

[Stinson 2006] Stinson, D. 2006. *Cryptography: Theory and practice*, Chapter 1. Boca Raton, FL: CRC Press.

[Young 2006] Young, A. 2006. *Mathematical ciphers: From Caesar to RSA*, Chapters 2–6. Providence, RI: American Mathematical Society.

[OR1] http://www.nytimes.com/2014/03/04/opinion/has-privacy-become-a-luxury-good.html (accessed May 27, 2014).

[OR2] http://oxforddictionaries.com/words/what-is-the-frequency-of-the-letters-of-the-alphabet-in-english (accessed May 27, 2014).

[OR3] http://www.cryptogram.org/solve_cipher.html (accessed May 27, 2014).

[OR4] http://www.mrc-cbu.cam.ac.uk/personal/matt.davis/Cmabrigde/ (accessed May 27, 2014).

[OR5] http://www.cryptograms.org/tutorial.php (accessed May 27, 2014).

[OR6] http://evergreen.loyola.edu/ayoung/www/prg/cp_ms.htm (accessed May 27, 2014).

Chapter 23

Binary Representation and Symmetric Cryptosystems

> There are 10 kinds of people in the world: those who understand binary numerals, and those who don't.
>
> —Ian Stewart, British mathematician, 1945–

Chapter Preview

The transition from historical to contemporary digital cryptosystems requires an understanding of binary representation. In current systems, plaintext messages first need to be converted into binary data before they are encrypted. Encryption and decryption keys are likewise binary, and so the encoding and decoding of messages involve the use of various logical operations, such as the *exclusive-or* (XOR). As computing power increases, security concerns have dictated that the encryption process needs to be more complex. Plaintext is therefore transformed by a series of operations that include not only substitution, but also the application of other mathematical transformations, such as permutation and matrix multiplication. Security for symmetric ciphers also involves the issue of how to transmit or share a private key, and a process for doing this is described.

23.1 Number Representations and Plaintext Encoding

Historical ciphers replace letters with other letters. By contrast, contemporary ciphers are designed to run on computers that replace binary numbers with other binary numbers. Plaintext and ciphertext messages, as well as encryption and decryption keys, are all processed as binary data consisting of sequences of 0s and 1s.

Section 10.3 discussed how computer signals can be represented by binary digits (bits) in the context of circuit design, and Section 10.4 extended the representation to bit strings. We pause now to explain more carefully the binary number system as well as the hexadecimal number system, also frequently used by computers.

Number Representations

The binary number system and the hexadecimal number system, like our decimal number system, are *positional number systems*, in which each symbol takes on meaning according to its position within the number.

DEFINITION **Positional Number System**
A system for representing numbers in which the numerical value depends on the position of the digits. For example, the number 123 has a very different value than the number 321, although the same digits are used in both.

HISTORICAL CONTEXT **Positional Number System**
The Babylonians are often credited as having developed the first positional number system, which appeared around 3100 BC; it used the number 60 as its base. Our common base-10 system, or Hindu-Arabic decimal system, was first used in India between the 1st and 4th centuries AD.

In the decimal system, the number 245 (two hundred and forty-five) is used to represent the quantity $(2 \times 100) + (4 \times 10) + (5 \times 1)$; namely, 2 is in the hundred's position, 4 is in the ten's position, and 5 is in the one's position. We see that the positional values grow larger by a factor of 10 as we move from right to left in the expression. Also, the available coefficients of the powers of 10 are the digits $0, 1, \ldots, 9$.

By contrast, the binary system uses just the coefficients 0 and 1 to multiply successive powers of two, again with powers increasing from right to left. For example, the binary representation 11010 corresponds to the decimal number $(1 \times 16) + (1 \times 8) + (0 \times 4) + (1 \times 2) + (0 \times 1) = 16 + 8 + 2 = 26$. The hexadecimal system uses the sixteen symbols $0, 1, \ldots, 9$, A, B, C, D, E, F to represent the coefficients $0, 1, \ldots, 15$. For example, the hexadecimal representation 5E3 corresponds to the decimal number $(5 \times 16^2) + (14 \times 16) + (3 \times 1) = 1280 + 224 + 3 = 1507$.

Table 23.1 summarizes these three positional systems; the decimal number system uses the base 10, the binary number system uses the base 2, and the hexadecimal system uses the base 16. It also shows what the digits *abc* would represent in each such system.

TABLE 23.1: Three positional number systems

	Decimal	Binary	Hexadecimal
Base	10	2	16
Digits	$0, 1, \ldots, 9$	$0, 1$	$0, 1, \ldots, 9$, A=10, B=11, C=12, D=13, E=14, F=15
Sample: *abc*	$(a \times 10^2) + (b \times 10^1) + (c \times 10^0)$	$(a \times 2^2) + (b \times 2^1) + (c \times 2^0)$	$(a \times 16^2) + (b \times 16^1) + (c \times 16^0)$

Example 23.1 Convert Decimal Numbers to Binary and Hexadecimal

(a) Find binary and hexadecimal representations of the decimal numbers 13, 17, 254, 255, 256, and 778. (b) Compare the binary and hexadecimal representations.

Solution

(a) To convert 13 into binary, we find the largest power of 2 that does not exceed 13 (namely, $8 = 2^3$) and then subtract that from 13, yielding $13-8 = 5$. The process is now repeated with 5: the largest power $4 = 2^2$ is subtracted from 5, leaving $5 - 4 = 1 = 2^0$. This produces the expansion $13 = 2^3 + 2^2 + 2^0$ and therefore the (unique) binary representation 1101. Table 23.2 records the results for all binary conversions. Notice that the largest number we can represent in a binary arrangement of eight positions is 255. The hexadecimal representations are found in a similar way, by successively processing remainders once the largest power of 16 is divided into the current value.

TABLE 23.2: Binary and hexadecimal representations of decimal numbers

Decimal	Binary Representation	Hexadecimal Representation
13	$8 + 4 + 1 = 2^3 + 2^2 + 2^0$ $\rightarrow 1101$	$13 = 13 \times 16^0$ $\rightarrow \text{D}$
17	$16 + 1 = 2^4 + 2^0$ $\rightarrow 10001$	$16 + 1 = (1 \times 16^1) + (1 \times 16^0)$ $\rightarrow 11$
254	$128 + 64 + 32 + 16 + 8 + 4 + 2 =$ $2^7 + 2^6 + 2^5 + 2^4 + 2^3 + 2^2 + 2^1$ $\rightarrow 11111110$	$240 + 14 = (15 \times 16^1) + (14 \times 16^0)$ $\rightarrow \text{FE}$
255	$128 + 64 + 32 + 16 + 8 + 4 + 2 + 1 =$ $2^7 + 2^6 + 2^5 + 2^4 + 2^3 + 2^2 + 2^1 + 2^0$ $\rightarrow 11111111$	$240 + 15 = (15 \times 16^1) + (15 \times 16^0)$ $\rightarrow \text{FF}$
256	$256 = 2^8$ $\rightarrow 100000000$	$256 = (1 \times 16^2)$ $\rightarrow 100$
778	$512 + 256 + 8 + 2 = 2^9 + 2^8 + 2^3 + 2^1$ $\rightarrow 1100001010$	$768 + 10 = (3 \times 16^2) + (10 \times 16^0)$ $\rightarrow 30\text{A}$

(b) In the binary representation of 254, we can group the bits into groups of four, starting from the right-hand side: 1111–1110. The first of these groups of bits encodes the decimal number 15, which is F in hexadecimal. The second group encodes the decimal number 14, which is E in hexadecimal. As another example, the binary representation of 778 can be arranged as 0011–0000–1010, where we have added the leftmost pair of 0s to create a group of four bits. These three groups encode the decimal numbers 3, 0, and 10, which correspond to 3, 0, and A in hexadecimal. This holds in general: we can easily move back and forth between binary and hexadecimal representations by replacing each consecutive group of four bits by a hexadecimal number or by replacing each hexadecimal number by four bits.

◁

Example 23.2 Convert Binary and Hexadecimal Numbers to Decimal

(a) Convert the hexadecimal number 203C to decimal. (b) Convert the binary number 01101100 to decimal.

Solution

(a) Recall that each position in a hexadecimal representation corresponds to a power of 16, increasing as we move from right to left. Consequently, 203C is an encoding of the decimal number $(2 \times 16^3) + (0 \times 16^2) + (3 \times 16^1) + (12 \times 16^0) = 8192 + 0 + 48 + 12 = 8252$.

(b) To convert the binary string 01101100 to decimal, we sum the appropriate powers of two that correspond to each 1 in the binary representation: $2^6 + 2^5 + 2^3 + 2^2 = 64 + 32 + 8 + 4 = 108$. Alternatively, we can first convert the binary string 01101100 to the hexadecimal number 6C, and then compute $(6 \times 16^1) + (12 \times 16^0) = 96 + 12 = 108$. ◁

(Related Exercises: 1–4)

Plaintext Encoding

Realizing that modern cryptographic methods are implemented on computers that deal with binary data, we discuss how plaintext messages in English are represented as plaintext messages in binary. Each letter, space, number, or punctuation mark in a message is associated with a number by means of agreed-upon conversion tables. These tables are referred to as *ASCII* or *Unicode*. The American Standard for Information Interchange (ASCII) system, based on the English language, converts letters, numbers, and other symbols into 7-bit binary numbers. Unicode extends the ASCII system to include symbols for all the languages in the world, including mathematical, scientific, and other symbols.

HISTORICAL CONTEXT **ASCII (1963)**
This is a character-encoding scheme for 128 specific characters (e.g., a space, the numbers 0–9, upper and lower case letters, punctuation marks, and certain control codes) using 7-bit binary numbers.

HISTORICAL CONTEXT **Unicode (1991)**
This scheme expanded the ASCII code to include symbols needed by other languages and by scientific disciplines. It now contains more than a million characters. Each character is represented using the hexadecimal number system because the compactness of hexadecimal numbers makes it easier to represent large numbers.

Table 23.3 displays the word **HELLO** in ASCII/Unicode in order to illustrate how each letter is represented by a number in the decimal, hexadecimal, and binary number systems. The native representation of a computer utilizes a binary representation; however, the hexadecimal representation (used by ASCII/Unicode) provides a more concise way of referring to long strings of binary symbols.

TABLE 23.3: Representation of text in ASCII/Unicode

	H	E	L	L	O
Decimal	72	69	76	76	79
ASCII/Unicode (Hexadecimal)	48	45	4C	4C	4F
Binary	01001000	01000101	01001100	01001100	01001111

Using similar conversions for English letters, it is straightforward to convert English plaintext into a form suitable for digital processing.

Example 23.3 Convert English Plaintext to Hexadecimal and Binary

Table 23.4 shows the ASCII/Unicode assignments for representing uppercase English letters by decimal numbers. We wish to encode the plaintext message **REPLY** into a string of binary digits. (a) Use Table 23.4 to convert each plaintext letter into hexadecimal. (b) Then convert each hexadecimal number into an 8-bit binary number. Add leading 0s if necessary to the binary representation to maintain the same length for each encoded binary string. (c) Produce the final (binary) plaintext message.

TABLE 23.4: ASCII/Unicode decimal assignments for uppercase letters

A	B	C	D	E	F	G	H	I	J	K	L	M
65	66	67	68	69	70	71	72	73	74	75	76	77
N	O	P	Q	R	S	T	U	V	W	X	Y	Z
78	79	80	81	82	83	84	85	86	87	88	89	90

Solution

(a) From Table 23.4, the letter R has the decimal representation 82. This is converted into hexadecimal by dividing 82 by 16 (the highest power of 16 that does not exceed 82), which gives a quotient of 5 with a remainder of 2: $82 = (5 \times 16^1) + (2 \times 16^0)$. As a result, R has the hexadecimal representation 52. In a similar way, E has the representation 45, P has the representation 50, L has the representation 4C, and Y has the representation 59.

(b) Translate the hexadecimal number 52 into binary by expanding each hexadecimal digit into its 4-bit binary equivalent. Namely, $5 = 4+1 \rightarrow 0101$ and $2 \rightarrow 0010$. This gives 01010010 as the binary representation of R. In a similar way, $4 \rightarrow 0100$ and $5 \rightarrow 0101$ so that E encodes as 01000101; P $\rightarrow 50 \rightarrow 01010000$; L $\rightarrow 4C \rightarrow 01001100$; and Y $\rightarrow 59 \rightarrow 01011001$.

(c) The final binary encoding of the English plaintext message is

$$01010010010001010101000001001100010110010.$$ ◁

(Related Exercise: 5)

Stream and Block Ciphers

Binary numbers can be encrypted one bit at a time, or by groups of bits at a time. This contrast is expressed by distinguishing *stream ciphers* from *block ciphers*. In Chapter 22, we explored the Vigenère Cipher, which involves a key that is repeatedly applied to blocks of letters the same size as the key.

> DEFINITION **Stream Cipher**
> A cipher that uses a key to encrypt the plaintext one symbol at a time.
>
> DEFINITION **Block Cipher**
> A cipher that divides the plaintext into groups of symbols (letters or bits). It uses the same key to encrypt each group of plaintext symbols. The number of symbols in each fixed group is the *block size*.

We will look at some block ciphers later in this chapter. First, we will look at a stream cipher, called the *Vernam Cipher*.

23.2   Vernam Cipher

The *Vernam Cipher* is a stream cipher that encrypts the plaintext by combining the key and the plaintext bit-by-bit. The key is a randomly chosen binary string; it is the same length as the plaintext and is only used to encrypt one plaintext. A logical operation called the *exclusive-or* (XOR) is performed to combine each pair of bits (one from the key and one from the plaintext) to produce a corresponding ciphertext bit. We now discuss this new operation, which can be very efficiently implemented in hardware.

Exclusive-Or(XOR)

The *exclusive-or* operation is a variant of the *or* connective introduced in Section 8.2. The *or* connective is labeled "inclusive" because it defines the compound statement $p \vee q$ to be true when both p and q are true. By contrast, the *exclusive-or* connective defines the compound statement to be false when both p and q are true. As a result, the XOR of two statements p and q is true only when p and q have different truth-values. We define the corresponding operation $p \oplus q$ on bits p and q in Table 23.5.

TABLE 23.5: Definition of *exclusive-or*

p	q	$p \oplus q$
1	1	0
1	0	1
0	1	1
0	0	0

Examination of Table 23.6 shows a relation between this operation and modular arithmetic. Specifically, $p \oplus q$ gives the same result as the modular remainder function $(p+q) \pmod 2$. Also, this table reveals an important property of the XOR operation: namely, $(p \oplus q) \oplus q = p$. This means that applying an XOR using the bit q twice in succession does not change the original bit p. This property means that we can carry out encryption and decryption with the same (binary) key, as will be next seen in the description of the Vernam Cipher.

TABLE 23.6: Modular arithmetic and *exclusive-or*

p	q	$p \oplus q$	$(p+q) \pmod 2$	$(p \oplus q) \oplus q$
1	1	0	0	1
1	0	1	1	1
0	1	1	1	0
0	0	0	0	0

The *Vernam Cipher* generates a random binary key and performs a bitwise XOR operation with the plaintext to produce the ciphertext. Table 23.7 provides an illustration of this encryption process. Notice that when two bits are the same, combining them generates a 0; when they are different, combining them generates a 1.

The Vernam Cipher is symmetric because the same key is used to encrypt and to decrypt. As seen in Table 23.8, forming the XOR of the ciphertext with the key retrieves the original plaintext. This occurs because of the property $(p \oplus q) \oplus q = p$ previously mentioned.

TABLE 23.7: Encryption using the Vernam Cipher

Plaintext	01011000100110
Random Key	10011100101010
Ciphertext	11000100001100

TABLE 23.8: Decryption using the Vernam Cipher

Ciphertext	11000100001100
Random Key	10011100101010
Plaintext	01011000100110

DEFINITION **Vernam Cipher**
This symmetric cipher converts plaintext into ciphertext (and vice versa) by processing one bit at a time. It forms the XOR of each text bit with each bit of the randomly chosen key of the same length as the text. This key is used only one time.

HISTORICAL CONTEXT **Vernam Cipher**
Gilbert Vernam (an engineer at AT&T) and Joseph Mauborgne (U.S. Army) co-invented this cipher in 1917 during WWI for use with telegraphs. This was the first time that electrical signal impulses were used in the encryption process and it became the basis for the voice encryption system called SIGSALY used in WWII.

There remains one final detail: how do we generate a random key of length n? This can be done by conducting a sequence of n tosses of a fair coin (see Chapter 12). Each time a head H appears, we generate a 1; each time a tail T appears, we generate a 0. So the sequence of coin tosses $HHHTTHT$ would generate the random key 1110010. There are other ways to generate random numbers: by using physical data or by computer. When random numbers are generated, for example, from atmospheric noise (radio noise caused primarily by lightning discharges in thunderstorms), these are considered true random numbers. When they are generated by computer, they depend on a formula that is initiated with a starting number, called a seed, and that uses a mathematical algorithm to generate successive bits. Computer-generated random numbers are called *pseudo-random* numbers. In a stream of truly random bits, it is not possible to predict the next number in the sequence from knowledge of the previous bits. By contrast, in a pseudo-random stream, there is a mathematical relation between the next bit generated and the previous bits; good pseudo-random number generators attempt to mimic the behavior of a truly random sequence.

The following example shows how we can produce a stream of pseudo-random bits by starting with an initial state (seed) and then moving to successive states by iteratively applying logical operations. This idea parallels that used in Section 11.2, where logical expressions were used to track the evolution of a biological system through successive time steps.

Example 23.4 Generate a Stream of Pseudo-Random Bits

Generate a stream of bits $x_1 x_2 x_3 x_4 \ldots$, starting with the seed $x_1 x_2 x_3 = 001$ by means of the linear recurrence relation $x_n = x_{n-2} + x_{n-3}$ (mod 2) for $n \geq 4$. (a) How many distinct 3-bit sequences are generated by this process? (b) What happens if you begin with the seed 111? (c) What happens if you begin with the seed 000?

Solution

(a) $x_4 = x_2 + x_1$ (mod 2) $= 0 + 0$ (mod 2) $= 0$; $x_5 = x_3 + x_2$ (mod 2) $= 1 + 0$ (mod 2) $= 1$; $x_6 = x_4 + x_3$ (mod 2) $= 0 + 1$ (mod 2) $= 1$; $x_7 = x_5 + x_4$ (mod 2) $= 1 + 0$ (mod 2) $= 1$; $x_8 = x_6 + x_5$ (mod 2) $= 1 + 1$ (mod 2) $= 0$; and so forth. This recurrence produces the stream 00101110010111100101110 ..., generating in succession the 3-bit sequences 001, 010, 101, 011, 111, 110, 100, 001, ... a pattern which is seen to repeat after 001 is generated again. This recurrence has generated all seven 3-bit strings except for 000.

(b) Beginning with the seed 111 produces the stream 11100101110010111... , generating in succession the 3-bit sequences 111, 110, 100, 001, 010, 101, 011, 111. This produces the same seven 3-bit strings found in (a), generated in the same order but starting with 111.

(c) When the seed 000 is used, we just keep generating over and over again the 3-bit sequence 000. Only one 3-bit sequence is produced. ◁

(Related Exercises: 6–9, 34)

What Example 23.4 has shown is that by choosing a recurrence and the initial seed wisely, we can generate all 3-bit strings (except one). The sequence produced has a *period* of $7 = 2^3 - 1$. Suppose that instead of using a recurrence that extends back three steps (to x_{n-3}), we use one extending back 17 steps. Then by using a suitably designed recurrence, it is possible to generate all $2^{17} - 1 = 131{,}071$ 17-bit strings before repeating. In this way, one can produce very long strings of bits that can be used in a stream cipher like Vernam.

Example 23.5 Encrypt Plaintext Using the Vernam Cipher

Suppose the message ART is to be transmitted. (a) Convert the letters in this message into decimal numbers using Table 23.4 and then convert each resulting decimal number into an 8-bit binary number. (b) Combine these three binary strings and then generate a random key of the same length. Encrypt the plaintext bit-by-bit by performing an XOR with the random key.

Solution

(a) Using Table 23.4, A is assigned the decimal number 65, R is assigned 82, and T is assigned 84. These decimal numbers are converted (respectively) into 01000001, 01010010, and 01010100.

(b) Combining these three 8-bit strings produces the following plaintext message of length of 24: 010000010101001001010100. Suppose we generate a random encryption key, also with 24 bits: 111001010111000101101001. Table 23.9 shows the result of combining these two binary strings bit-by-bit using the XOR operation, giving the ciphertext indicated. ◁

(Related Exercises: 10–11)

Example 23.6 Decrypt Vernam Ciphertext

Decrypt the ciphertext 101001000010001100111101 from Example 23.5. (a) XOR the ciphertext with the 24-bit key 111001010111000101101001. (b) Convert each 8-bit string to a decimal number, and then use Table 23.4 to determine the letters comprising the plaintext.

TABLE 23.9: Encryption using the Vernam Cipher

Plaintext	010000010101001001010100
Random Key	111001010111000101101001
Ciphertext	101001000010001100111101

Solution

(a) Table 23.10 shows the result of taking the XOR of the ciphertext with the key.

TABLE 23.10: Decryption using the Vernam Cipher

Ciphertext	101001000010001100111101
Random Key	111001010111000101101001
Plaintext	010000010101001001010100

(b) The binary number 01000001 is converted to the decimal number $(1 \times 64) + (1 \times 1) = 65$, which then corresponds to the plaintext letter A. Similarly, 01010010 is converted to the decimal number $(1 \times 64) + (1 \times 16) + (1 \times 2) = 82$, giving the plaintext letter R, and 01010100 is converted to the decimal number $(1 \times 64) + (1 \times 16) + (1 \times 4) = 84$, giving the plaintext letter T. As anticipated, the original plaintext ART is obtained. ◁

(Related Exercises: 12–13)

Another purpose of encryption, besides protecting passwords, communication, and personal data, could be to prevent the unauthorized use of a digital image. We might display a small image on a website that could be screen-captured, but could not be downloaded for higher resolution because it is stored in an encrypted form that alters the colors. In this way, we can prevent the unauthorized use of the image by others in its original high-resolution format.

Many image-editing tools use the RGB color model in which *red, green,* and *blue* light are added together in different mixtures (represented by numbers in the range 0–255) so it is possible to reproduce approximately $256 \times 256 \times 256 = 16,777,216$ different colors. The percentages of red, green, and blue for any pixel (a small element of the entire image) are listed as three numbers in these editing tools. When the red pixel is set to 0, the light is turned off. When the red pixel is set to 255, the light is turned fully on. Any value in between gives partial light emission. A black pixel would be coded as $(0, 0, 0)$ and a white pixel would be coded as $(255, 255, 255)$.

Example 23.7 Encrypt a Digital Image Using the Vernam Cipher

Suppose a pixel representing a particular shade of red has the RGB values $(209, 29, 29)$. We want to encrypt this pixel so that it becomes some other color. (a) Convert each of these three decimal numbers into an 8-bit binary number. (b) Combine these three binary strings and then generate a random key of the same length. Encrypt the plaintext bit-by-bit by performing an XOR with the random key. (c) Convert the resulting 24-bit string into the associated RGB values.

Solution

(a) To convert 209 into binary, we subtract the largest power of 2, giving $209 - 128 = 81$;

from 81 we subtract the largest power of 2, giving $81 - 64 = 17$; and so on. The result is the expression $209 = 128 + 64 + 16 + 1 = 2^7 + 2^6 + 2^4 + 2^0$, providing the binary representation 11010001. Similarly, $29 = 16 + 8 + 4 + 1 = 2^4 + 2^3 + 2^2 + 2^0$, giving the binary representation 00011101. Note that we have added leading 0s to obtain an 8-bit string.

(b) The three binary RGB numbers are combined into the 24-bit string 110100010001110100 011101. Suppose that we use the randomly generated 24-bit key 111100100100000111100000. The 24-bit plaintext and the 24-bit key are combined in Table 23.11 using the XOR operation.

TABLE 23.11: Encrypting RGB values

Plaintext	110100010001110100011101
Random Key	111100100100000111100000
Ciphertext	001000110101110011111101

(c) The ciphertext 00100011–01011100–11111101 is then translated into three decimal numbers:

$$00100011 \rightarrow 2^5 + 2^1 + 2^0, = 32 + 2 + 1 = 35,$$

$$01011100 \rightarrow 2^6 + 2^4 + 2^3 + 2^2 = 64 + 16 + 8 + 4 = 92,$$

$$11111101 \rightarrow 2^7 + 2^6 + 2^5 + 2^4 + 2^3 + 2^2 + 2^0 = 128 + 64 + 32 + 16 + 8 + 4 + 1 = 253.$$

As a result, the encrypted pixel has color $(35, 92, 253)$, which corresponds to a shade of royal blue. Without knowing the encryption key, unauthorized users will be unable to determine the true color of the pixel. Of course, this process can be applied to all pixels of the original image. ◁

(Related Exercises: 14, 35)

We summarize the steps of the Vernam Cipher as follows:

Vernam Encryption Algorithm

1. Using Table 23.4, convert each plaintext letter into a decimal number and then into an 8-bit binary number.

2. Randomly generate a binary key having the same length as the binary plaintext.

3. Combine the plaintext and the key using the XOR operation. This produces the ciphertext.

Vernam Decryption Algorithm

1. Produce the binary key.

2. Combine the ciphertext and the key using the XOR operation. This produces the binary plaintext.

3. Convert each 8-bit plaintext segment into a decimal number and then determine the associated plaintext letter using Table 23.4.

Security of the Vernam Cipher

Can the Vernam Cipher be broken if we don't know the key? The answer is no. The Vernam Cipher has *perfect security* because it satisfies the following three conditions:

1. The number of possible keys is equal to the number of possible plaintexts.
2. The key is randomly selected from the key space.
3. The key is only used once.

Recall that in Section 22.4, we mentioned that the only version of the Vigenère cipher that would resist attacks to find the decryption key had to meet these same three conditions.

DEFINITION **Perfect (Theoretical) Security**
A cryptosystem is perfectly secure when, without knowledge of the key, the ciphertext resulting from the encryption algorithm gives no information about the plaintext.

To appreciate why the Vernam Cipher has perfect security, suppose that the randomly generated key has length n, which is the same length as the plaintext message. Note that the number of possible plaintexts and the number of possible keys are the same: there are 2^n different ways to construct a binary plaintext of length n and 2^n different ways to construct a binary key of length n. The Vernam Cipher cannot be broken using brute force because, given enough computational power, brute force would eventually try every key of length n. When each such key is used to decrypt ciphertext of the same length, it will produce *every* possible sequence of n bits and (using Table 23.4) all English words of length $n/8$. There would be no way of knowing which resulting plaintext is the correct one. Since the actual key is randomly generated and then discarded after encrypting a single message, a frequency analysis is of no use in identifying this key.

The Vernam Cipher can be described as a *one-time pad*. This designation reflects the historic origins of the system in which both sender and receiver possess a booklet whose pages contain a stream of random digits. Each message is encrypted (and then decrypted) using the same sheet(s) from this pad. Once used, these sheets are discarded. In modern cryptography, a one-time pad is understood as a technique for encryption that has perfect security if the three conditions listed earlier are implemented correctly.

HISTORICAL CONTEXT **One-Time Pad**
The idea of using a random key (the same length as the message) only one time was first described in the context of telegraphy (1882) and in teleprinting (Vernam, 1917). It was subsequently used during World War I and World War II. In 1945, Claude Shannon, American mathematician, electrical engineer, and cryptographer, recognized and proved the theoretical significance (the perfect security) of the one-time pad.

Though the Vernam Cipher has perfect security, there are certain important objections to its use in practice. Such a system would be used only for the highest level of security for the following reasons:

1. Key storage and transmission: To decrypt ciphertext, the receiver must have access to the key. Since the key must be as long as the plaintext, this often requires very long keys. Also, since the sender generates the key, it must be transmitted secretly to the

receiver. The costs and inconvenience involved with storing and transmitting the key securely discourage its use.

2. One-time use: Repeatedly generating and storing a new key adds to the expense.

3. Randomness: The system assumes that the key used is randomly generated, so it cannot be guessed by any intruder. However, it is difficult to generate long streams of truly random numbers. Instead, pseudo-random number generators are commonly used, and as a result perfect security may be compromised.

23.3   Block Ciphers

Many stream and block ciphers have been developed to use the XOR operation. The stream ciphers that encrypt and decrypt one bit at a time are fast but not as secure as block ciphers, which manipulate an entire set of bits at one time. Modern block ciphers divide the plaintext into several equal-sized blocks of binary digits, and then perform a variety of mathematical transformations and substitutions on these blocks to produce the ciphertext. Also, unlike the Vernam Cipher, they do not generate a key that is as long as the plaintext. Rather, they use a shorter key and then use it, along with specific mathematical operations, to generate a longer key.

One important example is the block cipher DES (Data Encryption Standard), which divides a message into 64-bit blocks and transforms it through a series of *rounds*. Typical operations that may be applied in turn during a round are substitution (replacing groups of bits), permutation (shifting groups of bits within the block in a circular fashion), and then applying the XOR operation using the key assigned to the round. All these manipulations increase security because they make the relationship between the ciphertext and the key, as well as the ciphertext and the plaintext, very complex.

HISTORICAL CONTEXT **DES (1977)**
Developed through the efforts of IBM (International Business Machines), NBS (National Bureau of Standards), and NSA (National Security Agency), it became mandatory for federal agencies. It processed blocks of 64 bits and used an initial key of length 56 bits to generate various keys for use in each round.

The length of the DES key became too short for serious security operations; computers quickly became powerful enough to do a brute-force search for the key at a reasonable cost. Consequently, another block cipher called AES (American Encryption Standard) increased the block size and key length. It also added manipulations to increase complexity of the encryption process including *matrix multiplication*, which will be discussed in the next section.

HISTORICAL CONTEXT **AES (2001)**
The National Institute of Standards and Technology (NIST) recognized that DES was not adequately secure and called for a new algorithm. The result was AES, a symmetric block cipher using a block size of 128 bits and a key length of 128, 192, or 256 bits.

The underlying idea of AES is to carry out a series of transformations on each message block that then make it difficult to reconstruct. Figure 23.1 illustrates placing a single block MEET ME AT THE WALL of a larger plaintext message into successive columns of a 4×4 array. Each letter in the array (requiring 8 bits) is then converted using Table 23.4 into its hexadecimal representation (two hexadecimal digits); we use hexadecimal rather than binary to simplify the presentation. Thus, the 16 letters in the plaintext array represent a total of $16 \times 8 = 128$ bits. Spaces between words have been omitted and a 0 is added in the last cell as padding to ensure that we always work with a block of exactly 128 bits.

FIGURE 23.1: Converting a plaintext array into hexadecimal

AES Operations on a Block of Plaintext

The plaintext (hexadecimal) array is subjected to a series of transformations before it is finally encrypted. Here is an overview of the steps of the AES algorithm and its operations:

1. An initial round where the plaintext array is XORed with the initial key array.
2. Intermediate (typical) rounds where the array from the previous round is manipulated by *substitution, row shifting, column mixing*, and then XORing with a new key produced just for this round.
3. A final round where the array from the last intermediate round is subjected to substitution, row shifting (but not column mixing), and then XORing with the new key produced for the final round. The result is the final ciphertext.

Figure 23.2 illustrates the transformations applied to the current plaintext array during a typical round. The number of rounds depends on the length of the key; the longer the key, the more rounds required (e.g., 10 rounds are used for a 128-bit key). Since AES is a symmetric system, this entire process can be reversed for decryption.

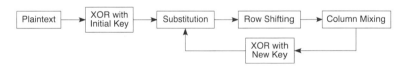

FIGURE 23.2: A typical AES round

We now discuss the four transformations (XOR, substitution, row shifting, and column mixing) that are applied during each round. Later we will discuss how the current key for a round is modified to produce a key for the next round.

XOR. The current 128-bit plaintext array is XORed with the current 128-bit key. Recall that the XOR (*exclusive-or*) operation plays an integral role in the Vernam Cipher (Section 23.2). Figure 23.3 shows how the initial plaintext array of Figure 23.1 is XORed with a particular 128-bit key, entry-by-entry. For example, when the upper left corner 4D = 01001101 of the plaintext array is XORed with the corresponding key element 42 = 01000010, the result is $01001101 \oplus 01000010 = 00001111 = $ 0F.

4D	4D	54	41
45	45	48	4C
45	41	45	4C
54	54	57	30

$\oplus$

42	40	4A	46
44	4C	51	41
43	6A	7C	44
50	49	4B	49

=

0F	0D	1E	07
01	09	19	0D
06	2B	39	08
04	1D	1C	79

FIGURE 23.3: Current array XORed with the current key

Substitution. This transformation involves a substitution rule, analogous to the monoalphabetic substitution discussed in regard to cryptograms (Section 22.3). In this case, however, each entry of the current array is replaced by a new value generated by a mathematical formula. The substitution is constructed so that multiple occurrences of the same hexadecimal pair are mapped to the same value, whereas different hexadecimal pairs are mapped to different values. For example, Figure 23.4 shows that multiple occurrences of the hexadecimal number 0D are all replaced by the same hexadecimal number D7.

FIGURE 23.4: Applying substitution to the current array

Row Shifting. This operation takes every row of the current array produced by substitution and shifts it left by a different amount. In fact, the first row remains unchanged, the second row is shifted left by one place, the third row by two places, and the fourth row by three places. Each shift is circular, meaning that the numbers on the left side of the row that have been displaced by the shift will move to the empty spaces on the right-hand portion of the row. For example, Figure 23.5 shows that the last two entries 12 and 30 of the third row of the current array become the first two entries of the transformed array; the first two entries 6F and F1 become, respectively, the last two entries of the shifted row.

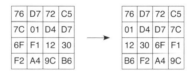

FIGURE 23.5: Applying row shifting to the current array

Column Mixing. In this transformation of the current array, each column is successively altered using the technique of matrix multiplication, which will be explained in Section 23.4.

We now illustrate how several of these transformations can be applied in a simplified setting.

Example 23.8 Transform a Block of Plaintext Using AES Operations

Suppose that the block size is 32 bits, and the first block of plaintext is TAKE. (a) Use Table 23.4 to find the decimal representation of each plaintext letter and then convert each decimal number into hexadecimal. Arrange these hexadecimals in a 2×2 array. (b) XOR this array with the initial key array shown in Figure 23.6(a). (c) Then apply substitution by referencing the partial substitution table (generated using a mathematical formula) in Figure 23.6(b). Note that the numbers in the left column of the table represent the first hexadecimal digit, the numbers in the top row represent the second hexadecimal digit, and the intersection of each row and column gives the replacement number in hexadecimal. (d) Finally, apply row shifting to the current array.

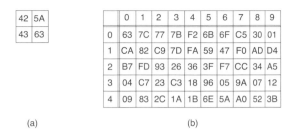

(a)   (b)

FIGURE 23.6: Initial key array and substitution table

Solution

(a) Using Table 23.4 we identify the decimal equivalent for each letter: $\mathsf{T} \to 84, \mathsf{A} \to 65, \mathsf{K} \to 75, \mathsf{E} \to 69$. Then these decimal numbers are converted to the corresponding hexadecimal numbers 54, 41, 4B, 45 resulting in the numerical array shown in Figure 23.7.

$$\begin{array}{|c|c|} \hline \mathsf{T} & \mathsf{K} \\ \hline \mathsf{A} & \mathsf{E} \\ \hline \end{array} \longrightarrow \begin{array}{|c|c|} \hline 54 & 4B \\ \hline 41 & 45 \\ \hline \end{array}$$

FIGURE 23.7: Plaintext array of hexadecimal numbers

(b) The initial plaintext array is XORed with the initial key. This is done by converting each hexadecimal number to binary and then carrying out the XOR operation bit-by-bit. For example, hexadecimal 54 is 01010100 in binary and hexadecimal 42 is 01000010 in binary; taking the *exclusive-or* of these two binary numbers results in $01010100 \oplus 01000010 = 00010110 = 16$ in hexadecimal. Figure 23.8 shows the result of applying the XOR operation to the initial plaintext array.

$$\begin{array}{|c|c|} \hline 54 & 4B \\ \hline 41 & 45 \\ \hline \end{array} \oplus \begin{array}{|c|c|} \hline 42 & 5A \\ \hline 43 & 63 \\ \hline \end{array} = \begin{array}{|c|c|} \hline 16 & 11 \\ \hline 02 & 26 \\ \hline \end{array}$$

FIGURE 23.8: Plaintext array XORed with the initial key

(c) Next, we apply to the current array the substitutions defined by Figure 23.6(b), which shows how to transform the first and second digits of each hexadecimal pair. Only a relevant portion of the larger 16×16 table is displayed in Figure 23.6(b). Specifically, 16 is replaced by 47, 02 by 77, 11 by 82, and 26 by F7. This results in the array shown in Figure 23.9.

$$\begin{array}{|c|c|} \hline 16 & 11 \\ \hline 02 & 26 \\ \hline \end{array} \longrightarrow \begin{array}{|c|c|} \hline 47 & 82 \\ \hline 77 & F7 \\ \hline \end{array}$$

FIGURE 23.9: Substitution applied to the current array

(d) The first row of the current array remains the same, while the second row is shifted (circularly) to the left by one place. This transformation is shown in Figure 23.10. ◁

(*Related Exercise: 15*)

FIGURE 23.10: Result of row shifting

Subsequent rounds apply in turn the four transformations (substitution, row shifting, column mixing, XOR) to the result of the previous round. What is new at each round is the key applied during the XOR transformation. We now briefly discuss how a sequence of keys $K_1, K_2, \ldots$ is generated from the initial key K.

AES Key Expansion

AES prescribes an algorithm for transformation of the initial 4×4 key array K into a new 4×4 key array K' for use in the next round. Here is an overview of the key expansion process and its constituent operations, also illustrated in Figure 23.11.

1. Create the first column of the new key array K': Starting with the last column of the previous key array K, apply *column shifting*, substitution, XOR this column with a column of predefined numbers, and finally XOR it with the first column of K. This produces the first column of K'.

2. Create the remaining columns of the new key array K': XOR the first column of K' with the second column of K to produce the second column of K'; XOR the second column of K' with the third column of K to produce the third column of K'; and XOR the third column of K' with the last column of K to produce the last column of K'.

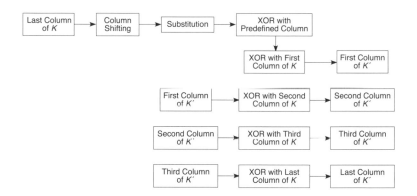

FIGURE 23.11: AES key expansion process

We now describe how the key shown in Figure 23.3 is transformed into a key for the next round, using substitution, column shifting, and XOR.

Substitution and Column Shifting. Figure 23.12(a) shows the key array from the first round. To begin, the last column of that array is shifted in a manner analogous to the row shift described earlier. Specifically, the first number in this column is moved to the end, and all remaining entries in the column shift up one place. Then a substitution is applied to this shifted column, again using a mathematical formula to generate the substitution ordering. Figure 23.12(b) shows the result of applying column shifting and substitution (using Figure 23.6(b)) to the last column of the previous key matrix.

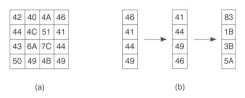

(a)   (b)

FIGURE 23.12: Key expansion through column shifting and substitution

XOR. In order to produce the first column of the new key array, the current column is XORed twice, first with a column of predefined numbers and then with the first column of the previous key array. Carrying out the two successive *exclusive-or* operations in our example yields

$$[83\ 1B\ 3B\ 5A] \oplus [02\ 3B\ 65\ 6C] \oplus [42\ 44\ 43\ 50] = [C3\ 64\ 1D\ 66].$$

To produce the second column of the new key array, the process is much simpler. We XOR the second column of the previous key array with the new first column just generated:

$$[40\ 4C\ 6A\ 49] \oplus [C3\ 64\ 1D\ 66] = [83\ 28\ 77\ 2F].$$

Then we XOR the third column of the previous key with the new second column just generated to produce the new third column:

$$[4A\ 51\ 7C\ 4B] \oplus [83\ 28\ 77\ 2F] = [C9\ 79\ 0B\ 64].$$

Finally, we XOR the fourth column of the previous key with the new third column to produce the new fourth column:

$$[46\ 41\ 44\ 49] \oplus [C9\ 79\ 0B\ 64] = [8F\ 38\ 4F\ 2D].$$

This process gives the new key array displayed in Figure 23.13.

C3	83	C9	8F
64	28	79	38
1D	77	0B	4F
66	2F	64	2D

FIGURE 23.13: New key array

Example 23.9 Expand a Key Using AES Operations

Figure 23.14 provides an initial 2×2 key array, a partial substitution table, and the column of predefined hexadecimal numbers. We want to produce the next key array, for use in the second round of the AES encryption process. (a) Extract the last column of the initial key array, carry out column shifting, and then apply the substitution defined by the given partial substitution table. (b) XOR the resulting column with the column of predefined numbers and then with the first column of the initial key array. This produces the first column of the next key array. (c) XOR this new first column with the second column of the initial key array to produce the second column of the next key array.

Solution

(a) As seen in Figure 23.15, the last column of the current key array is shifted up, and then substitutions are applied.

FIGURE 23.14: Initial key, substitution table, and predetermined column

FIGURE 23.15: Row shifting and substitution

(b) The new column is XORed with the column of predefined numbers and then with the first column of the initial key array: [52 BE] $\oplus$ [C4 44] $\oplus$ [42 43] = [D4 B9].

(c) XORing this new first column with the second column of the initial key array gives the new second column [5A 48] $\oplus$ [D4 B9] = [8E F1].

The new key array for use in round two is shown in Figure 23.16. ◁

(Related Exercise: 16)

D4	8E
B9	F1

FIGURE 23.16: Key array for the next round

AES is possible to break, and does not have perfect security. Depending on the complexity of the mathematical operations underlying the relationship between the key and the encryption algorithm, it can however take an impractical amount of time to break. Consequently, AES achieves *practical security*—at least until new developments in computational power or new decryption algorithms are able to shorten the amount of time required.

DEFINITION **Practical Security**
When a cryptosystem can be ultimately broken using brute force (to find the decryption key), but cannot be broken in a meaningful amount of time, it has practical security. That is, the level of required computational effort is too great to be handled effectively by the existing state of computing power.

Cryptanalysts rank the security of cryptosystems based on the level of their computational effort or complexity. With flexibility in key length, AES can protect itself against new hardware that performs faster and cheaper brute-force searches. In addition, its complicated mathematical operations confound statistical attacks.

We will now look at a fairly simple block cipher, the *Hill Cipher*, to explain one of the techniques used by AES: matrix multiplication. The Hill Cipher has an encryption algorithm that is easier to describe than that of either DES or AES.

23.4 Hill Cipher

The Hill Cipher is a block cipher that uses matrix multiplication to transform plaintext into ciphertext. It is not considered a secure cipher, but it illustrates the application of such a mathematical tool for encryption of messages.

HISTORICAL CONTEXT **Hill Cipher (1929)**
This polyalphabetic cipher was invented by Lester Hill (1891–1961), an American mathematician. It is an early block cipher, like the Vigenère Cipher, but it involves the arithmetic manipulation of matrices.

In describing matrix multiplication, we will return to using letters and their alphabetic position numbers to encrypt the plaintext, though we understand that AES would be applying matrix multiplication to blocks of binary data.

Matrix Multiplication

In discussing the Hill Cipher, the plaintext will be given as letters. For example, the plaintext TO THE CITY AT NIGHT has 16 letters once the spaces are removed. Each letter will be converted into a decimal number in the range 0–25 based on the position of each letter in the alphabet. Table 23.12 repeats the alphabetic position table previously used in Chapter 22.

TABLE 23.12: Alphabetic position numbers

A	B	C	D	E	F	G	H	I	J	K	L	M	N	O	P	Q	R	S	T	U	V
0	1	2	3	4	5	6	7	8	9	10	11	12	13	14	15	16	17	18	19	20	21
W	X	Y	Z																		
22	23	24	25																		

These 16 numbers 19 14 19 7 4 2 8 19 24 0 19 13 8 6 7 19 will be encrypted into blocks of a fixed size. Specifically, suppose we use a block size of two and group the above message into the blocks 19 14, 19 7, ..., 7 19. It is convenient to represent each of these blocks by a column:

$$\begin{pmatrix} 19 \\ 14 \end{pmatrix}, \begin{pmatrix} 19 \\ 7 \end{pmatrix}, \ldots, \begin{pmatrix} 7 \\ 19 \end{pmatrix}.$$

Next, how might we disguise the first block? One approach is to take a weighted sum of 19 and 14 (namely, $19a + 14b$) to replace 19, and then use a different weighted sum of 19 and 14 (namely, $19c + 14d$) to replace 14. This can be accomplished by *multiplying* the column on the left by a 2×2 *key matrix*. A 2×2 *matrix* is a just a grid of numbers arranged in two rows and two columns. The multiplication operation, described next for a 2×1 column s with elements x and y, accomplishes the task of disguising these two elements by taking weighted sums.

DEFINITION **Matrix Multiplication**

Multiplication of the 2×2 key matrix $K = \begin{pmatrix} a & b \\ c & d \end{pmatrix}$ and the 2×1 column $s = \begin{pmatrix} x \\ y \end{pmatrix}$ produces the 2×1 column

$$Ks = \begin{pmatrix} a & b \\ c & d \end{pmatrix} \begin{pmatrix} x \\ y \end{pmatrix} = \begin{pmatrix} ax + by \\ cx + dy \end{pmatrix}.$$

The first element of the resulting column is formed by applying the first row of multipliers $(a\ b)$ to the given column s, while the second element of the resulting column results from applying the second row of multipliers $(c\ d)$ to the given column s. All operations are carried out using modulo 26 arithmetic.

For example, suppose that the key matrix K is given by $K = \begin{pmatrix} 3 & 2 \\ 5 & 7 \end{pmatrix}$. Then the result of multiplying this matrix and the first block of our message is

$$\begin{pmatrix} 3 & 2 \\ 5 & 7 \end{pmatrix} \begin{pmatrix} 19 \\ 14 \end{pmatrix} = \begin{pmatrix} 3 \times 19 + 2 \times 14 \\ 5 \times 19 + 7 \times 14 \end{pmatrix} = \begin{pmatrix} 85 \\ 193 \end{pmatrix} \equiv \begin{pmatrix} 7 \\ 11 \end{pmatrix} \bmod 26.$$

Using Table 23.12 would result in the ciphertext letters H and L.

(Related Exercises: 17–19, 36)

It is important to point out that decryption can also be carried out by matrix multiplication, in this case by using the 2×2 *inverse matrix* M shown below:

$$M = \begin{pmatrix} 3 & 14 \\ 9 & 5 \end{pmatrix}.$$

To illustrate, multiplying the previously encrypted column on the left by M will restore the original plaintext column:

$$\begin{pmatrix} 3 & 14 \\ 9 & 5 \end{pmatrix} \begin{pmatrix} 7 \\ 11 \end{pmatrix} = \begin{pmatrix} 3 \times 7 + 14 \times 11 \\ 9 \times 7 + 5 \times 11 \end{pmatrix} = \begin{pmatrix} 175 \\ 118 \end{pmatrix} \equiv \begin{pmatrix} 19 \\ 14 \end{pmatrix} \bmod 26.$$

This is not an accident! The important feature of the inverse matrix M is that it reverses the scrambling achieved by applying the key matrix K. This definition parallels that given for the inverse of an element modulo n, defined in Section 22.2.

DEFINITION **Inverse Matrix**

The inverse of the 2×2 matrix $K = \begin{pmatrix} a & b \\ c & d \end{pmatrix}$ is a matrix $M = \begin{pmatrix} e & f \\ g & h \end{pmatrix}$ which, when multiplied by K, produces the identity matrix I:

$$MK = \begin{pmatrix} e & f \\ g & h \end{pmatrix} \begin{pmatrix} a & b \\ c & d \end{pmatrix} = \begin{pmatrix} ea + fc & eb + fd \\ ga + hc & gb + hd \end{pmatrix} = \begin{pmatrix} 1 & 0 \\ 0 & 1 \end{pmatrix} = I.$$

To produce MK, the matrix M is multiplied in turn by each column of K. All operations are carried out using modulo 26 arithmetic.

Example 23.10 Verify the Inverse of a Key Matrix

(a) Use the definition of matrix multiplication to verify that MK is the identity matrix I, where K and M are as previously defined. (b) What effect does multiplication by I have when it is applied to the column $s = \begin{pmatrix} x \\ y \end{pmatrix}$?

Solution

(a) $MK = \begin{pmatrix} 3 & 14 \\ 9 & 5 \end{pmatrix} \begin{pmatrix} 3 & 2 \\ 5 & 7 \end{pmatrix} = \begin{pmatrix} 3 \times 3 + 14 \times 5 & 3 \times 2 + 14 \times 7 \\ 9 \times 3 + 5 \times 5 & 9 \times 2 + 5 \times 7 \end{pmatrix} = \begin{pmatrix} 79 & 104 \\ 52 & 53 \end{pmatrix} = \begin{pmatrix} 1 & 0 \\ 0 & 1 \end{pmatrix}$ mod 26.

(b) $Is = \begin{pmatrix} 1 & 0 \\ 0 & 1 \end{pmatrix} \begin{pmatrix} x \\ y \end{pmatrix} = \begin{pmatrix} 1x + 0y \\ 0x + 1y \end{pmatrix} = s$. That is, multiplication by I does not change the column. Since $MK = I$, this means that multiplying a message encrypted by K by using the matrix M will produce the original plaintext: $M(Ks) = (MK)s = Is = s$. ◁

(Related Exercises: 20–21)

Example 23.11 Encrypt a Message Using the Hill Cipher

We wish to encrypt the plaintext message **HELP ME** using the Hill Cipher with a block size of two. (a) Use Table 23.12 to convert each block of two plaintext letters into (decimal) position numbers and place each in a column. (b) Apply matrix multiplication, modulo 26, to each column using the key matrix K given below. (c) Translate the resulting ciphertext numbers into ciphertext letters.

$$K = \begin{pmatrix} 4 & 1 \\ 5 & 3 \end{pmatrix}$$

Solution

(a) Using Table 23.12 we replace H and E by 7 and 4, giving the 2×1 column $\begin{pmatrix} 7 \\ 4 \end{pmatrix}$. Similarly, L and P result in the 2×1 column $\begin{pmatrix} 11 \\ 15 \end{pmatrix}$, while M and E result in the 2×1 column $\begin{pmatrix} 12 \\ 4 \end{pmatrix}$.

(b) $\begin{pmatrix} 4 & 1 \\ 5 & 3 \end{pmatrix} \begin{pmatrix} 7 \\ 4 \end{pmatrix} = \begin{pmatrix} 32 \\ 47 \end{pmatrix} \equiv \begin{pmatrix} 6 \\ 21 \end{pmatrix}$ mod 26.

$\begin{pmatrix} 4 & 1 \\ 5 & 3 \end{pmatrix} \begin{pmatrix} 11 \\ 15 \end{pmatrix} = \begin{pmatrix} 59 \\ 100 \end{pmatrix} \equiv \begin{pmatrix} 7 \\ 22 \end{pmatrix}$ mod 26.

$\begin{pmatrix} 4 & 1 \\ 5 & 3 \end{pmatrix} \begin{pmatrix} 12 \\ 4 \end{pmatrix} = \begin{pmatrix} 52 \\ 72 \end{pmatrix} \equiv \begin{pmatrix} 0 \\ 20 \end{pmatrix}$ mod 26.

(c) Again using Table 23.12, the ciphertext numbers 6, 21, 7, 22, 0, 20 are converted to the ciphertext letters **GVHWAU**. Notice that the two occurrences of E in the plaintext get encrypted first as V and then as U in the ciphertext. ◁

(Related Exercises: 22, 24–25)

The next example verifies that the ciphertext message just created can be decrypted using an appropriate inverse matrix.

Example 23.12 Decrypt a Message Using the Hill Cipher

We wish to decrypt the ciphertext message GVHWAU using the Hill Cipher. (a) Consult Table 23.12 to convert each block of two ciphertext letters into decimal numbers and place these in a column. (b) Apply matrix multiplication, modulo 26, to each column using the inverse matrix M given below. (c) Translate the resulting plaintext numbers into plaintext letters. (d) Verify that M is in fact the inverse of matrix K from Example 23.11.

$$M = \begin{pmatrix} 19 & 11 \\ 3 & 8 \end{pmatrix}$$

Solution

(a) Using Table 23.12 we replace G and V by 6 and 21, giving the 2×1 column $\begin{pmatrix} 6 \\ 21 \end{pmatrix}$.

Similarly, H and W result in the 2×1 column $\begin{pmatrix} 7 \\ 22 \end{pmatrix}$, while A and U result in the 2×1 column $\begin{pmatrix} 0 \\ 20 \end{pmatrix}$.

(b) $\begin{pmatrix} 19 & 11 \\ 3 & 8 \end{pmatrix} \begin{pmatrix} 6 \\ 21 \end{pmatrix} = \begin{pmatrix} 345 \\ 186 \end{pmatrix} \equiv \begin{pmatrix} 7 \\ 4 \end{pmatrix}$ mod 26.

$\begin{pmatrix} 19 & 11 \\ 3 & 8 \end{pmatrix} \begin{pmatrix} 7 \\ 22 \end{pmatrix} = \begin{pmatrix} 375 \\ 197 \end{pmatrix} \equiv \begin{pmatrix} 11 \\ 15 \end{pmatrix}$ mod 26.

$\begin{pmatrix} 19 & 11 \\ 3 & 8 \end{pmatrix} \begin{pmatrix} 0 \\ 20 \end{pmatrix} = \begin{pmatrix} 220 \\ 160 \end{pmatrix} \equiv \begin{pmatrix} 12 \\ 4 \end{pmatrix}$ mod 26.

(c) Again using Table 23.12, the ciphertext numbers 7, 4, 11, 15, 12, 4 are converted to the plaintext letters HELPME.

(d) Multiplying M by K results in the identity matrix I, verifying that M is the inverse of K, modulo 26:

$$\begin{pmatrix} 19 & 11 \\ 3 & 8 \end{pmatrix} \begin{pmatrix} 4 & 1 \\ 5 & 3 \end{pmatrix} = \begin{pmatrix} 131 & 52 \\ 52 & 27 \end{pmatrix} \equiv \begin{pmatrix} 1 & 0 \\ 0 & 1 \end{pmatrix} \text{ mod } 26. \quad \lhd$$

(Related Exercises: 23, 25)

While we have worked with a block size of two, and accordingly 2×2 matrices K and M, the encryption and decryption processes can be extended to use a block size k and therefore larger $k \times k$ matrices. The concept of an inverse matrix extends as well, although its calculation becomes more laborious. You will not be asked to determine inverses of matrices, but rather to appreciate how important they are in the process of decryption.

A formal statement of the Hill encryption/decryption procedure follows:

Hill Encryption Algorithm

1. Using Table 23.12, convert each block of plaintext into (decimal) alphabetic position numbers, and form a column containing these numbers.
2. Agree on a key matrix K of the appropriate dimensions that has an inverse.
3. Multiply the plaintext column by the matrix K using modulo 26 arithmetic.

---

4. Use Table 23.12 to convert the resulting column entries into letters. These form one block of the ciphertext.

Hill Decryption Algorithm

1. Using Table 23.12, convert each block of ciphertext into (decimal) alphabetic position numbers, and form a column containing these numbers.
2. Determine the inverse M of the key matrix K.
3. Multiply the ciphertext column by the inverse matrix M using modulo 26 arithmetic.
4. Use Table 23.12 to convert the resulting column entries into letters. These form one block of the original plaintext.

Security of the Hill Cipher

Our purpose in discussing the Hill Cipher is to see how matrix multiplication can be used for encryption in a block cipher. The AES block cipher also uses matrix multiplication among other mathematical transformations.

While the Hill Cipher is difficult to break using only intercepted ciphertext messages, it is vulnerable to a *known-plaintext attack*. Namely, if an attacker is able to intercept several pairs of plaintext and ciphertext, it is possible to determine the key matrix by solving certain systems of equations.

The security of the Hill Cipher, and the AES block cipher, increase with the block size used, and consequently the size of the matrix K. Security also depends, as it does for all symmetric ciphers, on the security of the key. This consideration leads us to discuss methods for ensuring that the key is delivered without unauthorized access.

23.5 Diffie-Hellman Key Exchange

Key management is a major problem in all symmetric cryptosystems. How can everyone in the communication loop acquire the secret key, yet no one else? Hiring a secure courier to deliver it physically isn't practical or secure. Nor is placing the key within an email sent to all authorized persons.

The *Diffie-Hellman Key Exchange* introduces a way to solve the problem of transmitting private keys. Two individuals, two businesses, or two computers far away from each other, who don't know anything about each other, can create a *shared private key* without transmitting anything secret. Moreover, they can do this over insecure communication channels where unauthorized parties might also be viewing the transmission. This shared private key can be used as the key to any symmetric encryption/decryption algorithm: for example, the Caesar, Vigenère, Vernam, Hill, AES, and DES systems discussed in Chapters 22 and 23.

HISTORICAL CONTEXT **Diffie-Hellman Key Exchange (1976)**
This scheme was first published by the American cryptographers Whitfield Diffie and Martin Hellman, although it had also been invented within the British intelligence agency but was kept classified.

Here is a general outline of how this exchange protocol works. Suppose B is a business and S is one of its subsidiaries. These entities wish to communicate sensitive corporate information with one another using one of the encryption/decryption algorithms previously discussed, but a shared private key is required. Both B and S agree in an email on two numbers (which can be made public). Each also chooses its own private number, which is held securely and never transmitted publicly. Each entity then uses the two public numbers and its own private number to calculate another public number, which is then sent (over an insecure communication channel) to the other entity. This new public number is subsequently used by each entity to determine the shared private key needed for encryption and decryption.

The Diffie-Hellman procedure involves the use of exponentiation and modular arithmetic; this process can itself be made public. Specifically, suppose that a and n are the public numbers (agreed upon in advance by the entities), whereas b and s are the private numbers for B and S, respectively. These private numbers are shown in blue in the boxes for B and S located at the top of Figure 23.17. The public number b^* for B is obtained using the formula $b^* = a^b \pmod{n}$, while the public number s^* for S is obtained using the formula $s^* = a^s \pmod{n}$. It should be emphasized that a and n are publicly available numbers and need not be kept secret. However, B and S maintain securely their private numbers b and s.

At the next step of the Diffie-Hellman procedure, B sends b^* to S and S sends s^* to B, as indicated by the horizontal arrows in Figure 23.17. Then B, which knows its private number b as well as s^*, computes $k = (s^*)^b \pmod{n} = (a^s)^b \pmod{n}$. Likewise S, which knows its private number s as well as b^*, computes $k = (b^*)^s \pmod{n} = (a^b)^s \pmod{n}$. Notice that the same shared private key $k = a^{sb} \pmod{n} = a^{bs} \pmod{n}$ is now computed by both B and S, being some integer in the range $0, 1, \ldots, n-1$. This key now can be used to encrypt and decrypt messages between B and S in a secure way. There are certain technical restrictions on the choices for a, n, b, and s and these will be discussed in Chapter 24.

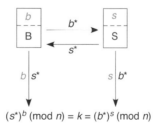

$$(s^*)^b \ (\text{mod } n) = k = (b^*)^s \ (\text{mod } n)$$

FIGURE 23.17: Diffie-Hellman key exchange

The following example illustrates the mechanics of exchanging a key using the above procedure.

Example 23.13 Establish a Shared Key Using the Diffie-Hellman Procedure

Business B and Subsidiary S wish to use a symmetric cipher system to encrypt and decrypt messages that they will send to one another. They agree on two public numbers: the prime modulus $n = 17$ and the base $a = 3$. Suppose that B maintains the private number $b = 15$, known only to B, and that S maintains the private number $s = 13$, known only to S. (a) Calculate the public number b^* for B, which will be sent to S. (b) Calculate the public number s^* for S, which will be sent to B. (c) Using the known public numbers and its own private number, determine the shared key k calculated by B. (d) Using the known public numbers and its own private number, determine the shared key k calculated by S.

Solution

(a) $b^* = a^b \pmod{n} = 3^{15} \pmod{17} = 14{,}348{,}907 \pmod{17} = 6$.

(b) $s^* = a^s \pmod{n} = 3^{13} \pmod{17} = 1{,}594{,}323 \pmod{17} = 12$.

(c) $k = (s^*)^b \pmod{n} = 12^{15} \pmod{17}$. Since 12^{15} is too large to display in a calculator, we rewrite this as $k = 12^{15} \pmod{17} = 12^5 \times 12^5 \times 12^5 \pmod{17} = 248{,}832 \times 248{,}832 \times 248{,}832 \pmod{17} \equiv 3 \times 3 \times 3 \pmod{17} = 27 \pmod{17} = 10$. So B calculates the shared private key as 10.

(d) $k = (b^*)^s \pmod{n} = 6^{13} \pmod{17}$. Since 6^{13} is too large to display in a calculator, we rewrite this as $k = 6^{13} \pmod{17} = 6^5 \times 6^5 \times 6^3 \pmod{17} = 7776 \times 7776 \times 216 \pmod{17} \equiv 7 \times 7 \times 12 \pmod{17} = 588 \pmod{17} = 10$. So S also calculates the shared private key as 10. ◁

(Related Exercises: 26–33)

What has happened in Example 23.13 is that without transmitting the private number $s = 13$ of S or without transmitting the private number $b = 15$ of B, both entities are able to use these private numbers indirectly and create the shared private key $k = 10$. No one else will be able to discover this private key because the private numbers $s = 13$ and $b = 15$ are never transmitted publicly.

Security of Diffie-Hellman

The main purpose of the Diffie-Hellman Key Exchange is to establish a secure shared private key, which it does. It must be emphasized that the Diffie-Hellman procedure does not encrypt any messages. It is a numerical algorithm that allows two people (or computers) to securely exchange a cryptographic key in a symmetric system.

The Diffie-Hellman system is secure in that there is no easy way to obtain the private numbers used in the presented formulas because these formulas involve a "one-way function" that is easy to compute but difficult to reverse. Namely, consider the formula that B uses to compute its private number b^*: $b^* = a^b \pmod{n}$. Since a, n, and b^* are public numbers, couldn't someone solve this modular equation for the unknown (private) b and then obtain access to the private shared key $k = (s^*)^b \pmod{n}$? In other words, how easy is it to solve an equation of the form $b^* = a^x \pmod{n}$ for the unknown value x? Here x can be considered as the logarithm of b^* relative to the base a, using modular arithmetic. It turns out that this *Discrete Logarithm Problem* appears to be quite difficult to solve when the modulus n is a large prime. More details about this important problem will be presented in Section 24.2.

One difficulty with the Diffie-Hellman Key Exchange is that if you wish to communicate with many people, you have to send many emails and perform several computations to establish these shared keys. Consider the situation where a group of 10 people all want to be able to send encrypted information to each person in the group, without others in the group being able to read it. This means that every two people in the group would need their own shared key. This then would require $C(10, 2) = 45$ secure key exchanges because there are 45 different pairs of people in a group of 10; we have applied here the

Combinations Formula of Chapter 13. In a group of 1000 people, there would need to be $C(1000, 2) = 499{,}500$ key exchanges. This can be expensive and time-consuming.

Also, since the Diffie-Hellman Key Exchange was intended to be used with a symmetric encryption algorithm, the same vulnerabilities exist that allow symmetric systems to be broken without knowing the key (these were discussed in Chapter 22).

An additional serious difficulty with the Diffie-Hellman system is that some unauthorized third party could manipulate the transmissions. Namely, two entities could believe that they are communicating with each other, when actually someone has intercepted the transmission and placed himself in the middle of the transmission stream. Consequently, two people A and B may unknowingly establish a shared private key with a third person C. This is referred to as the *man-in-the-middle attack*. Authentication techniques are then needed so A and B can be assured that they are communicating directly with each other. When Diffie-Hellman is strengthened by authentication techniques (which are discussed in Section 24.5), it can avoid this criticism.

Chapter Summary

Plaintext messages can be converted into binary data using an ASCII/Unicode conversion table before they are encrypted. This binary data may encrypted one bit at a time using a stream cipher such as the Vernam cipher, or one block at a time using block ciphers such as DES or AES. Matrix multiplication (as explained using the Hill Cipher) is one operation that helps AES increase the complexity of the relationship between the key and the ciphertext, thus making it harder to discover the key and to decrypt the ciphertext. There is a difference between perfect security (using a one-time pad where a brute-force search for the key will not be successful) and practical security (where a brute-force search for the key takes so long that it is impractical). The problem of transmitting a secret key between individuals sharing a symmetric cipher can be solved using the Diffie-Hellman Key Exchange. This procedure relies heavily upon the use of modular arithmetic and exponentiation. The benefits of asymmetric cryptosystems will be explored in Chapter 24.

Applications, Representations, Strategies, and Algorithms

Applications				
Security (5–6, 8–9, 11–13)			Graphics (7)	

Representations				
Tables		Symbolization		
Data	Decision	Numerical		Logical
3	5–9	1–3, 10–13		4–9

Strategies	
Rule-Based	Composition/Decomposition
1–13	7

Algorithms				
Exact	Sequential	Numerical	Logical	Recursive
1–13	1, 4, 8–9	1–3, 10–13	4–9	4

Exercises

Decimal, Hexadecimal, and Binary Number Systems

1. Convert the following decimal numbers to binary and to hexadecimal:

   (a) 5

   (b) 22

   (c) 139

2. Convert the following decimal numbers to binary and to hexadecimal:

   (a) 313

   (b) 579

   (c) 1000

3. Convert the following binary numbers to decimal:

   (a) 1100

   (b) 101010

   (c) 11010011

4. Convert the following hexadecimal numbers to decimal:

   (a) B3

   (b) 2A7

   (c) D4C

5. We want to convert the plaintext message PM into binary form. (a) Use Table 23.4 to represent each uppercase letter by a decimal number. (b) Convert each decimal number to hexadecimal. (c) Convert each hexadecimal number into an 8-bit binary number.

Pseudo-Random Numbers

6. Generate a stream of bits $x_1 x_2 x_3 x_4 \ldots$ by means of the linear recurrence relation $x_n = x_{n-1} + x_{n-2} \pmod 2$ for $n \geq 4$. (a) Starting with the seed $x_1 x_2 x_3 = 001$, how many distinct 3-bit sequences are generated by this process? (b) Starting with the seed $x_1 x_2 x_3 = 101$, how many distinct 3-bit sequences are generated by this process?

7. Generate a stream of bits $x_1 x_2 x_3 x_4 x_5 \ldots$, starting with the seed $x_1 x_2 x_3 x_4 = 0011$ by means of the linear recurrence relation $x_n = x_{n-3} + x_{n-4} \pmod 2$ for $n \geq 5$. How many distinct 4-bit sequences are generated by this process?

8. Generate a stream of bits $x_1 x_2 x_3 x_4 x_5 \ldots$, starting with the seed $x_1 x_2 x_3 x_4 = 0011$ by means of the linear recurrence relation $x_n = x_{n-2} + x_{n-3} \pmod 2$ for $n \geq 5$. How many distinct 4-bit sequences are generated by this process?

9. Generate a stream of bits $x_1 x_2 x_3 x_4 x_5 \ldots$, starting with the seed $x_1 x_2 x_3 x_4 = 0010$ by means of the linear recurrence relation $x_n = x_{n-2} + x_{n-3} + x_{n-4} \pmod 2$ for $n \geq 5$. How many distinct 4-bit sequences are generated by this process?

Vernam Cipher

10. Encrypt the plaintext message RUN using the Vernam Cipher. (a) Convert the uppercase letters in this message to decimal numbers using Table 23.4. (b) Convert each decimal number into an 8-bit binary number. (c) Combine these binary numbers into a 24-bit number and then encrypt using the random key 100001010111000101101001.

11. Encrypt the plaintext message KGB using the Vernam Cipher. (a) Convert the uppercase letters in this message to decimal numbers using Table 23.4. (b) Convert each number into an 8-bit binary number. (c) Combine these binary numbers into a 24-bit number and then encrypt using the random key 000001000011000101001001.

12. Decrypt the ciphertext 110100100010101000110000 obtained from application of the Vernam Cipher. (a) Use the 24-bit key 100000010110000101101001 to convert the ciphertext into a 24-bit plaintext. (b) Convert each successive 8-bit string of this plaintext to a decimal number, and then use Table 23.4 to determine the letters comprising the plaintext.

13. Decrypt the ciphertext 010001110111000000001100 obtained from application of the Vernam Cipher. (a) Use the 24-bit key 000001010011000101001011 to convert the ciphertext into a 24-bit plaintext. (b) Convert each successive 8-bit string of this plaintext to a decimal number, and then use Table 23.4 to determine the letters comprising the plaintext.

14. A digital image is to be encrypted using the Vernam Cipher. In particular, we want to encrypt a green-shaded pixel having the RGB values $(79, 139, 26)$ so that it becomes

some other color. (a) Convert each of the three RGB decimal numbers into an 8-bit binary number and then combine to form a 24-bit plaintext. (b) Then generate a random 24-bit key by tossing a coin 24 times, using H and T to designate 1 and 0, respectively. Use this random key to encrypt the 24-bit plaintext. (c) Convert the 24-bit ciphertext into the corresponding three RGB decimal numbers.

AES Transformations

15. Suppose that the block size is 32 bits, and the first block of plaintext is SYNC. (a) Use Table 23.4 to find the decimal representation of each plaintext letter and then convert each decimal number into hexadecimal. Place these hexadecimals in a 2×2 array. (b) XOR this array with the 2×2 key array shown below. (c) Then apply substitution by referencing the partial substitution table shown below. Note that the numbers in the left column represent the first hexadecimal digit, the numbers in the top row represent the second hexadecimal digit, and the intersection of each row and column gives the replacement number in hexadecimal. (d) Apply row shifting to the current array.

42	5A
43	48

	0	1	2	3	4	5	6	7	8	9	A	B
0	63	7C	77	7B	F2	6B	6F	C5	30	01	67	2B
1	CA	82	C9	7D	FA	59	47	F0	AD	D4	A2	AF
2	B7	FD	93	26	36	3F	F7	CC	34	A5	E5	F1

16. Below are the initial key array, a partial substitution table, and the column of predefined hexadecimal numbers. Produce the key array for use in the next round of the AES encryption process. (a) Extract the last column of the initial key array, carry out column shifting, and then apply the substitution defined by the given partial substitution table. (b) XOR the resulting column with the column of predefined numbers and then with the first column of the initial key array. This produces the first column of the next key array. (c) XOR this new first column with the second column of the initial key array to produce the second column of the next key array.

43	4A
5B	59

	8	9	A	B	C
4	52	3B	D6	B3	29
5	6A	CB	BE	39	4A

C4
25

Hill Cipher and Matrix Multiplication

17. Apply matrix multiplication, modulo 26, to the column of decimal numbers $\begin{pmatrix} 2 \\ 5 \end{pmatrix}$ using the key matrix $K = \begin{pmatrix} 15 & 0 \\ 12 & 3 \end{pmatrix}$.

18. Apply matrix multiplication, modulo 26, to the column of decimal numbers $\begin{pmatrix} 1 \\ 5 \end{pmatrix}$ using the key matrix $K = \begin{pmatrix} 6 & 8 \\ 12 & 4 \end{pmatrix}$.

19. Apply matrix multiplication, modulo 26, to the column of decimal numbers $\begin{pmatrix} 1 \\ 5 \\ 3 \end{pmatrix}$

    using the key matrix $K = \begin{pmatrix} 4 & 20 & 11 \\ 6 & 3 & 10 \\ 2 & 25 & 7 \end{pmatrix}$.

20. Multiply the key matrix $K = \begin{pmatrix} 2 & 1 \\ 3 & 4 \end{pmatrix}$ and the matrix $M = \begin{pmatrix} 6 & 5 \\ 15 & 16 \end{pmatrix}$. Verify that MK produces the identity matrix (modulo 26) and so M is the inverse of K.

21. Multiply the key matrix $K = \begin{pmatrix} 19 & 7 \\ 7 & 4 \end{pmatrix}$ and the matrix $M = \begin{pmatrix} 4 & 19 \\ 19 & 19 \end{pmatrix}$. Verify that MK produces the identity matrix (modulo 26) and so M is the inverse of K.

22. Suppose you have one block of plaintext **ES** of size two. (a) Convert the letters to decimal numbers using Table 23.12 and place them in a column. (b) Given the key matrix $K = \begin{pmatrix} 3 & 2 \\ 5 & 7 \end{pmatrix}$, use matrix multiplication to encrypt the plaintext. (c) Use Table 23.12 to convert the resulting ciphertext numbers into letters.

23. Suppose you have one block of ciphertext **GC** of size two. (a) Convert the letters to decimal numbers using Table 23.12 and place them in a column. (b) Given the inverse matrix $M = \begin{pmatrix} 6 & 5 \\ 15 & 16 \end{pmatrix}$, use matrix multiplication to decrypt the ciphertext. (c) Use Table 23.12 to convert the resulting plaintext numbers into letters.

24. Encrypt the plaintext message **GOBACK** using the Hill Cipher with a block size of two. (a) Use Table 23.12 to convert each block of two plaintext letters into (decimal) position numbers and place each in a column. (b) Apply matrix multiplication to each column using the key matrix K given below. (c) Translate the resulting ciphertext numbers into letters.

$$K = \begin{pmatrix} 4 & 1 \\ 5 & 3 \end{pmatrix}$$

25. Suppose the key matrix is $K = \begin{pmatrix} 3 & 25 \\ 24 & 17 \end{pmatrix}$ and its inverse matrix is $M = \begin{pmatrix} 3 & 17 \\ 8 & 25 \end{pmatrix}$. We wish to encrypt the message **MICHIGAN** using the Hill Cipher with block size two. (a) Convert each plaintext letter using Table 23.12 and then encrypt each block of size two using matrix multiplication with the key K. (b) Decrypt the resulting six decimal numbers using matrix multiplication with the matrix M. (c) Verify that M is in fact the inverse of matrix K, modulo 26.

Diffie-Hellman Key Exchange

Business B and Subsidiary S wish to use a symmetric cipher system to encrypt and decrypt messages that they will send to one another. They agree on two public numbers: the prime modulus $n = 47$ and the base $a = 5$. Suppose that B maintains the private number $b = 18$, known only to B, and that S maintains the private number $s = 22$, known only to S.

26. Calculate the public number b^* for B, which will be sent to S.

27. Calculate the public number s^* for S, which will be sent to B.

28. Using the known public numbers and its own private number, determine the shared key k calculated by B.

29. Using the known public numbers and its own private number, determine the shared key k calculated by S.

Business B and Subsidiary S wish to use a symmetric cipher system to encrypt and decrypt messages that they will send to one another. They agree on two public numbers: the prime modulus $n = 23$ and the base $a = 5$. Suppose that B maintains the private number $b = 6$, known only to B, and that S maintains the private number $s = 15$, known only to S.

30. Calculate the public number b^* for B, which will be sent to S.

31. Calculate the public number s^* for S, which will be sent to B.

32. Using the known public numbers and its own private number, determine the shared key k calculated by B.

33. Using the known public numbers and its own private number, determine the shared key k calculated by S.

Projects

34. Pseudo-random number generators are extremely useful in many applications. One method for generating a sequence of pseudo-random integers (not just bits) is called a *linear congruential generator*. Further investigate this topic, using for example [OR3] as a starting point.

35. Example 23.7 illustrates how to disguise the color of a single pixel. Apply this technique to a grid containing several shades of a particular color and then observe the resulting encoded grid of colors. You might consult the site [OR4], which enables one to visualize colors defined by their specified RGB values.

36. Develop a spreadsheet that can carry out matrix multiplication, modulo 26. Notice that the integer k can be converted to a corresponding value modulo 26 by using the spreadsheet function MOD($k, 26$).

Bibliography

[Bauer 2013] Bauer, C. 2013. *Secret history: The story of cryptology*, Chapters 7, 12, 13, 19. Boca Raton, FL: CRC Press.

[Lewand 2000] Lewand, R. 2000. *Cryptological mathematics*, Chapter 3. Washington, D.C.: Mathematical Association of America.

[Sinkov 1966] Sinkov, A. 1996. *Elementary cryptanalysis*, Chapter 4. Washington, D.C.: Mathematical Association of America.

[OR1] `http://www.mathsisfun.com/binary-decimal-hexadecimal-converter.html` (accessed June 12, 2014).

[OR2] `http://www.mathcelebrity.com/bitwise.php` (accessed June 12, 2014).

[OR3] `http://www.www.phy.ornl.gov/csep/rn/node6.html` (accessed June 12, 2014).

[OR4] `http://www.rapidtables.com/web/color/RGB_Color.htm` (accessed June 12, 2014).

Chapter 24

Prime Numbers and Public-Key Cryptosystems

A very little key will open a very heavy door.

—Charles Dickens, British novelist, 1812–1870

Chapter Preview

In the private-key systems discussed in the previous chapter, the shared private key for encryption and decryption had to be transmitted between sender and receiver, and security was an issue in transmitting the key. The Diffie-Hellman Key Exchange introduces the notion of a public-key system, in which public information enables the sender and receiver each to generate their own copy of the shared private key. This private key can then be used to encrypt and decrypt messages in any symmetric system. In this chapter we will discuss in detail two public-key systems (RSA and ElGamal) that extend the Diffie-Hellman idea so that different keys can be used for encryption and decryption; a publicly available key is used for encryption, whereas a private key is used for decryption. The RSA and ElGamal systems, and a variation of ElGamal based on elliptic curves, are all public-key asymmetric systems achieving a high level of security. These systems owe their security to their mathematical underpinnings, based on one-way functions, modular arithmetic, and prime numbers.

Public-Key Cryptosystems

When discussing symmetric cryptosystems, in which encryption and decryption use the same key, we learned that there are some drawbacks to using the same key: it must be kept secret, it must somehow be securely transmitted between the entities, and different keys are required for each sender/recipient pair. In a symmetric cryptosystem, knowledge of the encryption process immediately reveals how to carry out decryption. Unlike a symmetric cryptosystem, a *public-key cryptosystem* uses two keys: one key that is kept secret and known to only one person (the private key) and another key that is public and available to everyone (the public key). Someone who wants to receive ciphertext will publish his own encryption key and keep his own decryption key private.

DEFINITION **Public-Key Cryptosystem**
An asymmetric system in which the key used for encrypting plaintext is different from the key used for decrypting ciphertext. The encryption key is public and the decryption key is private.

A useful analogy can help to explain the underlying idea of a public-key system. Suppose that Anthony wants to receive secure messages from his friends. He provides each one with an unlocked box into which the friend can place a message. Everyone knows where to obtain such a box: they are *publicly* available. Moreover, each box is supplied with a combination lock provided by Anthony and he instructs each friend to lock the box before sending. Since only Anthony knows the lock's combination (his *private key*), only he can open each box sent to him.

This chapter discusses two important public-key cryptosystems, the RSA scheme and the ElGamal scheme. In these systems, knowledge of the encryption key does not easily reveal the decryption key. The encryption and decryption keys are mathematically interrelated in such a way that it is extremely difficult to derive one key from the other.

The RSA and ElGamal schemes use modular arithmetic and prime numbers to carry out their calculations, and their security depends upon certain computational problems involving these concepts. The next two sections present some relevant background on these topics.

24.1   Prime Numbers

We introduced prime numbers briefly in Section 9.3 when discussing mathematical induction. Recall that a *prime number* is an integer greater than 1 that can be divided evenly only by itself and 1. Otherwise, such an integer is called *composite*. For example, 2, 3, 5, 7, 11, 13 are primes whereas 4, 6, 8, 9, 10, 12 are composites. There are an infinite number of primes and an infinite number of composites. What makes primes useful is that they are the building blocks of the integers: every integer greater than 1 can be represented as a unique product of prime numbers. This discovery is called the *fundamental theorem of arithmetic*. For example, 18 can be represented uniquely as the following product of prime factors: $18 = 2 \times 3 \times 3$.

DEFINITION **Prime Number**
A prime number is an integer greater than one that has only itself and 1 as factors.

RESULT **Fundamental Theorem of Arithmetic**
Any integer greater than 1 can be expressed as a unique product of primes.

Another important concept for the construction of public-key cryptosystems is the notion that two integers are *relatively prime*. This means that they share no prime factors. For example, while neither 18 nor 35 is prime, they share no common prime factors: only 2 and 3 divide (evenly) 18, while only 5 and 7 divide (evenly) 35. By contrast, 51 and 15 are not relatively prime, since the prime 3 divides both of them.

DEFINITION **Relatively Prime**
Two integers are relatively prime when they share no prime factor.

The concept of relatively prime integers is quite important in solving modular equations—in particular, in determining if the integer b has an inverse, modulo n. Recall from Section 22.2 that the inverse of b modulo n is an integer a in the range $0, 1, 2, \ldots, n-1$ that satisfies $ab \equiv 1 \pmod{n}$. The modular inverse of b exists, and is unique, precisely when b and n are relatively prime. In this case, we write the inverse of a as a^{-1}.

RESULT **Modular Inverse**

The inverse of $b \pmod{n}$ exists and is unique when b and n are relatively prime.

The following example provides cases in which the inverse does and does not exist.

Example 24.1 Determine When Modular Inverses Exist

Define the composite modulus $n = 2 \times 5 = 10$. (a) Verify that 7 has an inverse, modulo 10. (b) Verify that 9 has an inverse, modulo 10. (c) Verify that 6 does not have an inverse, modulo 10. (d) What other numbers in the range $1, 2, \ldots, 9$ have an inverse, modulo 10?

Solution

(a) We keep trying multiples of 7 until we find a remainder of 1, modulo 10: $7 \equiv 7 \pmod{10}$; $2 \times 7 = 14 \equiv 4 \pmod{10}$; $3 \times 7 = 21 \equiv 1 \pmod{10}$. This means that 3 is the inverse of 7 modulo 10. Since $7 \times 3 = 3 \times 7 \equiv 1 \pmod{10}$, 7 is the inverse of 3 modulo 10. Notice that both 3 and 7 are relatively prime to 10; neither is divisible by 2 or 5.

(b) We keep trying multiples of 9 until we find a remainder of 1, modulo 10: $9 \equiv 9 \pmod{10}$; $2 \times 9 = 18 \equiv 8 \pmod{10}$; $3 \times 9 = 27 \equiv 7 \pmod{10}$; $4 \times 9 = 36 \equiv 6 \pmod{10}$; $5 \times 9 = 45 \equiv 5 \pmod{10}$; $6 \times 9 = 54 \equiv 4 \pmod{10}$; $7 \times 9 = 63 \equiv 3 \pmod{10}$; $8 \times 9 = 72 \equiv 2 \pmod{10}$; $9 \times 9 = 81 \equiv 1 \pmod{10}$. This means that 9 is its own inverse modulo 10.

(c) Trying multiples of 6 produces $6 \equiv 6 \pmod{10}$; $2 \times 6 = 12 \equiv 2 \pmod{10}$; $3 \times 6 = 18 \equiv 8 \pmod{10}$; $4 \times 6 = 24 \equiv 4 \pmod{10}$; $5 \times 6 = 30 \equiv 0 \pmod{10}$; $6 \times 6 = 36 \equiv 6 \pmod{10}$; $7 \times 6 = 42 \equiv 2 \pmod{10}$; $8 \times 6 = 48 \equiv 8 \pmod{10}$; $9 \times 6 = 54 \equiv 4 \pmod{10}$. This means that 6 does not have an inverse modulo 10. Notice that 6 is not relatively prime to 10, as they both share the factor 2.

(d) The numbers that have inverses modulo 10 are exactly the ones that are relatively prime to 10: 1, 3, 7, 9. $\triangleleft$

As seen in Example 24.1, there are four numbers with inverses, modulo $10 = 2 \times 5$. We can write $4 = (2-1) \times (5-1)$. This is no accident, as there is a nice formula that tells us how many numbers in the range $1, 2, \ldots, n-1$ have an inverse, modulo $n = pq$, where p and q are distinct primes.

RESULT **Number of Modular Inverses**

There are $(p-1)(q-1)$ numbers relatively prime to $n = pq$, where p and q are distinct primes. Consequently, $(p-1)(q-1)$ numbers in the range $1, 2, \ldots, n-1$ have inverses, modulo n.

(Related Exercises: 1–2)

The composite $n = pq$ will be important when we discuss the RSA cryptosystem in Section 24.3. Another concept involving primes, important to the development of the ElGamal cryptosystem in Section 24.4, involves exponentiation, modulo n. We have previously used exponentiation in the Diffie-Hellman Key Exchange procedure of Chapter 23. Now we discuss whether it is possible to invert the exponentiation process when using modular arithmetic.

In standard arithmetic, exponentials and logarithms are inverses of one another. Since 10 raised to the power 2 equals $10^2 = 100$, the logarithm to the base 10 of 100 is 2: $\log_{10} 100 = 2$. In words, the logarithm gives the exponent (2) to which the base (10) is raised to produce the given value (100). We can also talk about logarithms in the context of modular arithmetic. For example, 2^{11} (mod 13) $= 2048$ (mod 13) $= 7$, so that $\log_2 7$ (mod 13) $= 11$. The following example provides cases when the discrete logarithm does and does not exist.

Example 24.2 Find Discrete Logarithms When They Exist

Define the prime modulus $p = 7$. (a) Determine (if it exists) $\log_2 6$ (mod 7) by computing successive powers of 2, modulo 7. (b) Determine (if it exists) $\log_3 6$ (mod 7) by computing successive powers of 3.

Solution

(a) Taking successive powers of 2, modulo 7 produces: 2^0 (mod 7) $= 1$; 2^1 (mod 7) $= 2$; 2^2 (mod 7) $= 4$; 2^3 (mod 7) $= 8$ (mod 7) $= 1$; 2^4 (mod 7) $= 16$ (mod 7) $= 2$; 2^5 (mod 7) $= 32$ (mod 7) $= 4$; 2^6 (mod 7) $= 64$ (mod 7) $= 1$. Notice that only the remainders 1, 2, 4 are produced but the remainder 6 is not. As a result, $\log_2 6$ (mod 7) does not exist: there is no exponent e so that 2^e (mod 7) $= 6$.

(b) Taking successive powers of 3, modulo 7 produces: 3^0 (mod 7) $= 1$; 3^1 (mod 7) $= 3$; 3^2 (mod 7) $= 9$ (mod 7) $= 2$; 3^3 (mod 7) $= 27$ (mod 7) $= 6$. So the exponent 3 works. To see whether this is the only solution exponent in the range $0, 1, \ldots, 6$ we continue the process: 3^4 (mod 7) $= 81$ (mod 7) $= 4$; 3^5 (mod 7) $= 243$ (mod 7) $= 5$; 3^6 (mod 7) $= 729$ (mod 7) $= 1$. So the remainder 6 is only produced using the exponent 3, meaning that $\log_3 6$ (mod 7) $= 3$. Notice that the successive remainders $1, 3, 2, 6, 4, 5$ produced are all the nonzero remainders, modulo 7. ◁

(Related Exercises: 3–4)

As seen in Example 24.2, taking successive powers of 3, modulo the prime 7, produced all of the nonzero remainders $1, 2, \ldots, 6$ for that modulus. We say then that 3 is a *generator*, modulo 7.

DEFINITION **Generator Modulo p**
The number b between 1 and $p - 1$ (where p is a prime) is generator modulo p when successive powers of b generate every remainder from 1 to $p - 1$.

If b is a generator modulo p, then every nonzero remainder will have a logarithm to the base b, modulo p. Generators play an important role in the ElGamal cryptosystem, discussed later. For now we state a result that indicates how many generators there are for a given prime p.

RESULT **Number of Generators Modulo p**
If p is a prime, then the number of generators modulo p is equal to the number of integers in the range $1, 2, \ldots, p - 2$ that are relatively prime to $p - 1$.

We verify this formula in the following case, which is based on Example 24.2.

Example 24.3 Determine All Generators for a Modulus

(a) Determine all generators b modulo $p = 7$ by finding successive powers of b, modulo p.
(b) Verify that the number you find agrees with the formula stated earlier.

Solution

(a) Since all powers of 1 equal 1, $b = 1$ is not a generator. Example 24.2 demonstrated that $b = 2$ is not a generator modulo 7, but $b = 3$ is a generator modulo 7. Table 24.1 calculates the successive powers b^k for $b = 4, 5, 6$. As a result, we see that 3 and 5 are the only generators modulo 7.

TABLE 24.1: Finding generators for the modulus 7

k	$b = 4$	$b = 5$	$b = 6$
1	$4^1 \pmod 7 = 4$	$5^1 \pmod 7 = 5$	$6^1 \pmod 7 = 6$
2	$4^2 \pmod 7 = 2$	$5^2 \pmod 7 = 4$	$6^2 \pmod 7 = 1$
3	$4^3 \pmod 7 = 1$	$5^3 \pmod 7 = 6$	$6^3 \pmod 7 = 6$
4	$4^4 \pmod 7 = 4$	$5^4 \pmod 7 = 2$	$6^4 \pmod 7 = 1$
5	$4^5 \pmod 7 = 2$	$5^5 \pmod 7 = 3$	$6^5 \pmod 7 = 6$
6	$4^6 \pmod 7 = 1$	$5^6 \pmod 7 = 1$	$6^6 \pmod 7 = 1$

(b) Let's determine which numbers in the range $1, 2, \ldots, 5$ are relatively prime to $p - 1 = 6$. Only 1 and 5 are relatively prime to 6, so there are two such numbers. This agrees with the number of generators b found in part (a). ◁

(*Related Exercises: 5–6*)

24.2 One-Way Functions

One of the features of a secure public-key cryptosystem is that it should be easy to encrypt messages to a particular receiver, who can easily decrypt them, but it should be computationally very difficult for someone else to decrypt the messages. This can be accomplished by using *one-way functions*, which are easy to implement in one direction but difficult to implement in the reverse direction without knowledge of additional information (a private key). One-way functions slow down the decryption of a ciphertext by anyone who does not possess the key.

DEFINITION **One-Way Function**

A function which can be easily computed for every possible input but which is difficult to invert. That is, if we are given the output from applying the function, it is computationally difficult to recover the inputs that produced the given output.

We will look at two problems that involve one-way functions. The *Prime Factorization Problem* provides the basis for the security of the RSA cryptosystem (Section 24.3), while the *Discrete Logarithm Problem* provides the basis for security of the ElGamal cryptosystem (Section 24.4).

Prime Factorization Problem

Given two large prime numbers p and q, it is easy to multiply these two factors together (even if they are hundreds of digits long) and produce the product $n = pq$ (several hundred digits long). But given n, can we find its factors p and q in a meaningful amount of computing time?

DEFINITION **Prime Factorization Problem**
Given the composite number $n = pq$, find its prime factors p and q.

While progress has been made in factoring such large composite numbers n, it becomes increasingly difficult for instances in which n is several hundred digits long. This factoring problem then provides a suitable one-way function for use in public-key cryptosystems, as it is easy to compute the product $n = pq$, but quite difficult to retrieve the factors p and q directly from n.

HISTORICAL CONTEXT **Factoring Challenges**
From 1991 to 2007, RSA Laboratories offered a monetary prize to the first group able to factor a large product of two primes. These products ranged from a length of 100 decimal digits up to 617 decimal digits. The largest product successfully factored was 232 decimal digits.

We will now look at a very simple, but not very efficient, factoring technique applied to a composite number n; here n is the product of a number of primes (not necessarily two). To begin, we divide n by each prime up to $\sqrt{n}$ until one is found (say p) that evenly divides n. Then we repeat the process for the integer n/p. This continues until a complete factorization is obtained. There are tables of primes [OR1] that we can consult that list the primes $2, 3, 5, 7, 11, \ldots$ used as trial divisors. For example, if $n = 90$, find the smallest prime that divides 90 evenly; it is $p_1 = 2$, giving the quotient $90/2 = 45$. Next, find the smallest prime that divides 45 evenly; it is $p_2 = 3$, giving the quotient $45/3 = 15$. Then the smallest prime that divides 15 evenly is $p_3 = 3$, giving the quotient $15/3 = 5$. Since $p_4 = 5$ is prime we are done, resulting in the prime factorization $p_1 p_2 p_3 p_4 = 2 \times 3^2 \times 5 = 90$. The problem with this prime division method is that for large composites there are too many primes to check. For example, if n has 100 digits, then there are over 8×10^{47} primes less than $\sqrt{n}$ that might need to be checked. More complicated factoring algorithms have been developed that will factor numbers larger than 100 digits (such as the *General Number Field Sieve*), but they still consume significant computer storage and time.

DEFINITION **Prime Division Method**
Divide the given number (evenly) by its smallest prime factor. Then divide the resulting quotient by its smallest prime factor. Continue in this manner until the resulting number is prime. The product of all these prime divisors gives the unique factorization.

Example 24.4 Use the Prime Division Method to Obtain the Prime Factorization

Apply the prime division method to obtain the (unique) factorization of 23,100.

Solution

The square root of 23,100 is approximately 152, so we only need to consider primes less than 152. The smallest prime that divides 23,100 is $p_1 = 2$, giving the quotient $23{,}100/2 = 11{,}550$. We continue to divide each quotient by the smallest prime that divides it evenly:

$$p_2 = 2 \text{ gives the quotient } 11{,}550/2 = 5775;$$
$$p_3 = 3 \text{ gives the quotient } 5775/3 = 1925;$$
$$p_4 = 5 \text{ gives the quotient } 1925/5 = 385;$$
$$p_5 = 5 \text{ gives the quotient } 385/5 = 77;$$
$$p_6 = 7 \text{ gives the quotient } 77/7 = 11.$$

We stop here because 11 is prime. The prime factorization of 23,100 is $2^2 \times 3 \times 5^2 \times 7 \times 11$.

◁

The next example shows that the process can be more time consuming if $n = pq$, where p and q are similar in magnitude.

Example 24.5 Use the Prime Division Method to Obtain Another Prime Factorization

Apply the prime division method to obtain the (unique) factorization of 22,499.

Solution

The square root of 22,499 is approximately 150 so we only need to consider primes less than 150. The smallest prime that divides 23,100 turns out to be $p_1 = 149$, found after trying 35 different trial prime divisors. Once we have found $p_1 = 149$, the quotient is $22{,}499/149 = 151$ which is a prime. So the prime factorization of 22,499 is 149×151. ◁

(Related Exercises: 7–8)

Discrete Logarithm Problem

Suppose that p is a prime and b is a generator modulo p. Then it is fairly easy to compute the result $r = b^a \pmod{p}$ for a specified integer a. However, how easy is it to reverse this process, to compute the logarithm to the base b of r: $a = \log_b r \pmod{p}$? Since b is a generator modulo p, we are assured that this logarithm exists and is unique. Finding the solution a in the range $1, 2, \ldots, p - 1$ is termed the *Discrete Logarithm Problem*.

DEFINITION **Discrete Logarithm Problem**
Suppose p is a prime, b is a generator modulo p, and r is an integer in the range $1, 2, \ldots, p-1$. Determine x in the range $1, 2, \ldots, p - 1$ that satisfies $r = b^x \pmod{p}$.

Solving this problem in general appears to be quite challenging, especially when p is a large prime. The Discrete Logarithm Problem then provides a suitable one-way function for use in public-key cryptosystems, as it is easy to compute the exponential b^a (mod p), but quite difficult to compute the logarithm $\log_b r$ (mod p). The different computational requirements of both problems are illustrated in the next example.

Example 24.6 Compute Exponentials and Logarithms Relative to a Prime Modulus

It is known that 3 is a generator modulo 31. (a) Compute 3^{22} (mod 31). (b) Compute $\log_3 30$ (mod 31).

Solution

(a) Because 3^{22} is itself very large, we express it as the product $3^{22} = 3 \times (3^3 \times 3^4)^3$. Now $3^4 = 81 \equiv 19$ (mod 13) so that $(3^3 \times 3^4)^3 \equiv (27 \times 19)^3 = 513^3 \equiv 17^3 = 4913 \equiv 15$ (mod 31). Then $3^{22} = 3 \times (3^3 \times 3^4)^3 \equiv 3 \times 15 = 45 \equiv 14$ (mod 31).

(b) The brute-force solution is to try all possible exponents $0, 1, \ldots, 30$: $3^0 = 1$ (mod 31), $3^1 = 3$ (mod 31), $3^2 = 9$ (mod 31), $3^3 = 27$ (mod 31), $3^4 = 81 \equiv 19$ (mod 31), $3^5 = 3^4 \times 3 \equiv 19 \times 3 = 57 \equiv 26$ (mod 31), $\ldots, 3^{15} = 3^5 \times 3^5 \times 3^5 \equiv 26 \times 26 \times 26 \equiv 30$ (mod 31). So after much effort, we finally discover that $\log_3 30$ (mod 31) = 15. ◁

(Related Exercises: 9–12)

24.3   RSA Cryptosystem

The RSA cryptosystem is a well-known public-key cipher whose security rests on the Prime Factorization Problem.

DEFINITION **RSA Cryptosystem**
An asymmetric public-key system in which a public key is used for encryption and a different private key is used for decryption. It relies on carrying out modular arithmetic operations.

HISTORICAL CONTEXT **RSA (1977)**
The RSA cryptosystem was developed by Ron Rivest, Adi Shamir, and Leonard Adleman. It is owned by RSA Security (a subsidiary of EMC Corporation since 2009), which licenses the algorithm technologies and also sells development kits.

Suppose someone (a sender) wants to send encrypted information to someone else (a receiver). With a symmetric cryptosystem, a single private key is used to encrypt as well as decrypt the information, and this key must be in the possession of both parties. The Diffie-Hellman Key Exchange devised a way that both parties involved in this symmetric system could acquire the private key through calculations, yet never have to risk sending the actual private key to each other. By contrast, the RSA cryptosystem uses a public key for encryption and a different private key for decryption; the receiver never has to reveal or send his private decryption key to anyone.

The receiver makes public his RSA encryption key and encryption formula so that any sender can encrypt information destined for the receiver. The public and private numbers are chosen carefully so that only the receiver's private key will decrypt messages from any sender using the posted encryption key and formula. Because the RSA cryptosystem involves a one-way function based on the Prime Factorization Problem using very large primes, it would be difficult computationally for anyone else to discover the receiver's private key when given only the public numbers and formulas.

Here is an overview of the RSA procedures for creating keys, encrypting plaintext, and decrypting ciphertext.

Key Generation: The calculation of the public encryption key e and the private decryption key d depends on two very large private prime numbers p and q, each composed of hundreds of digits. The *public* encryption key e is chosen so that it is relatively prime to the product $(p-1)(q-1)$. Then the *private* decryption key d is chosen as the inverse of e modulo $(p-1)(q-1)$. Results from Section 24.1 ensure that the inverse of e will exist because e and this modulus are relatively prime. The values e and $n = pq$ are made available publicly.

Encryption: To encrypt the given plaintext, it is first broken into blocks of numerical values in the range $0, 1, \ldots, n - 1$. The sender of the message uses the public formula $C = P^e \pmod{n}$ to obtain the encrypted C from the block of plaintext P.

Decryption: To decrypt the ciphertext C, the receiver of the message uses his private key d to obtain the original plaintext P by means of the formula $P = C^d \pmod{n}$.

Figure 24.1 provides a diagram of the sequence of steps taken, showing in blue the information that is private.

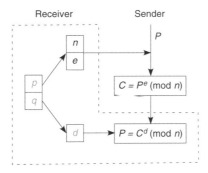

FIGURE 24.1: Components of the RSA cryptosystem

It is helpful to illustrate how this process works by introducing specific numerical values. The receiver wants to receive encrypted messages from a number of senders, and selects the two private primes $p = 3$ and $q = 11$. Then the public value n is computed as $n = 3 \times 11 = 33$. The receiver also chooses the public encryption key $e = 3$, which is relatively prime to the product $(p-1)(q-1) = 2 \times 10 = 20$; note that 3 is relatively prime to 20 because they only share the common factor 1. Next, the receiver computes his private decryption key $d = 7$, which is the inverse of $e = 3$ modulo 20, since $7 \times 3 \equiv 1 \pmod{20}$.

A particular sender wishes to transmit the numerical plaintext $P = 8$ to the receiver, so this plaintext value is encrypted as $P^e \pmod{n} = 8^3 \pmod{33} = 512 \pmod{33} = 17$ and sent to the receiver. The receiver then decrypts the ciphertext $C = 17$ by computing C^d

$(\bmod\ n) = 17^7\ (\bmod\ 33) = 17^3 \times 17^3 \times 17\ (\bmod\ 33) \equiv 29 \times 29 \times 17\ (\bmod\ 33) = 8$. Table 24.2 summarizes the calculations performed in this numerical example.

TABLE 24.2: Sample RSA calculations

Key Generation	$p = 3$, $q = 11$, $n = pq = 33$ $e = 3$: e and $(p-1)(q-1) = 20$ are relatively prime $d = 7$: d is the inverse of e, modulo 20
Encryption of Plaintext $P = 8$	$P^e\ (\bmod\ n) = 8^3\ (\bmod\ 33) = 17$
Decryption of Ciphertext $C = 17$	$C^d\ (\bmod\ n) = 17^7\ (\bmod\ 33) = 8$

Example 24.7 Encrypt Plaintext Using the RSA System

Suppose the two private prime numbers $p = 3$ and $q = 23$ are chosen by the receiver. (a) Calculate the public modulus n. (b) Verify that the public encryption key $e = 5$ is relatively prime to $(p-1)(q-1)$. (c) A sender wants to encrypt the plaintext message $P = 10$. Calculate the ciphertext C by using the encryption formula.

Solution

(a) The public modulus is $n = 3 \times 23 = 69$.

(b) The public key $e = 5$ is relatively prime to $(p-1)(q-1) = 2 \times 22 = 44$ because the only common factor of 5 and 44 is 1.

(c) The ciphertext is $C = P^e\ (\bmod\ n) = 10^5\ (\bmod\ 69) = 100{,}000\ (\bmod\ 69) = 19$. ◁

(Related Exercises: 13–15)

Example 24.8 Decrypt Ciphertext Using the RSA System

Suppose the two private prime numbers $p = 7$ and $q = 19$ are chosen by the receiver. (a) Calculate the public modulus n. (b) Verify that the public encryption key $e = 5$ is relatively prime to $(p-1)(q-1)$. (c) Show that the private decryption key $d = 65$ is the inverse of $e = 5$ modulo $(p-1)(q-1)$. (d) The receiver wants to decrypt the ciphertext message $C = 62$. Calculate the plaintext P using the decryption formula $P = C^d\ (\bmod\ n)$.

Solution

(a) The public modulus is $n = 7 \times 19 = 133$.

(b) The public key $e = 5$ is relatively prime to $(p-1)(q-1) = 6 \times 18 = 108$ because the only common factor of 5 and 108 is 1.

(c) To verify that $d = 65$ is the (unique) inverse of $e = 5$ modulo $(p-1)(q-1)$, we calculate $de\ (\bmod\ (p-1)(q-1)) = 65 \times 5\ (\bmod\ 108) = 325\ (\bmod\ 108) = 1$.

(d) The plaintext P is calculated using $P = C^d\ (\bmod\ n) = 62^{65}\ (\bmod\ 133)$. Since 62^{65} is too large to calculate directly, we rewrite it as $62^{65} = 62 \times 62^{64} = 62 \times (62^2 \times 62^2)^{16} \equiv 62 \times (120 \times 120)^{16}\ (\bmod\ 133) \equiv 62 \times 36^{16}\ (\bmod\ 133) = 62 \times (36^4)^4\ (\bmod\ 133) \equiv 62 \times 92^4$ $(\bmod\ 133) = 62 \times 92^2 \times 92^2\ (\bmod\ 133) \equiv 62 \times 85 \times 85\ (\bmod\ 133) = 6$. This gives the plaintext $P = 6$. ◁

(Related Exercises: 16–18)

Example 24.9 Encrypt and Decrypt Using the RSA System

Suppose the two private prime numbers $p = 5$ and $q = 11$ are chosen. Carry out the following steps of the RSA cryptosytem. (a) Calculate the public modulus n. (b) Explain why the public encryption key $e = 7$ is relatively prime to $(p-1)(q-1)$. (c) To send the plaintext message MOP, first use Table 22.7 to convert each plaintext letter into a decimal number in the range $0, 1, \ldots, 25$. Then calculate the ciphertext using the encryption formula for each decimal. [In actual practice, the letters would be converted to binary via an ACSCI/Unicode table and subdivided into blocks of the same size.] (d) Use the private decryption key $d = 23$ and the decryption formula to recover the plaintext. (e) Verify that d and e are inverses modulo $(p-1)(q-1)$.

Solution

(a) The public modulus is $n = 5 \times 11 = 55$.

(b) The public key $e = 7$ is relatively prime to $(p-1)(q-1) = 4 \times 10 = 40$ because the only common factor of 7 and 40 is 1.

(c) From Table 22.7, $\mathsf{M} \to 12, \mathsf{O} \to 14, \mathsf{P} \to 15$. Each block P of the plaintext is encrypted using $C = P^e \pmod{n}$ so that the plaintext message 12-14-15 is transformed as follows: $12^7 \pmod{55} = 23, 14^7 \pmod{55} = 9$, and $15^7 \pmod{55} = 5$. The ciphertext 23-09-05 will be sent.

(d) Each block C of the ciphertext 23-09-05 is decrypted using $P = C^d \pmod{n}$: $23^{23} \pmod{55} = 23^6 \times 23^6 \times 23^6 \times 23^5 \pmod{55} \equiv 34 \times 34 \times 34 \times 23 \pmod{55} = 903{,}992 \pmod{55} = 12; 9^{23} \pmod{55} = 9^{10} \times 9^{10} \times 9^3 \pmod{55} \equiv 1 \times 1 \times 14 \pmod{55} = 14; 5^{23} \pmod{55} = 5^{10} \times 5^{10} \times 5^3 \pmod{55} \equiv 45 \times 45 \times 15 \pmod{55} = 15$. So the plaintext recovered is 12-14-15, which corresponds to the plaintext message MOP.

(e) To verify that $d = 23$ is the (unique) inverse of $e = 7$ modulo $(p-1)(q-1)$, we calculate $de \pmod{(p-1)(q-1)} = 23 \times 7 \pmod{40} = 161 \pmod{40} = 1$. ◁

(*Related Exercises: 19–20*)

We now describe more formally the general RSA procedure used above:

RSA Key Generation

1. Choose two large prime numbers p and q, which are kept private. Compute the public modulus $n = pq$.
2. Calculate the public encryption key e by selecting a number that is relatively prime to the product $(p-1)(q-1)$.
3. Calculate the private decryption key d that is the inverse of e modulo $(p-1)(q-1)$.

RSA Encryption Algorithm

1. Using the public key e and the public modulus n, encrypt the plaintext P using the public formula $C = P^e \pmod{n}$.

RSA Decryption Algorithm

1. Using the private key d and the public modulus n, decrypt the ciphertext C using $P = C^d \pmod{n}$.

Practical Security and the RSA Cipher

Without knowledge of the two large primes p and q, it is computationally challenging to determine the private decryption key d. Even though the product $n = pq$ of these two large primes is public, it is difficult to recover the product $(p - 1)(q - 1)$ and thus determine the inverse d of the public key e. The security of RSA is based on the difficulty of solving the Prime Factorization Problem for large composite numbers. The size of the numbers currently used in the RSA cryptosystem are chosen to ensure *practical security* so that hacking the decryption key, though theoretically possible, could range from six months to hundreds of years depending on the computing power available. Note that currently there is no theoretical result showing that factoring is as difficult as the NP-hard problems encountered in Chapter 5.

RSA continues to be used in most web servers and browsers, e-commerce applications, and in commercially available security products. However, it is not alone in terms of cryptosystems currently used to maintain commercial and personal information. The alternative ElGamal cryptosystem is another public-key system that is widely used and it is discussed in the next section.

24.4 ElGamal Cryptosystem

The ElGamal scheme is the best-known alternative to RSA.

DEFINITION **ElGamal Cryptosystem**
An asymmetric public-key system that also uses modular arithmetic. Its security rests on the difficulty of solving the Discrete Logarithm Problem.

HISTORICAL CONTEXT **ElGamal Public-Key System (1985)**
This encryption/decryption scheme was described by the Egyptian cryptologist Taher ElGamal.

Just as with the RSA cryptosystem, the ElGamal system enables a receiver to publish certain public information so that any sender can transmit a message, encrypted using this public information, with a high degree of security. In the ElGamal system, the receiver creates three public numbers together with a private key. Any sender uses these three public numbers together with his own random key to encrypt a plaintext message, which is then sent to the receiver along with another disguised piece of information (derived from the public information and the sender's random key). Using both of these items and his own private key, the receiver is able to piece together the original plaintext.

The public and private numbers are chosen carefully in conjunction with the encryption formulas so that only the receiver's private key will be able to decrypt messages from the sender. Because the ElGamal cryptosystem involves a one-way function based on the Discrete Logarithm Problem, it is difficult computationally for anyone else to deduce the receiver's private key from the available public numbers and formulas.

Here is an overview of the El Gamal procedures for creating keys, encrypting the plaintext, and decrypting the ciphertext.

Key Generation: The public key consists of three numbers (p, g, y) created by the receiver. The receiver chooses a prime modulus p, a number g that is a generator modulo p, and computes another public number y from his private key x using the formula $y = g^x \pmod{p}$. The receiver's private key x is a randomly chosen integer in the range $2, 3, \ldots, p - 2$ and it is used for decrypting ciphertext.

Encryption: To encrypt the plaintext, the sender uses the receiver's published key (p, g, y) and a randomly chosen integer k in the range $0, 1, \ldots, p - 2$. We assume that the plaintext has been divided into blocks P of size less than p. He sends the receiver two encrypted numbers (C_1, C_2): C_1 is computed from k using the formula $C_1 = g^k \pmod{p}$ and the plaintext block P is encrypted using the formula $C_2 = Py^k \pmod{p}$.

Decryption: In essence, the receiver uses the encrypted C_1 and C_2 together with his private key x to solve the modular equation $C_1^x P \equiv C_2 \pmod{p}$ and thus retrieve the plaintext P. To do this, he first computes $z = C_1^x \pmod{p}$ and then calculates z^{-1}, the inverse of z modulo p. To obtain the original plaintext P, the receiver simply computes $z^{-1} C_2 \pmod{p}$.

Figure 24.2 provides a diagram of the sequence of steps taken from key creation, to encryption, to decryption. The private information is shown in blue.

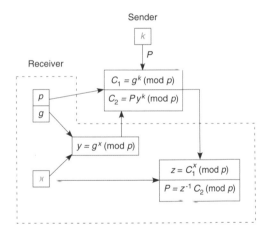

FIGURE 24.2: Components of the ElGamal cryptosystem

It is helpful to illustrate how this process works by introducing specific numerical values. Suppose that the receiver has chosen the prime modulus $p = 23$ and the generator $g = 11$ modulo 23. The receiver also selects the private key $x = 6$, which is then used to calculate y using the formula $y = g^x \pmod{p} = 11^6 \pmod{23} = 9$. He publishes these three numbers $(p = 23, g = 11, y = 9)$. The sender now uses these three public numbers and his randomly chosen number $k = 3$ to encrypt the plaintext $P = 10$. Namely, he sends the receiver the ciphertext pair (C_1, C_2): $C_1 = g^k \pmod{p} = 11^3 \pmod{23} = 20$, $C_2 = Py^k \pmod{p} = 10 \times 9^3 \pmod{23} = 22$. The ciphertext pair transmitted is then $(20, 22)$. Upon receiving this pair, the receiver calculates $z = C_1^x \pmod{p} = 20^6 \pmod{23} = 16$ and then finds the inverse of $z = 16$ modulo 23, which turns out to be $z^{-1} = 13$. Finally, the receiver decrypts the original message by computing $z^{-1} C_2 \pmod{p}$: $13 \times 22 \pmod{23} = 10$. The receiver now knows that the plaintext sent was 10.

Table 24.3 summarizes the calculations performed in this numerical example where the sender transforms the plaintext $P = 10$ into the ciphertext pair $(20, 22)$. It also indicates how the receiver reverses the process and recovers the original plaintext $P = 10$.

TABLE 24.3: Sample ElGamal calculations

Private numbers	Receiver's private key: $x = 6$ Sender's random number: $k = 3$
Public numbers	Prime modulus: $p = 23$ Generator modulo p: $g = 11$ Transformed private key: $\quad y = g^x \pmod{p} = 11^6 \pmod{23} = 9$
Plaintext $P = 10$ Send (C_1, C_2)	$C_1 = g^k \pmod{p} = 11^3 \pmod{23} = 20$ $C_2 = Py^k \pmod{p} = 10 \times 9^3 \pmod{23} = 22$
Receiver decrypts Ciphertext	Solve $C_1^x P \equiv C_2 \pmod{p}$ $z = C_1^x \pmod{p} = 20^6 \pmod{23} = 16$ $z^{-1} \pmod{23} = 13$ $P = z^{-1} C_2 \pmod{p} = 13 \times 22 \pmod{23} = 10$

To use the ElGamal procedure for the example in Table 24.3, the generator g modulo 23 had to be specified. As shown in Section 24.1, we can demonstrate that 11 is a generator modulo 23 by verifying that successive powers of 11 produce all the nonzero remainders (mod 23). Also, the inverse of $z = C_1^x \pmod{p} = 16$ needed to be found. Since $13 \times 16 \equiv 1$ (mod 23) holds, 13 is in fact the inverse of 16, modulo 23. Modular inverses can be easily found using online tools such as [OR2].

Example 24.10 Encrypt Plaintext Using the ElGamal System

A sender wishes to transmit plaintext to a receiver whose public information $(p, g, y) = (31, 6, 26)$ is available. Here the plaintext has already been converted to the numerical value $P = 15$. The sender's randomly chosen number is $k = 2$. (a) Compute the ciphertext pair (C_1, C_2) that is sent to the receiver. (b) Since the values (p, g, y) are publicly known, what equation would an intruder have to solve to retrieve the private key x of the receiver?

Solution

(a) $C_1 = g^k \pmod{p} = 6^2 \pmod{31} = 5$ and $C_2 = Py^k \pmod{p} = 15 \times 26^2 \pmod{31} = 10{,}140 \pmod{31} = 3$. The pair $(5, 3)$ is sent to the receiver.

(b) The equation used to define the public number y is $y = g^x \pmod{p}$. Using the known published numbers, this becomes the modular equation $26 = 6^x \pmod{31}$. In other words, we would need to find the discrete logarithm $\log_6 26 \pmod{31}$. ◁

(Related Exercises: 21–22)

As seen in the solution to part (b) of Example 24.10, the security of the ElGamal system depends on the difficulty of solving a discrete logarithm problem. In practice, the prime p is chosen to be several hundred digits long so that this task is computationally infeasible.

Example 24.11 Decrypt Ciphertext Using the ElGamal System

The receiver has been sent the ciphertext pair $(C_1, C_2) = (38, 80)$. His public key is $(p, g, y) = (139, 3, 44)$ and his private key is $x = 12$. (a) Calculate $z = C_1^x \pmod{p}$. (b) Verify that the inverse (modulo 139) of z is 36. (c) Obtain the original plaintext by solving the modular equation $zP \equiv C_2 \pmod{p}$.

Solution

(a) $z = C_1^x \pmod{p} = 38^{12} \pmod{139} = 38^6 \times 38^6 \pmod{139} \equiv 116 \times 116 \pmod{139} = 112$.

(b) To verify that 36 is the inverse (modulo 139) of $z = 112$, we calculate the product $36 \times 112 \pmod{139} = 4032 \pmod{139} = 1$.

(c) To solve $zP \equiv C_2 \pmod{p}$, we compute $z^{-1}C_2 \pmod{p} = 36 \times 80 \pmod{139} = 100$. So the plaintext is 100. ◁

(Related Exercises: 23–24)

Example 24.12 Encrypt and Decrypt Using the ElGamal System

A receiver selects the public modulus $p = 13$, the public generator $g = 2$ modulo 13, and the private key $x = 7$. (a) Calculate the receiver's transformed key y. (b) A sender wishes to transmit the plaintext $P = 3$ using the random number $k = 5$. Calculate the ciphertext pair (C_1, C_2) that should be sent to the receiver. (c) Calculate $z = C_1^x \pmod{p}$ and verify that $z^{-1} = 2$. (d) Using the inverse found in the previous step together with C_2, determine the plaintext sent.

Solution

(a) The receiver's transformed key y is found using $y = g^x \pmod{p} = 2^7 \pmod{13} = 11$.

(b) $C_1 = g^k \pmod{p} = 2^5 \pmod{13} = 6$. $C_2 = Py^k \pmod{p} = 3 \times 11^5 \pmod{13} = 8$. So the ciphertext pair $(C_1, C_2) = (6, 8)$ is sent.

(c) $z = C_1^x \pmod{p} = 6^7 \pmod{13} = 7$; since $2 \times 7 \pmod{13} = 1$, we have verified that $z^{-1} = 2$.

(d) We multiply C_2 by the modular inverse to obtain $z^{-1}C_2 \pmod{p} = 2 \times 8 \pmod{13} = 3$. ◁

(Related Exercises: 25–26)

We now describe more formally the general ElGamal procedure used above:

ElGamal Key Generation

1. The receiver chooses a large prime as the public modulus p, a public number g that is a generator modulo p, and a random private number x in the range $2, 3, \ldots, p - 2$.

2. The receiver computes the transformed key $y = g^x \pmod{p}$.

3. The receiver publishes the three numbers (p, g, y) as his public key.

ElGamal Encryption Algorithm

1. The sender looks up the receiver's public key (p, g, y).

2. The sender breaks up the plaintext into blocks P of size less than p.

---

3. The sender selects a random number k in the range $0, 1, \ldots, p - 2$.

4. The sender transmits (C_1, C_2) to the receiver, where $C_1 = g^k \pmod{p}$ and $C_2 = Py^k \pmod{p}$.

ElGamal Decryption Algorithm

1. The receiver computes $z = C_1^x \pmod{p}$ and then finds the inverse of z modulo p.

2. The receiver computes $z^{-1}C_2 \pmod{p}$ to determine the plaintext sent.

Security of ElGamal

The security of ElGamal rests on the belief that solving the Discrete Logarithm Problem is difficult. No attacker can determine the private key x from the available public information because this requires an efficient algorithm for solving the Discrete Logarithm Problem; such an algorithm does not currently exist. By contrast, there are efficient techniques for exponentiation and for finding inverses in a modular system.

Each time a sender encrypts plaintext, a new value k is randomly generated. This is an advantage, since the same plaintext sent at different times will be encrypted differently.

An undesirable feature of the ElGamal procedure is that it requires sending a ciphertext pair, which is twice as long as the plaintext. By contrast, the RSA procedure sends ciphertext that is the same size as the plaintext.

Elliptic Curve Cryptography

An alternative form of the ElGamal scheme is based on the mathematics of *elliptic curves*. Basing ElGamal on elliptic curves allows the size of the keys to become smaller so this system is preferred over RSA for efficiency reasons as well as increased security.

DEFINITION **Elliptic Curve Cryptography**
This type of cryptography uses mathematical objects called elliptic curves in conjunction with modular arithmetic. Instead of working with the integers modulo p, it uses numbers associated with points on an elliptical curve.

DEFINITION **Elliptic Curve Modulo p**
An elliptic curve is the set of points (x, y) that satisfy the congruence $y^2 \equiv x^3 + ax + b \pmod{p}$.

Example 24.13 Determine All Points on an Elliptic Curve

Find all pairs (x, y) that satisfy $y^2 \equiv x^3 + x + 6 \pmod{11}$ by trying in turn each value of $x = 0, 1, \ldots, 10$ and then solving the associated modular equation for y. Generally, such a quadratic equation either has two or no solutions.

Solution

When $x = 0$ the equation $y^2 \equiv x^3 + x + 6 \pmod{11}$ becomes $y^2 \equiv 6 \pmod{11}$. By examining all possible values for $y = 0, 1, \ldots, 10$ we find that there is no value for y that will work.

On the other hand, when $x = 8$ the equation becomes $y^2 \equiv 8^3 + 8 + 6 \pmod{11}$ or $y^2 \equiv 9 \pmod{11}$. Trial-and-error produces two solutions, $y = 3$ and $y = 8$; note that $3^2 \pmod{11} = 9$ and $8^2 \pmod{11} = 64 \pmod{11} = 9$. Similar calculations result in the 12 solutions listed in Table 24.4. Notice that if y is a solution for a particular x, then so is $11 - y$. $\triangleleft$

(Related Exercise: 27)

TABLE 24.4: Solutions to an elliptic curve

x	2	2	3	3	5	5	7	7	8	8	10	10
y	4	7	5	6	2	9	2	9	3	8	2	9

It turns out that we can define the addition of two points (x_1, y_1) and (x_2, y_2) on the elliptic curve. The precise definition is somewhat complicated, so we just illustrate the use of this operation. For example, it turns out that $(2, 7) + (3, 6) = (7, 9)$. Notice that all three of these points lie on our particular elliptic curve $y^2 \equiv x^3 + x + 6 \pmod{11}$, as seen in Table 24.4. In general, the addition of two points on the curve produces a point that is also on the curve. Moreover, we can find a generator g of all points on this elliptic curve, once we remember that now our operation is addition and not multiplication (as it was for the ElGamal scheme). One such generator is $g = (5, 2)$. Then $g + g = 2g$ turns out to be $(10, 2)$ and $g + g + g = 3g$ turns out to be $(7, 9)$. Table 24.4 shows that both of these points remain on our particular elliptic curve. We have all the ingredients to define an "additive" version of ElGamal, defined with respect to any elliptic curve.

Suppose that the receiver uses a generator g of the points on an elliptic curve and also selects a private key x. Then the public y is defined as xg rather than g^x. A sender then selects a random k and uses this to send a ciphertext pair (C_1, C_2) to the receiver. Keeping in mind our additive structure, the sender computes $C_1 = kg$ rather than g^k and transforms the plaintext P using $C_2 = P + ky$ rather than $C_2 = Py^k$. Finally, the receiver can recover the plaintext P by solving $xC_1 + P = C_2$ instead of the congruence $C_1^x P \equiv C_2$. The following example provides a numerical illustration. Again, since the operation of addition is fairly tedious, we will only indicate the final results of certain computations.

Example 24.14 Encrypt and Decrypt Using an Elliptic Curve

Suppose that the receiver uses the generator $g = (5, 2)$ for the points on $y^2 \equiv x^3 + x + 6 \pmod{11}$ and selects the private key $x = 3$. Then the public $y = xg = 3g = (7, 9)$. (a) Suppose a sender selects $k = 5$ and wishes to send the plaintext message $P = (2, 7)$, which is a point on the elliptic curve. What ciphertext pair (C_1, C_2) will be sent? (b) Show the steps used by the receiver to decrypt the message received.

Solution

(a) $C_1 = kg = 5g = (8, 8)$. This value can be determined by using the online calculator [OR3], which easily enables the calculation of multiples and addition for elliptic curves. Also, the sender calculates $C_2 = P + ky = (2, 7) + 5(7, 9) = (2, 7) + (10, 2) = (3, 6)$.

(b) Given these values for C_1 and C_2, the receiver then needs to solve the equation $xC_1 + P = C_2$ or $3(8, 8) + P = (3, 6)$, which simplifies to $(10, 2) + P = (3, 6)$. To solve this equation we add the (additive) inverse of (10,2) to both sides of this equation. It turns out that (10,9) is the additive inverse of (10,2): $(10,9) + (10,2) = (0,0)$ using the elliptic curve arithmetic. The solution steps are as follows:

$$(10, 2) + P = (3, 6)$$
$$(10, 9) + (10, 2) + P = (10, 9) + (3, 6)$$
$$(0, 0) + P = (2, 7)$$
$$P = (2, 7).$$

So we are able to recover the original message $P = (2, 7)$. ◁

(Related Exercise: 28)

How secure is the elliptic curve cryptosystem? Similar to the ElGamal cryptosystem, its security rests on the difficulty of solving an appropriate discrete logarithm problem. Namely, given y and generator g, find the multiple x that satisfies $y = xg \pmod{p}$. Solving this problem is believed to be computationally infeasible for suitably chosen primes p.

In addition to its use by governments to ensure security, elliptic curve cryptography is rapidly gaining usage in commerical applications such as banking, telecommunications, the use of digital postmarks, and the production of Blu-Ray discs.

24.5   Message Authentication

Encryption and decryption of a message provides privacy, but alone these do not provide authentication that the source of the message is who you think it is, or that the message has not been intentionally manipulated by a *man-in-the-middle attack* (see Section 23.5). Recall that a man-in-the-middle attack occurs when an intruder inserts himself between the authorized sender and receiver and, without their knowledge, alters messages or creates new ones. This section will discuss a system of authentication that can prevent this kind of invasive eavesdropping.

Also, encryption and decryption of a message does not guarantee *message integrity*, ensuring that the message has not been altered accidentally, for example, by signal interference in the communications channel that can distort the digital data being transmitted.

DEFINITION **Message Authentication**
A process that guarantees a message's origin and provides assurance that it is free from intentional manipulation by some unauthorized third party.

DEFINITION **Message Integrity**
The goal of preventing, detecting, or correcting unintended changes to a message, resulting from noise in the communications channel, hardware failure, human error, etc.

We will concentrate here on authentication because it connects directly with cryptography. Data integrity is addressed using *coding theory*, where methods for detecting and correcting unintentional errors are studied and evaluated.

Definition **Coding Theory**
A discipline that develops methods for error detection, error correction, and data compression. For example, error-correcting codes are used to detect and repair errors on CDs and DVDs affected by scratches or dust, and to correct for noise in cell phone transmission.

Digital Signatures

In cryptography, one way to provide an assurance of authenticity is through the use of *digital signatures*, which make use of the private keys involved in a public-key cryptosystem. In 2000, President Clinton signed into law the E-SIGN Act that made digital signatures legally binding.

Definition **Digital Signature**
This provides assurance of the identity of the sender and an indication that the document has not been altered after it was signed.

Digital signatures often involve using a *hash function*, which is simply a compressed version of a longer message. Its purpose is not to hide data, but rather to accompany a message so that if someone alters the message, the hashed value will very likely reveal an alteration.

Definition **Hash Function**
A mathematical function that operates on binary data (a message) and compresses it to reduce its size for transmission. Any change to the message will (with very high probability) change the hashed value and alert the receiver of the message of some alteration.

There are different ways to implement hash functions. We will illustrate this concept using modular arithmetic.

Example 24.15 Apply a Modular Hash Function to a Given Message

Suppose the message M to be transmitted is 4578230945122344, a numerical representation of a bank account. (a) Divide this 16-digit message into four parts M_1, M_2, M_3, M_4 each with four digits. Use the modulus $n = 6101$ to compute the hashed value of M as $h(M) = (M_1 + M_2 + M_3 + M_4) \pmod n$. (b) What would be the hashed value if the first part was incorrectly given as 4587?

Solution

(a) $h(M) = M_1 + M_2 + M_3 + M_4 \pmod n = (4578 + 2309 + 4512 + 2344) \pmod{6101} = 13{,}743 \pmod{6001} = 1541$.

(b) $h(M) = M_1 + M_2 + M_3 + M_4 \pmod n = (4587 + 2309 + 4512 + 2344) \pmod{6101} = 13{,}752 \pmod{6101} = 1550$. The (inadvertent) error has caused the hashed value to change. ◁

(Related Exercise: 29)

Before we use hash functions in conjunction with digital signatures, we discuss how a public-key system can easily support the inclusion of a digital signature to verify the sender and the integrity of a transmitted message.

RSA and Digital Signatures

In a public-key system, a public key e is used to encrypt a message, while a private key d is used to decrypt a message. This is summarized for the RSA method in the first column of Table 24.5. This process can be reversed in order to produce a digital signature. Namely, the sender can encrypt data with his own private key d, and the receiver can then apply the sender's public key e. In this way, the receiver can verify the identity of the sender, as no one else has access to the sender's private key d. The second column of Table 24.5 indicates how the sender produces the signature S and the receiver verifies it by computing V.

TABLE 24.5: Public/private keys and digital signatures

Public key e used to encrypt message: $C = P^e \pmod n$	Private key d used to sign data: $S = P^d \pmod n$
Private key d used to decrypt message: $P = C^d \pmod n$	Public key e used to verify signature: $V = S^e \pmod n$

To implement the RSA system with digital signatures, both parties A and B need their own set of private/public keys as well as a selected modulus. So A makes available the public numbers (e_A, n_A) but maintains the private key d_A; B makes available the public numbers (e_B, n_B) but maintains the private key d_B.

Below is a specific example in which A wishes to send a plaintext message securely to B, and B wants to be confident that this message did in fact come from A. To do this, A will send two items to B, who will then subsequently process them to determine the plaintext as well as verify that it originated from A.

Example 24.16 Send Plaintext and Verify a Signature Using the RSA Cryptosystem

Suppose A publishes the two numbers $e_A = 3$ and $n_A = 55$, and maintains the private key $d_A = 27$; B publishes the two numbers $e_B = 5$ and $n_B = 69$, and maintains the private key $d_B = 9$. Entity A wishes to send the plaintext message $P = 6$ to B, after first adding a verifiable signature. (a) If A encrypts the plaintext with B's public key, what ciphertext C will be sent to B? (b) If A also signs the plaintext with his own private key, what signature S will be sent to B? (c) If B decrypts the ciphertext C using his own private key, what plaintext will result? (d) Using S, show how B can verify that the plaintext came from A.

Solution

(a) A encrypts $P = 6$ using B's public data, obtaining the ciphertext $C = P^{e_B} \pmod{n_B} = 6^5 \pmod{69} = 48$.

(b) A signs the plaintext $P = 6$ using his own private key, creating the signature $S = P^{d_A} \pmod{n_A} = 6^{27} \pmod{55} = 41$.

(c) Now B decrypts the ciphertext $C = 48$ using his own private key to obtain the plaintext $P = C^{d_B} \pmod{n_B} = 48^9 \pmod{69} = (48^3)^3 \pmod{69} \equiv 54^3 \pmod{69} = 6$.

(d) B verifies the signature S by computing $V = S^{e_A} \pmod{n_A} = 41^3 \pmod{55} = 6$. Since the decrypted signature V matches the plaintext P, the signature is verified. ◁

(Related Exercise: 30)

Figure 24.3 shows how this combined process of encryption, decryption, and signature verification is carried out using the RSA cryptosystem. The two private keys are shown in blue. It should be noted that the two encrypted messages C and S need to be sent from A to B; by contrast, the RSA scheme only requires the transmission of C.

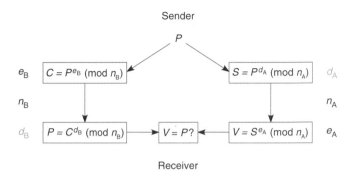

FIGURE 24.3: RSA with signatures

Since the purpose of a signature is not the content of the message, but just to verify the identity of the sender, we can use a hash function to reduce the size of the signature S sent, as well as the overall computational effort. The real benefit of hashing comes not from hashing a single plaintext block, but from combining several blocks $P_1, P_2, \ldots$ of plaintext, as illustrated in Example 24.15. In the next example, we apply a hashing function, but since we are working with small numbers and a single plaintext block, it is important to remember that efficiency is achieved through the hashing (compression) of multiple message blocks.

Example 24.17 Send Plaintext and Verify a Hashed Signature Using RSA

Suppose A publishes the two numbers $e_A = 3$ and $n_A = 55$, and maintains the private key $d_A = 27$; B publishes the two numbers $e_B = 5$ and $n_B = 69$, and maintains the private key $d_B = 9$. Entity A wishes to send the plaintext message $P = 53$ to B, after first adding a verifiable signature. In this case, rather than signing the plaintext P, it is first hashed using the modulus $n = 11$ to reduce its size. (a) If A encrypts the plaintext P with B's public key, what ciphertext C will be sent to B? (b) If A signs the hashed plaintext with his own private key, what signature S will be sent to B? (c) If B decrypts the ciphertext C using his own private key, what plaintext will result? (d) Using S and the hash function h, show how B can verify that the plaintext was sent from A.

Solution

(a) A encrypts $P = 53$ using B's public data, obtaining the ciphertext $C = P^{e_B} \pmod{n_B} = 53^5 \pmod{69} = 17$.

(b) First, A computes the hashed plaintext $H = h(P) = P \pmod{n} = 53 \pmod{11} = 9$. Then H is signed using $S = H^{d_A} \pmod{n_A} = 9^{27} \pmod{55} = 4$.

(c) Now B decrypts the ciphertext $C = 17$ to obtain the plaintext $P = C^{d_B} \pmod{n_B} = 17^9 \pmod{69} = (17^3)^3 \pmod{69} \equiv 14^3 \pmod{69} = 53$.

(d) First, B computes the hashed plaintext, given the numerical value $P = 53$ just determined: $H = h(P) = P \pmod{n} = 53 \pmod{11} = 9$. Next, B verifies the hashed signature S by computing $V = S^{e_A} \pmod{n_A} = 4^3 \pmod{55} = 9$. Since the decrypted signature V matches the hashed plaintext H, the signature is verified. ◁

(Related Exercise: 31)

The general process of using a hashed signature within the RSA framework is shown in Figure 24.4. The two private keys are shown in blue. The sender A encrypts the plaintext message the same way as done with RSA, using B's published number e_B. In addition, A computes the hashed value $H = h(P)$ from the plaintext, using a suitable hash function h. This hashed value H is then signed and also sent to B. As with RSA, B determines the plaintext P from the ciphertext C using his private key d_B. From P, it is then easy for B to determine $H = h(P)$ and also to compute the decrypted signature V from the signed hash value S. The signature is verified if $V = H$.

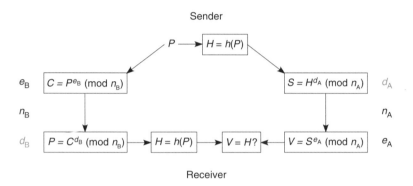

FIGURE 24.4: RSA with a signed hash value

We have expressed the signature scheme in Figure 24.4 using a general hash function h. It turns out that in order to be useful in cryptography and be resistant to attacks, the function h should satisfy several conditions:

1. Calculation of the hashed value $h(P)$ should be quick.
2. The hash function h should be a one-way function: that is, given the hashed value $h(P)$ it should be computationally infeasible to obtain the corresponding message P.
3. It should be computationally infeasible to find two messages that produce the same hashed value.

To support a system of authenticated digital signatures, a Public Key Infrastructure has been developed that provides reliable credentials. When the receiver of a digital message can be confident of the identity of the sender and that the document has not been altered since it was signed, then *authentication* has taken place. There are many Certification Authorities that issue *digital certificates*: electronic documents used to verify that a public key really belongs to a specific individual. There are also *digital timestamps* which substantiate when a message was created or modified. These procedures are important to ensure the safe and accurate transmission of electronic information.

Chapter Summary

The real contribution of RSA and ElGamal (as well as ElGamal based on elliptic curves) takes place during key generation where the relationship between the encryption and decryption keys is designed so that one key can unlock the effect of the other, using certain secret information. Consequently, the encryption key can be publicly known, whereas the decryption key is kept secret. This allows encryption information to be publicized (posted on a website) so a sender can encrypt a message for an intended receiver, without allowing unauthorized parties to discover the private decryption key. No transmission of a private key is required. Since the security of these systems relies on our inability to efficiently solve reasonably large prime factorization or discrete logarithm problems, potential attackers would require an impractical amount of time or computing power. We have introduced the underlying number theory that makes the mathematical relationship between the keys possible: primes, modular arithmetic, discrete logarithms, and one-way functions. These concepts also enable the authentication of messages transmitted between entities. Choices between competing public-key systems will continue to be made based on security needs compared with the ever-improving skills and computing power of potential attackers.

Applications, Representations, Strategies, and Algorithms

Applications				
Security (7–12, 14, 16–17)		Business (15)		

Representations				
Symbolization				
Numerical		Algebraic		
1–13, 15–17		13–14		

Strategies		
Brute-Force	Rule-Based	Composition/Decomposition
1–3, 6, 13	7–12, 14–17	4–5

Algorithms				
Exact	Sequential	Numerical	Algebraic	Recursive
1–17	1–3, 6	1–3, 4–13, 15–17	14	4–5

Exercises

Modular Inverses

1. Define the composite modulus $n = 3 \times 5 = 15$.

(a) Verify that 7 has an inverse, modulo 15.

(b) Verify that 11 has an inverse, modulo 15.

(c) Verify that 10 does not have an inverse, modulo 15.

(d) What other numbers in the range $1, 2, \ldots, 14$ have an inverse, modulo 15?

2. Define the composite modulus $n = 2 \times 7 = 14$.

(a) Verify that 7 does not have an inverse, modulo 14.

(b) Verify that 11 has an inverse, modulo 14.

(c) Verify that 10 does not have an inverse, modulo 14.

(d) What other numbers in the range $1, 2, \ldots, 13$ have an inverse, modulo 14?

Discrete Logarithms

3. Define the prime modulus $p = 5$.

(a) Determine (if it exists) $\log_3 2 \pmod 5$ by computing successive powers of 3, modulo 5.

(b) Determine (if it exists) $\log_4 2 \pmod 5$ by computing successive powers of 4, modulo 5.

4. Define the prime modulus $p = 11$.

(a) Determine (if it exists) $\log_2 6 \pmod{11}$ by computing successive powers of 2, modulo 11.

(b) Determine (if it exists) $\log_3 6 \pmod{11}$ by computing successive powers of 3, modulo 11.

Generators Modulo p

5. Define the prime modulus $p = 5$.

(a) Find all generators b modulo 5 by computing successive powers of b, modulo 5.

(b) Verify that the number of generators you find agrees with the formula stated in Section 24.1.

6. Define the prime modulus $p = 11$.

(a) Verify that $b = 6$ is a generator modulo 11.

(b) How many generators are there modulo 11?

Prime Division and Prime Factorization

7. Apply the prime division method to obtain the (unique) factorization of 5670.

8. Apply the prime division method to obtain the (unique) factorization of 52,745.

Exponentials and Logarithms Modulo p

9. It is known that 2 is a generator modulo 13.

   (a) Compute 2^{25} (mod 13).
   (b) Compute $\log_2 11$ (mod 13).

10. It is known that 5 is a generator modulo 23.

   (a) Compute 5^{42} (mod 23).
   (b) Compute $\log_5 8$ (mod 23).

11. Determine the last two digits of 27^{27}. [*Hint:* Consider using modulo 100 arithmetic.]

12. Determine the last two digits of 19^{26}. [*Hint:* Consider using modulo 100 arithmetic.]

RSA Encryption

13. Suppose the two private prime numbers $p = 5$ and $q = 17$ are chosen by the receiver.

   (a) Calculate the public modulus n.
   (b) Verify that the public encryption key $e = 13$ is relatively prime to $(p-1)(q-1)$.
   (c) A sender wants to encrypt the plaintext message $P = 9$. Calculate the ciphertext C by using the encryption formula $C = P^e$ (mod n).

14. Suppose two private prime numbers $p = 7$ and $q = 11$ are chosen by the receiver.

   (a) Calculate the public modulus n.
   (b) Verify that the public encryption key $e = 13$ is relatively prime to $(p-1)(q-1)$.
   (c) A sender wants to encrypt the plaintext message $P = 5$. Calculate the ciphertext C by using the encryption formula $C = P^e$ (mod n).

15. Suppose the two private prime numbers $p = 61$ and $q = 53$ are chosen by the receiver.

   (a) Calculate the public modulus n.
   (b) Verify that the public encryption key $e = 17$ is relatively prime to $(p-1)(q-1)$.
   (c) A sender wants to encrypt the plaintext message $P = 123$. Calculate the ciphertext C by using the encryption formula $C = P^e$ (mod n).

RSA Decryption

16. Suppose that two private prime numbers $p = 5$ and $q = 7$ are chosen by the receiver.

   (a) Calculate the public modulus n.
   (b) Explain why the public encryption key $e = 5$ is relatively prime to $(p-1)(q-1)$.
   (c) Verify that the private decryption key $d = 29$ is the inverse of $e = 5$ modulo $(p-1)(q-1)$.

(d) The receiver wants to decrypt the ciphertext message $C = 18$. Calculate the plaintext P using the decryption formula $P = C^d \pmod{n}$.

17. Suppose that two private prime numbers $p = 3$ and $q = 17$ are chosen by the receiver.

    (a) Calculate the public modulus n.

    (b) Explain why the public encryption key $e = 7$ is relatively prime to $(p-1)(q-1)$.

    (c) Verify that the private decryption key $d = 23$ is the inverse of $e = 7$ modulo $(p-1)(q-1)$.

    (d) The receiver wants to decrypt the ciphertext message $C = 26$. Calculate the plaintext P using the decryption formula $P = C^d \pmod{n}$.

18. Suppose that two private prime numbers $p = 47$ and $q = 59$ are chosen by the receiver.

    (a) Calculate the public modulus n.

    (b) Explain why the public encryption key $e = 157$ is relatively prime to $(p-1)(q-1)$.

    (c) Verify that the private decryption key $d = 17$ is the inverse of $e = 157$ modulo $(p-1)(q-1)$.

    (d) The receiver wants to decrypt the ciphertext message $C = 187$. Calculate the plaintext P using the decryption formula $P = C^d \pmod{n}$.

RSA Encryption and Decryption

19. Suppose the two private prime numbers $p = 17$ and $q = 11$ are chosen. Carry out the following steps of the RSA cryptosystem.

    (a) Calculate the public modulus n.

    (b) Explain why the public encryption key $e = 7$ is relatively prime to $(p-1)(q-1)$.

    (c) To send the plaintext message RUN, first use Table 22.7 to convert each plaintext letter into a decimal number in the range $0, 1, \ldots, 25$. Then calculate the ciphertext using the encryption formula for each decimal number.

    (d) Use the private decryption key $d = 23$ and the decryption formula to recover the plaintext.

    (e) Verify that d and e are inverses modulo $(p-1)(q-1)$.

20. Suppose the two private prime numbers $p = 23$ and $q = 41$ are chosen. Carry out the following steps of the RSA cryptosystem.

    (a) Calculate the public modulus n.

    (b) Explain why the public encryption key $e = 7$ is relatively prime to $(p-1)(q-1)$.

    (c) To send the plaintext message $P = 35$, calculate the ciphertext using the encryption formula.

    (d) Use the private decryption key $d = 503$ and the decryption formula to recover the plaintext.

    (e) Verify that d and e are inverses modulo $(p-1)(q-1)$.

ElGamal Encryption

21. A sender wishes to transmit plaintext to a receiver whose public information $(p, g, y) = (11, 2, 3)$ is available. Here the plaintext has already been converted to the numerical value $P = 5$. The sender's randomly chosen number is $k = 9$.

    (a) Compute the ciphertext pair (C_1, C_2) that is sent to the receiver.

    (b) Since the values (p, g, y) are publicly known, what equation would an intruder have to solve to retrieve the private key x of the receiver?

22. A sender wishes to transmit plaintext to a receiver whose public information $(p, g, y) = (29, 2, 7)$ is available. Here the plaintext has already been converted to the numerical value $P = 26$. The sender's randomly chosen number is $k = 5$.

    (a) Compute the ciphertext pair (C_1, C_2) that is sent to the receiver.

    (b) Since the values (p, g, y) are publicly known, what equation would an intruder have to solve to retrieve the private key x of the receiver?

ElGamal Decryption

23. The receiver has been sent the ciphertext pair $(15, 9)$. His public key is $(p, g, y) = (17, 6, 7)$ and his private key is $x = 5$.

    (a) Calculate $z = C_1^x \pmod{p}$.

    (b) Verify that the inverse (modulo 17) of z is 9.

    (c) Obtain the original plaintext by solving the modular equation $zP \equiv C_2 \pmod{p}$.

24. The receiver has been sent the ciphertext pair $(26, 12)$. His public key is $(p, g, y) = (107, 2, 22)$ and his private key is $x = 23$.

    (a) Calculate $z = C_1^x \pmod{p}$.

    (b) Verify that the inverse (modulo 107) of z is 38.

    (c) Obtain the original plaintext by solving the modular equation $zP \equiv C_2 \pmod{p}$.

ElGamal Encryption and Decryption

25. A receiver selects the public modulus $p = 13$, the public generator $g = 2$ modulo 13, and the private key $x = 3$.

    (a) Calculate the receiver's transformed key y.

    (b) A sender wishes to transmit the plaintext $P = 7$ using the random number $k = 5$. Calculate the ciphertext pair (C_1, C_2) that should be sent to the receiver.

    (c) Calculate $z = C_1^x \pmod{p}$ and verify that $z^{-1} = 5$.

    (d) Using the inverse found in the previous step together with C_2, determine the plaintext sent.

26. A receiver selects the public modulus $p = 167$, the public generator $g = 4$ modulo 167, and the private key $x = 37$.

    (a) Calculate the receiver's transformed key y.

    (b) A sender wishes to transmit the plaintext $P = 65$ using the random number $k = 71$. Calculate the ciphertext pair (C_1, C_2) that should be sent to the receiver.

    (c) Calculate $z = C_1^x \pmod{p}$ and verify that $z^{-1} = 66$.

    (d) Using the inverse found in the previous step together with C_2, determine the plaintext sent.

Elliptic Curve Cryptography

27. Find all pairs (x, y) that lie on the elliptic curve $y^2 \equiv x^3 + x + 4 \pmod 5$ by trying in turn each value of $x = 0, 1, \ldots, 4$ and then solving the associated modular equation for y.

28. The elliptic curve $y^2 \equiv x^3 + 2x + 1 \pmod 5$ contains the points $(0,1)$, $(0,4)$, $(1,2)$, $(1,3)$, $(3,2)$, $(3,3)$. The table below shows the result of adding any two of these points; we also include the point $(0,0)$, which serves as the "zero" for the addition operation.

    (a) Using this table, verify that $g = (1, 2)$ is a generator for all seven points.

    (b) Suppose a receiver selects the private key $x = 4$. Calculate the public y produced from x and g.

    (c) Suppose a sender selects $k = 3$ and wishes to transmit the plaintext message $P = (3, 2)$, a point on the elliptic curve. What ciphertext pair (C_1, C_2) will be sent?

    (d) Show the steps needed by the receiver to decrypt this ciphertext and recover the original plaintext.

+	(0,0)	(0,1)	(0,4)	(1,2)	(1,3)	(3,2)	(3,3)
(0,0)	(0,0)	(0,1)	(0,4)	(1,2)	(1,3)	(3,2)	(3,3)
(0,1)	(0,1)	(1,3)	(0,0)	(0,4)	(3,3)	(1,2)	(3,2)
(0,4)	(0,4)	(0,0)	(1,2)	(3,2)	(0,1)	(3,3)	(1,3)
(1,2)	(1,2)	(0,4)	(3,2)	(3,3)	(0,0)	(1,3)	(0,1)
(1,3)	(1,3)	(3,3)	(0,1)	(0,0)	(3,2)	(0,4)	(1,2)
(3,2)	(3,2)	(1,2)	(3,3)	(1,3)	(0,4)	(0,1)	(0,0)
(3,3)	(3,3)	(3,2)	(1,3)	(0,1)	(1,2)	(0,0)	(0,4)

RSA, Digital Signatures, and Hashing

29. Suppose the message M to be transmitted is 7865345662115249, a numerical representation of a bank account.

    (a) Divide this 16-digit message into four parts M_1, M_2, M_3, M_4 each with four digits. Use the modulus $n = 5011$ to compute the hashed value of M as $h(M) = (M_1 + M_2 + M_3 + M_4) \pmod{n}$.

    (b) What would be the hashed value if the first part was incorrectly given as 7685?

30. Suppose A publishes the numbers $e_A = 7$ and $n_A = 51$, and maintains the private key $d_A = 23$; B publishes the numbers $e_B = 5$ and $n_B = 35$, and maintains the private key $d_B = 29$. Entity A wishes to send the plaintext message $P = 11$ to B, after first adding a verifiable signature.

    (a) If A encrypts the plaintext with B's public key, what ciphertext C will be sent to B?

    (b) If A also signs the plaintext with his own private key, what signature S will be sent to B?

    (c) If B decrypts the ciphertext C using his own private key, what plaintext will result?

    (d) Using S, show how B can verify that the plaintext was sent from A.

31. Suppose A publishes the numbers $e_A = 7$ and $n_A = 55$, and maintains the private key $d_A = 23$; B publishes the numbers $e_B = 3$ and $n_B = 33$, and maintains the private key $d_B = 7$. Entity A wishes to send the plaintext message $P = 17$ to B, after first adding a verifiable signature. In this case, rather than signing the plaintext P, it is first hashed using the modulus $n = 7$ to reduce its size.

    (a) If A encrypts the plaintext P with B's public key, what ciphertext C will be sent to B?

    (b) If A signs the hashed plaintext with his own private key, what signature S will be sent to B?

    (c) If B decrypts the ciphertext C using his own private key, what plaintext will result?

    (d) Using S and the hash function h, show how B can verify that the plaintext was sent from A.

Projects

32. Investigate another public-key encryption method called the *knapsack encryption algorithm*, developed by Ralph Merkle and Martin Hellman. It is based on the knapsack problem discussed in Chapter 2, where the challenge is how best to fill a knapsack given certain constraints on the knapsack size. For example, suppose a knapsack must be filled with exactly 30 lb and items available to pack in the knapsack weigh $1, 6, 8, 15$, and 24 lb, respectively. One choice would be to select the items with weights $1, 6, 8, 15$; another choice would be the items with weights $6, 24$. The idea behind knapsack cryptography is that the message is encrypted using binary digits, where 1 indicates an item is included and 0 means it is not. A private key is used to disguise a solvable instance of the knapsack problem so that without the key it will be difficult to solve.

33. When using RSA, we were given a value for the decryption key d and confirmed that it was the multiplicative inverse of the encryption key e. Given e, there is an efficient way to determine its inverse d using the *Extended Euclidean Algorithm*. Investigate this algorithm and apply it to the problems in this chapter.

34. A cryptosystem called *Pretty Good Privacy* (PGP) is a popular program used to encrypt and decrypt email. It is available both as freeware and as a commercial version that will connect it with RSA or the Diffie-Hellman formulas. PGP was developed in 1991 and has an interesting history. Investigate this history and describe the basic design of the method. Possible sources are [Bauer 2013, Ch. 17] and [Singh 1999, Ch. 7].

35. It has been demonstrated that a quantum computer (based on quantum mechanics) will be able to vastly increase the speed of computing, but a practical quantum computer has not yet been developed. Quantum cryptography is a new area of cryptography that takes advantage of *quantum states*, in which two separate objects become linked in such a way (entangled) that they must be described together, not individually. This has implications for shared keys between two parties. If the bits of the key are encoded as quantum data between receiver and sender, then if some third party tries to intercept the data, the data will be disturbed and the presence of the third party will be revealed. Investigate this area of quantum cryptography, called *quantum key distribution*.

Bibliography

[Bauer 2013] Bauer, C. 2013. *Secret history: The story of cryptology*, Chapters 14, 17. Boca Raton, FL: CRC Press.

[Lewand 2000] Lewand, R. 2000. *Cryptological mathematics*, Chapter 4. Washington, D.C.: Mathematical Association of America.

[Singh 1999] Singh, S. 2002. *The code book*, Chapters 6 and 7. New York: Anchor Books.

[Young 2006] Young, A. 2006. *Mathematical ciphers: From Caesar to RSA*, Chapters 20, 21, 23. Providence, RI: American Mathematical Society.

[OR1] http://www.math.niu.edu/~richard/Math420/primelist.pdf (accessed June 25, 2014).

[OR2] http://www.math.brown.edu/~jhs/MA0158/MathCryptoCalc.02.html (accessed June 25, 2014).

[OR3] http://www.christelbach.com/ECCalculator.aspx (accessed June 25, 2014).

Appendix A

Applications

TABLE A.1: List of applications developed in chapter examples

	Unit I	Unit II	Unit III	Unit IV	Unit V
Archaeology	5.10				
Biology	3.1–3.2; 3.4–3.6; 4.9	7.6; 11.4–11.5	15.6		
Business	3.10; 5.4		12.8–12.10; 14.3; 15.1; 15.8; 15.10; 16.11	19.1–19.2; 19.4–19.6; 20.8; 20.10; 21.3–21.6	22.1–22.2; 22.5; 24.15
City Planning		11.1–11.3		17.9–17.10	
Communication	3.11; 6.8				
Education			16.2	19.9; 20.2–20.3; 20.7; 21.1	
Finance	2.1–2.4; 2.6	9.8; 9.11			
Gambling			12.1; 13.2–13.4; 14.13; 16.3–16.4		
Geography	6.1				
Graphics					23.7
Language		8.2–8.7			
Law Enforcement	3.12	7.3–7.4; 11.8–11.9	15.7; 15.9; 15.11–15.13		
Manufacturing	5.8				
Marketing		11.12			
Medicine	6.14	7.1–7.2; 7.7–7.9	14.10; 15.2; 16.10		
Navigation	3.7–3.8				
Psychology			15.4		
Routing	4.4–4.8; 5.2				
Scheduling	3.9; 6.9–6.10				
Security					22.6–22.11; 22.13–22.15; 23.5–23.6; 23.8–23.9; 23.11–23.13; 24.7–24.12; 24.14; 24.16–24.17
Society/Politics	1.1	10.16; 11.10	14.5–14.6; 14.9	17.1–17.8; 18.1–18.14; 19.3; 19.7–19.8; 19.10; 20.1; 20.4–20.6; 20.9; 20.11; 21.2; 21.7–21.9	
Sports			14.4; 14.15		
Technology		10.1; 10.11–10.15	15.3		
Time Measurement	1.10				22.3
Transportation		10.17		20.12	

Appendix B

Classifications

We describe here the terms used in the classification tables found at the end of each chapter. Examples appearing in the chapter are first categorized by their use of certain Representations, Strategies, and Algorithms; these categories are then further refined.

Representations

These are structures that help us visualize, model, or describe an unfamiliar problem in a more familiar form in order to expose the relationships between its components. Each representation is intended to capture the most important properties and relationships in the original problem. The common representations encountered involve graphs, tables, sets, and symbolic forms.

1. *Graphs.* These are structures that show the relationships between elements (vertices), which are interconnected using lines (edges). They are used to model many situations, including communication and transportation networks, as well as personal relationships. The type of graph representation can be further classified as follows.

    - Unweighted: neither edges nor vertices have associated weights.
    - Weighted: edges and/or vertices can have weights, possibly reflecting a cost, distance, or other quantitative measure.
    - Undirected: the graph edges have no specific orientation, indicating a mutual relationship between the corresponding vertices.
    - Directed: the graph edges have a specific orientation, possibly indicating the allowable direction of travel or an allowable transformation.
    - Path: a sequence of edges and vertices that join two particular vertices.
    - Tree: a special type of graph in which there is a unique path between any two vertices.

2. *Tables.* These are arrays of individual cells, organized into rows and columns. Two particular types of tables are encountered here.

    - Data: problem data are organized in such a way that promotes further analysis of the problem.
    - Decision: table cells and their relationships enable (optimal) actions to be carried out.

3. *Sets.* These are well-defined collections of objects that can be presented in several ways.

    - Lists: a collection of objects ordered in a specific way.
    - Venn diagrams: visual depictions that show the logical relationships between various sets of objects.

4. *Symbolization.* These representations use sets of symbols and certain rules to establish relationships between mathematical or non-mathematical objects.

    - Numerical: numbers and their standard arithmetic operations are used.
    - Logical: a set of symbols and operations are used to express logical relationships.
    - Algebraic: unknown or unspecified symbols are used to model relationships.
    - Geometric: relationships between objects are displayed visually.

Strategies

These are higher-level plans or approaches that guide the design of effective techniques to find problem solutions. Four such strategies are frequently encountered in devising context-sensitive solutions.

1. *Brute-Force.* A method of systematically checking all possible candidates for a solution in order to determine whether any candidate satisfies the problem requirements.

2. *Solution-Space Restrictions.* In contrast to using brute-force enumeration, the set of possible candidates is reduced by applying conditions that must be satisfied by a solution to the problem.

3. *Rule-Based.* A particular set of problem-specific rules is applied to transform the given information into the required solution.

4. *Composition/Decomposition.* An approach that involves breaking the given problem up into smaller subproblems and solving each one separately, sometimes sequentially and sometimes in parallel. These component solutions are then combined to produce a solution to the original problem.

Algorithms

These are specific step-by-step procedures for finding a solution to the given problem. They can be categorized in several ways.

1. *Solution Type.* The algorithm can produce either an exact or an approximate solution.

   - Exact: the algorithm is guaranteed to deliver an exact solution to any problem.
   - Approximate: the algorithm is not guaranteed to deliver an exact solution to the problem, only a solution that is typically close to optimal.

2. *Greedy.* Such an algorithm makes the locally optimal (best) choice at each stage in the hope that this will result in an optimal solution. While such a method is typically fast, it may or may not be guaranteed to produce an optimal solution.

3. *Trial-and-Error Approaches.* These methods are based on implicit or explicit evaluation of all alternatives.

   - Inspection: a solution is obtained by a straightforward application of simple techniques, without recourse to involved calculations.
   - Enumeration: all feasible solutions to the problem are generated and the best is selected.

4. *Search Techniques.* These algorithms apply standard procedures to solve problems of a certain type.

   - Sequential search: a prescribed set of steps is repeatedly applied until a solution is identified.
   - Breadth-first search: the search proceeds level by level, starting at an initial point, exploring all options available from the initial point, and continuing until a solution is reached.

- Depth-first search: the search starts at an initial point, explores an option available from the initial point, and continues to explore an available option (backtracking if necessary) until a solution is reached.

5. *Computational Approaches.* These algorithms determine a solution by carrying out well-defined computations.

   - Numerical: arithmetic operations are applied to find a solution.
   - Logical: logical (Boolean) operations or rules of inference are applied to find a solution.
   - Algebraic: Equations or relations involving the unknown variables are solved to determine a solution.
   - Geometric: spatial relationships between objects are used to determine a solution.
   - Classification: objects are allocated to different categories based on their characteristics.

6. *Decomposition Approaches.* These methods are based on decomposing a problem and then assembling the solutions to its subproblems.

   - Dynamic programming: optimal solutions to various subproblems are determined and then combined to reach an overall optimal solution.
   - Recursive: these algorithms repeatedly apply the same procedure to smaller instances of the same problem and thus build up a solution to the original problem.

Index